Evaluation Guide for Principles of Science, Book One

Principles of Science, Book One, is a textbook that discusses the most modern principles and concepts of science. Recent developments and basic ideas in science are explained in language suited to the abilities of the student. To strengthen the presentation of the textual material and to increase the student's understanding and interest in science, the authors and editors have included many special text features.

To examine examples of the format and the features of *Principles of Science, Book One,* please turn to the following pages:

(pp. 2–3)	A unit **photograph** in color, and a **brief paragraph** provide a visual and conceptual introduction to each of the six units.
(pp. 4–5)	A one-page **color photograph** and a **brief paragraph** introduce each chapter.
(p. 45)	A **goal statement** at the beginning of each chapter tells your students what they will learn from each chapter.
(pp. 74–75)	**Questions** printed in the margins are useful in guiding the student's thinking while reading and reviewing.
(p. 155)	**New terms** are printed in boldface type where they are defined. Many new terms are spelled phonetically where they first appear.
(pp. 196–197)	**Photographs** and clearly-labeled **illustrations,** most in full color, help students visualize ideas presented in the accompanying text.
(p. 298)	**Tables** appear in the text where they are relevant to the material being presented.
(p. 312)	**Perspectives: People and Frontiers** present new developments in science and current information on people and careers in science.
(pp. 330–331)	**Activities** immediately reinforce scientific principles. Activity materials are inexpensive and readily available. **Drawings** accompany many of the activities to help the student in preparing laboratory materials.
(pp. 400–401)	**Making Sure** questions and problems are presented at the end of sections to reinforce the preceding material.
(p. 486)	**Perspectives: Skills** pages present additional aids geared toward reading comprehension.
(pp. 490–491)	**Side Roads** are two-page features which discuss topics of current scientific interest. Each unit includes one Side Roads feature which is related to the material in the unit.
(pp. 505–507)	A summary of **Main Ideas,** a list of **Vocabulary** words, a variety of **Study Questions,** and suggestions for further **Challenges** and **Interesting Reading** appear at the end of each chapter.
(pp. 508–513)	**Appendices.**
(pp. 514–522)	**Glossary** with pronunciation key on each page.
(pp. 524–533)	**Index.**

Teacher's Annotated Edition

principles of SCIENCE Book One

Author

Charles H. Heimler is a Professor of Science Education and Director of Teacher Education at California State University, Northridge, CA. He received his B.S. degree from Cornell University and his M.A. and Ed.D. degrees from Columbia University and New York University. He has 32 years of teaching experience at the junior high, high school, and university levels. Dr. Heimler is a member of the American Association for the Advancement of Science, National Science Teachers Association, and the National Association of Biology Teachers. He is currently a consultant in science education to several California school districts. Dr. Heimler is author of the Merrill *Focus on Life Science* program and co-author of the *Focus on Physical Science* program.

Consultant

Charles D. Neal is Professor Emeritus of Education at Southern Illinois University, Carbondale, IL. Dr. Neal has taught at the junior high school level and has lectured extensively on the teaching of science in the elementary and junior high school. He received his B.S. degree from Indiana University, his M.A. degree from the University of Illinois, and his M.S. and Ed.D. degrees from Indiana University. He is a member of various professional organizations and is the author of numerous science books for children and adults.

Content Consultants
Earth Science: Dr. Jeanne Bishop, Parkside Junior High School, Westlake, OH
Life Science: Lucy Daniel, Rutherfordton-Spindale High School, Rutherfordton, NC
Physical Science: Dr. Richard H. Moyer, University of Michigan-Dearborn

Reading Skills Consultant
Dr. David L. Shepherd, Hofstra University

Charles E. Merrill Publishing Co.
A Bell & Howell Company
Columbus, Ohio
Toronto, London, Sydney

A Merrill Science Program

Principles of Science, Book One Program
  Principles of Science, Book One and Teacher's Annotated Edition
  Principles of Science, Book One, Teacher Resource Book
  Principles of Science, Book One, Activity-Centered Program, Teacher's Guide
  Principles of Science, Book One, Evaluation Program, Spirit Duplicating Masters
Principles of Science, Book Two Program
(This program contains components similar to those in the Book Two Program.)

Series Editor: Terry B. Flohr; *Project Editor:* Francis R. Alessi, Jr.; *Editors:* Janet Helenthal, Peter R. Apostoluk; *Book Design:* Patricia Cohan; *Project Artist:* Dick Smith; *Illustrators:* Don Robison, Jim Shough, Bill Robison; *Photo Editor:* Lindsay Gerard; *Production Editor:* Joy E. Dickerson

ISBN 0-675-07081-3

Published by
Charles E. Merrill Publishing Company
A Bell & Howell Company
Columbus, Ohio 43216

Copyright 1986, 1983, 1979, 1975, 1971, 1966 © by Bell & Howell

All rights reserved. No part of this book may be reproduced in any form, electronic or mechanical, including photocopy, recording, or any information storage or retrieval system without written permission from the publisher.
Printed in the United States of America

Table of Contents

Philosophy	4T	Unit 2 Mechanics	55T
Basic Themes	5T	Chapter 4: Work and Energy	55T
Scope and Sequence	5T	Chapter 5: Machines	58T
Earth Science	6T	Chapter 6: Motion	62T
Life Science	8T	Chapter 7: Pressure and Fluids	66T
Physical Science	10T		
Text Design	12T	Unit 3 Earth Science	70T
Teacher Resource		Chapter 8: The Earth	70T
Book, Activity-		Chapter 9: Geology	73T
Centered Programs	21T	Chapter 10: Earth History	78T
Using Principles of Science	22T	Chapter 11: The Atmosphere	81T
Reading Science Information	26T	Chapter 12: Weather and Climate	85T
Planning the Course	28T	Chapter 13: Oceanography	89T
Planning Guide	29T		
Equipment and Supplies for		Unit 4 Living Things	93T
Activities	30T	Chapter 14: Life and the Cell	93T
Suppliers	35T	Chapter 15: Plants	97T
Preparation of Solutions	35T	Chapter 16: Animals	102T
Teaching Aids	36T	Chapter 17: Animals with Backbones	105T
Teacher References	36T		
Student References	37T	Unit 5 Environment and	
Films and Filmstrips	37T	Heredity	109T
Film Distributors	38T	Chapter 18: Populations and	
Careers Related to Science	39T	Communities	109T
Performance Objectives	41T	Chapter 19: Food and Energy	112T
Answers to Textbook Inventory	42T	Chapter 20: Heredity	115T
		Chapter 21: Descent and Change	118T
Unit I Matter and Energy	43T	Unit 6 Conservation	121T
Chapter 1: Science and Measurement	43T	Chapter 22: Soil, Water, and Air Conservation	121T
Chapter 2: Matter	46T	Chapter 23: Forest and Wildlife Conservation	124T
Chapter 3: Energy and Changes in Matter	51T	Teacher Questionnaire	126T

Philosophy

Principles of Science is a two-book program that emphasizes scientific inquiry and interrelationships between different areas of science. Major attention is given to presenting a balanced approach to life, earth, and physical sciences and to everyday practical applications in each of these areas. As students progress from chapter to chapter, they realize that the technological advances they take for granted are merely applications of basic science principles.

One advantage of this balanced approach is that it provides a variety of science topics that stimulate interest and motivate learning. Students are taught the essential unity of science in terms of the scientific methods and principles that pervade all science fields. This learning is important for all students as future adult citizens.

In familiarizing yourself with the many features of *Book I*, you will see that a successful learning experience for students is the first priority. The sequence of presentation is designed to allow students to move from simple, concrete concepts to more difficult and abstract ideas.

Unit 1 concerns the definition of science, basic principles of measurement, and the study of matter, energy, and their interrelationship. This unit reviews basic physical science concepts studied at the elementary level. It is the foundation on which the rest of the text builds. The International System of Measurement is presented. SI units are then used throughout both texts as the primary system of measurement.

Unit 2 expands the topic of energy. Controlled use of energy to do work and the physical laws that govern the behavior of matter and motion are discussed.

Unit 3 presents a comprehensive study of earth science concepts. The earth is presented first because this material is most familiar to students. After studying the overall characteristics of our planet, students move on to study its internal structure and the processes by which it has evolved. They then cover the characteristics and behavior of the area around our earth, the atmosphere. The unit concludes with the study of the characteristics and behavior of the oceans.

Unit 4 begins the study of life. The unit begins with a study of the features of living things. These features are then used as a basis in covering the characteristics of representative members of each of the five kingdoms.

In Unit 5, the study of individual organisms is expanded to cover the characteristics of large numbers of interacting organisms. The unit then covers elementary genetics and its applications in covering how organisms have changed over time.

Unit 6 considers the important resources required to maintain life as it exists today. Students will be made aware of the advantages, disadvantages, and trade-offs to be considered in altering our environment. The knowledge gained from this unit should serve as a foundation in helping students make informed decisions concerning the future uses of resources.

You can see from the unit organization that topics are presented in a sequence that builds in complexity. Basic principles are stressed. Then applications are presented as a means of emphasizing basics. When students begin with familiar ideas, they develop a sense of confidence and more readily move on to more complex ideas. Thus, a successful learning experience is assured.

Basic Themes

Several unifying themes provide a conceptual framework for **Principles of Science, Book I.**
1. Observation and experimentation are the bases of scientific inquiry and discovery.
2. Matter and energy are interrelated. Changes in matter cause a gain or release of energy.
3. Properties of matter depend upon the kinds of atoms it contains and the order and arrangement of the atoms.
4. Energy can be changed from one form to another.
5. The classification of objects, organisms, and events simplifies the study of science and allows us to make predictions.
6. The features and behavior of Earth and the universe can be explained through the study of matter and energy.
7. The earth is constantly changing through the action of forces within the earth and surrounding atmosphere.
8. Technological advancements are the result of applying scientific principles.
9. Organisms use matter and energy for growth and maintenance.
10. Organisms are interrelated with each other and with their physical environments.
11. Organisms inherit traits from their parents and these traits are modified by environment.
12. Changes have occurred in living things through time. Survival of a species depends on adaptation to its environment.
13. The wise use of resources affects the quality of life on our planet.

Scope and Sequence

The **Principles of Science** program is organized in a manner by which students progress from simple or familiar concepts to abstract, complex, or unfamiliar ones. Note the sequence of presentation in each of the areas (life, earth, and physical science). The life science units of *Book I* begin with the basic characteristics of life. Students then study organisms beginning with the most simple, single-celled organisms and progressing to mammals, the most complex. The life science units of *Book II* are then an extension of this theme in covering complex, yet familiar organisms—humans. Students then move on to learn practices that will keep their bodies in good working order.

In looking at the earth science units, note that most of the material in *Book I* concerns the earth itself. *Book II* then expands on this foundation when students look at other parts of the universe in studying astronomy, space, galaxies, and the solar system.

The physical science units of *Book I* cover the most basic principles of chemistry and physics in a descriptive manner. The material in *Book II* builds on the concepts presented in *Book I* in looking at reaction chemistry and interactions between matter and energy by studying heat, light, sound, electricity, magnetism, and nuclear energy.

The scope and sequence is divided into two areas: content and problem solving/process skills. The table enables you to see at a glance the interrelationships between the two books as well as the balanced approach to life, earth, and physical science.

Book I

Content	Problem solving	Process
8 The Earth Earth's Shape and Size; Earth's Motions; Seasons; Latitude and Longitude; Mapping the Earth's Surface; Time Zones; The Earth as a Magnet Calculating scale equivalents/Using angles		8-1 Observing changing shadows throughout the day 8-2 Demonstrating Earth's motions 8-3 Locating cities on a map 8-4 Constructing a map scale 8-5 Using topographic maps
9 Geology Structure of the Earth; Plate Tectonics; Rocks and Minerals; Weathering and Erosion; Igneous Rock; Sedimentary Rock; Metamorphic Rock; Rock Formations; Earthquakes; Volcanoes Calculating density		9-1 Comparing the shapes of continents 9-2 Identifying rocks 9-3 Observing the expansion effects of ice 9-4 Observing the texture of igneous rocks 9-5 Testing for the presence of carbonate ion 9-6 Determining the composition of sediment 9-7 Observing the characteristics of anticlines and synclines
10 Earth History The Grand Canyon; Age of Rocks; Fossils; Fossil Records; Geologic Time; Precambrian Era; Paleozoic Era; Mesozoic Era; Cenozoic Era; Earth's Future Calculating density		10-1 Observing the characteristics of petrified wood and oak 10-2 Making a cast of a leaf 10-3 Predicting fossil formations
11 The Atmosphere Air; Parts of the Atomosphere; Air Pressure; Heating the Atmosphere; Winds; Local Winds; Moisture in Air; Clouds; Precipitation		11-1 Observing the effects of air pressure 11-2 Using a barometer 11-3 Observing the movement of warm air 11-4 Observing the path of water on a spinning globe 11-5 Recording wind direction 11-6 Determining dew point 11-7 Using a sling psychrometer 11-8 Observing a cloud in a bottle
12 Weather and Climate Observing the Weather; Air Masses; Fronts; Thunderstorms; Highs and Lows; Weather Maps; Weather Forecasting; Climate Using precentages/Using angles		12-1 Recording time, temperature, pressure, relative humidity, cloud cover, wind speed and direction, and precipitation 12-2 Locating fronts on a weather map 12-3 Comparing forecasts to actual weather
13 Oceanography Seawater; Temperature and Density; The Seafloor; Sea Life; Ocean Currents; Waves; Tides; Seismic Sea Waves and Surges; Ocean Resources Using percentages/Calculating density/Calculating wave period and time intervals		13-1 Making a brine solution 13-2 Determining the density of seawater 13-3 Observing convection currents in water 13-4 Observing wave motion 13-5 Observing wave refraction 13-6 Comparing the time between high and low tide
22 Soil, Water, and Air Conservation Conservation of Natural Resources; Soil; Soil Erosion and Mineral Loss; Soil Conservation; Water Resources; Water Conservation; Water Sources for the Future; Air Pollution		22-1 Observing the effects of soil type of plant growth 22-2 Observing the effects of leaching on soil type 22-3 Constructing a contour plowing model 22-4 Detecting air pollution
23 Forest and Wildlife Conservation Forest Resources; Forest Conservation; Forest Conservation Practices; Vanishing Wildlife; Wildlife Resources; Wildlife Conservation; Wildlife Refuges		23-1 Constructing and observing the characteristics of a watershed 23-2 Listing wood products 23-3 Observing the characteristics of birds at a feeder

Earth Science

BOOK II

Content	Problem solving	Process

18 Astronomy and the Moon Astronomy; Optical Telescopes; Astronomical Observations; Radio Astronomy and Satellites; The Moon; Moon Phases; Eclipses; Lunar Surface and Composition; Origin of the Moon Calculating scale conversions/Calculating time intervals	18-1 Making a refracting telescope 18-2 Determining an approximate distance to the moon 18-3 Recording the positions and shapes of the moon 18-4 Observing a model of a solar eclipse	
19 Our Solar System Motion of the Planets; The Sun; Sunspots, Solar Flares, and Solar Wind; Mercury; Venus; Earth; Mars; Asteroids and Meteors; Jupiter; Saturn; Uranus, Neptune, and Pluto; Comets; Origin of the Solar System Calculating period of revolution	19-1 Observing the characteristics of an ellipse 19-2 Determining the period of revolution	
20 Stars and Galaxies Stars and Stellar Distances; Classifying the Stars; Early Stages of a Star; Final Stages of a Star; Life Cycle of Our Sun; Interstellar Space; Galaxies; Origin of the Universe	20-1 Determining the relationship between distance and illumination 20-2 Determining the amount of parallax	
21 Space Exploration Gravity and Space Flight; Types of Rocket Engines; Rocket Guidance Systems; Satellites; The Apollo Program; Space Stations; Space Shuttle; Space Training and Survival; Space Colonies	21-1 Observing action and reaction forces 21-2 Observing a model for rocket thrust 22-3 Observing the behavior of a gyroscope	
22 People and Resources People; Resources; Food; Increasing Food Production; Minerals; Electricity; Fossil Fuels; Hydroelectricity and Geothermal Energy; Nuclear Energy; Energy from Tides, Waves, and Wind; Solar Energy Calculating population growth	22-1 Studying the effects of increased population 22-2 Classifying resources 22-3 Comparing food prices 22-4 Comparing evaporation rates 22-5 Classifying mineral and living resources	
23 People and Their Enrivonment Environment; Air Pollution; Water Pollution; Radiation Pollution; Pesticides; Benefits and Risks; Solid Wastes; Conservation Calculating percentages	23-1 Testing acidity of water 23-2 Testing water for oxygen loss 23-3 Comparing contents of trash containers in cafeteria and home	

BOOK I

Content	Problem Solving	Process

Life Science

Content	Problem Solving / Process
14 Life and the Cell Features of Living Things; The Cell; Cell Activities; Classification; Bacteria; Fungi; Amoebas and Paramecia; Flagellates; Sporozoans; Viruses Calculating percentages and population	14-1 Determining the characteristics of life 14-2 Observing osmosis 14-3 Observing bacteria using a microscope 14-4 Observing yeast and testing for CO_2 14-5 Observing amoeba movement 14-6 Observing organisms in a hay infusion
15 Plants Algae; Mosses and Liverworts; Ferns; Seed Plants; Roots; Stems; Leaves; Photosynthesis; The Flower; Seeds; Seed Dispersal	15-1 Growing and observing liverworts 15-2 Observing the characteristics of roots 15-3 Observing the response of plant roots 15-4 Observing transport in a celery stalk 15-5 Counting the age of a tree 15-6 Observing the growth of a sweet potato 15-7 Observing water loss in plants 15-8 Observing the construction of a leaf skeleton 15-9 Determining the relationships among sunlight, chlorophyll, and photosynthesis 15-10 Testing for starch 15-11 Determining the factors affecting seeds 15-12 Observing the seed germination process
16 Animals Tissues, Organs, and Systems; Sponges, Jellyfish and Their Relatives; Flatworms and Roundworms; Earthworms and Their Relatives; Spiny-skinned Animals; Mollusks; Arthropods; Insects	16-1 Observing the characteristics of hydra 16-2 Identifying the parts of an earthworm 16-3 Identifying insects
17 Animals with Backbones Backbones and Skeletons; Fish; Amphibians; The Frog; Reptiles, Birds; Mammals	17-1 Observing circulation in goldfish 17-2 Observing the characteristics of fish 17-3 Dissecting a frog 17-4 Observing metamorphosis of a frog 17-5 Identifying the internal organs of a chicken
18 Populations and Communities Communities; Populations; Population Change; Competition; The Climax Community; The Boundary Community; Cooperation in Populations; Ecology	18-1 Observing life in a community 18-2 Comparing aquarium and terrarium environments 18-3 Observing the effect of competition of population size 18-4 Observing succession in a hay infusion
19 Food and Energy Food Chains; Producers and Consumers; Energy Pyramids; Predators; Parasites; Symbiosis	19-1 Indentifying producers and consumers in pond water 19-2 Constructing an energy pyramid 19-3 Comparing the structure of algae and fungi
20 Heredity Inherited Traits; Law of Dominance; Crossing Hybrids; Blending; Chromosomes and Genes; Reduction Division and Fertilization; Sex Determination; Reproduction; Mutations; Plant and Animal Breeding	20-1 Observing the traits of fruit flies 20-2 Determining probabilities with coins 20-3 Observing the color of hybrid offspring 20-4 Comparing body cell division and reduction division 20-5 Observing albanism in tobacco plants
21 Descent and Change Origin of Living Things; Darwin's Theory; Fossil Evidence; Natural Selection; Changes in Species; Mutations and Change	21-1 Counting the number of seeds produced by a plant

Book II

Content	Problem solving / Process
2 The Human Body Human Beings; Cells, Tissues, and Organs; Skeletal System; Bone Tissue; Muscles Calculating averages	2-1 Measuring lung capacity 2-2 Observing cells under a microscope 2-3 Observing the effects of heat on bone
3 Circulatory Systems Blood, Blood Cells; Blood Vessels; The Heart; Plumonary and Coronary Circulation; Systemic Circulation; Blood Types; Rh Factor; Diseases of the Circulatory System; Lymphatic System Using ratios/Calculating averages	3-1 Recording the time required for blood to clot 3-2 Measuring pulse rate 3-3 Comparing heart beats
4 Internal Body Processes Metabolism; Respiration; Respiratory System; Digestion; The Digestive System; Digestion in the Digestive Tract; Absorption of Food; Excretion and Excretory System Using ratios/Calculating averages	4-1 Measuring rate of breathing 4-2 Observing the function of enzymes 4-3 Observing protein digestion 4-4 Testing for starch digestion 4-5 Observing the function of bile in fat digestion 4-6 Observing diffusion through a membrane
5 Nervous and Endocrine Systems Behavior and Its Regulation; The Central Nervous System; Neurons; Sense Organs; Reflex; Conditioned Reflex; The Autonomic Nervous System; Loewi's Experiment; The Endocrine System; Behavior and Habit	5-1 Observing the sensation of touch 5-2 Observing the sensation of taste 5-3 Observing reflex actions 5-4 Determining reflex actions 5-5 Producing conditioned responses
6 Nutrition Food and Health; Water; Minerals; Carbohydrates; Fats and Oils; Proteins; Vitamins; Energy from Food; Food Additives and Food Labels	6-1 Classifying foods 6-2 Testing foods for starch and sugar 6-3 Testing foods for the presence of fat 6-4 Testing foods for the presence of Vitamin C 6-5 Calculating caloric intake 6-6 Comparing information on food labels
7 Disease Disease and Microbes; Koch's Postulates; Microbe Carriers; Cancer; Defenses Against Disease; Vaccination; Allergy; Diseases of the Immune System	7-1 Growing microbes on agar
8 Drugs Classification of Drugs; Antibiotics; Stimulants; Tobacco; Depressants; Depressants: Alcohol; Depressants: Narcotics; Hallucinogens; Marijuana	8-1 Recording the functions of nonprescription drugs 8-2 Comparing the concentration of penicillin with microbe growth
9 Human Reproduction and Heredity Human Reproduction; Menstrual Cycle; Pregnancy; Traits and Genes; Human Pedigrees; Sex-Linked Traits; Multiple-Gene Inheritance; Genetic Diseases and Disorders Using ratios/ Calculating averages/Determining probabilities	91- Recording the presence of a dominant trait 9-2 Determining the factors associated with hand size

Scope and Sequence

Book I

Content	Problem solving	Process
1 Science and Measurement What is Science?; Careers in Science; Using Science Skills; Experiments; Metric Measurement; Area and Volume; Mass Measurement conversions/Calculating area and volume		1-1 Reading about science 1-2 Researching science careers 1-3 Heating air 1-4 Finding area 1-5 Measuring length, area, and volume
2 Matter Properties; States of Matter; Density; Elements; Atoms; Compounds; Formulas; Mixtures Calculating density/Using ratios in chemical formulas		2-1 Determining a property of matter 2-2 Finding the density of wood 2-3 Finding density using a graduated cylinder 2-4 Using molecular models 2-5 Separating mixtures 2-6 Observing the properties of solutions
3 Energy and Changes in Matter Energy; Heat and Temperature; Freezing and Melting; Evaporation and Boiling; Chemical Change; Chemical Equations; Nuclear Change Sources of Energy Calculating rate of heat gain or loss		3-1 Observing temperature changes 3-2 Measuring temperature changes to determine heat movement 3-3 Separating water from saltwater 3-4 Graphing evaporation rate 3-5 Observing a burning candle 3-6 Observing energy from the sun 3-7 Collecting solar energy
4 Work and Energy Force; Weight; Friction; Work; Work and Energy; Power; Engines Calculating work and power		4-1 Measuring forces 4-2 Measuring weight 4-3 Comparing friction forces 4-4 Finding the power used in climbing stairs
5 Machines Machines; Levers; Mechanical Advantage; Pulleys; Wheel and Axle; Inclined Plane; Wedge; Screw; Efficiency; Compound Machines Calculating I.M.A. and A.M.A./Calculating work input and output/Calculating efficiency		5-1 Using a ruler as a lever 5-2 Measuring the I.M.A. of three pulleys 5-3 Measuring the I.M.A. and A.M.A. of inclined planes
6 Motion Inertia; Speed; Change in Speed; Action and Reaction Forces; Falling Objects; Circular Motion; Momentum Calculating speed, acceleration, and deceleration		6-1 Demonstrating inertia 6-2 Observing inertia 6-3 Measuring action and reaction forces 6-4 Observing an action and reaction 6-5 Graphing time vs. speed 6-6 Measuring the speed of falling objects 6-7 Observing circular motion 6-8 Observing the effect of inertia
7 Fluids and Pressure Fluids; Pressure in Liquids; Gas Laws; Buoyancy; Bernoulli's Principle; Lift; Flight Calculating pressure		7-1 Observing changes in water levels 7-2 Measuring pressure in a liquid 7-3 Observing the relationship between buoyance and density 7-4 Measuring displacement 7-5 Demonstrating Bernoulli's principle

Physical Science

Book II

Content	Problem solving	Process

1 Science and Technology Scientists and Science; Observations; Measurement; Units of Measurement; Technology Calculating differences in measurements		1-1 Making observations of a photograph 1-2 Measuring length, mass, and volume 1-3 Observing science/technology programs on TV
10 Matter Chemistry: The Science of Matter; Properties of Matter; Elements and Compounds; Mixtures; Solutions, Suspensions, and Colloids; Solubility; Atoms Using ratios in chemical formulas		10-1 Determining physical properties of matter 10-2 Observing the characteristics of a chemical change 10-3 Separating mixtures 10-4 Determining solubility
11 Elements Elements and the Periodic Table; Metals, Nonmetals, and Metalloids; Alkali Metals; Alkaline Earth Metals; Transition Metals and Synthetic Elements; Boron and Aluminum; Carbon Family; Nitrogen Family; Oxygen Family; Halogens; Noble Gases		11-1 Identifying elements using flame tests 11-2 Identifying transition metal compounds by color 11-3 Determining a product of burning 11-4 Identifying sulfide compounds
12 Reactions and Solutions Bonding; Oxidation Numbers; Naming Compounds; Equations for Chemical Reactions; Types of Chemical Reactions; Chemical Change and Energy; Acids and Bases; Ions in Acids and Bases; Indicators and pH Scale; Neutralization and Salts		12-1 Observing the properties of acids and bases 12-2 Testing substances to determine pH 12-3 Observing the process of neutralization Using ratios in chemical formulas/Using proportions
13 Chemical Technology Chemical Technology; Carbon and Chemical Technology; Petroleum; Plastics; Synthetic Fibers; Soap and Detergent; Water Purification; Hard Water		13-1 Classifying plastics 13-2 Determining the properties of synthetic fabrics and comparing them to natural fibers 13-3 Observing a precipitation reaction
14 Heat Energy; Temperature and Heat; Effects of Heat and Matter; Heat and Changes of State; Specific Heat; Measuring Heat Changes; Heat Transfer; Heating and Insulation; Central Heating Systems; Cooling Systems Calculating energy released and absorbed/Calculating temperature intervals		14-1 Observing the effects of heat on metals 14-2 Recording temperature during a change of state 14-3 Comparing specific heats 14-4 Using a calorimeter to determine energy transfer 14-5 Comparing conduction properties 14-6 Comparing insulating properties
15 Sound and Light Waves; Characteristics of Waves; Sound Waves; Volume and Pitch of Sound; Reflection and Refraction; Electromagnetic Waves; Mirrors and Reflection; Refraction of Light; Lenses, Prisms, and Color; Polarized Light; Sources of Light		15-1 Comparing reflection and refraction 15-2 Locating a Virtual Image 15-3 Observing refraction through water Calculating frequency and speed
16 Magnetism and Electricity Magnets; Theory of Magnetism; Electromagnets; Static Electricity; Electric Potential Energy; Electric Current; Electric Generators; Resistance and Ohm's Law; Types of Circuits; Electric Energy and Power Calculating, current, voltage, resistance, power		16-1 Observing the properties of magnets and magnetic fields 16-2 Observing the characteristics of electromagnets 16-3 Observing the characteristics of static charge 16-4 Determining the relationship between magnetic field and current 16-5 Comparing series and parallel circuits
17 Nuclear Energy Radioactive Elements; Nuclear Radiation; Half-Life and Decay Series; Radiation Biology; Detecting Radiation; Fission; Fusion		17-1 Measuring radiation Calculating numers of neutrons/Calculating half-life

principles of SCIENCE

Content that focuses on the basic principles of life, earth, and physical science so that students have a better understanding of their world.

Written by experienced educators in science, **Principles of Science** reflects the preferences of middle school and junior high teachers throughout the country.

7	**Fluids and Pressure**	126	10:6	Precambrian Era	205
7:1	Fluids	127	10:7	Paleozoic Era	206
7:2	Pressure in Liquids	129	10:8	Mesozoic Era	210
7:3	Gas Laws	133	10:9	Cenozoic Era	212
7:4	Buoyancy	134	10:10	Earth's Future	214
7:5	Bernoulli's Principle	138		Perspectives	
7:6	Lift	139		Mike Hansen: Earth Scientist	216
7:7	Flight	140			
	Perspectives		**11**	**The Atmosphere**	220
	First Flight	142	11:1	Air	221
			11:2	Parts of the Atmosphere	222
unit **3**			11:3	Air Pressure	224
Earth Science		146	11:4	Heating the Atmosphere	226
			11:5	Winds	227
		148	11:6	Local Winds	229
		149	11:7	Moisture in Air	231
		151	11:8	Clouds	234
		152	11:9	Precipitation	236
		155			
		156	**12**	**Weather and Climate**	240
		159	12:1	Observing the Weather	241
		161	12:2	Air Masses	243
			12:3	Fronts	245
		164	12:4	Thunderstorms	246
			12:5	Highs and Lows	247
		168	12:6	Weather Maps	251
		169	12:7	Weather Forecasting	252
		171	12:8	Climate	254
		174		Perspectives	
		177		Weather Watching	256
		179			
		180	**13**	**Oceanography**	260
		183	13:1	Seawater	261
		184	13:2	Temperature and Density	264
		186	13:3	The Seafloor	266
		189	13:4	Sea Life	269
			13:5	Ocean Currents	271
			13:6	Waves	273
			13:7	Tides	277
		194	13:8	Seismic Sea Waves and Surges	279
		195	13:9	Ocean Resources	280
		198		Perspectives	
		200		Commercial Diving	282
		202		Side Roads	
		204		The Sandy Seashore	286

ix

Environment is all the surroundings of an organism. The environment includes the plants, animals, climate, and nonliving parts of an area. Only the plants and animals that are suited to a particular environment live in it. How are these big horn sheep suited to their environment? Why is a water buffalo unsuited to this environment?

Rick McIntyre for Tom Stack & Assoc.

The **sequence** of chapters allows students to begin with topics which are familiar and progress within a framework of concepts suitable for their level of development.

The **unit organization** allows students to see the relationships among the concepts presented over many chapters.

12T Text Design

A Comprehensive and Balanced Approach to the Study of Science

A format designed to streamline the job of teaching, motivate students, and provide a successful learning environment.

Many ocean studies take place below the surface. A deep-sea diver must wear a heavy diving suit in order to withstand the pressures at great depths in the ocean. While a diver is beneath the ocean's surface, there are opportunities for observation and study. What forms of life are present in the ocean? What causes currents? What resources are found within the ocean?

260

The Pupil's text is overprinted in red with **Teacher Annotations**. Answers, points of emphasis, teaching strategies, and demonstrations are located with the text material.

Full color **chapter openings** are designed to motivate students, stimulate thought processes, and provide a purpose for studying the chapter.

Energy and Changes in Matter — chapter 3

Introducing the chapter: Place an ice cube in water in a beaker. Heat the water with a Bunsen burner until the ice melts and the water begins to boil. Discuss the use of heat energy to melt ice and boil water. Have students make a list of different examples of energy such as heat, light, electricity, and the kinetic energy of running water. Discuss the examples with the class and use this introduction as a springboard to the material in section 3:1.

3:1 Energy

Everything you do requires energy. Reading, running, throwing, climbing, and eating are all activities that use energy. Energy is also needed for travel and industry and to warm our homes, schools, and factories. Energy is used in refrigeration, lighting, and cooking. **Energy is the ability to do work.** In science, work means to move something.

GOAL: You will learn the sources and forms of energy and how energy is related to changes in matter.

What is energy?

Table 3–1.
Major Uses of Energy in the United States

Use	Percent
Industry	36
Transportation	27
Homes	21
Business	16
Total	100

Go over the scientific definition of energy and explain the meaning of work in a scientific sense.

Ask students where the energy they used yesterday has gone.

Energy can be divided into two kinds—potential and kinetic. **Potential energy** is energy of position, or stored energy. **Kinetic energy** is energy of motion. For example, moving water in a river has

How is potential different from kinetic energy?

45

A **Goal**, found in statement form, at the beginning of each chapter identifies the major objective to be accomplished in studying the chapter.

Each chapter is divided into **short numbered sections**. The section titles can be used as a framework for an outline of the chapter. Thus, students are better able to see the relationships among topics within a chapter.

Text Design **13T**

Reading and study aids designed to remove any

Readability is enhanced by using a single column format. Thus, the margin area is available for student notes and annotations to highlight important information.

Phrases and **sentences to be emphasized** are printed in italic type.

New terms are highlighted in boldface type. They are then used repeatedly in the following discussion to allow students to establish a familiarity with the term and its definition. All new terms are listed in the Vocabulary section at the end of the chapter and in the Glossary.

rapidly the rock is broken down. Sedimentary rock is worn away faster than metamorphic rock. The average rate of wear is about 1 m of rock for every 15 000 years.

Many fossils have been found in the rocks of the Grand Canyon. These fossils indicate that the land once lay under a shallow sea. It was at this time that the layers of sedimentary rock were formed. Later, over long periods of time, the land was uplifted from the sea. Desert conditions were present and land animals roamed the area. Fossils of animals, including reptiles are preserved in sand sediments formed by the blowing desert winds.

The rock history of the Grand Canyon is not complete. There are some time periods from which no rocks remain. The absence of rock layers is the result of erosion. A surface of eroded rock that separates younger rock layers from older layers is called an **unconformity.** Erosion took place when layers of rock were uplifted from beneath the sea and exposed to rain and wind. Sometimes a whole layer of rock was eroded away. See Figure 10-3.

According to the **principle of uniformity,** the processes that changed the earth in the past still exist today. The processes of weathering and erosion, which are evident today, have produced many of the earth's features.

Define unconformity.

Explain the principle of uniformity.

FIGURE 10-3. Sedimentary deposits beneath the ocean formed sedimentary rocks. Magma cut through the rock layers and formed a granite intrusion (a). When the rock layers were uplifted, erosion removed the upper layers. As ocean waters covered the rocks again, new sedimentary rock layers were deposited (b).

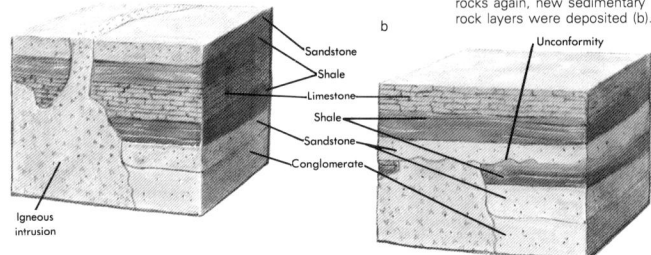

10:1 The Grand Canyon 197

14T   Text Design

obstacles students may encounter in learning.

Example problems are used to integrate math skills with science content. The format used enables students to develop logical problem solving skills.

Deceleration (dee sel uh RAY shun) means negative acceleration. You can use the acceleration equation to calculate deceleration. The deceleration you calculate is always a negative number. Why?

How is deceleration calculated?

Example
A sled traveling at 6 m/s slows to a stop in 3 s. Find the deceleration.
Solution
Step 1: Write the equation for average acceleration.
$$a = \frac{S_2 - S_1}{t}$$
Step 2: Substitute the values for final speed, initial speed, and time given in the problem.
$$a = \frac{0 - 6 \text{ m/s}}{3 \text{ s}}$$
Step 3: Subtract, then divide to find the answer.
$$\frac{0 - 6 \text{ m/s}}{3 \text{ s}} = -2 \text{ m/s}^2 \quad a = -2 \text{ m/s}^2$$

To stop or slow a moving object, force must be applied. The rate of deceleration depends on mass and force. The greater the force applied in stopping something the faster it decelerates. The greater the mass of an object, the more slowly it stops. In the example problem above, friction and gravity were the two forces that caused the sled to stop.

making sure
6. In which of the following does acceleration occur? Explain your answers.
   (a) Air rushes out of an inflated balloon.
   (b) A clock's second hand moves in a circle.
   (c) An ocean current flows at 3 km/h.
7. What is the acceleration of a skateboard whose speed changes from 5 km/h to 10 km/h in 5 s?
8. A parachute on a drag racer is opened. It slows the racer's speed from 260 km/h to 130 km/h in 10 s. Find the deceleration.

6:3 *Change in speed*   113

Guide Questions in the margin, printed in blue, are designed as reading comprehension checks. Answering these questions will highlight, for the student, the important points in the text.

Groups of questions and problems are located throughout each chapter under the heading **Making Sure**. They can be used for immediate review of the material just studied.

Text Design   15T

A program in which students expand and reinforce their knowledge of science through concrete laboratory experiences.

BUOYANCY AND DENSITY

Pour 15 mL of water into a 25-mL graduated cylinder. Tie one end of a piece of string tightly around a small stone. Hold the opposite end of the string and lower the stone into the water until it is completely submerged (Figure 7–12). What is the volume of the stone? Find the mass of the stone on a balance and then calculate its density. Compare it to the density of water (1 g/mL). Did the stone sink because it was more dense or less dense than water?

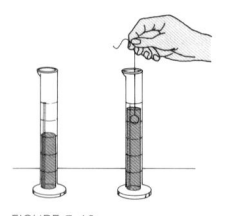
FIGURE 7–12.

FLOATING OBJECTS

(1) Place a displacement can on a table or desk top. Set a small beaker under the arm of the can. (2) Fill the can with water until water begins to run into the beaker. Empty water from the beaker and set it back under the arm of the filled can. (3) Fold a piece of heavy aluminum foil into the shape of a saucer. Place several small coins or washers in the saucer. (4) Find the mass of the s[...] and contents on a beam balance. (5) Then put the [...] in the displacement can (Figure 7–13). What happe[...] the saucer? Catch the displaced water in the smal[...] Measure the volume of the displaced water in a g[...] cylinder. (6) Then find the mass of the water (1 m[...] (7) Now fold the aluminum saucer tightly around [...] Put it into the refilled displacement can. What hap[...] (8) Find the volume and mass of the displaced wa[...] Compare the mass of the water to the mass of th[...] saucer.

making sure
10. Would a block of wood float higher or l[...] fresh water than in ocean water?
11. The density of alcohol is about 0.8 g/mL[...] a block of wood float higher or lower in [...] than in water?
12. Explain why a steel washer will sink in w[...] float in mercury.

Over 120 **Activities** are included throughout the text to reinforce science concepts and teach scientific inquiry. These Activities develop skills in simple observation, data collection, data organization, and data interpretation to form valid conclusions. Data collection Activities are organized into a numbered step format. This design makes for a more readable format and allows students to develop organizational skills in collecting data and formulating a report.

GAINING AND LOSING HEAT

Objective: To determine the movement of heat by measuring temperature

Materials
balance
beaker, 250 mL
ice cube
stirring rod
stopwatch or watch with second hand
thermometer
tongs

Procedure
1. Determine the mass of the empty beaker.
2. Fill the beaker with 125 mL of water. Find the total mass of the beaker and the water. Determine the mass of the water by subtracting the mass of the empty beaker from the total mass of the beaker and water. Record the mass of the water.
3. Carefully place the thermometer into the water and record the temperature of the water. Note: Be certain that the thermometer does not touch the sides or bottom of the beaker.
4. Add the ice cube to the water and stir for 1 min. CAUTION: Use the stirring rod and not the thermometer to stir the water. After 1 min, measure and record the temperature of the water.
5. Continue stirring the water and ice and measuring the temperature for 5 min. Record the temperature after each minute.
6. After 5 min, remove any ice that is left. Stir the water gently again and measure the temperature each minute until the water returns to room temperature.
7. Find and record the mass of water in the beaker.

Observations and Data

Mass of the water before ice was added: ____ g

Mass of the water after the ice melted: ____ g

Time (min)	Temperature (°C)
0	
1	
2	
3	

Questions and Conclusions
1. What was the lowest temperature to which the water cooled in the beaker?
2. How was the temperature of the water changed by the loss and gain of heat?
3. Where did the heat go when the water was cooled and warmed?
4. How did the rate in which the water cooled compare to the time it took for the water to return to room temperature?
5. How did the change in the mass of water affect the rate in which the water returned to room temperature?

3:2 Heat and Temperature 49

A strong graphic presentation to help students visualize new concepts and relate science principles to their everyday lives.

Illustrations provide the visual models needed to fully understand the abstract concepts and descriptions on the microscopic level.

9:10 Volcanoes

Volcanoes form when magma is squeezed from the earth's interior to the surface. Volcanoes form on continents as well as on the ocean floor. Most volcanic activity occurs where plates collide or move apart. Volcanism refers to the process by which magma is produced and moved to the surface. Magma that reaches the surface is called **lava.** Volcanism also produces and expels hot gases and solid debris from openings in the crust.

Where do volcanoes form?

What is lava?

FIGURE 14–3. There are some differences between plant and animal cells.

How is a plant cell different from an animal cell?

What are protoplasm and cytoplasm?

In animal cells, the cell membrane is the outer layer of the cell (Figure 14–3). Plant cells, however, have an outer layer called a cell wall which surrounds the cell membrane. The cell wall forms a stiff case around the cell. It is made mostly of a material called cellulose (SEL yuh lohs). Cellulose gives strength to the cell wall.

Protoplasm is the "living material" of the cell. Each cell is a unit or mass of protoplasm. Protoplasm is 70 percent water. Proteins, fats, carbohydrates, and minerals make the remaining 30 percent. With the exception of the cell wall in plant cells, all parts of a cell are made of protoplasm. Cytoplasm, for example, is the protoplasm that lies outside the nucleus of a cell.

...sometimes lava ...surface. The ...Instead, the ...high plateau. ...in this manner.

FIGURE 9–32. Locations of the world's mountain ranges, volcanoes, and earthquake zones are related to the movements of Earth's crustal plates (a). This volcano in Iceland erupted along the Mid-Atlantic Ridge where two crustal plates are moving apart (b).

FIGURE 9–33. The Columbia Plateau in northwestern United States formed when basalt lava erupted from large cracks in the earth's surface.

Table 14–1.
Elements in Human Protoplasm

Element	Percent
Oxygen	65
Carbon	18
Hydrogen	10
Nitrogen	3
Calcium	2
Phosphorus	1
Trace elements	1

Tables are used to organize text information and quantitative data. The use of a tabular format allows students to see relationships among ideas and data at a glance.

Additional special features provide for remedial and enrichment activities thereby completing the program.

PERSPECTIVES
people
A Dedicated Microbiologist

Microbiology is the study of small organisms, also called microorganisms. Bacteria, viruses, fungi, and some algae are common examples of microorganisms. Natalie Jones graduated from college with a degree in microbiology. Numerous careers that were open to her included work in hospitals, private laboratories, food industries, and environmental agencies.

Natalie Jones decided to work with more than just microorganisms. Her job at the Ohio State University microbiology laboratories is to help over 1000 students perform experiments. The students that she works with are involved in pre-medical studies, pharmacy, veterinary medicine, optometry, dentistry, nursing, and other sciences. Microbiology is an important aspect of each of these sciences.

As an instructional assistant, her job is to make sure that each student receives the necessary equipment and supplies to perform a required experiment. The job is not easy, as Ms. Jones must know every aspect of the experiment. Many of the microorganisms that she works with are pathogenic, or disease-causing. She has to be very organized to make necessary materials available to students and researchers. All supplies, equipment, and specimens must be ordered well in advance of an experiment.

Cultures, or containers with living organisms, are maintained under the supervision of Natalie Jones. Such cultures cannot be contaminated with other microorganisms that are floating in the air. Special techniques are used when working with these microorganisms, and are monitored by Ms. Jones.

Doug Wynn

312

Perspectives on **People** and **Frontiers** are magazinelike pages featuring articles on current science topics and applications. **Careers** are highlighted by looking at the activities of people in various disciplines related to science.

PERSPECTIVES
skills
Using Graphs

Graphs are used often to show and explain scientific information. The study of science involves your understanding many facts and seeing how they are related. The ability to interpret graphs makes it possible for you to acquire much information quickly. The *bar graph* below shows how land is used in a typical community.

1. What part of the community uses more land than any other. How many hectares are used?
2. What part of the community uses the least amount of land?
3. Determine the total number of hectares used for all activities.

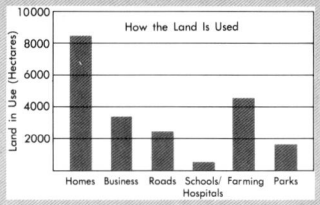

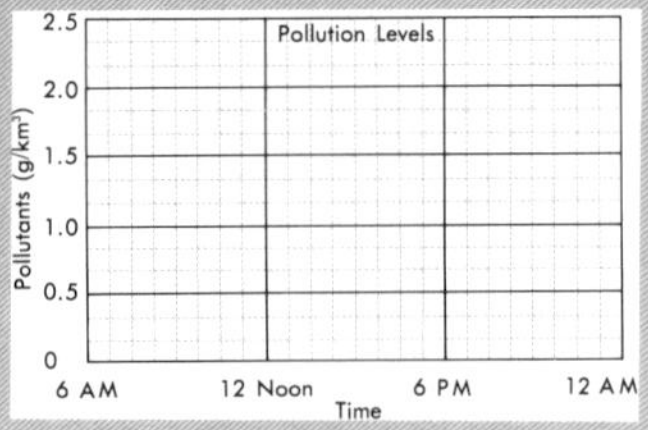

Circle graphs show the parts of a whole in the form of percentages. Even without percentages given, you can see the relative areas given to each part of the circle graph. The circle graph below shows water usage for a typical household.

1. In a household, what two uses of water amount to about two thirds of the total?
2. If a typical household uses 1020 L of water per day, how many liters would be used for bathing?
3. Suppose you had to conserve water and you were requested to cut the amount you used by 40 L. Using the graph, list all the ways by which you might be able to decrease the amount of water you use.

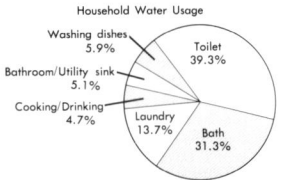

Line graphs show change such as growth, development over a time period, or a process. The following graph shows changes in air pollution levels for a time period of a typical day in a city.

1. How would you account for the highest amount of pollution to occur around 5:00 P.M.?
2. What do you think causes the first peak of pollution around 9:00 A.M.?
3. When does the lowest level of pollution occur? Why?

484

Skills pages feature activities designed to improve reading comprehension and study skills.

Photographs are used to illustrate the text information or present additional applications and analogies to help students relate science to everyday life. The captions reinforce the text or provide additional relevant information.

21:3 Fossil Evidence

Fossils provide evidence supporting the theory of evolution by natural selection. Fossils show that life in the past was very different from today. At one time there were huge forests of giant ferns, and dinosaurs roamed the earth. Fossils of one reptile species show it had wings and could fly. Woolly mammoths have been discovered preserved in ice in Siberia.

Fossil records show that species have changed during millions of years. In some cases these changes appear to be sudden and sharp. But most changes in plants and animals were slow and gradual. The changes in species over time caused most species to die off and new species to take their place. Scientists believe that the new species were descended from the old species. According to one theory, birds and reptiles are descended from the same ancestors.

Trilobites are an example of an animal for which there are many fossils. Trilobites lived about 500 million years ago. Trilobites no longer exist on the earth. They became **extinct** many years ago. However, species that resemble trilobites are alive today. Among these species are crabs and lobsters. It is possible that the trilobites that lived long ago are the distant ancestors of crabs and lobsters.

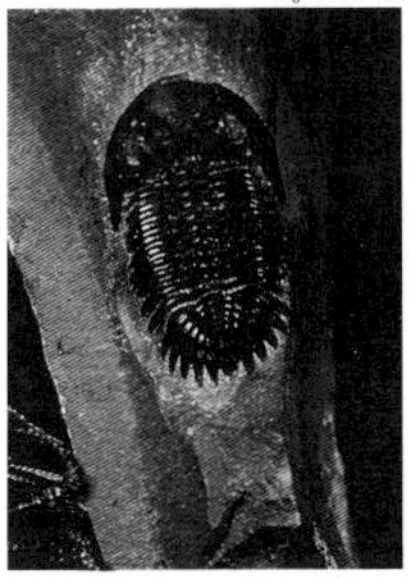

FIGURE 21-7. Trilobites became extinct millions of years ago.

What do fossils show about past life?

FIGURE 21-8. Fossil records show that tropical vegetation (a) once covered Greenland (b).

456 Descent and Change

SIDE ROADS
The Sandy Seashore

Unit 3 describes many physical conditions that are found on Earth. These conditions determine the manner in which plants and animals survive in a given area. In the marine environment, some factors include salinity, temperature, density, currents, waves, and tides. Each plant and animal reflects the physical conditions of its individual habitat. By examining a typical sandy beach in the Caribbean, we can observe the relationships between life forms and the physical conditions.

If one stands in shallow water and walks toward the shore, different zones of life are encountered. The first zone is called the subtidal zone. This area is always covered with water. It usually has a sandy bottom and very little plant growth. Most food in the subtidal zone comes in from outside the area. Predominant animals are burrowers that feed on plankton or small particles of food called detritus. Fan worms spread brightly colored tentacles into the water to collect plankton. When disturbed, these tentacles quickly retreat into tubes that are buried in the sand.

Sea urchins, protected by hard spines, move slowly across the bottom in search of food. Sand dollars also slide their thin bodies through the sand while feeding on detritus. Occasionally, a sting ray will swim in from outside the subtidal zone. Crabs are usually buried in the sand and are very difficult to see. When looking for food, a crab blends in with the color of the sand.

The next area encountered inland is the intertidal zone. It is covered with water during high tide and exposed during low tide. Intertidal animals must be able to breathe under water and also survive without water. Worms are the most common type of animal, living in

286

Side Roads features give students the opportunity to look at some practical application of science in more depth. These features are designed as an enrichment section for the unit.

A complete range of chapter review materials to review science concepts and diagnose weaknesses.

main ideas   Chapter Review

1. Animals can be divided into two groups, vertebrates and invertebrates. — 17:1
2. Fish are vertebrates that live in water, obtain oxygen through gills, and are covered with scales. — 17:2
3. Amphibians are vertebrates with moist skin and no scales. They live part of their lives in water and part on land. — 17:3
4. A frog's body is a good illustration of the systems of organs present in vertebrate animals. — 17:4
5. Fish, amphibians, and reptiles are cold-blooded animals. Birds and mammals are warm-blooded. — 17:2–17:7
6. Most species of reptiles live in tropical or subtropical regions — 17:5
7. Birds are vertebrates that have wings, two legs, feathers, and reproduce by laying eggs. — 17:6
8. Some birds stay in the same area all year and others migrate when the seasons change. — 17:6
9. Mammals are vertebrates that have hair, are warm-blooded, and feed their young on milk.
10. Most animals reproduce by laying eggs.
11. Fertilization and development in most s occur inside the mother's body.

Main Ideas lists the major points presented in the chapter. Each statement is annotated for students allowing them to refer back to the section where the statement is presented.

vocabulary
Define each of the following words or terms.
mammal   reptile   amphibian

Vocabulary is a list of important science terms presented in the chapter.

study questions
DO NOT WRITE IN THIS BOOK.
A. True or False
Determine whether each of the following sente If the sentence is false, rewrite it to make it true
1. The fertilization of eggs in a reptile oc female's body.
2. Bones protect internal body organs.

4. ____, ____, and ____ are three kinds of earthquake waves.
5. Most earthquakes are located on the ____ of the plates.
6. Quartz is a ____ found in rocks.
7. ____ is molten rock that flows from a volcano and turns into solid rock when it cools.
8. Bedrock is usually covered with a layer of ____.
9. A ____ is a break in the rocks, within the earth's crust.
10. ____ is the breaking down of the earth's crust by wind, water, ice, plants, animals, and chemical changes.

D. How and Why
1. How could you identify a rock or mineral?
2. How is sedimentary rock formed?
3. Why is the earth's crust constantly changing?
4. Why is it not possible to locate the center of an earthquake with a single seismograph?
5. How were the Alps and Himalaya mountains formed?

challenges
1. Have you ever thought about becoming a rock hound? Obtain a book about rock and mineral collecting. Make a collection of different rocks and minerals. Find out about chemical tests for identifying minerals.
2. Obtain information about research that scientists are doing to predict the occurrence of earthquakes. Prepare a report for your class.
3. Obtain a book from the library on crystal growing. Learn how to grow the crystals of several kinds of minerals. Prepare an exhibit of the crystals along with a description of the process you used to grow them.

interesting reading
Deeker, Robert and Barbara, *Volcanoes.* San Francisco, CA: Freeman, 1980.
Dietrich, R. V., *Stones: Their Collection, Identification, and Uses.* San Francisco, CA: Freeman, 1980.

Study Questions contains a range of evaluation materials—True-False questions, Multiple Choice, Completion, and short answer applications questions listed under How and Why.

Challenges is a set of questions and activities that take students beyond the material presented in the chapter.

Interesting Reading is a set of suggested references for further study of the chapter material or related topics.

Ancillary materials designed to provide for hands-on learning, development of reading and mathematics skills, and evaluation of student progress.

Evaluation Program

The **Evaluation Program for Principles of Science** is designed to reinforce student learning, to evaluate student progress, and to discover areas of weakness in instruction. The program consists of a four-part test for each of the 23 chapters of **Principles of Science, Book I.** Test items are based on the Performance Objectives for each chapter listed in this Teacher's Guide. The program is available as Spirit Duplicating Masters and Reproducible black line Masters in the Teacher Resource Book.

The four sections of each chapter test include:
Understanding Concepts —A set of multiple choice questions in which students demonstrate their ability to recall specific facts from the chapter.
Interpreting Concepts—A set of completion and matching items in which students demonstrate their ability to interpret basic concepts.
Using Concepts—A set of applications items in which students demonstrate their ability to apply their text knowledge to new situations.
Completing Concepts—A set of completion items in which students demonstrate their understanding of science terms used in context.

The major focus of each test is to help students improve their knowledge of science. Using these tests as a diagnostic tool, you can decide what supplemental materials may be needed to overcome any weaknesses in presentation and interaction that may become apparent. Annotations are used to answer all questions on the tests.

Teacher Resource Book

The **Teacher Resource Book** is designed as a time-saving teaching aid with the purpose of assisting the teacher in all areas of science instruction. It is organized with tabbed dividers that correspond to each text chapter. The materials provided for each chapter include a Chapter Planning Guide, Teaching Masters, Review Guide Masters, Challenge Masters, and Evaluation Masters.

Chapter Planning Guide
The Chapter Planning Guide lists all resources available for the chapter in the pupil's text, the Teacher's Guide, the Teacher Resource book, and the Activity-Centered Program Teacher's Guide. It aids in planning for daily instruction.

Teaching Masters
Teaching Masters are designed to be used as transparencies and/or student reference sheets. They facilitate class discussion and aid in the reinforcement of basic concepts.

Review Guide Masters
The Review Guide Masters contain pencil and paper exercises designed to aid student mastery of reading and mathematics skills simultaneously with science concepts. The exercises stress the areas of word-processing skills, context-processing skills, study skills, and problem-solving skills.

Challenge Masters
Challenge Masters are designed to challenge the more motivated students. The exercises emphasize inference and other higher-level critical thinking skills. Students are often asked to apply concepts learned in the chapter to new situations. Expansion of chapter concepts may be a goal of other exercises.

Evaluation Masters
The Evaluation Program described above is included as reproducible blackline masters in the Teacher Resource Book.

Activity-Centered Program Teacher's Guide

Principles of Science: Activity-Centered Program Teacher's Guide is designed to assist in planning a general science course using a laboratory approach. This guide is correlated to the activities that appear in the pupil's text. It provides additional questions and hints for directing students in making observations and interpreting data. The **Activity-Centered Program** contains detailed instructions for carrying out the paragraph format activities in the text and additional activities, and challenging follow-up procedures and questions. The additional activities require inexpensive and readily available materials. Suggestions for the use of alternative supplies and for organizing laboratory materials in the classroom are also provided. *Challenge: Applying Scientific Methods* exercises are designed to challenge more motivated students. These are laboratory activities that stress application of scientific methods to the concepts learned in the chapter.

Using Principles of Science

Sequence

Principles of Science is designed for maximum flexiblity in teaching. The text is divided into six units, which are further subdivided into 23 chapters. Each chapter is then divided into numbered sections, each being a specific topic of study. The listing of topics in the Table of Contents can be used as a unit outline for students. This format also aids you in planning lessons and making assignments. The Planning Guide on page 29T will further assist you in planning and organizing your lessons to meet the needs of your students.

In examining the Scope and Sequence on pages 6T-11T, you can see how the two **Principles of Science** books relate to each other. You should notice the balanced approach in that life, earth, and physical science are given equal treatment.

The sequence of units in both books follows the philosophy of having students progress from simple or familiar topics to more complex or abstract topics. This philosophy is clearly demonstrated in looking at the Scope and Sequence. The material in *Book 1* is basic enough to be a starting point for those students who have had a minimum of science at the elementary level yet challenging enough to stimulate students with good science backgrounds.

The unit overviews provide additional rationale for why topics are organized in a particular sequence.

Teacher's Guide
Unit Overview and Teaching Tips

The Overview gives you background information concerning the relationship among the chapters in the unit. The unit opening paragraph and photograph in the text are designed to be used to begin class discussions concerning the material in the unit. This approach gives you an opportunity to determine what prior knowledge your students have concerning particular topics. Thus, you are better able to structure your lessons to fit the needs of your students.

The Teaching Tips listed beside each unit overview provide you with insights on how you can present the unit. Ideas for long term projects can also be found here.

Demonstrations should be an essential part of your teaching strategy. The demonstration as a teaching technique allows you to present a maximum amount of text material in a concrete and visual manner. Good demonstrations also allow for the development of observational skills and proper lab techniques. You can use the demonstrations in this text to emphasize safety rules and procedures. Demonstrations can be found as red annotations on the Pupil's pages, as part of the Teaching Tips for the unit, and as part of the Section background for each chapter. Additional demonstrations can be found in the **Principles of Science Activity-Centered Program Teacher's Guide.**

Demonstrations may be approached in one of the following ways:

1. Begin with a clearly defined problem and have students develop a solution.
2. Silently perform the demonstration so students can sharpen their abilities to observe and draw conclusions.

Allowing students to develop skills via both of these approaches will contribute to their success in the laboratory portion of the course. These approaches also allow students to recognize and verbalize relationships based on their level of experience.

Review lessons are also an essential teaching strategy. Review before chapter or unit exams helps students establish an overall picture for the purpose of developing relationships. The Skills pages aid students in establishing models for gathering and summarizing information and discriminating between important and unimportant material.

The review lesson can be teacher or student directed. For example, the class can be divided into teams. Each team can prepare questions for the others. The questions can be a homework assignment. Another alternative would be to have a team of "experts" for each unit. These experts would be responsible for preparing the questions for the unit review. Students can be graded on the quality and quantity of the questions they generate.

You could also use the format from any question and answer game show on TV as the basis for your review lesson.

Chapter Overview

A brief summary of the major themes in the chapter are given. Suggested discussion questions, points of emphasis, and additional teaching strategies are often provided.

Performance Objectives

The Goal for each chapter is listed with an expanded list of objectives. These objectives highlight the important concepts for you. When designing your chapter tests or course of study for general science, use these objectives as a framework. The **Evaluation Program for Principles of Science** is written based on the Performance Objectives listed in this Teacher's Guide. An explanation of the coding system used with these objectives is found on page 41 T.

Section Background

Teaching strategies for presenting each section are given. Information concerning the authors' approach to a particular topic will be stated. For example, the background for Section 2:2 describes how to introduce the section with illustrations and a demonstration. This feature helps you by providing a number of alternative strategies throughout the text. This section also provides additional science background for you if it is appropriate, thus eliminating the need to research supplementary texts. Discussion questions and additional information concerning text illustrations may also be included.

Pupil's Text
Guide Questions in the Margins

These questions assist students by identifying the major principles, concepts, and ideas that are important in reading and later reviewing the chapter. In assigning a section for study, you may wish to read these questions to the class or have students read them silently in order to clarify the nature and purpose of the assignment. These questions provide students with the opportunity to check their learning as they proceed through an assignment. You can use the questions as a homework assignment. Then, an oral review of the questions and answers makes an excellent summary for the chapter.

Activities

Each Activity in this text is numbered for you using a red annotation. Use these numbers when referring to this Teacher's Guide for background information. The Activities in this text are of two types. For example, turn to the Activity on page 15. Note that it is a different format from the Activity on page 16. The short paragraph Activities require little preparation and are usually observational in nature. The procedures are generally finished quickly allowing ample time for post lab discussion.

The long-form Activities (like that on page 16) require more time for data collection and interpretation. The equipment for these Activities is still relatively simple but may require more preparation time. The primary objective of the Activity is listed under the title. These Activities are organized in a manner that enables students to generate organized, coherent laboratory reports. Since the communication of knowledge is an important aspect of science, you should give students frequent opportunities to express the results of their work in the laboratory. It is also vitally important that students make the distinction between observations and conclusions. The organization of the long-form Activities makes this distinction clear.

You will note that, for the most part, all Activities are placed toward the end of a text section. This placement should not imply that the Activities are to be performed after the reading. We are aware that teachers use Activities at many different points in presenting a concept. By consistently placing them toward the end of a section, you can be more flexible in introducing them at the time that best suits your purposes. The insertion of Activity copy within the text copy may distract some students and hamper readability. In reading a section, you may want students to establish the logical flow of ideas. Keeping the sections short and uninterrupted increases comprehension.

Additional information on structuring an activity-centered program is found in the **Principles of Science Activity-Centered Program Teacher's Guide.** This guide furnishes ideas on how to set up the classroom for efficient use of work space, safety considerations, and how to evaluate laboratory work.

Laboratory Equipment and Safety

Student laboratory work must always be supervised. All students should wear safety glasses when conducting any Activities that require the use of chemicals and any procedure that requires students to heat a substance. Clothing protection in the form of laboratory aprons is also recommended. Encourage students to develop serious attitudes toward laboratory work.

Specific safety precautions necessary for individual Activities are given in the Activity itself and as teacher annotations. Note these precautions and explain them to students. As part of the introduction to your course, be sure to cover the Safety and First Aid information in Appendix B of the text. For safety reasons, it is important that you supervise the dispensation of all chemicals for Activity work.

At the beginning of the year, students should be given proper instructions in the use of the microscope, balance, and burner. Correct techniques are the key to success in the lab.

The master equipment list begins on page 30T. This list is broken down on a chapter-by-chapter basis in the sections following the performance objectives for each chapter. Specific equipment for the paragraph format Activities is listed under the heading for the Activity number in this Teacher's Guide only. The equipment is not listed in the pupil's edition.

Additional information concerning ways to construct your own equipment is given in the **Principles of Science Activity-Centered Program Teacher's Guide.**

Keeping a Science Notebook

Keeping a notebook is important in communicating the idea to students that records are essential to promote scientific inquiry. Data is used to discover patterns and to draw conclusions. Observations about events can be recorded in diverse ways such as simple drawings, graphs, or lengthy reports. Data should be recorded accurately in a notebook as soon as possible. Use the notebook as a tool to help students sharpen their observational skills as well as their verbal and written communication skills.

You can vary the type of records your students keep. The following are some ideas.

1. labeling objects in a collection—name, location found, composition, and so on
2. making drawings, diagrams, graphs, and charts
3. writing short formal reports for work in the lab
4. class notes from lectures, movies, and discussions
5. homework, quizzes, and tests

Photographs and Illustrations

Today's students rely on visual presentations to help clarify and reinforce verbal presentations. We maintain that a visual presentation not only motivates but adds to their level of understanding. The chapter and unit opening presentations are designed to stimulate student interest by using photographs to provide the visual link between science and the real world. It is important to use the graphics throughout the text as a functional part of your lesson presentation. The interpretation of graphic aids as a reading skill is discussed on page 26T.

The Skills pages cover some of these imporant skills. Interpreting graphs, tables, and charts, and using maps are essential methods of gaining scientific information. Take the time to review the kinds of information that can be gained from graphic aids.

Examples

Selected mathematics concepts from the middle/junior high school curriculum are integrated into the text. This approach provides meaningful illustrations of practical applications of mathematics in science. It also makes mathematics more relevant and interesting for students. The use of selected mathematics concepts in science also reinforces learning.

Specific mathematics skills covered throughout the text are listed in the Scope and Sequence beginning on page 6T. You may want to review these math concepts before beginning the particular chapter so students can be ready to apply this material to the science concepts presented.

Calculations for working each kind of problem are presented in step-by-step detail. Encourage students to adopt this method in working problems on their own.

Making Sure

All questions in these sections are answered in detail in this Teacher's Guide. Note that these questions are placed at the end of a section for consistency. You may wish to assign them individually at different points in the reading.

Many of these questions inolve more than simple recall. Problems are included as a quick test to see whether students can apply math concepts to science. Students should be encouraged to answer these questions and problems on an individual basis. Reviewing the answers and solutions should then be a group activity. This approach should help clarify concepts and guard against a vague acceptance of statements in the text without having had to apply these ideas.

Perspectives/Skills Pages

These pages are designed to present topics in science in a form more like a magazine article. Thus, students are given a break from the intense, precise style of scientific writing. These pages are especially useful in planning career education materials for your science class. Many careers are the focus of an article or discussed subtly in covering an interesting topic related to science. The following should be some of your goals in presenting the Perspectives pages.
1. The activities associated with science are not relegated to the lab and classroom.
2. Those working in science-related fields are not all classified as scientists. People with varying educational backgrounds work in science-related fields.
3. Not all scientific breakthroughs are the result of research teams of scientists. Many ideas and inventions are the result of human ingenuity rather than scientific expertise.

For additional information concerning careers related to science, see pages 39T-41T.

The Skills pages are designed to give students additional practice in sharpening their reading, math, and study skills. In planning your lessons for a particular chapter, you may want to have students turn to the Skills page as one of your introductory lessons. Many of the techniques they will learn can be applied to the material throughout the whole chapter. The Skills pages are placed at the end of the chapter for convenience.

Chapter-End Materials
Main Ideas

This section is a concise review of the major concepts in the chapter. Each Main Ideas statement is keyed back to the chapter section by a blue student note. You can begin the chapter using the Main Ideas to determine how much your students know about a topic from previous science courses. This technique is especially important when your class is composed of students with a variety of backgrounds in elementary school science. The use of the chapter preview as a reading comprehension aid is illustrated on page 18 of **Principles of Science, Book One.** The Main Ideas not only help students discern the important points in the chapter, but give you insight as to the authors' points of emphasis.

The Vocabulary section contains the boldfaced science terms for the chapter in alphabetical order. We strongly urge you to make vocabulary development an integral part of your lesson planning. The following are some suggestions for ways to vary your approach in reviewing vocabulary terms.

1. Students should be given the opportunity to use new terms as often as possible in writing and in class discussion. Asking them to write a short story using the vocabulary words for the chapter can show you their ability to use terms correctly in context.
2. Students can be asked to make an outline of the chapter using the vocabulary words. An example of this technique is shown on page 100 of **Principles of Science, Book One.**
3. A number of other vocabulary development activities and techniques are found in **Principles of Science Teacher Resource Book.** The evaluation of students' grasp of new terms and their definitions is Part D of the **Evaluation Program for Principles of Science.**

The Study Questions are progressively more difficult as students move from Parts A to D. You should keep this fact in mind if you are trying to individualize, or are teaching low-ability students. Some of the How and Why questions may require paragraph-type answers. These questions generally involve the application and interpretation of knowledge. Parts A and C are generally simple recall questions.

The questions and projects under Investigations can be used for individualization or enrichment activities. The reports required give students practice in using the library and researching information. The experiments encourage an inquiring attitude about science. Many times these items provide the link between science principles and everyday applications.

Students should be encouraged to read materials related to science other than their text. The popularity of the new science magazines shows a demand for information in a format that can be used by the nonscientist. Readings are suggested at the end of each chapter and in this Teacher's Guide in the hope that you will encourage students to read materials in science. Assigning reports and research projects is another way to get them to read scientific materials.

Appendices

The Appendices are a functional part of the text in that they contain supplementary, review, and enrichment information. Having the information in the Appendices eliminates the need to buy a number of different handbooks. Scientific notation is reviewed in *Book I*. Safety guidelines and first aid procedures appear in both books. It is very important that you review this material before beginning any of the Activities in the text. Supplementary information concerning SI measurement is found in both books. The other reference information and data tables are pertinent to the text material in that book. For example, *Book II* contains a Calorie chart that is used throughout the Nutrition chapter. Give students as much practice as possible in deriving information from the Appendices. The ability to use tables and charts is a reading comprehension skill.

Glossary

All the boldfaced science terms in the text are defined in the glossary. Vocabulary building is an important part of teaching science. Be sure students are familar with the Glossary and encourage them to get into the habit of looking up words that are unfamiliar.

Reading Science Information

Though we think of science study as one in which students conduct experiments and discover scientific data, we find that the dissemination and acquisition of data among scientists is through the reading of printed materials. Thus, by reading, scientists are able to know the information their colleagues have discovered.

Science writing can be described as intense, precise, specific, and jammed with facts, among which are many interrelationships. This compact writing often presents difficulties to students who, in most cases, have learned to read by using more casual narrative material. Specifically, students must master technical vocabulary representing scientific concepts. These concepts are in ascending order with new knowledge building on what has been learned. The use of the context and the knowledge of word parts enables students to determine the meaning of technical words. Then, students must not only note the facts but see the interrelationships among them—discern the organization of the information leading toward a critical interpretation and the application of the data. Further, students must have competence in the various reading-study skills: attacking a reading assignment, using an appropriate study procedure, and using graphic aids such as graphs, pictures, tables, and charts effectively.

Being able to note the author's organization of data and to organize this information for oneself is a most important requisite for adequate comprehension of science materials. Specific techniques would include:

1. Taking notes that will give students practice in noting main ideas and related details and in noting the interrelationships of the main ideas and details.
2. Noting the author's organization (outline) through the topical headings.
3. Previewing the reading assignment with students to acquaint them with the material as well as to point out the organization of the information.
4. Giving students, as necessary, a skeletal plan of the information that would show the organization of concepts.
5. Helping students to evolve purposes for reading that will guide them toward the specific understandings you wish them to acquire.

Two other major areas in which students need instruction are in the understanding of technical terms and in the use of graphic aids. Definitions of technical terms can be obtained from the glossary. Meanings can also be gained from the context (a new term is often defined when it is introduced) and by knowing the meanings of word parts (roots, prefixes, and suffixes). Graphic aids seem simple to use, but students need instruction in reading and interpreting the information contained in tables, charts, graphs, diagrams, and pictures.

The text, **Principles of Science,** is designed to assist you in helping your students to read competently in the following ways:

1. The topical headings in each chapter are the author's organization of the information. Look at these in a preview before reading, and aid students in noting the development and progression of information.
2. The text has numerous graphic aids. Discuss information from these. This will give students practice in noting the information. These aids can be very helpful to a slow reader.
3. Note the series of "Making Sure" questions throughout each chapter. Alert your students to the purpose of these questions so that an understanding of data can build progressively.
4. Each text has six Skills pages emphasizing student previewing of an assignment, seeing the organization through ways of taking notes, using graphic aids, and getting the meaning of technical terms through the context and etymology. These Skills pages are designed to involve the student in the appropriate techniques in the chapter where they are found. However, either give preparatory instruction in the skill *before* assigning the page or work through the page with the students in class with a follow-up of your own design. DO NOT merely assign the pages—they will prove to be too difficult for most of the students.
5. Use the end-of-chapter features for review, extension, enrichment, and skill application. The "Main Ideas" are listed with the paragraph coding showing where they are located. The "Study questions" are excellent for review. The "Challenges" and "Interesting Reading" are suggestions for extension and enrichment.

If we think of science as learning through discovery, then as teachers we must make that possible by aiding student development of a basic conceptual background and competence in the reading-study skills.

Bibliography

Dupuis, Mary M. and Askov, Eunice N. *Content Area Reading, An Individualized Approach.* Prentice-Hall, Inc., Englewood Cliffs, NJ 1982.

Lamberg, Walter J. and Lamb, Charles E. *Reading Instruction in the Content Areas.* College Publishing Company, Chicago, 1980.

Shepherd, David L. *Comprehensive High School Reading Methods,* 3rd ed. Charles E. Merrill Publishing Co., Columbus, OH 1982.

Planning the Course

Principles of Science provides an in-depth study of selected topics in the life, earth, and physical sciences. Each book may be used as the basis of a one-year general science course. However, this program is flexible enough to be used in a variety of ways. You may only need a single semester course. This Planning Guide can be used to help you design a course which covers only basic principles from life, earth, and physical sciences.

The guide is organized to present a suggested number of class sessions for each chapter. The entire course is based on 180 class sessions in the school year. The sections of each chapter are classified as being of primary or secondary importance. A single semester course should include all sections designated as primary in importance. If you are teaching a full-year course to lower ability students you may want to use only the sections of primary importance. Then, if time allows, you can select additional topics from the secondary importance sections. The number of class sessions suggested includes time for activities, laboratory experiments, and testing. You may find that some classes will take longer to cover a chapter than what is projected in the planning guide. Use this guide as a framework and not as a rigid schedule to which you must adhere.

Principles of Science lends itself to a variety of teaching methods—activity-centered, demonstration, group reading of text material, lecture, activity reporting, class discussion, homework review, viewing films and filmstrips, and preparation for tests and quizzes. Different methods reinforce each other and provide a change of pace thereby increasing student motivation and success.

Principles of Science may also be used as the basis for a course in which the emphasis is on some aspect of life, earth, or physical science. The following are some options.

Single semester course in Earth and Physical Science
Unit 1: Matter and Energy
Unit 2: Mechanics
Unit 3: Earth Science
Unit 6: Chapter 22: Soil, Water, and Air Conservation

Single semester course in Life Science, Ecology, and Conservation
Unit 1: Chapter 1: Science and Measurement
Unit 4: Living Things
Unit 5: Environment and Heredity
Unit 6: Conservation

Single semester Introduction to Science course
Unit 1: Matter and Energy
Unit 3: Earth Science
Unit 4: Living Things

Planning Guide for Principles of Science

Chapter	Class Sessions		Emphasis Level		Teacher Resource Book Pages
	Semester Course	Full Year Course	Primary	Secondary	
1	5	5	All sections		1–1 to 1–18
2	6	8	All sections		2–1 to 2–18
3	6	8	3:1–3:5	3:6–3:8	3–1 to 3–18
4	6	10	4:1, 4:2, 4:4–4:6	4:3, 4:7	4–1 to 4–18
5	6	10	5:1–5:4	5:5–5:10	5–1 to 5–18
6	5	8	6:1, 6:2, 6:4, 6:5	6:3, 6:6, 6:7	6–1 to 6–18
7	5	8	7:1–7:4	7:5–7:7	7–1 to 7–18
8	4	6	8:1–8:4, 8:7	8:5, 8:6	8–1 to 8–18
9	5	8	9:1, 9:3–9:8	9:2, 9:8, 9:10	9–1 to 9–20
10	4	7	10:1–10:5	10:6–10:10	10–1 to 10–20
11	5	8	11:1–11:5, 11:7	11:6, 11:8, 11:9	11–1 to 11–18
12	—	8	12:1–12:3, 12:5, 12:6	12:4, 12:7, 12:8	12–1 to 12–18
13	—	8	13:1–13:5	13:6–13:9	13–1 to 13–18
14	5	8	All Sections		14–1 to 14–20
15	5	8	All Sections		15–1 to 15–18
16	5	10	All Sections		16–1 to 16–18
17	5	10	All Sections		17–1 to 17–18
18	4	8	18:1–18:4	18:5–18:8	18–1 to 18–18
19	—	8	19:1–19:3	19:4–19:6	19–1 to 19–18
20	5	10	20:1–20:9	20:10	20–1 to 20–18
21	4	6	All Sections		21–1 to 21–20
22	—	5	22:1–22:6, 22:8	22:7	22–1 to 22–18
23	—	5	23:1–23:6	23:7	23–1 to 23–18

Equipment and Supplies for Activities

A complete list of equipment and supplies for the entire course is given below for your convenience in ordering materials. The amounts listed are approximate values for one class of 30 students working in pairs. The materials have been divided into laboratory equipment, biological supplies, and chemical supplies. Most biological and chemical supplies will need to be replaced each school year. Most laboratory equipment may be used many years without replacement. The activities are designed to use simple apparatus. Use materials that are available locally whenever possible. A list of supply houses follows the equipment list.

Item	Amount	Activity used in
Laboratory Equipment		
anemometer	1	12–1
aquarium tank (with air pump, filter, heater, and gravel)	1	17–2, 17–4, 18–2, 23–1
baking dish	15	22–3
ball, small rubber	15	6–7
large playground	15	8–2
balloons	60	1–3, 3–6, 6–4 14–4
barometer	1	11–2, 12–1
battery (1 and ½ or 6 volt)	15	6–6
beakers, 50 mL	45	3–4, 15–10, 18–3
250 mL	15	3–2, 7–4, 15–9, 22–2
1 L	15	2–1, 7–2, 13–1, 13–2, 14–2, 14–4, 14–6, 17–1, 18–4
beam balance	15	2–1, 2–2, 3–2, 7–3, 7–4, 10–1, 13–2
bell, electric	15	6–6
bell jar	15	15–10
blotters	40	15–8, 15–11
bottles, clear and colored	45	3–6, 3–7, 14–4
boxes, assorted sizes	30	1–5
cardboard	30	11–3
brick or wooden block	15	4–1, 4–3, 5–3
Bunsen burner/hot plate	15	1–3, 2–5, 13–3, 14–6, 15–9, 18–4
can (to collect rain)	15	12–1
candle	15	3–5, 14–1
cardboard	60 sheets	9–1, 16–3, 19–2
clay, modeling	15 packages	2–4, 9–8
cloth, dark color	15	20–1
coins	30	6–1, 7–4, 9–3, 20–2
construction paper,		
black	1 ream	15–9
manila	1 ream	9–1, 19–1
container, clear plastic rectangular	1	13–5
cord or heavy string	5 balls	1–3, 4–1, 4–2, 5–2, 5–3, 6–2, 6–7, 7–3, 15–7, 18–1
corks	15	2–2, 3–7, 13–4
cotton	2 packages	16–3, 17–1, 22–2
coverslips	1 package	14–5, 14–6, 18–3, 18–4, 19–1, 19–3

Item	Quantity	References
dishpan, plastic	15	18–1
displacement can	15	7–4
dissecting needle	15	16–1, 16–2, 17–3, 19–3
dissecting pan	15	16–2, 17–3
drinking glass, plastic	15	1–5, 3–3, 6–1
dropper pipet	30	1–5, 2–6, 11–4, 14–5, 16–1, 17–1, 18–3, 18–4, 19–1, 19–3
egg carton	15	9–2
electric lamp	1	3–3, 15–9, 15–11, 15–12
facial tissue	1 box	2–1
flashlight	15	8–2
flask, 500 mL Florence	15	1–3
flower pot	45	20–3, 22–1
funnel	45	7–1, 7–5, 22–2
glass baby food jars	15	15–9, 20–1
glass baking dish	15	20–5, 22–3
glass jar (varied sizes specified)	15	3–3, 3–5, 11–6, 11–8, 14–6, 15–6, 16–1, 16–3, 18–4, 19–1, 22–4
glass plate	15	9–3, 13–5, 15–1
glass slide	30	14–5, 14–6
glass tubing,		
5 mm/100 cm long	15	1–3
5 mm/small length	15	11–1, 14–2, 14–4
globe	1	8–2, 11–4
graduated cylinders,		
25 mL	15	7–3
50 mL	15	1–5, 2–3
100 mL	15	3–4, 7–4, 22–2
graph paper	30 sheets	6–5, 9–7
hammer	1	2–5, 20–1, 22–4
hand lens	15	9–7, 15–2, 16–1, 18–1, 20–1
hygrometer (or psychrometer)	1	12–1
ink,		
blue	2 jars	7–2, 13–3
red	1 jar	7–2, 15–4
insect catching net	1	16–3
insect field guide	1	16–3
knife	15	9–3, 9–8
liquid solder	1 package	6–6
litmus paper (blue/red)	2 vials	13–1
magazines	30	1–1, 19–2
magnet	15	2–5
magnifying glass	15	3–7
maps		
world	1	9–1, 9–7
U. S.	15	10–3
weather	30	12–2
marble, large glass	15	6–8
marking pencil	15	15–11
marking pen	15	16–3, 19–2, 22–1
matches, wooden	1 box	1–3, 2–5, 3–5, 11–8, 13–3, 14–1, 14–6

materials for bird feeder	15 sets	23–3
metal ball	15	6–6
metal shears	1	6–8
metal washers	40	13–5
meter stick	30	1–5, 5–3, 18–1
metric ruler	30	1–4, 2–2, 6–5, 12–1, 15–12, 19–2, 22–3, 23–1
microscope	15	14–3, 14–5, 14–6, 15–2, 17–1, 18–3, 18–4, 19–1, 19–3, 20–4, 22–4
microscope slide	1 package	14–5, 14–6, 17–1, 18–3, 18–4, 19–1, 19–3
milk carton, paper	15	9–4
plastic	15	11–1
nails	75	6–6, 20–1, 22–4
pan, flat	15	13–3, 13–4, 17–5, 18–4
paper cup	45	2–1, 3–1
paper towel	10 rolls	2–5, 15–2, 17–3, 20–5
paste	15 containers	6–6, 19–2
pebbles		8–1
petri dish	15	15–2, 17–1
pie pan, thin metal	15	6–8, 10–2, 14–1
Ping-Pong ball	15	7–5
plastic bags,		
large clear	15	3–3
small clear	15	15–7, 15–10
plastic sheet, clear	1 roll	11–3
playing cards	1 deck	6–1
pulleys,		
double	15	5–2
single	30	5–2
putty	1 can	23–1
ring stand/clamp	15	1–3, 5–2, 7–2
rubber bands	1 box	14–6, 18–4
rubber gloves	1 pair	15–10
rubber sheets	15	7–2
rubber stoppers,		
1 hole (to fit bottle)	15	14–4
(to fit flask)	15	1–3, 14–4
(to fit milk container)	15	11–1
solid (to fit test tube)	45	2–3, 2–6, 15–11
rubber tubing,		
2 m	15	7–1, 7–2
1 m	15	11–1
30 cm	15	14–4
ruler, wooden	15	5–1
scalpel	15	16–2, 17–3
scissors	15	6–7, 17–3, 19–2
sealing wax	1 block	14–2
sheet metal shears	1 pair	6–8
shovel	1	18–1
sling psychrometer	15	11–7, 12–1
soda straw	15	2–6

sponge	15	3–3
spoon	15	13–4, 18–3
spray bottle	15	15–3
spring scale	30	4–1, 4–2, 4–3, 5–2, 5–3, 6–3
sprinkling can	1	22–1, 22–3, 23–1
stirring rod	15	3–2, 11–6
stopwatch or clock with second hand	1	3–2, 4–4, 6–6, 11–8
straight pins	1 box	16–2, 16–3, 17–3, 20–1
tape, masking	2 rolls	3–1, 20–1, 22–4
clear	2 rolls	19–2
teacup	15	1–5
terrarium,		
with cover	15	15–1
with variety of populations	15	18–2
test tubes	270	2–5, 2–6, 14–5, 15–11
test tube holder	15	2–5, 15–11
test tube rack	15	2–6
thermometer, Celsius	15	3–1, 3–2, 11–3, 11–6, 12–1, 13–2, 15–2
thistle tube	15	7–2
thread	1 spool	3–7
thumbtack	15	3–7
tongs	1 pair	3–2
toothpicks	3 boxes	2–4, 15-6
tray or pan	15	15–8, 15–12
tripod for heating	15	13–3
tweezers	15	17–3
U-tube	15	7–2
watch glass	15	3–5, 16–1
wind vane	1	11–5, 12–1
wire, bell type	10 m	6–6
wire screen	15	1–3
wood (for bird feeder)	15 sets	23–2
wood blocks, pine and oak	15 sets	2–2, 4–1, 10–1, 22–4
wooden board, smooth		
(1 m × 20 cm)	15	4–3, 5–3
(4 m × 15 cm)	15	6–6
wooden dowels	90	4–3, 13–5
wooden stakes	60	18–1
wooden sticks, 1 m long	15	8–1
Biological Supplies		
Amoeba culture		14–5
bean seeds	40 packages	15–11, 15–12, 22–1
carrot	15	14–2
celery stalks	15	15–4
coleus leaf	15	15–9
corn syrup	1 bottle	14–2
Drosophila fruit fly culture	15	20–1
earthworm, preserved	15	16–2
egg	15	13–2
fish, different live varieties	10–20	8–2
frog, preserved	15	17–3

geranium plant	30	15–7, 15–9, 15–10
goldfish, live	15	17–1
hay or dried grass	for hay infusion	14–6, 18–4
Hydra culture	1	16–1
lichen	15	19–3
magnolia leaf	15	10–2, 15–8
Marchantia (liverwort)	15	15–1
milkweed pods	15	21–1
oat seed	1 box	15–3
pine cones	15	21–1
plants, small	1 dozen	23–1
pond water, with infusorians	1 L	14–6, 15–8, 18–4, 19–1,
radish seeds	4 packages	15–2, 15–3
rice	1 box	14–6, 18–4
slides, prepared		
animal testis	15	20–4
bacteria	15 sets	14–3
onion root tip	15	20–4
whitefish eggs	15	20–4
sod or moss	15	23–1
soybean seed	15 packages	20–3
sphagnum moss	15	15–3
sugar, granular	11 kg	14–4
cube	1 box	18–3
sweet potato	15	15–6
tadpole		17–4
tobacco seeds, albino	15 packages	20–5
tree trunk, cross-section	1	15–5
wild bird seed	as needed	23–3
wood		

Chemical Supplies

alcohol,		
ethanol	500 mL	2–6, 15–9
isopropyl	500 mL	3–4
aluminum foil,		
regular	1 roll	6–6
heavy duty	1 roll	7–4
aquarium water	10 mL	16–1
bromothymol blue	5 g	2–6
calcium chloride	20 g	13–1
calcium oxide	50 g	2–6
glycerol	100 mL	1–3, 11–1, 14–2, 14–4
hydrochloric acid, dilute	50 mL	9–6
ice cubes	3 trays	3–2, 11–6, 13–3
iodine solution	1 bottle	15–9, 15–10
iron filings	500 g	2–5
limewater	500 mL	14–4
mineral oil	25 mL	2–6

mineral samples, including calcite, corundum, fluorite, gypsum, orthoclase, quartz, talc, topaz	15 sets	9–3
nail polish remover	1 bottle	16–3
petrified wood	15	10–1
petroleum jelly	1 jar	14–5, 15–10
phenol red	100 mL	2–6
plaster of paris	2 kg	10–2, 22–3
rock samples	15 sets	9–2, 9–5, 9–6, 23–1
saline solution	1 L	13–1, 13–2
salt	1 box	3–3
salt compounds,		
sodium chloride	23 g	13–1
magnesium chloride	100 g	13–1
sodium sulfate	4 g	13–1
calcium chloride	1 g	13–1
potassium chloride	0.7 g	13–1
sand	5 kg	3–1, 9–7, 22–1, 22–2, 23–1
sodium hydroxide pellets	50 g	15–10
soil, clay	10 kg	22–2
humus	10 kg	15–1
loam	1 kg	22–2
subsoil	5 kg	22–1
topsoil	15 kg	2–6, 15–12, 20–3, 22–1
stone or small rock	45	7–3, 8–1
sulfur, powdered	500 g	2–5
yeast	45 packages	14–4, 18–3

Suppliers

Central Scientific Co.
2600 S. Kostner Ave.
Chicago, IL 60623

Fisher Scientific Co.
4901 W. LeWoyne St.
Chicago, IL 60651

Hubbard Scientific Co.
2855 Shermer Road
Northbrook, IL 60062

Sargent-Welch Scientific Co.
7300 N. Linder Ave.
Skokie, IL 60076

Ward's Natural Science Establishment, Inc.
P. O. Box 1712
Rochester, NY 14603
or
P. O. Box 1749
Monterey, CA 93940

Preparation of Solutions

Hydrochloric acid
Use 258 mL of $11.6\,M$ concentrated HCl. Dilute to 1 L with distilled water. Caution: Add acid slowly to water.

Iodine
Dissolve 3 g potassium iodide and 6 g iodine in 200 mL water. Dilute with enough water to form a very pale yellow solution.

Limewater
Add 2 g of calcium hydroxide to 1 L water. Filter and protect from the air.

Phenol Red
Add 0.1 g phenol red to 28 mL of $0.01\,M$ sodium hydroxide and 222 mL water.

Bromothymol Blue
Mix 0.5 g bromothymol blue powder with 500 mL distilled water. Dilute 10 mL stock solution with 500 mL distilled water.

Teaching Aids

Teacher References
General—Science Education
Bartholomew, Rolland B. and Frank E. Crawley. *Science Laboratory Techniques.* Reading, MA: Addison-Wesley Publishing Co., 1980

Bork, Alfred. *Learning With Computers.* Bedford, MA: Digital Press, 1981.

Collette, Alfred T. and Eugene L. Chiappetta. *Science Instruction in the Middle and Secondary Schools.* St. Louis: Times Mirror/Mosby, 1984.

Educators Guide to Free Films. Educators Progress Service, Inc. Randolph, WI 53956. (published yearly)

Educators Guide to Free Science Materials. Progress Service, Inc., Randolph, WI 53956. (published yearly)

Farmer, Walter A. and Margaret A. Farrell. *Systematic Instruction in Science for the Middle and High School Years.* Reading, MA: Addison-Wesley Publishing Co., 1980.

How To Do It Series. National Science Teachers Association, 1742 Connecticut Ave., NW, Washington, DC 20009.

Safety in the Science Classroom. National Science Teachers Association, 1742 Connecticut Ave. NW, Washington, DC 20009, 1983.

TenBrink, T. D. *Classroom Teaching Skills.* Lexington, MA: DC Heath & Co., 1982.

The Science Teacher. National Science Teachers Association, 1742 Connecticut Ave., NW, Washington, DC 20009.

UNESCO *Handbook for Science Teachers.* New York: Uni Publishing, Inc., 1980.

Earth Science
Bascom, W. *Waves and Beaches.* Garden City, NY: Anchor Books, 1980.

Critchfield, Howard J. *General Climatology,* 4th ed. Englewood Cliffs, NJ: Prentice-Hall, 1983.

Dietrich, R. V. and E. Read Wicander. *Minerals, Rocks and Fossils.* New York: Wiley, 1983.

Hendrickson, Robert. *The Ocean Almanac.* Garden City, NY: Doubleday & Company Inc., 1984.

Ludlum, David. *The American Weather Book.* Boston: Houghton Mifflin Co., 1982.

Niesen, Thomas. *The Marine Biology Coloring Book.* New York: Harper & Row Publishing Co., 1982.

Schaefer, Vincent and John A. Day. *A Field Guide to the Atmosphere.* New York: Houghton Mifflin, 1981.

Spencer, Edgar W. *Physical Geology.* Reading, MA: Addison-Wesley, 1983.

Strahler, Arthur H. and Alan H. Strahler. *Elements of Physical Geography.* New York: John Wiley & Sons, 1984.

Life Science
Claflin, William E. *Collecting, Culturing and Caring for the Teacher, Student, and Hobbyist.* Saratoga, CA: Century Twenty-One, 1981.

Curtis, Helena. *Biology* 4th ed. New York: Worth, 1983.

Guiding Principles for the Use of Animals by Secondary School Students and Science Club Members. American Humane Association, 1980. (Available from the American Humane Association, P. O. Box 1266, Denver, CO 80201).

Kimball, John W. *Biology* 5th ed. Reading, MA: Addison-Wesley, 1983.

Lewin, Roger. *Thread of Life: The Smithsonian Looks at Evolution.* Washington, DC: Smithsonian Books, 1982.

Nilsson, Greta. *The Endangered Species Handbook.* Washington, DC: Animal Welfare Institute, 1983.

Purdom, P. Walton and Stanley H. Anderson. *Environmental Science.* Columbus, OH: Charles E. Merrill Publishing Co., 1983.

Ray, Peter M., Taylor A. Steeves, and Sara A. Fultz. *Botany.* New York: Saunders College Publishing, 1983.

Rowland-Entwistle, Theodore. *Illustrated Facts and Records Book of Animals.* New York: Arco, 1983.

Sterba, Gunther. *The Aquarium Encyclopedia.* Cambridge, MA: MIT Press, 1983.

The Audubon Society's Encyclopedia of Animal Life. New York: Distributed by Crown Publishing Co. Inc., 1982.

Physical Science
Beiser, Arthur. *Physics, 3rd ed.* Menlo Park, CA: The Benjamin/Cummings Publishing Co., Inc., 1982.

Dickerson, Richard E., Harry B. Gray, Marcetta Y. Darrensbourg, and Donald J. Darrensbourg. *Chemical Principles* 4th ed. Menlo Park, CA: The Benjamin Cummings Publishing Co., 1984.

Epstein, Lewis C. *Thinking Physics is Gedanken Physics.* San Francisco: Insight Press, 1983.

Herold, Kenneth and Frank Walmsley. *Chemical Principles, Properties and Reactions.* Reading, MA: Addison-Wesley Publishing Co., 1984.

Hewitt, Paul B. *Conceptual Physics . . . A New Introduction to Your Environment,* 4th ed. Boston: Little, Brown and Company, 1981.

Student General References

The following references should be accessible to students in the classroom or library for supplemental reading and research.

Current Energy and Ecology
Current Health
Current Science
Curriculum Innovations Inc.
P. O. Box 310
Highwood, IL 60040

Discover
Time-Life Building
541 N. Fairbanks Court
Chicago, IL 60672

National Geographic
National Geographic Society
1155 Sixteenth St., NW
Washington, DC 20036

National/International Wildlife
National Wildlife Federation
1412 Sixteenth St., NW
Washington, DC 20036

Oceans
The Oceanic Society
Membership Services
P. O. Box 10167
Des Moines, IA 50340

Popular Science
Subscription Department
P. O. Box 2873
Boulder, CO 80321

Science Digest
P. O. Box 10076
Des Moines, IA 50350

Sci Quest
Circulation Manager
American Chemical Society
P. O. Box 2895
Washington, DC 20013

Films and Filmstrips
Unit 1
Byron B. Blackbear and the Scientific Method. Learning Corporation of America
Make Mine Metric. Pyramid Films
The Nature of Matter. McGraw-Hill Text Films
The New Industrial Revolution. National Audio Visual Center
The Philosophy of Science I. The Media Guild
What Is Science? (2nd ed). Coronet Instructional Media

Unit 2
Airplanes and How They Fly. Centron Films
Attraction of Gravity. BFA
Pedal Power. Bullfrog Films
Robots: Intelligent Machines Serving Mankind. Science Screen Report
Simple Machines: Inclined Planes and Levers, 3rd ed. EBEC
Simple Machines—Using Mechanical Advantage. Barr Films
Thonk! Science and Hitting a Ball. Centron Films

Unit 3
The Beach: A River of Sand. EBEC
Challenge of the Oceans. McGraw-Hill Text Films
Continental Drift. Green Mountain Post Films
The Earth Beneath the Sea. McGraw-Hill Text Films
The Earth: Its Atmosphere, revised ed. Coronet Films & Video
The Legacy of L. S. B. Leakey. National Geographic
Life in the Sea. EBEC
Message In The Rocks. Time-Life Video
Weather Forecasting. EBEC
What Makes Clouds. EBEC
Why The Earth Moves. Modern Talking Pictures Service, Inc.

Unit 4
Amoeba: One Celled Organism. International Film Bureau
Animals Without Backbones. Coronet Instructional Media
Animals With Backbones. Coronet Instructional Media
Diffusion and Osmosis. Coronet Instructional Media
Flowering Plants and Their Parts. EBEC
Flowers At Work, 3rd ed. EBEC
Insect Life Cycles. Centron Films
The Living Cell: An Introduction. EBEC
Plants Make Food (2nd ed) Churchill Films.
Poison Plants. EBEC
The Protist Kingdom. BFA

Unit 5
Food Chains—A Bond of Life. Barr Films
Genetics: Mendel's Laws. Coronet Instructional Media
Genetics: Improving Plants and Animals. Coronet Instructional Media
Genetics and Plant Breeding. BFA
Introducing Ecology. Coronet Instructional Media
John Muir's High Sierra. Churchill Films
Rain Forest. National Geographic

The Galapagos: Darwin's World Within Itself. EBEC
What Ecologists Do. Centron Films

Unit 6
Backyard Wildlife. Marty Stouffer Productions Ltd.
Choices. The Conservation Foundation.
Drip Drop. Alden Films
Erosion. BFA
Environment. BFA
Flight to Survival. Churchill Films.
National Parks: Promise and Challenge. National Geographic
Problems of Conservation: Wildlife. EBEC
The Water Crisis. Time-Life Video
Watching Wildlife. Marty Stouffer Productions Ltd.
We Are Of The Soil. International Film Bureau, Inc.

Film Distributors

Alden Films
7820 20th Ave.
Brooklyn, NY 11214

Barr Films
P. O. Box 5667
3490 East Foothill Blvd.
Pasadena, CA 91107

Benchmark Films, Inc.
145 Scarborough Rd.
Briarcliff Manor, NY 10510

BFA Educational Media
P. O. Box 1795
2211 Michigan Ave.
Santa Monica, CA 90404

Bullfrog Films
Oley, PA 19547

Centron Educational Films
1621 West 9th Street
Lawrence, KS 66044

Churchill Films
662 West Robertson Blvd.
Los Angeles, CA 90069

The Conservation Foundation
1717 Massachusetts Ave. NW
Washington, D.C. 20036

Coronet Instructional Media
65 East South Water Street
Chicago, IL 60601

Educational Media International
P. O. Box 328
14740 Jordan La.
Haymarket, VA 22069

Encyclopedia Britannica
Educational Corporation
425 North Michigan Ave.
Chicago, IL 60611

Green Mountain Post Films
P. O. Box 229
Turner Falls, MN 01376

International Film Bureau
332 South Michigan Ave.
Chicago, IL 60604

Learning Corporation of America
1350 Avenue of the Americas
New York, NY 10019

Marty Stouffer Productions Ltd.
300 S. Spring St.
Aspen, CO 81611

The Media Guild
11526 Sorrento Valley Rd., Suite J
San Diego, CA 92121

McGraw-Hill Text Films
110 Fifteenth Street
Del Mar, CA 92014

Modern Talking Picture Service, Inc.
5000 Park St.
St. Petersburg, FL 33709

National Audio Visual Center
National Archives and Records Service
General Services Administration
Washington, DC 20409

National Geographic Society
17th and M Streets, NW
Washington, DC 20036

Pyramid Films
P. O. Box 1048
Santa Monica, CA 90406

Science Screen Report
P. O. Box 691
New York, NY 10003

Time-Life Films
100 Eisenhower Dr.
Paramus, NJ 07652

Careers Related to Science

Career exploration must begin early in the formal educational process. Young people should be exposed to the vast possibilities that are open to them. Emphasis on career education throughout the year adds relevance and a special significance to the topics presented in your science course. Scientific and science-related careers provide challenges, stimulation, satisfaction, and many opportunities for creativity. For students who may not have aspirations toward a career in science, it is important to make them aware of the significance of science in other fields.

Use the Perspectives pages to give students an insight into science-related careers. Use these pages as the starting point for additional reports concerning a career or the training required to pursue a particular line of work.

Field trips to nearby facilities such as water treatment plants, hospitals, food processing plants, weather stations, and police crime labs can be excellent sites for learning. If possible, document these trips using photographs and slides which can then be used for displays and projects.

The following information is included to show you the variety of careers that require some knowledge of science. Students should be exposed to jobs at all levels from Ph.D. geologists to nursing aids needing only a high school diploma. You can use this information to assign reports on various careers. A short bibliography and list of addresses is included to provide you with sources for more specific job descriptions and training requirements. Try to obtain catalogs from colleges, technical schools, and vocational schools in your area as they are excellent sources of information.

Earth Science Related Occupations
Geology

Workers in the field of geology must have an in-depth knowledge of earth's structure, composition, and evolution. Occupations in this field are not limited to the study of earth science but require some background knowledge of physics, chemistry, and biology. A thorough understanding of earth's structure and composition is based on physical and chemical theories and laws. The study of fossil remains requires a knowledge of biology to form valid conclusions concerning the events in earth's history. Occupations in the petroleum industry involve locating new oil and gas supplies and developing the technology to make existing wells more efficient. Most professional positions require at least a four year college degree with some post graduate work required in order to specialize in a field. The following list covers some of the various occupations in the field of geology—crystallographer, geologist, geophysicist, mineralogist, paleontologist, seismologist, and soil engineer.

Meteorology

Occupations in meteorology require a knowledge of the composition and behavior of the atmosphere. People working in this field use a variety of instruments and maps to predict changes in weather and climate. Map interpretation is an essential skill in forecasting. Accurate weather information is essential to other fields such as air and sea transportation, fire prevention, and air quality control. A four year degree or equivalent military training in meteorology is required. People trained in this field are generally employed as local forecasters for the media and airports or by the National Weather Service.

Oceanography

Professionals in the field of oceanography are well versed in the composition and behavior of seawater. Water chemistry, the geology of coastal areas, and ocean life are some aspects of oceanographic study. In studying ocean currents, waves, and tides a background in elementary physics is required. A four year degree including extensive fieldwork is usually essential. This field overlaps meteorology in that some oceanographers are involved in weather prediction.

Environmental Science

Occupations in environmental science cover the prevention and control of pollution. Pollution level data from soil, air, and water samples is compared to government standards. New equipment and procedures are developed to solve pollution problems that may arise. An in-depth study of earth, physical, and life science is necessary to adequately study the effects of pollution on the environment. The following are some occupations in this field—air pollution, soils, and water quality analyst; environmental researcher; design engineers for pollution control; and government positions in setting pollution standards.

Technical Fields

Technical assistants in earth science-related occupations provide support services for professionals such as meteorologists, geologists, environmental scientists, mining engineers, and oceanographers. Most positions in these areas require a high school diploma with technical school or on-the-job training. Technicians are usually more responsible for data collection rather than interpretation. They are also generally more familiar with the instrumentation needed in collecting data for a particular use. For example, a weather observer is trained by the National Weather Service in recording meteorological data. Analytical technicians involved in testing soil, air, and water samples may have two year degrees from a vocational or technical school.

Life Science Related Occupations

Biology

Most professional biological occupations would involve research work. Biological scientists are actively involved in studying life processes, structure, growth, behavior, and reproduction. These studies can have a variety of applications from disease control to the impact of health hazards on the environment. Most professional biological positions require at least a four-year degree with post graduate work. The following are some fields of specialization for the biological scientist: anatomist, aquatic biologist, biochemist, entomologist, botanist, geneticist, microbiologist, mycologist, pharmacologist, and zoologist.

Health and Medicine

The occupations available in the fields of medicine, veterinary medicine, dentistry, and pharmacy are many and diverse. Most all who work in these fields at the professional level need a basic background in life science with additional coursework in the physical sciences. Professional fields such as physicians, dentists, optometrists, pharmacists, podiatrists, opthalmologists, and veterinarians require a college degree with post graduate training. Professionals in these fields must also pass a state licensing exam.

Careers in nursing, physical and occupational therapy, dietetics, environmental health, radiation medicine, and medical technology usually require a four-year degree. These people are also generally licensed by the state.

A two-year associate degree or on-the-job vocational training is required for most assistant and technician jobs. The following are some examples of careers in this area: EKG and EEG technicians, medical and dental assistants, operating room technicians, optometric assistants, X-ray technicians, and practical nurses. As with other technical positions these generally involve collecting data using diagnostic equipment or simple patient care.

Health and Safety

Workers in this field are primarily concerned with quality control functions. Jobs are varied with many options available in the food industry. Food processors have many quality control personnel to assure the safety and freshness of their products. As a further check, the government employs a number of health inspectors who handle on-site inspections of food processors, groceries, and restaurants to assure that food is being handled under safe conditions.

Those employed by the Occupational Health and Safety Administration are involved in inspections at industrial sites to see that proper safety measures have been implemented and workers are in a safe environment.

Jobs in these fields would generally require a 2-4 year degree or on-the-job technical training. Any analysis type work would require a background in life science with some additional knowledge of chemistry.

Agricultural

Occupations in this field involve workers who are primarily involved with developing new procedures to help solve problems related to growing plants and breeding animals. Jobs in this field require varied amounts of training from post graduate work down to assistant positions requiring only a high school diploma. Some professional positions requiring at least a four-year degree are agronomist, animal scientist, dairy scientist, forester, poultry scientist, soil scientist, and seed analyst.

Careers in dairy technology, fiber technology, horticulture, and wood technology would require a four-year degree.

Lab technicians involved in breeding and hybrid studies may require only a high school diploma with on the job training or a two year technical degree.

Physical Science Related Occupations

Chemistry/Physics

Most professional scientific occupations in the fields of chemistry and physics require at least a four-year degree with post graduate work. These occupations deal primarily with scientific research. Extensive coursework in mathematics is also required. The following are some professional occupations in the chemistry and physics area: chemist, physicist, food chemist, industrial chemist, nuclear physicist, water purification chemist, biochemist, biophysicist, and environmental chemist.

Engineering

All aspects of engineering require a knowledge of design, properties of materials, and machines. Engineers must have an extensive background in physical science with additional course work in mathematics. A four-year degree is required and the option of taking a state licensing exam. Engineering is primarily concerned with the application of science principles to the development of new products or more efficient ways of manufacturing existing products. The following list covers some branches of engineering: aerospace, agricultural, biomedical, ceramic, chemical, civil, electrical, industrial, mechanical, metallurgical, mining, and nuclear.

Technical Fields

Technicians and assistants provide support services for scientists and engineers. Their jobs also involve the development and production of new products. However, these jobs are usually more specific and limited in scope than those at the professional level. Some technical positions require only a high school diploma. However, many technicians and assistants have two-year degrees or have been trained at post secondary technical schools. Many companies will provide on-the-job training for interested applicants. Some branches of technical work related to physical science are broadcasting, drafting electronics, heating and cooling systems, nuclear energy, and surveying.

Construction and Machine Trades

Construction jobs include those involved in the initial design as well as the structural workers. Designers and architects have four-year degrees in these fields. Most construction and machine trades workers have been trained through apprenticeship programs after high school graduation. Examples of jobs in these fields are foundry workers, machinists, tool and die makers, millwrights, instrument makers, electroplaters, waste water treatment workers, electricians, bricklayers, welders, iron workers, painters, plumbers, and pipefitters.

Additional Information

Keys to Careers in Science and Technology
National Science Teachers Association
1742 Connecticut Ave. NW
Washington, DC 20009

Dictionary of Occupational Titles
U.S. Department of Labor
U.S. Employment Service

Occupational Outlook Handbook
U.S. Department of Labor
Bureau of Labor Statistics

Performance Objectives

Performance Objectives are listed for each chapter of **Principles of Science.** They are related to the four broad goals of science education.

A **Attitudes:** to develop students' attitudes of curiosity and involvement with occurrences in their environment; to develop an appreciation for the contributions of science; and to recognize the value of solving problems in a scientific manner.

P **Processes:** To develop those intellectual processes of inquiry by which scientific problems and occurrences are explained, predicted, and/or controlled.

K **Knowledge:** To develop knowledge of facts, terminology, concepts, generalizations, and principles which help students confront and interpret occurrences in their environment.

S **Skills:** To develop students' ability to handle, construct, and manipulate materials and equipment in a productive and safe manner; and to develop the ability to measure, organize, and communicate scientific information.

The following are some examples of each type of performance objective.

Attitudes: Students will request the use of science equipment during free periods.

Processes: Given data about weather fronts, temperature, and barometer readings, students will predict the weather conditions in their area.

Knowledge: Students will identify correctly the roots, stem, leaves, and flowers of a plant.

Skills: Using a stopwatch, students will count the number of swings a pendulum makes in one minute.

Bibliography

Doran, R. L. *Basic Measurement and Evaluation of Science Instruction.* Washington, DC: National Science Teachers Association, 1980.

Gronlund, Norman E. *Constructing Achievement Tests.* 3rd ed. Englewood Cliffs, NJ: Prentice-Hall, 1982.

Sullivan, Howard and Norman Higgins. *Teaching for Competence.* New York: Teachers College Press, 1983.

TenBrink, T. D. *Classroom Teaching Skills.* Lexington, MA: DC Heath & Co., 1982.

Wall J. *Compendium of Standardized Science Tests.* Washington, DC: National Science Teachers Association, 1981.

Textbook Inventory

Answers to questions:

A. Introduction
1. The name of the textbook is *Principles of Science, Book One.*
2. The author's name is Charles H. Heimler.
3. Science is important because it explains the world in which we live; science affects how we live our daily lives.
4. There are 23 chapters in this book; sedimentary rocks are studied in section 9:6.

B. Graphic Aids
5. Unit I begins on page 2; the photograph and opening statement illustrate how science affects our everyday lives.
6. The photograph of an industrial arts class is filled with examples of machines; chapter 5 is about machines.
7. Title: composition of air; the table shows a list of gases that make up air and the percent by volume in which the gases are present in air.
8. The information could be indicated in name words and numerals only; sometimes pictures convey information more quickly than words.

C. Study Aids
9. A student will gain an understanding of fluids and how they are related to forces and motion.
10. Section 7:2 discusses pressure in liquids.
11. The blue margin notes highlight the main points of page 135.
12. The example problem is about work output; example problems can be used as samples for setting up equations.
13. Making sure questions are useful for review and reinforcement of main ideas.
14. Terms include latitude, longitude, meridians, and prime meridian.

D. Activities
15. The activity on page 330 is a long-form activity.
16. The purpose of the activity is to determine if sunlight and chlorophyll are needed for photosynthesis.
17. A coleus leaf and a geranium plant are used in the activity on page 330; the iodine solution is an indicator of the presence of starch in a plant.
18. Touching sodium hydroxide can cause severe burns.
19. A line graph showing the relationship between the time and the number of yeast cells in the population is to be drawn.

E. Chapter Review
20. Centripetal force is discussed in section 6:6.
21. There are twelve new terms.
22. True or false, multiple choice, completion, and how and why questions are included.
23. The Challenges section contains activity ideas.
24. The chapter reading list contains sources of additional information.

F. Other Features
25. Special features highlight current science topics, applications, and careers.
26. Skills include reading for meaning, understanding science words, using tables, organizing notes in outline form, reading maps, and using graphs.
27. Appendix B contains information about safety and first aid.
28. Vocabulary terms are defined in the glossary.
29. Mineral is discussed as a term on page 175; page 299 shows a photograph of a cougar.

Unit 1 Matter and Energy

Overview

Unit 1 covers the nature of science, matter, energy, and changes in matter. In Chapter 1 science is defined and measurement skills using SI units are introduced. A section on careers in science is included to motivate students towards thinking about jobs in science and related fields. Chapter 2 continues the themes from Chapter 1, explaining the nature of matter and how density is calculated. Activities for scientific inquiry and discovery and model making are included. Chapter 3 introduces the concept of energy and the relationships between matter and energy. Emphasis is given to the forms of energy, sources of energy, and energy transformations. The sum total of the content of Unit 1 provides the basic background for the study of other topics in the units that follow.

Teaching Tips

Science is in the news. Ask five students each week to bring in a newspaper article on a science topic. Every Friday devote time to science current events, allowing each student to give a short science current event report. Assign extra credit for the reports.

Select one or two able students to assist you in handling science supplies for demonstrations and activities.

Start making a collection of activity cards. For each activity, list the title and the equipment needed on a 3" × 5" index card. This system will make it easier to collect, store, and retrieve equipment.

Have students make a display of everyday items found at home, the grocery, or hardware store that have quantities labeled in metric units.

---

1 Science and Measurement

Overview

Chapter 1 explains the use of science skills, including experiments and metric measurement. It describes the methods scientists use to solve problems and develop new knowledge by forming hypotheses and theories. One way to introduce a chapter is to direct students to the chapter opening photo and ask them to explain how the picture is related to science. Then ask students the question that is the topic for Section 1:1, What Is Science? You may then want students to complete the Textbook Inventory in the front of the book. Go over the answers and acquaint students with the features that will aid their learning and increase achievement.

Use the guide questions in the margins and the Making Sure questions to introduce and summarize the learning for each section. Students will get more out of their reading if they review the questions first because the questions guide the students to important material in each section. Students enjoy doing science activities; they will be more motivated and learn more if they participate in these activities. Use the chapter-end material for review and reinforcement.

Performance Objectives

Goal: Students will gain an understanding of scientific methods and the use of SI units.

Objectives: Upon completion of the reading and activities and when asked to diagram, demonstrate, or respond either orally or on a written test, students will:

1:1	K	differentiate the product and process of science.
1:2	A/K	seek information related to science career interests.
1:3	K	list and explain the methods scientists use to solve problems.
	A/S	use scientific methods in the solution of everyday problems.
1:4	K	identify and explain the parts of an experiment.
	K	explain why a control is used in an experiment.

	S	perform an activity that shows the main parts of an experiment.
1:5	K	list the most frequently used metric (SI) units.
	K/S	list the most frequently used metric prefixes and explain their meaning. Change metric units to smaller or larger units.
1:5–7	S	use metric units in measuring length, area, volume, and mass.
	A/S/K	carry out one Challenge at the end of the chapter.
	A/K	read a science article or book and prepare a short report.

Equipment for Activities

boxes, 2 different sizes
burner or hot plate
drinking glass
dropper
flask, Florence, 500 mL
glass tubing, 10 cm
glycerol
graduated cylinder, 50 mL
matches
meter stick
metric ruler
rag or towel
ring stand/clamp
rubber balloon
rubber stopper, 1-hole
science magazines, 5
string
teacup

1:1 What Is Science?

Explain the difference between the process and product of science. The facts and theories in this textbook are good examples of science product. Activities, experiments, and measuring are examples of science process.

Activity 1–1

In this activity the student surveys five science magazines and writes a short report describing the contents of the magazines. Check with the librarian before giving this assignment. Explain the assignment and exactly what students are to do. Explain how students will be graded in your class and how the activities and assignments are counted in the grading system. The report should be at least one page in length.
Materials: five science magazines.

Making Sure, page 6

1. By reading, a person can learn what other scientists have discovered.

1:2 Careers in Science

Career goals can motivate your students towards higher achievement. Poll the class to discover individual career interests and the reasons for these interests. Stress the value of science classes in helping to achieve career goals.

Activity 1–2

In this activity the student obtains information about a career or job related to science. Invite the school guidance counselor to class to talk about science careers and sources of information in this area.
Materials: career information sources

1:3 Using Science Skills

Have students read this section and answer the guide questions. Then use the Making Sure questions for discussion by applying scientific methods to each problem. Explain the use of tools such as a microscope, telescope, and space satellite in scientific problem solving. Stress the fact that scientific research is not cut and dried, but involves skill, luck, trial and error, and creativity on the part of scientists.

Making Sure, page 9

2. Answers will vary. One possible answer to (a) is given below.
   (1) Make a clear statement of the problem: Engine is overheating.
   (2) Obtain information: Thermostat controls flow of coolant in the engine.
   (3) Form a hypothesis: Thermostat is not working properly.
   (4) Test the hypothesis: Replace old thermostat with new one.
   (5) Accept or reject hypothesis: Engine does not overheat; faulty thermostat was cause of the problem.

1:4 Experiments

Stress the value in keeping records of experiments. You may wish to have your students record all activities and demonstrations in their notebooks using the outline given on page 11. Make certain students understand the use of a control.

Activity 1–3

In this activity the student performs an experiment that illustrates the parts of an experiment. A

flask connected to a balloon is heated and the warmed air expands and inflates the balloon. Have students write a hypothesis for the experiment before they begin work. Example: The balloon will expand when the flask is warmed.
Materials: 500-mL Florence flask, 1-hole stopper, glass tubing (10 cm), balloon, burner or hot plate, glycerol, rag or towel, string, matches
Questions: As the air is warmed it expands and inflates the balloon to a larger volume. If the flask did not contain air there would be no change in the balloon. Ask students to describe a control for this experiment. A control would be the same apparatus used in the experiment without the flask being heated.

1:5 Metric Measurement

You will note throughout the text that there are no English-metric conversions. Metric units are used exclusively to encourage thinking in SI. Students are likely to be familiar with metric units, but SI will be a new term. Explain that SI is the modern, up-to-date form of the metric system. Although the calorie, liter, milliliter, and Celsius temperature scale are not SI, they are all used extensively in science and industry.

The Celsius temperature scale is used exclusively throughout the text and no Celsius-Fahrenheit conversions are given. This is to encourage student thinking in Celsius temperatures.

Note that commas are not used to group digits in large or small numbers used in this text. In some countries a comma indicates a decimal point. Because of the international use of metric units (SI), spaces are used to separate groups of three digits. For example:
  10 000   100 000   1 000 000   0.000 01

Making Sure, page 14
3. 100
4. 300 cm

1:6 Area and Volume

Use Figure 1–11 to teach the area of 1 cm^2 and 1 mm^2. Centimeter-lined and millimeter-lined graph paper can also be used to reinforce this concept. You can demonstrate metric volume using 10-mL, 25-mL, and 50-mL graduates. Show that it takes two and one-half 10-mL graduates of water to fill the 25-mL graduate and five 10-mL graduates of water to fill the 50-mL graduate.

Activity 1–4
In this activity the student uses the equation for finding the areas of two rectangles. Write the equation *(Area = length × width)* on the chalkboard and explain its use. Remind students that 1 cm^2 = 100 mm^2.
Materials: Figure 1–14, page 15, metric ruler
Questions: 7.5 cm^2 (750 mm^2), 3.75 cm^2 (375 mm^2)

Activity 1–5
In this activity the student uses metric units to measure length, area, and volume. Have students work in groups of two to four. Each student should make his or her own measurements. Students should record their data in their notebooks. One possible format for writing up activities is given below. Remind students not to write in their textbooks.
(1) Write the title and objective for the activity at the top of the notebook page.
(2) List the text page number for the activity.
(3) Make tables and charts as necessary to record data. (Students will need help in recording and organizing the data they obtain.)
(4) Answer Questions and Conclusions.
Materials: 2 boxes (different sizes), drinking glass, dropper, graduated cylinder (50 mL), meter stick, teacup
Questions and Conclusions: **1.** Answers will vary. **2.** $A = l \times w$. **3.** Answers will vary. **4.** Divide volume (cm^2) by 10. **5.** Use the equation $V = l \times w \times h$.

Making Sure, page 17
5. 1000 mL or 1 L
6. Answers will vary.
7. Area of the page is 348.5 cm^2.

1:7 Mass

Explain the concept of mass as different from weight. Mass is the amount of matter and weight is the force of the earth's gravity on an object. Further discussion of weight is found in Chapter 4. Demonstrate the use of a beam balance (Figure 1–17) by finding the mass in grams of selected small objects. Select three or four additional objects and have students find the mass of each.

Making Sure, page 17
8. Use a balance and find the mass of the beaker with the water. Empty the water from the

beaker and determine the mass of the dry beaker. Subtract the mass of the empty beaker from the mass of the beaker with water.
9. 40.58 kg

Perspectives

Reading for Meaning

Reading for meaning involves alerting students to words which indicate the relationship between facts and ideas. The text aids in this by indicating important new words in bold-faced type. The bold-faced new term is linked to its definition by a verb.

Questions: area—(1) definition: the number of square units required to cover a surface, (2) verb: is, (3) section 1:6; control—(1) definition: a part of an experiment that is held constant, (2) verb: is, (3) section 1:4; cubic meter—(1) definition: one unit for volume, (2) verb: is, (3) section 1:6; experiment—(1) definition: a method of testing a hypothesis, (2) verb: is, (3) section 1:3; hypothesis—(1) definition: a possible solution for a problem, (2) verb: is, (3) section 1:3; International System of Units SI—(1) definition: the modern form of the metric system, (2) verb: is called, (3) section 1:5; kilogram—(1) definition: the basic unit of mass in SI, (2) verb: is, (3) section 1:7; liter—(1) definition: a unit of volume often used for measuring liquids, (2) verb: is, (3) section 1:6; mass—(1) definition: the amount of matter in an object, (2) verb: is, (3) section 1:7; meter—(1) definition: basic unit for length in SI, (2) verb: is, (3) section 1:5; volume—(1) definition: the space occupied or filled by an object, (2) verb: is, (3) section 1:6.

Study Questions

A. True or False
1. True **2.** True **3.** True **4.** False (Scientists always report their discoveries after they do an experiment.) **5.** False (A control group is the group that is held constant.) **6.** False (There are 1/100 g in 1 cg.) **7.** False (A cubic centimeter is smaller than a cubic meter.) **8.** True **9.** False (A paper clip has a mass of about 1 g.) **10.** True

B. Multiple Choice
1. procedure **2.** problem **3.** meter **4.** divide by **5.** cubic millimeter **6.** gram **7.** mass **8.** increases **9.** decimeters **10.** kg

C. Completion
1. kilogram **2.** 1/100 **3.** 1/100 **4.** 2500 **5.** 1 **6.** graduated cylinder **7.** balance **8.** 50 **9.** 100 cm^2 **10.** 10

D. How and Why
1. A control group is used as a standard of comparison for the results.
2. A welder needs knowledge about the physical and chemical properties and behavior of various metals and alloys.
3. Science is not always a step by step process. Scientists are also creative people. Skill and imagination are major aspects of creativity.
4. An object is placed on the left pan and objects of known or standard masses are placed on the right pan. The pans will balance when the masses on both pans are equal.
5. 100 cm^3, 4

Challenges
1. You may also wish to have students prepare a science career bulletin board for your classroom. Have a committee of students obtain materials from a library, guidance office, local industry, or government offices.

2   Matter

Overview

Chapter 2 describes the major properties of matter and the forms in which it occurs. Introduce this material by pointing out the different kinds of matter in the classroom or in a home. Then ask students to explain how the kinds of matter are different and how they are alike. Use of the guide questions will aid student reading comprehension. Making Sure questions provide review and application of material that is learned.

Performance Objectives

Goal: Students will learn the properties and composition of different kinds of matter.

Objectives: Upon completion of the reading and activities and when asked to diagram, demonstrate, or respond either orally or on a written test, students will:

2:1   K   list examples of matter and their properties.

	K	explain that every substance has properties that identify it and determine its use.
	K/S	do an activity to show that air takes up space.
2:2	K	list the four states of matter and give examples of each state.
	K	explain how each state of matter is different from the other states.
2:3	K	define density.
	P/S	calculate density when given the mass and volume.
2:4	K	define element and give examples of elements.
	K/S	write symbols for elements.
2:5	K	describe the structure of an atom.
	K	list the main characteristics of a proton, neutron, and electron.
2:6	K/P	explain how elements combine to form compounds.
	K	define and explain chemical bonding.
	K	explain how an atom becomes an ion and give examples of ions.
2:7	K	explain what a formula is and what information it contains.
	K/S	use symbols to write formulas for simple compounds.
2:8	K	define mixture and give examples of mixtures.
	K/S	explain how mixtures can be separated by physical means.
	P/S	separate a mixture of iron filings and sulfur.

Equipment for Activities

alcohol, rubbing	metric ruler
balance	mineral oil
beaker	paper cup
bromothymol blue	paper tissues
burner	paper towels
clay, modeling (different colors)	phenol red
cork	rubber stopper
dropper	soda straw
graduated cylinder, 25 mL	soil
hammer	sulfur, powdered
iron filings	test tubes/stoppers, 5
lime (calcium oxide)	test tube holder
magnet	test tube rack
matches	toothpicks
	wooden blocks (1 pine and 1 oak)

2:1 Properties

Have students list all the properties of water they can think of. Point out that properties are used to identify a substance and how properties affect the use of a substance. Display examples of substances such as copper, iron, sulfur, oil, alcohol, ammonia water, sand, and salt. Have students list the properties of each. You may wish to distinguish between physical and chemical properties. Physical properties depend only on the matter itself and include color, taste, smell, melting point, etc. Chemical properties depend on the reaction of a substance with other substances. For example, some substances may burn when combined with oxygen.

Activity 2-1

In this activity the student submerges in a container of water a paper cup containing a crushed paper tissue. The cup is submerged twice. A hole is made in the bottom of the cup before it is submerged the second time.
Materials: paper cup, paper tissue, large beaker or jar, water
Questions: The first time the air in the cup prevents the water from entering all the way. The second time, the hole allows air to escape and the water rises to wet the tissue. The property of matter observed is that matter (air) takes up space. Two volumes of matter cannot occupy the same place at the same time.

Making Sure, page 24

1. Answers will vary; answers may include:
   (a) Rubber is elastic; aluminum is a shiny metal that can be bent or shaped.
   (b) Chalk is powdery and leaves a mark; snow is cold and melts at room temperature.
   (c) Milk is white; water is clear.
   (d) Salt and sugar taste different.
   (e) Wood burns; some plastics melt.
2. Possible answers may include:
   (a) rubber balls, tires; aluminum pans, boats
   (b) writing (on chalkboard); surface for sledding and skiing
   (c) drink as a food; washing and cooling
   (d) melts ice, seasons food; used to make food
   (e) construction materials; making of sporting equipment.

2:2 States of Matter

Plasma is a state that will be unfamiliar to students. You can explain plasma by reviewing the fact that all matter is made of atoms and that each atom is made up of particles. When matter is sufficiently heated to be in the plasma state, the electric particles in atoms separate. Review Table 2–1 and have students list additional examples of solids, liquids, and gases.

Making Sure, page 26
3. (a) gas (b) liquid (c) gas (d) liquid (e) gas (f) solid
4. Plasma contains electrically charged particles; gas does not.

2:3 Density

Stress the fact that mass and volume are not always directly related. In other words a large object may contain less matter than a smaller object. Referring to Table 2–2, have students explain why 10 g of lead has a smaller volume than 10 g of gasoline. Note that density can be expressed using g/mL or g/cm^3. Go over the example problem on page 27 and explain how the equation is used to calculate density.

Activity 2–2

In this activity the student calculates the volume, mass, and density of two different wooden blocks. You can use a block of Styrofoam and a paperback book as substitutes for the wooden blocks. A paperweight (rectangular-prism shape) can also be used.
Materials: wooden blocks (pine and oak), metric ruler, balance
Questions: Oak is more dense than pine. Students should use the equation on page 27 to calculate density.

Activity 2–3

In this activity the student measures the mass and volume of a cork and a rubber stopper and calculates the density of each. Before beginning the activity be sure students are able to correctly determine the water level in the graduated cylinder. You may also wish to demonstrate how to find the volumes of the stopper and the cork by water displacement.
Materials: rubber stopper, cork, graduated cylinder (25 mL), balance
Questions: Use the equation on page 27 to calculate density. The rubber stopper is more dense than the cork.

As an additional activity, suggest that students devise an experiment to measure the densities of two liquids such as cooking oil and rubbing alcohol.

Making Sure, page 28
5. In order for an airplane to fly, it must be constructed of lightweight materials such as aluminum.
6. 100 g/cm^3 ÷ 0.0013 g/cm^3 = 769.2 cm^3
7. density = 7.9 g/cm^3; iron
8. density = 16.7 g/cm^3; not pure gold because density is too low

2:4 Elements

Elements are the building blocks of all matter. Every substance consists of one or more of the 106 known elements. If possible, display examples of elements such as carbon, iron, lead, sulfur, iodine, magnesium, and nickel. Have students compare and contrast properties. After students have completed this section, you may want to give a short quiz in which students write the symbols for ten elements listed in Table 2–3. Stress the fact that the first letter of a symbol is a capital and the second is lowercase.

Making Sure, page 29
9. Al, Au, C, Pb, O
10. Al—silver color, lightweight; Au—yellow color, bends easily; C—black color, may be soft; Pb—high density, soft metal; O—colorless, odorless gas

2:5 Atoms

Point out that the diagrams and pictures of atoms that students see in books and magazines represent visual models. Since atoms are very small and cannot be seen, scientists formulate conceptual models based on observable information and data. Remind students that these conceptual models may change as scientists gather additional information about the nature of the atom. You may wish to display pictures of the Dalton, Thomson, and Bohr models of the atom to demonstrate how the concept of the atom has changed through time.

Diagram different atoms on the chalkboard to illustrate their structure. Two-dimensional diagrams that show protons and neutrons in the nucleus and electrons in energy levels help develop the concept of atomic structure. The electron cloud model of the atom is discussed in this text. Point out that the electron cloud that surrounds the nucleus represents the probability where electrons are located within the atom. Stress the fact that there is a connection between the structure of the element's atom and the properties of the element. The atom of each element is different, having a different number of protons and neutrons. Atoms are made of electric particles and most of the volume is space. Go over Table 2-4 and answer questions students may have to clarify the idea of energy levels.

Making Sure, page 31
11. The number of electrons is equal to the number of protons.
12. 1, 8, 16
13. two

2:6 Compounds

Begin by going over the compounds and elements in Table 2-5. Review the definition of element and then define compound as a substance containing two or more elements held together by a chemical bond. Explain how atoms of different elements bond together to form compounds. Take the time to show how an atom becomes an ion when it gains or loses an electron. For example:
$$Na \longrightarrow Na^+ + e \text{ and } Cl + e \longrightarrow Cl^-$$

Be sure students are able to differentiate between ionic and covalent bonds. On the chalkboard, use diagrams to show bonding between potassium and iodine to form potassium iodide and bonding between carbon and oxygen to form carbon dioxide. Have students determine which compound is formed by ionic bonding and which is formed by covalent bonding. Stress that the properties of a compound are usually different from the properties of the elements from which it is made.

2:7 Formulas

Begin by directing students to Table 2-6 showing the formulas for different compounds. Point out that each formula contains the symbols for the elements in a compound. Stress the importance of subscripts in formulas. Explain that the components of a compound exist in a definite ratio. Display samples and have students compare the properties of hydrogen peroxide (H_2O_2) with those of water (H_2O). Both compounds have different properties even though they consist of hydrogen and oxygen.

Activity 2-4

In this activity the student, using modeling clay, constructs models of molecules. Use three different colors of clay to represent the three kinds of atoms in sugar and sulfuric acid. Remind students that they should use a different color for each element. Select compounds such as calcium carbonate, potassium nitrate, and aluminum chloride for which students can construct additional models.
Materials: modeling clay, toothpicks
Questions: The models will show the number and kinds of atoms. The models do not show, however, the differences in size (diameter) of atoms.

Making Sure, page 35
14. (a) 2 nitrogen (b) 2 chlorine (c) 1 hydrogen, 1 chlorine (d) 2 hydrogen, 1 sulfur (e) 1 carbon, 4 chlorine (f) 2 nitrogen, 5 oxygen (g) 2 aluminum, 3 oxygen (h) 1 carbon, 4 hydrogen
15. (c) 1:1 (d) 2:1 (e) 1:4 (f) 2:5 (g) 2:3 (h) 1:4

2:8 Mixtures

Students should be aware that the components of a mixture retain their physical and chemical properties and that the components can be mixed in any proportion. Go over the examples of the kinds of solutions given in the section. Have students list additional examples. Discuss those substances not commonly thought of as solutions such as coins, whipped cream, or fog.

Activity 2-5

For safety and practicality, this activity is best done as a teacher demonstration. In this activity the student observes the separation of a mixture of sulfur and iron filings. In the second part of the activity, the student observes the attempted separation of the same mixture after it has been heated. Have all students record the parts of the activity in their notebooks. If a hood is not available, do the activity near an open window or door. If possible, use a large fan to blow the fumes out of the room.
Materials: iron filings, powdered sulfur, magnet, test tubes, test tube holder, paper towel or old newspaper, hammer, burner
Questions: The filings are attracted by the magnet the first time the magnet is passed over the mixture. The filings do not separate after the

mixture has been heated. The iron and sulfur formed the compound iron sulfide.

Activity 2–6

In this activity the student observes properties of oil and water, oil dissolved in alcohol, soil and water, limewater solution, and exhaled air dissolved in water. Have students check to be certain that the test tubes are properly sealed before shaking them.
Materials: rubbing alcohol, bromothymol blue, dropper, lime (calcium oxide), mineral oil, phenol red, soda straw, soil, test tubes (5) with stoppers, test tube rack
Questions and Conclusions: **1.** A, C **2.** E **3.** Possible answers may include adding the phenol red and bromothymol blue solutions to the test tubes and filtering and evaporating the solution.

Perspectives

Using Tables

Tables are aids that supplement and illustrate the ideas presented within the text. They are of value to all students, but particularly to the student who has difficulty in reading and understanding words. The student may not give attention to the tables unless their importance is shown and the student can understand them. It is important to stress how the tables supplement text material.

1. general properties for the states of matter
2. properties and examples
3. Yes, plasma is similar to a gas but is composed of electrical particles.
4. 1.0
5. lowest density to highest density; Yes, the densities of substances can be easily compared.
6. a list of some elements and their symbols; to provide more information in less space
7. The table—the information is more easily located and identified when organized in columns.
8. Table does not include information about specific atoms/elements.
9. information provided in the second and third paragraphs of Section 2:5 and Figure 2-9
10. 4
11. The first, second, and third energy levels contain the remaining 28 electrons.
12. The third column combines the number of electrons from the second column with number of electrons from lower energy levels.
13. The table shows the arrangement of electrons in energy levels as introduced in the third paragraph of Section 2:5.
14. The table shows the name of the compound and the elements of which it is composed.
15. The same compounds are listed in both. Table 2-6 shows the number of atoms and ratio of how the elements are combined.
16. hydrogen; Section 2:7

Study Questions
A. True or False
1. False (The properties of solids and liquids are different.) **2.** True **3.** True **4.** True **5.** False (A compound is a substance in which two or more elements are chemically joined.) **6.** True **7.** True **8.** True **9.** False (A proton has a positive charge.) **10.** True

B. Multiple Choice
1. solids **2.** equal to **3.** Fe_2O_3 **4.** air **5.** mixture **6.** O_2 **7.** 3 **8.** 45 **9.** electrons **10.** proton

C. Completion
1. mass **2.** protons, electrons **3.** elements **4.** H_2 **5.** mixture **6.** iron **7.** ammonia **8.** nitrogen, hydrogen **9.** gold **10.** two

D. How and Why
1. oxygen—element; soil—mixture; nitrogen—element; water—compound; carbon dioxide—compound; mercury—element
2. Liquid has no definite shape but has a definite volume. A gas has neither a definite volume nor a definite shape. A solid has a definite volume and a definite shape.
3. NaCl—sodium, chlorine; H_2O_2—hydrogen, oxygen; CaO—calcium, oxygen; $ZnSO_4$—zinc, sulfur, oxygen
4. 19 (9 protons and 10 neutrons)
5. An atom is the smallest piece of an element which retains the element's properties. A molecule is the smallest part of a compound which retains the compound's properties.

Side Roads
Superconductors
Students can get a better understanding of the usefulness of superconductors by preparing a bulletin board display showing the many uses of electricity. A simple exercise dealing with how a student's lifestyle would change without electricity would be interesting also.

3 Energy and Changes in Matter

Overview

Chapter 3 introduces the concept of energy, defines potential and kenetic energy, and discusses forms of energy such as light and heat. The relationship between matter and energy in physical changes such as melting and boiling, and in chemical changes such as burning, are explained. Thus, the chapter extends the concepts related to matter that were developed in Chapter 2. Nuclear fission and fusion are introduced in describing the release of nuclear energy from atoms. Sources of energy that people use are named and described.

Performance Objectives

Goal: Students will learn the sources and forms of energy and how energy is related to changes in matter.

Objectives: Upon completion of the reading and activities and when asked to diagram, demonstrate, or respond either orally or on a written test, students will:

3:1	K	define energy and explain the differences between potential and kinetic energy.
	K	list different forms of energy.
	S/P	do an activity that shows the relationship between kinetic energy and heat.
3:2	K	define heat and differentiate heat and temperature.
	S	use the Celsius temperature scale to measure temperatures.
	S/P	do an activity in which the transfer of heat is observed by measuring changes in temperature.
3:3	K	describe the changes matter undergoes during melting, freezing, and sublimation.
3:4	K	describe the changes matter undergoes during evaporation and boiling.
	S/P	demonstrate and explain how salt water is purified by distillation.
	S/P	do an activity that compares the rate of evaporation under different conditions.
3:5	K	define chemical change and give examples of a chemical change.
	S/P	do an activity that shows water is produced during burning.
3:6	K/S	write balanced chemical equations for simple reactions.
3:7	K	explain what occurs in a nuclear change.
	K	explain the difference between nuclear fission and fusion.
	K/A	describe the difference between nuclear energy and other forms of energy.
3:8	K	list the major sources of energy.
	K	describe the advantages and disadvantages of solar energy.
	S/P	do an activity to show how solar energy can be collected.

Equipment for Activities

alcohol, rubbing
balance
beakers, 250 mL
bottle, clear glass, narrow neck
bottle, dark glass, narrow neck
cork
drinking glass
graduated cylinder, 10 mL
ice cubes
jar, large glass
lamp
magnifying glass
matches
paper cups
pencil
plastic bag, large, clear
rubber balloons
sponge
salt
sand
stirring rod
stopwatch or watch with second hand
tape, masking
thermometer
thumbtack
thread
tongs
watch glass
water

3:1 Energy

Energy is defined as the ability to do work; that is, to move something through a distance. Explain how walking, lifting, writing, and other human activities are work and how we obtain energy to do the work from the food we eat. Then discuss other forms of energy familiar to students such as heat, light, and electricity. Remind students that most forms of energy are derived from the sun. Demonstrate or list examples of energy such as a burning match, flashlight, rubber-band powered model airplane, wind up spring toy, pinwheel, and toaster. Ask students to describe the form of energy associated with each.

Activity 3-1

In this activity the student shakes sand vigorously in a container and observes that the

temperature of the sand increases. Caution students to be careful in using the thermometer to avoid breakage. Seal the two cups together so they will not come apart when shaken. After shaking the sand, puncture a hole in the cup with a pencil first and then insert the thermometer.
Materials: 2 paper cups, dry sand, masking tape, thermometer
Questions: After shaking the sand, the temperature of the sand is higher because the kinetic energy of the sand particles increased.

Making Sure, page 47
1. (a) PE (b) KE (c) PE (d) KE

3:2 Heat and Temperature

You may wish to begin this section by having a student hammer a large nail into a block of wood and then touch the nail to feel that it is hot. Then have your students read the section and answer the guide questions. Be certain students understand the difference between heat and temperature. Explain that temperature of a substance is a measure of the average kinetic energy of the particles (atom, molecules) of the substance. Heat, on the other hand, is the total kinetic energy of a substance. Heat energy is the measure of the internal kinetic energy transferred from objects of higher internal energy (temperature) to objects of lower internal energy. The Celsius temperatures in Table 3–2 are useful for students to learn and use as reference points for other Celsius temperatures. You may wish to introduce the term absolute zero. When a substance approaches absolute zero (–273°C), the average kinetic of its particles approaches zero. Absolute zero provides the basis for the kelvin temperature scale. See Appendix F, page 512.

Activity 3–2

In this activity the student observes that ice cools water by absorbing heat from the water. The water warms after the ice is removed because it absorbs heat from the air. You may wish to have students construct a graph using their data to show the rate at which the water temperature changed.
Materials: balance, beaker (250 mL), stop watch or watch with second hand, ice cubes, stirring rod, thermometer, tongs
Questions and Conclusions: **1.** The water temperature may decrease to within a few degrees of freezing (0°C). **2.** The water cools when the ice absorbs the heat from the water. The water warms when it absorbs heat from the air. **3.** Eventually, all heat is transferred to the atmosphere. **4.** The water cools much faster than it warms because the ice is so much colder than the water. **5.** The increased mass of the water from the melted ice causes the water to warm more slowly.

3:3 Freezing and Melting

The process by which a solid substance becomes a liquid is called fusion and the temperature at which fusion occurs is called the melting point. The melting and freezing points of pure crystalline substances are usually the same. Students should be aware that water is one of the few substances that expands when it freezes. The volume occupied by ice is almost 1.1 times greater than the volume occupied by the water from which it was formed. Stress that changes in state such as freezing, melting, and sublimation are physical changes and be certain that students understand that heat is lost or gained during a change in state.

Making Sure, page 51
2. lose heat
3. Freezing is a change from a liquid to a solid. Sublimation occurs when a gas changes directly to a solid (or solid to a gas).
4. Boiling involves a change of state but not in chemical composition.
5. Through sublimation, water vapor in the air forms ice crystals on cold surfaces. Remind students that frost is not frozen dew.

3:4 Evaporation and Boiling

A liquid evaporates when a molecule at the surface of the liqiud gains sufficient kinetic energy and escapes from the surface. If the temperature of the liquid increases, the kinetic energy of the particles increases. Thus more of the liquid evaporates. During condensation the particles of the vapor or gas lose kinetic energy. Have students give examples of condensation and evaporation they observe every day.

Activity 3–3

In this activity the student observes the natural distillation of salt water through evaporation and condensation.
Materials: sponge, salt, water, drinking glass, glass jar, clear plastic bag, lamp

Questions: The sponge absorbs the light and is warmed, causing the water in the salt water mixture to evaporate. Condensation of the water occurs on the cooler inner surface of the bag.

Activity 3–4

In this activity the student compares the rates of evaporation of liquids in three containers. You may wish to have students state a hypothesis first and then accept, reject, or amend the hypothesis based on their observations. Show students how to draw a graph for their data and supply graph paper if available. The volume of liquid that remains each day should be measured with the graduated cylinder and then returned to the beaker. Subtract the amount remaining from the amount present the previous day to find the amount that evaporated during the past 24 hours.
Materials: 3 beakers (250 mL), graduated cylinder, water, rubbing alcohol
Questions: Evaporation will be the greatest in the beaker that contains the alcohol because alcohol evaporates faster than water. Evaporation of the water outdoors is faster because air currents, sunlight, and lower humidity increase the rate of evaporation.

Making Sure, page 52
6. loses
7. As alcohol evaporates, it absorbs heat from the skin. (Evaporation of a substance requires energy.)

3:5 Chemical Change

To illustrate an example of chemical change, you may wish to do the following demonstration. Place some moistened steel wool in a test tube and invert the test tube in a pan of water. Fasten the test tube with a clamp. Invert a control test tube in the pan and mark the water level in each test tube with a piece of tape. After one day have students note the change in water level in the test tubes. Have students offer an explanation for the change in water level as well as the color change in the steel wool.

Activity 3–5

In this activity the student observes the changes that occur when a candle burns in an enclosed inverted jar. The activity shows that a candle needs air (oxygen) to burn and water is produced in burning.

Materials: small candle, watch glass, glass jar
Questions: When the jar is placed over the candle, the air supply is cut off and the flame goes out. Water is produced by the combining of hydrogen in the candle with oxygen from the air. The black material is carbon from the candle that remains unburned. Burning no longer continues when all of the oxygen from the air combines with the hydrogen from the candle.

Making Sure, page 53
8. (a) Heat and light are released. (b) Heat is released. (c) Electric energy is released. (d) Energy is not released.

3:6 Chemical Equations

In this section a procedure for writing chemical equations is explained. Be certain students understand the following terms: word equation, chemical equation, reactants, products, coefficients, and balanced. Show the writing of a chemical equation through several steps: (1) writing a word equation, (2) writing a chemical equation by using symbols and formulas for words, and (3) adding coefficients to balance the equation. The word equation for a reaction must be known before a chemical reaction can be written.

Making Sure, page 55
9. (a) $C + O_2 \longrightarrow CO_2$
   (b) $2H_2O \longrightarrow 2H_2 + O_2$
   (c) $Fe + S \longrightarrow FeS$
   The equations need not be balanced.

3:7 Nuclear Change

Begin this section by introducing the term isotope. Isotopes are two or more atoms of the same element having different numbers of neutrons in their nuclei. Explain that isotopes of some elements are unstable. Unstable elements contain atoms that have too many neutrons for the number of protons. Unstable elements are radioactive. The nucleus of a radioactive element emits radiation in the form of alpha particles, beta particles, and gamma rays. An alpha particle consists of two protons and two neutrons. A beta particle is an electron and gamma rays are a type of electromagnetic energy. Refer students to Figure 3–14 and explain that during fission the larger nucleus is split into two or more equal fragments. The combined mass of the fragments, however, is less than the mass of the original nucleus. The difference in mass is converted to energy in the form of radiation.

3:8 Sources of Energy

Begin this section with a discussion of the sources of the energy your students used today. Such examples may include food for body energy, fuel for cars and buses, electricity to heat water and cook food, etc.

You may wish students to research and report some of the following topics related to energy: (1) Using wind power to produce electricity (2) Types of nuclear reactors and their operations (3) Coal gasification (4) Gasohol (5) Possible use of passive solar heating for your school

Activity 3–6

In this activity the student learns that a dark-colored bottle absorbs more energy than a clear bottle.
Materials: 2 bottles having the same size and a narrow neck, one clear and the other dark, 2 balloons (same size)
Questions: The air in the dark bottle is heated to a higher temperature and expands more than the air in the clear bottle. In the dark bottle, the heated air expands and inflates the balloon.

Activity 3–7

In this activity the student observes the heating effect of solar energy.
Materials: thread (25 cm), clear glass bottle and cork, thumbtack, magnifying glass
Questions: The thread will burn and split. Solar energy that reaches the earth is diffuse. The lens concentrates or focuses the energy and the thread will not burn if the magnifying glass is not used.

Making Sure, page 59

10. Some examples may include the sun, natural gas, oil, and coal.
11. gasoline and diesel fuel
12. They do not pollute.
13. Methods may include driving less, turning off lights, etc., when not in use, lower thermostat, and so on.

Perspectives

Fluidized Bed Combustion

Have students do research on the coal industry and the use of nuclear power for energy. The nuclear energy issue is controversial. Have students debate the use of nuclear energy over alternate energy sources.

Study Questions

A. True or False

1. True   2. False (In a chemical change, new and different substances are formed.)   3. True   4. True   5. True   6. False (Kinetic energy is energy of motion.)   7. True   8. True   9. False (Water boils at 100°C.)   10. False (When iron rusts it combines with oxygen.)

B. Multiple Choice

1. oxygen   2. yield   3. chemical   4. uranium   5. fusion   6. nuclear   7. solar   8. electrical   9. coal   10. hydroelectric

C. Completion

1. chemical   2. physical   3. loss   4. symbols, formulas   5. chemical   6. nuclear change   7. water   8. coal, oil, natural gas   9. earth   10. kinetic

D. How and Why

1. A physical change occurs when a substance changes from one form to another without changing its composition. In a chemical change, new substances with different properties are produced. In a nuclear change, an element is changed into a different element.
2. The iron in the furniture combines with oxygen and forms rust.
3. One explanation is that water is trapped behind a dam. The water is allowed to fall through a shaft in the dam to a stream below. As it falls, the water flows through a turbine. The falling water has kinetic energy that is transferred to the moving turbine blades. As the turbine rotates, it turns an electric generator that produces electricity.
4. Potential energy is stored energy or energy of position. Examples include wound spring and a car on top of a hill. Kinetic energy is energy of motion. Examples include water falling over a waterfall and a spring unwinding.
5. We have a limited supply of fossil fuels.

Challenges

1. You may also wish for students to find out what kinds of energy sources your community uses and where these sources are derived.

Unit 2 Mechanics

Overview

Unit 2 covers force, work, power, machines, and motion. In Chapter 4, the use of a force to do work is explained, and the relationships between work, power, and engines are discussed. Simple and compound machines, and their applications are presented in Chapter 5. In Chapter 6, the use of forces to produce and stop the motion of moving objects is explained. Chapter 7 discusses fluids, pressure in gases and liquids, and related concepts such as buoyancy and the lift force on airplanes.

Teaching Tips

Obtain a used gasoline engine from a discarded lawn mower or other power machine. Have students study and dismantle the engine and save the parts in a large cardboard box. A local lawn mower repair shop may be able to supply the engine.

With the assistance of your students, make a collection of small, simple machines of all kinds for study in Chapter 5. Old toys, tools, and various kitchen utensils are useful here.

Relate the study of forces and motion to sports. Games such as billiards, tennis, baseball, Ping-Pong, basketball, golf, and soccer offer many concrete examples having meaning for students.

Ask your students to contribute to a collection of old toys that illustrate motion—cars, windup toys of all kinds, electric trains. Use the toys to motivate interest in forces and motion.

---

4 Work and Energy

Overview

Chapter 4 explains how forces are produced and work is done as an object is moved. Friction force is defined and explained, and the use of lubricants and bearings to reduce friction is discussed. The law of conservation of energy is covered in Section 4:5. The concept of power is presented in Section 4:6, and the use of engines to generate power and do work is described in Section 4:7.

Performance Objectives

Goal: Students will gain an understanding of forces, work, and the use of energy to do work.

Objectives: Upon completion of the reading and activities and when asked to diagram, demonstrate, or respond either orally or on a written test, students will:

4:1	K	define force and list examples of force.
	S/P	measure forces with a spring scale.
4:2	K	explain how weight is different from mass.
	S/P	measure weight in newtons.
4:3	K	list and describe three kinds of friction.
	K	explain how friction is related to motion, machines, and lubricants.
	S/P	do an activity that compares the force of sliding and rolling friction.
	A	understand the value of lubricants in reducing the wear of machine parts.
4:4	K	explain how forces produce work.
	S	calculate the work done when given the amount of force and distance through which the force is exerted.
4:5	K	explain how energy is changed from one form to another when work is done.
	K	state the law of conservation of energy and explain its meaning.
4:6	K	define power and watt, the unit used to measure power.
	P/S	calculate power when given the amount of work done and the time in which the work is done.
4:7	K	explain how a gasoline engine produces power.
	K	explain how a diesel engine is different from a gasoline engine.
	A/S/K	carry out one Challenge at the end of the chapter.
	A/K	read a science article or book and prepare a short report.

Equipment for Activities
brick or wooden block
cord or heavy string
spring scale
stopwatch or clock with second hand
wooden board, smooth,
   1 m × 20 cm
wooden dowels,
   10 cm

4:1 Force
In this section force is introduced and explained by giving examples of forces related to students' everyday experiences. Measurement of forces using newtons is discussed. The newton is a derived unit and is based on the force needed to accelerate a standard mass. In other words, a newton is defined as the force that will cause a 1-kg mass to accelerate at the rate of one meter per second squared (1 N = 1 kg • m/s²). A discussion of force in relation to motion (Newton's laws) is found in Chapter 6.

Demonstrate forces by having students move a variety of objects such as empty cardboard boxes, balls, and small cartons of books. Students should observe that the force needed to move the object is related to the object's mass.

Activity 4–1
In this activity the student measures the forces needed to drag and lift an object (brick or wooden block) and compares the magnitude of the forces.
Materials: spring scale, cord or heavy string, brick or wooden block
Questions: More force is needed to lift the object than to start dragging the object along the desk top. It takes more force to start than to drag the object once it is moving. Students should realize the amounts of forces will vary depending on the object's mass as well as the surfaces of the desk top and the brick. Friction forces and weight are discussed later in this chapter.

Making Sure, page 68
1. Answers will vary. Some examples may include the pull of a car on a trailer and the push of your hand against a ball when it is thrown.

4:2 Weight
In this section the concept of weight is presented as the force of the earth's gravity. You may wish to introduce the section with Activity 4–2 on page 70.

Then refer students to Table 4–1 for additional examples of weights measured in newtons. Be certain that students understand the distinction between weight and mass. Mass depends upon the amount of matter in an object. It is related to the actual number of protons, electrons, and neutrons that make up the object. Weight refers to the gravitational force exerted by the earth.

Activity 4–2
In this activity the student estimates the weights of several objects in newtons and then uses a spring scale to determine the actual weights. The student gains experience in measuring forces and develops a sense or "feel" of the weight of the objects in newtons.
Materials: spring scale, cord or string, five objects such as a book, baseball, shoe, notebook, tennis ball, etc.
Questions: Students determine their weight in newtons by multiplying their weight in pounds by 4.5.

Making Sure, page 70
2. Weight is determined by gravity and is measured in newtons. Mass is the amount of matter in an object and is measured in grams or kilograms.
3. The amount of matter in an object does not change regardless of location. The weight of an object depends on the force of gravity that is exerted on the object. The weight of an object varies according to its distance from Earth's center.
4. 486.02 kg

4:3 Friction
Friction exists when two surfaces slide or roll against each other. Friction is a force that opposes motion. Sliding friction has the following characteristics:
(1) Frictional force is parallel to the surfaces sliding over one another.
(2) Friction depends upon the materials and their surfaces.
(3) Frictional forces are roughly independent of the area of the surface of contact.
(4) The frictional force is roughly independent of the speed of sliding.
(5) Friction is directly proportional to the force pressing the two surfaces together.

These characteristics may be demonstrated by using various sizes of wooden blocks and bricks and different surfaces over which to slide them. A variety of lubricants such as motor oil, water, grease, and alcohol may also be used to demonstrate the various properties and effectiveness of lubricants.

Activity 4–3
In this activity the student pulls a brick over a rough concrete surface, over a smooth board, and on wooden dowels. Sliding and rolling friction are compared. The forces needed to pull the brick is measured wth a spring scale.
Materials: brick, smooth board (1 m × 20 cm), spring scale, wooden dowels (10 cm)
Questions and Conclusions: **1.** More force is needed to keep the brick sliding over the rough surface than the smooth surface. **2.** Answers will vary. More force was needed for sliding. **3.** The force needed to move the brick using two dowels is the same as the force for six dowels.

Making Sure, page 73
5. (a) fluid (b) rolling (c) fluid (d) rolling (e) sliding
6. The sand particles on the surface of the paper exerts a friction force that rubs off the surface particles of the rough wood.

4:4 Work
The scientific definition of work introduces the content of this section and a specific example—a girl lifting a box—is given. Stress the fact that no work is done if a force does not move an object. For example, although a person exerts a force by pushing against a wall, no work is done. It is a good idea to go over the example problem with the class to be certain they understand how work is calculated and how the unit for work, the joule, is derived. Explain how the equation for work is used to calculate the work done in Questions 8 and 9 of Making Sure.

Making Sure, page 74
7. (b), (d), (e) Work is done only when something is moved.
8. 1300 J (65 N × 20 m)
9. 300 J (30 N × 10 m)

4:5 Work and Energy
Begin this section by reviewing the definitions of kinetic and potential energy (Chapter 3) and discuss how the concepts of work and energy are related. For example, water at the top of a dam has the ability to do work. The potential energy of the water is determined by the product of the water's weight and height and is measured in joules. A moving object also has the ability to do work. The kinetic energy of a moving object is related to the mass and the velocity of the object and is also measured in joules.

Discuss the concept of conservation of energy. Have students give examples of energy that changes from one form to another. Some examples may include the changing of chemical energy to light and heat as wood burns. Remind students that in nuclear reactions, matter is changed to energy.

Making Sure, page 75
10. Kinetic energy is changed to heat.
11. The bulldozer does more work.
12. Chemical energy is changed to light and heat when coal burns. Thus, it changes forms but is not destroyed.

4:6 Power
Power refers to the speed at which work is done. The SI unit of power is the watt. Have students make a list of the appliances in their homes and determine the power requirements for each. Suggest students construct a bar graph comparing the watts used by each appliance. Be certain students can work through the steps in the example problem on page 76.

Activity 4–4
In this activity the student calculates the power used to climb a flight of stairs. Caution students to avoid running on the stairs for safety reasons. Students should do this activity in pairs, taking turns timing each other.
Materials: a flight of stairs, stopwatch or clock with second hand
Questions: Use the equation on page 76 to calculate power. The force used is the weight of the student in newtons (1 lb = 4.5 N) and the distance is the total height of the flight of stairs. Time is measured in seconds.

Making Sure, page 77
13. 1000 W (Be certain to calculate work done before determining the power.)

4:7 Engines
This section covers a practical application of work and power. Most of the world's work is done

by engines. You may want to assign a library report on the jet engine or steam engine as an enrichment activity. Give specific directions for the format of the report and the kind of information, including a diagram of the engine, that should be in the report. Point out that engines require a fuel such as gasoline to supply the energy needed to do work. The chemical energy in the fuel is changed to heat when the fuel burns, and the heat is changed into the kinetic energy of the moving parts of the engine.

Study Questions

A. True or False
1. False (A person does no work by pushing against a solid brick wall.) 2. True 3. True 4. False (If a 5-N force moves a box 10 m, 50 J of work are done.) 5. False (Weight and mass are not the same.) 6. True 7. False (A rolling object has less friction than a sliding object.) 8. True 9. False (Adding oil to an engine decreases the friction among moving parts.) 10. False (Power is the amount of work per unit of time.)

B. Multiple Choice
1. force and distance 2. friction 3. fluid 4. work 5. 400 6. newtons 7. weight 8. piston 9. changed 10. fluid

C. Completion
1. weight, force 2. decreases 3. friction 4. friction 5. gasoline, air 6. force 7. $W = F \times d$ 8. work 9. force 10. conservation of energy

D. How and Why
1. Force is any push or pull on an object, but work is a force that moves through a distance.
2. Sliding friction is the sliding of surfaces of solid objects past each other such as a sled or skis moving over snow. Rolling friction occurs when an object rolls across a surface such as a ball or wheel. Fluid friction occurs when objects move across or through a fluid—submarine through water for example.
3. to be sure that moving parts are sufficiently lubricated
4. Answers will vary. Forces may include push of feet on pedals, chain turning wheel, friction on brakes, etc.
5. The law states that energy can change forms, but energy cannot be created or destroyed under ordinary conditions. When wood burns, the energy originally derived from the sun becomes light and heat.

5 Machines

Overview
Chapter 5 covers machines, both simple and compound. Details of a lever, pulley, wheel and axle, inclined plane, wedge, and screw are given, and the use of these simple machines to do work is explained. The calculation of ideal and actual mechanical advantage is presented and illustrated with examples and problems. The percent efficiency of a machine based on work input and work output is discussed and the calculation is explained, step by step. In Section 5:10, a pencil sharpener, can opener, and typewriter are used to show how simple machines are combined together in a compound machine to do work.

Performance Objectives
Goal: Students will gain an understanding of the use of simple and compound machines to do work.

Objectives: Upon completion of the reading and activities and when asked to diagram, demonstrate, or respond either orally or on a written test, students will:

5:1	K	define a machine as something that changes the amount, speed, or direction of a force.
	K	explain the difference between a simple and a compound machine.
5:2	K	describe the three classes of levers and explain that levers may use small forces to overcome large forces.
	K/P	explain how a lever works.
5:3	K	define the actual mechanical advantage of a machine as the number of times a machine multiplies the effort force.
	K/P	explain the difference between actual mechanical advantage and ideal mechanical advantage.
	K/P	calculate the A.M.A. of a machine when given the resistance force and effort force.
	K/P	calculate the I.M.A. of a lever when given the length of the lever arm and the effort arm.

5:4	K	define the I.M.A. of a pulley as equal to the number of support strands.
	S/P	do an activity that demonstrates the I.M.A. and A.M.A. of three-pulley systems.
5:5	K	explain how a wheel and axle does work and that the I.M.A. of a wheel and axle is the diameter of the wheel divided by the diameter of the axle.
5:6	K	define the I.M.A. of an inclined plane as its length divided by its height.
	P/S	demonstrate how the I.M.A. and A.M.A. of an inclined plane is calculated.
5:7	K	explain how a wedge is used to do work and define the I.M.A. of a wedge as its length divided by its thickness.
5:8	K	explain how a screw is a circular inclined plane and how a screw does work.
5:9	K	define percent efficiency as the work output divided by the work input, multiplied by 100%.
	P/S	calculate the percent efficiency of a machine using the equation.
	K	explain why the work output of a machine is always less than the work input.
5:10	K	define compound machine as a machine that contains two or more simple machines.
	K/P	explain how a mechanical pencil sharpener, can opener, and typewriter use simple machines combined together to do work.
	A/S/K	carry out one Challenge at the end of the chapter.
	A/K	read a science article or book and prepare a short report.

Equipment for Activities

books
brick or wooden block
meter stick
metric ruler
object of known weight
pulleys,
   1 double and 2 single
ring stand/clamp
spring scale
string
wooden board,
   smooth,
   1 m × 20 cm

5:1 Machines

Begin this section with a discussion of machines that students know and the uses of these machines. Refer the students to the chapter photograph, page 82, and ask them to identify the tools shown in the picture. Then have students read the section. Review the definition of machine, compound machine, and the six simple machines. Relate these concepts to the photograph by pointing out the examples of simple machines and compound machines in the picture.

5:2 Levers

Begin this section with Activity 5–1. Be certain students know the parts of a lever shown in Figure 5–1. Have students read the section and answer the guide questions and Making Sure questions, then review these questions with your students. Point out how we use our body limbs as levers in walking, throwing, writing, and other activities. Long arms and legs provide leverage, which is valuable in sports such as basketball and ice hockey. A tennis racket, golf club, hockey stick, and baseball bat are all levers used in sports.

Activity 5–1

In this activity the student uses a ruler as a lever to raise a book set on a table or desk. The action of a first- and a second-class lever is demonstrated.
Materials: book, table or desk, wooden ruler, pencil
Questions: When the ruler is pulled up to raise the book, the fulcrum is the end of the ruler under the book. The book is the resistance, and the effort force is applied to the opposite end of the ruler (second-class lever). When the ruler is on the pencil, the pencil is the fulcrum, the book is the resistance, and the effort force is applied by the hand (first-class lever).

Making Sure, page 85

1. (a) between the ends, (b) between the ends, (c) lower hand on handle, (d) screw attaching the bars, (e) where it touches the top of the cap
2. crowbar, bottle opener, shovel
3. crowbar, bottle opener, shovel, scissors
4. crowbar—separate things; bottle opener—remove bottle caps; shovel—move soil

5:3 Mechanical Advantage

The mechanical advantage of some machines makes it possible to do work with less effort force. For example, power steering on a car makes it possible to turn the steering wheel with less effort force. In this section, the concept of mechanical advantage is introduced and applied to the workings of a lever. The fact that friction and weight of a machine affect the mechanical advantage, causing the actual mechanical advantage to be less than the ideal mechanical advantage is explained.

Making Sure, page 89
5. 5 (325/65)
6. 6 N (150/25)
7. 3 (3/1)

5:4 Pulleys

Some students may be familiar with a system of pulleys called a block and tackle that is used to lift the engine out of an automobile when it is to be rebuilt. Pulleys are also evident in factories, ship-loading docks, and on farms where they are used with ropes or chains to lift heavy objects. Certain kinds of machinery contain pulleys connected together by a belt where rotation of one pulley turns one or more other pulleys.

Activity 5-2

In this activity the student has an opportunity to experiment with different pulley systems and to observe the effects of mechanical advantage on effort force.
Materials: double pulley, object (rock or large bolt), ring stand, 2 single pulleys, spring scale, string (1 m)
Questions and Conclusions: **1.** Friction increases the amount of effort force needed to raise the weight. **2.** The system with an I.M.A. of 4 uses two double pulleys, one set fixed and one set moveable, connected by a string that pulls down.

5:5 Wheel and Axle

All students are familiar with a wheel and axle. Have them read the section and make a list of wheels and axles they know. Ask students to explain how the wheel and axle act like a first-class lever. Explain how the radius of the wheel is the effort arm and the radius of the axle is the resistance arm. The fulcrum is the center of the axle. Go over the I.M.A. of the wheel and axle to be certain students understand its calculation.

Making Sure, page 91
8. At the axle, a greater force moves through a small distance. At the wheel, a smaller force moves through a larger distance.
9. The wheel turns faster than the axle, making the bicycle move faster.

5:6 Inclined Plane

Have students give examples of inclined planes in or near your school. Expain that the I.M.A. of an inclined plane is found by dividing the length of the plane by the height of the plane.

Activity 5-3

In this activity the student uses a spring scale to measure the forces needed to lift and pull a brick up an inclined plane. Five different elevations of the inclined plane are used. The student uses the data to calculate the I.M.A. and A.M.A. of the inclined plane.
Materials: smooth wooden board (1 m × 20 cm), 5 books, meter stick, spring scale, string
Questions and Conclusions: **1.** Answers will vary. **2.** Answers will vary. **3.** The greater the angle, the closer the A.M.A. is to the I.M.A. The A.M.A. of the inclined plane at each elevation will be greater than the I.M.A. due to friction between the board and the brick.

Making Sure, page 93
10. I.M.A. = 2, 4; F = 60 N, 30 N

5:7 Wedge

Begin by illustrating the principle of the wedge by placing two inclined planes together base to base. Review the examples of wedges given in this section. Provide examples of different wedges made from scrap lumber. Have students measure the thickness and length of these wooden wedges with a metric ruler and calculate the I.M.A. of each.

Making Sure, page 93
11. (a) 36/5 = 7.7 (b) 42/3 = 14
12. Wedge (b) has a greater mechanical advantage.

5:8 Screw

Begin by having students wrap a wedge-shaped piece of paper around a pencil. See figure below. Ask students to explain the similarities between a screw and an inclined plane. Have students give

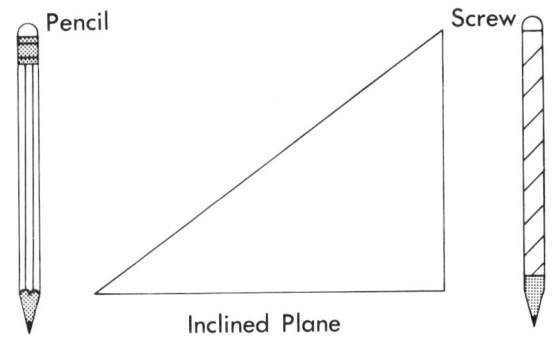

[Figure: Pencil and Screw shown with Inclined Plane forming a right triangle, illustrating that a screw is an inclined plane wrapped around a cylinder.]

examples of items that are held together with screws and bolts. Suggest that students assist in collecting discarded screws of all kinds. Examples may include sheet metal screws, wood screws, and bolts. Making Sure Question 13 provides a short review of the practical uses of simple machines.

Making Sure, page 94
13. (a) screw, lever (b) wheel and axle (c) lever, wedge (d) lever (e) inclined plane, wedge (f) pulley (g) wheel and axle (h) inclined plane (i) wedge (j) lever, wheel and axle (k) wedge, lever, wheel and axle (l) lever

5:9 Efficiency

Some machines are more efficient than others. Percent efficiency is the work output divided by the work input multiplied by 100%. More efficient machines use less energy in doing work. Relate this idea to energy conservation. Example: A small car is a machine that is more efficient than a large car in transporting a person because the small car uses less fuel (energy) per kilometer traveled. No machine is 100% efficient. Friction is one factor that reduces the efficiency of all machines. Hard metal parts that resist wear, roller and ball bearings, and proper lubrication reduce friction and increase efficiency. Example: Ask students to compare the rolling of a roller skate wheel with ball bearings and one that has lost its ball bearings. With the bearings the rotation of the wheel is faster and smoother. The calculation of work input and work output is explained in an example problem using a set of pulleys. These values are then used to show the calculation of efficiency.

Making Sure, page 97
14. A 30 J, 24 J, 80%
    B 2000 J, 1000 J, 50%
    C 50 J, 40 J, 80%
    D 500 J, 300 J, 60%
    E 3360 J, 2520 J, 75%
    F 400 J, 212 J, 53%
15. Lubrication reduces friction.
16. Less energy is needed to do work.
17. Friction is always present between surfaces in contact.

5:10 Compound Machines

Students will need help in identifying the simple machines in a compound machine. In this section the simple machines in a mechanical pencil sharpener, can opener, and typewriter are identified and the way the simple machines in each of these compound machines operate to do work is explained. If possible, display examples of other types of compound machines and have students identify the simple machines.

Making Sure, page 99
18. wheel and axle, wedge
19. wheel and axle, lever
20. When one simple machine does not function, the machines connected to it cannot function.

Perspectives

Organizing Notes in Outline Form

Outlining is a form of note-taking that is effective in helping the student to sort out information in order to show the relationships between main ideas and related details. Students will need guidance in correct outline form, which is not difficult if they are able to note the main ideas and the facts relating to it. Clarity and the proper relationships must be apparent to the student or the outline will be of little value to them.

(1) the friction and weight of the lever are not considered
(2) I.M.A. = $\dfrac{\text{effort arm}}{\text{resistance arm}}$

   I.M.A. = $\dfrac{\text{effort movement}}{\text{resistance movement}}$

1. a wheel that turns on an axle that is used with a rope, chain, or belt
2. a. fixed and moveable
   b. single and double
3. the total number of support strands of a moveable pulley
C. 1. Definition: works on the same principle as the lever and pulley

2. Mechanical advantage:
   Divide the diameter of the wheel by the diameter of the axle
D. Inclined Plane
   1. Definition: a slanted surface which may be used for raising objects to a higher place
   2. Mechanical advantage: divide the length of the plane by the height of the plane

Study Questions

A. *True or False*

1. False (Machines have practical value.) **2.** True **3.** True **4.** True **5.** True **6.** True **7.** False (A short, thick wedge has a smaller mechanical advantage than a long, narrow wedge.) **8.** False (The work output of a machine is usually less than the work input.) **9.** True **10.** False (Mechanical advantage has no units.)

B. *Multiple Choice*

1. two **2.** wheel and axle **3.** first class **4.** unchanged **5.** friction **6.** pulley **7.** screw **8.** efficiency **9.** percent **10.** 4 times

C. *Completion*

1. fulcrum **2.** 2 **3.** 4 **4.** 20 **5.** lever **6.** inclined plane **7.** never **8.** less than **9.** 1 **10.** lever

D. *How and Why*

1. All three are forms of a lever.
2. Lever—A crowbar separates things.
   Pulley—A fixed pulley changes the direction of force to make lifting easier. A flagpole would be an example.
   Wheel and axle—A bicycle wheel and axle makes it move.
   Inclined plane—A ramp allows less force to be used in going to different levels.
   Wedge—A sharp knife can cut things with little force.
   Screw—Requires a small effort force to overcome a large resistance force; automobile jack.
3. This system requires two single pulleys—one moveable, the other fixed, with three support strands. The resistance force will move 0.67 m.
4. $E = W_o/W_i \times 100\%$; $E = 50\%$
5. Possible answers may include:
   mechanical pencil sharpener—wedge; can opener—wheel and axle; crane—pulley; typewriter—lever; meat grinder—screw;

Challenges

2. Have students obtain information about the designs of different automobiles; make a poster listing the car weight, engine power, and km/L (miles/gallon).

Side Roads

The World of Robots

Students can do research and report on technical advances that make robots useful in areas such as medicine, space travel, research, transportation, and industry. Students might contact movie studios that produce science fiction movies such as "Star Wars" and report on the construction of the robots used in the movies. There is controversy over robots replacing humans on the assembly lines of factories. One side suggests that workers will lose their jobs; the other side suggests that robots will create even more jobs in the future. Students can do research on both sides of the issue and present a debate.

6 Motion

Overview

Chapter 6 begins with the concept of inertia as it relates to objects at rest and in motion. Speed and changes in speed are discussed in two sections in which acceleration and deceleration are explained. Examples of objects in motion, their speed, and changes in speed are given throughout the chapter. The acceleration of falling objects and the effect of air pressure on falling objects is explained in Section 6:5. Centripetal force and circular motion is explained using examples such as amusement park rides and water extraction in a washing machine. The last section in the chapter covers momentum of moving objects and the law of conservation of momentum. Chapter 6 builds on the preceding chapters in this unit and extends the student's knowledge of the action of forces on objects.

Performance Objectives

Goal: Students will gain an understanding of how forces produce, change, and stop the motion of objects.

Objectives: Upon completion of the reading and activities and when asked to diagram, demonstrate, or respond either orally or on a written test, students will:

6:1	K	define inertia.
	P/S	do two activities that illustrate the inertia of objects when at rest.
6:2	K/P	explain the difference between constant speed and average speed.
	P/S	calculate the average speed of an object.
6:3	K	explain how a force starts and stops an object in motion.
	K	explain the difference between acceleration and deceleration.
	P/S	calculate average acceleration and deceleration when given the change in speed and time.
6:4	K	define action and reaction forces.
	K	describe examples of action and reaction forces.
	P/S	use a spring scale to measure action and reaction forces.
6:5	K	explain the acceleration of a falling body and state the value for the rate of acceleration.
	K	define terminal speed and give one example of terminal speed.
	P/S	make a graph of the acceleration of a falling object given the time of fall in seconds and the speed in meters per second.
	P/S	use the graph for acceleration of a falling body to extrapolate the speed at times not plotted on the graph.
6:6	K	explain how centripetal force keeps an object moving in a circular path.
	K	give examples of circular motion.
	P/S	demonstrate two activities that illustrate circular motion and explain each activity in terms of speed, acceleration, centripetal force, and inertia.
6:7	K	explain how momentum is affected by mass and velocity.
	K	give examples of momentum and compare the amount of momentum in different objects.
	A/S/K	carry out one Challenge at the end of the chapter.
	A/K	read a science article or book and prepare a short report.

Equipment for Activities

aluminum foil
battery, 1½ or 6 volt
bell, electrical
books
coins
drinking glass, plastic
graph paper, metric-ruled
hammer
liquid solder or tape
metal ball
marble, large glass
metric ruler
nails
paste
pie pan, thin metal
playing card
rubber ball, small
rubber balloon
scissors
sheet metal shears
spring scale
stopwatch or watch with second hand
string
wire, electrical
2 wooden boards, 15 cm × 4 m

6:1 Inertia

Many students are interested in the United States' space shuttle program. Explain that launching a spacecraft requires a huge force to overcome the inertia of the spacecraft at liftoff. The path of the shuttle orbiter into space and back to earth again involves motion and force. Rockets are used to produce motion and guide the path of the space shuttle. You can demonstrate inertia by placing checkers in a stack on a desk and quickly knocking out the bottom checker with a ruler. Each time you knock out the bottom checker, the rest of the stack will drop down.

Activity 6–1

In this activity the student observes the inertia of a coin. The trick in this activity is to pull the card in a sudden swift movement. The students may have to practice the activity several times before they are successful.
Materials: coin, playing card, drinking glass, plastic
Questions: The inertia of the coin is illustrated by the fact the coin remains in place for a moment over the glass when the card is removed. There is not sufficient friction between the coin and the card to carry the coin along with the card. Gravity finally pulls the coin down into the glass.

Activity 6–2

In this activity the student demonstrates the inertia of a stack of books.

Materials: string (thin), 5 books

Questions: When the string is pulled slowly, the books are raised off the floor because the string overcomes the inertia of the books. When the string is pulled suddenly, it breaks due to the sudden large force exerted. The inertia of the books keeps them in place.

Making Sure, page 109

1. If a car were to stop suddenly, a person would keep moving forward because of inertia. A seat belt prevents this movement.
2. A golf ball has more mass, therefore more inertia.
3. The person would land next to the first seat. Because of inertia, the person moves at the same speed as the train while in the air.

6:2 Speed

Students may be familiar with the speed of an automobile or bicycle recorded on a speedometer. Other examples of speed familiar to your students are fast pitch and slow pitch softball, speed of a major league baseball pitcher, speed of runners in a track meet, and the different speeds at which a kitchen blender and mixer may be operated. Review the difference between constant speed and average speed. The example problem on page 110 will help students understand constant speed.

Making Sure, page 111

4. 860 km/h
5. 5 km/min

6:3 Change in Speed

Although some moving objects travel at constant speed, the speed of most moving objects is increasing or decreasing at times. Acceleration is any increase or decrease in speed and any change in direction of a moving object. Deceleration is negative acceleration. A force must be applied for acceleration to occur. Discuss examples of acceleration and deceleration and go over the two example problems which show the calculation of acceleration and deceleration. Explain that the value m/s² is read as "meters per second squared" and refers to the change in speed (m/s) per second.

Making Sure, page 113

6. (a) Yes, if a filled balloon is suddenly opened, pressure difference causes the air to move out of the balloon.
   (b) Yes, the clock's hand is constantly changing direction.
   (c) No, if the current moves in one direction. (Yes, if the current changes direction.)
7. 1 km/h/s
8. –13 km/h/s

6:4 Action and Reaction Forces

Begin this section with Activities 6–3 and 6–4. Then have students read the section and answer the guide questions. Allow students interested in model rocketry to display and demonstrate their models.

Activity 6–3

In this activity the student measures action and reaction forces by hooking together two spring scales. A force is exerted on one spring scale and the other spring scale is attached to a desk or table leg.

Materials: 2 spring scales, string

Questions: Both spring scales give the same reading, illustrating that the action and reaction forces are the same.

Activity 6–4

In this activity the student inflates a rubber balloon and releases it.

Materials: rubber balloon

Questions: When the balloon is released, it is propelled forward. The escaping air is the action force; the movement forward is the reaction force.

Making Sure, page 115

9. (a) The reaction force on the boat is opposite to the person's direction.
   (b) The reaction force on the sprinkler makes it turn in the opposite direction from the water.

6:5 Falling Objects

Some students have difficulty understanding that all objects fall at the same rate in a vacuum, accelerating at a constant rate of 9.8 m/s². Air pressure on falling objects in the atmosphere slows down the speed of fall depending upon the amount of surface area exposed to the air. The amount of gravitational force acting on a falling object varies directly with the mass of the object ($F = m \times a$). Although objects differ in mass, the amount of gravitational force per unit of mass is the same. Therefore, the acceleration of each object falling in a vacuum is the same.

Activity 6–5

In this activity the student constructs a graph plotting time and speed of an object falling at 9.8 m/s²—Earth's rate of acceleration due to gravity. Review the steps in making a graph by drawing the horizontal and vertical axes on the chalkboard and illustrating the plotting of time vs. speed. If graph paper is not available, have students draw the graph in their notebooks using a metric ruler.
Materials: graph paper (millimeter- or centimeter-lined), Table 6–1, page 115
Questions: The speed at 8 seconds (78.4 m/s) is found by continuing the plotted line and reading the speed on the vertical axis. The speed at 10 seconds is 98 m/s.

Activity 6–6

In this activity the student observes the uniform acceleration of a metal ball rolling down an inclined plane. Your school shop teacher or a student who is handy in construction may assist in making the apparatus.
Materials: 2 wooden boards (15 cm × 4 m), aluminum foil, paste, electrical wire, scissors, battery, electric bell, liquid solder or tape, stopwatch or clock with second hand, nails, hammer
Questions: The bell rings more often as the ball proceeds down the ramp because the ball is accelerating. The speed of the ball is the greatest at the end of the ramp.

Making Sure, page 117

10. Air friction decreases the speed.
11. There is less air resistance (drag) when the paper is crumpled.
12. The speed will be less because the acceleration due to gravity is less.

6:6 Circular Motion

Students have experienced circular motion riding in trains, cars, buses, and on amusement rides. Go over the definition and explanation of centripetal force. Some students will ask about centrifugal force, a term they have heard associated with circular motion. Centrifugal force is an apparent force and does not exist in reality. The tendency of an object moving in a circle to travel off in a straight line is due to inertia.

Activity 6–7

In this activity the student observes circular motion by attaching a ball to a string and swinging it in a circle. The string exerts a centripetal force on the ball that keeps it moving in a circle. This activity should be done where there is ample space or out-of-doors.
Materials: string, small rubber ball
Questions: The student must pull on the string and at the same time move the ball forward around the circle. If the string were cut, the ball would fly off in a straight line.

Activity 6–8

In this activity the student observes circular motion by moving a marble around the inner edge of a pie pan.
Materials: thin metal pie pan, marble, sheet metal shears
Questions: Centripetal force exerted by the edge of the pie pan acts on the marble towards the center of the pan. When the centripetal force is absent (cut-out part of the pan), the marble travels away from the circle and out of the pan in a straight line.

Making Sure, page 120

13. The road increases centripetal force which pushes inward on the car as the car moves around the curve.
14. There is a lack of friction due to the ice, and there is not sufficient centripetal force exerted on the tires (car).
15. the gravity of Earth
16. Motion is in a straight line unless acted upon by a force.

6:7 Momentum

Begin this section by introducing the definition of momentum. Then go over the Making Sure questions, which give concrete examples. Point out that a small object such as a bullet has great momentum when traveling at a high speed. A huge steam roller has a large momentum when traveling at a slow speed. Ask students to describe other examples of momentum.

Making Sure, page 121

17. The tennis ball has more momentum because it has more mass and velocity.

18. The bullet has a very high velocity, therefore more momentum.
19. A loaded truck has more mass and more momentum.
20. Its great mass gives it a large momentum.

Perspectives
Understanding Science Words
A student gathers a great deal of scientific knowledge by reading. For this reason, language and reading are vital tools for gathering scientific data. At times, science vocabulary can be a problem. Students can help themselves by learning to analyze words. One long word may be a compilation of strung-together word parts, each of which has a simple meaning. Combining the parts often gives meaning to the whole word.

Study Questions
A. True or False
1. True 2. True 3. True 4. True 5. True 6. False (Banked roads increase the centripetal force acting on cars.) 7. True 8. False (On a trip the car's speed varies.) 9. True 10. True

B. Multiple Choice
1. 0 2. on a merry-go-round 3. time 4. increases 5. 5 km/h 6. force 7. increases 8. increases 9. remains the same 10. no longer increases in speed

C. Completion
1. force 2. mass 3. higher 4. 10 km/s 5. centripetal 6. centripetal 7. 60 km/h 8. decelerating 9. force 10. speed

D. How and Why
1. A person riding in a car has inertia; therefore, the person is moving forward with the car. When the car is stopped suddenly, as in an accident, the person continues to move forward and may collide with the dashboard, windshield, etc.
2. Because of inertia, the water goes out through the holes in the tub as the clothes are spun around.
3. The speed of the jumper will first increase due to acceleration of Earth's gravity. Terminal speed will be reached when the air resistance equals the pull of gravity on the jumper. The jumper will decelerate when the chute is opened.
4. (a) When a driver steps on the gas pedal, the car accelerates. (b) As a football player kicks the ball, the ball accelerates. (c) Air rushing out of a balloon accelerates the balloon.
5. Although objects differ in mass, the amount of gravitational force per unit of mass is the same. All falling objects near the earth's surface accelerate at the same rate in a vacuum.

7 Fluids and Pressure
Overview
Chapter 7 discusses the nature of liquids and gases. General properties of fluids are presented and then the special properties of gases and liquids are discussed. The pascal is introduced as a unit of pressure, and the calculation of pressure is explained. Charles' law and Boyle's law are explained in the presentation of material on gases. Buoyancy, density, and floating are covered as these concepts pertain to gases and liquids. Bernoulli's principle is explained and illustrated with a simple activity and is related to the lift force that makes an airplane fly.

Performance Objectives
Goal: Students will gain an understanding of fluids and how they are related to forces and motion.

Objectives: Upon completion of the reading and activities and when asked to diagram, demonstrate, or respond either orally or on a written test, students will:

7:1	K	define fluid and list examples of gases and liquids.
	K	describe the particle structure of gases and liquids and explain how this structure is related to the properties of fluids.
	K/P	define surface tension and describe an example.
	P/S	use materials to demonstrate that water in a rubber tube will "seek its own level."
7:2	K	define pressure and pascal.
	P/S	calculate pressure when given the force and area.
	K	describe how a hydraulic lift works.
	P/S	do an activity which shows that pressure in a liquid increases as the depth increases.

7:3	K	explain how a confined gas exerts pressure within its container.
	K	state Charles' law and describe one example of its application.
	K	state Boyle's law and describe one example of its application.
7:4	K	explain how buoyancy is created in a liquid and gas.
	K/P	explain why certain objects float in water.
	K/P	explain why a balloon filled with hydrogen or helium will float in the air.
7:5	K/S	state Bernoulli's principle and demonstrate a simple activity that illustrates this principle.
	K/P	explain how an atomizer works.
7:6	K	explain how a lift force is produced on the wing of an airplane.
7:7	K	explain how an airplane flies.
	K	describe the action of an airplane's ailerons, elevators, and rudder.
	K	explain the effect of drag on an airplane and how it is designed to reduce drag.
	A/S/K	carry out one Challenge at the end of the chapter.
	A/K	read a science article or book and prepare a short report.

Equipment for Activities

aluminum foil, heavy
balance
beaker, 100 mL
coins or washers, small
displacement can
funnels
graduated cylinder, 25 mL
ink
Ping-Pong ball
ring stand/clamp
rubber band
rubber sheet, thin
rubber tubing, 2 m
stone or small rock
string
thistle tube
U-tube

7:1 Fluids

In this section the properties of gases and liquids are introduced. Compare the structure of gases, liquids, and solids and explain how the properties are related to differences in structure. You may wish to introduce this section by doing the demonstration described in paragraph 1, page 128. The blowing of soap bubbles illustrates surface tension, the molecules in the bubbles being held together by forces between them.

Activity 7-1

In this activity the student observes the water levels in two funnels connected by a rubber tube. The properties of a liquid are examined. The water in the two funnels flows so that the surfaces of the liquid in the two funnels are at the same level.
Materials: rubber tubing (2 m), 2 funnels
Questions: When the funnels are lowered or raised the water moves up and down so that the water levels are equal.

Making Sure, page 129

1. Some examples are water, oil, and steam.
2. Water flows through the shafts and turns turbines.
3. Through surface tension, the water molecules pull into a sphere. The water molecules do not adhere to the wax.

7:2 Pressure in Liquids

Distinguish between force and pressure, which is the force per unit area. The SI unit of pressure, pascal (Pa), is introduced and defined as N/m^2. Review the newton as a unit of force and the square meter as a unit of area. Use a meter stick to draw a square meter on the chalkboard. Because the pascal is such a small unit of force, the kilopascal (kPa) is often used to express the amount of pressure. Review the workings of a hydraulic lift by going over the diagram on page 131 with your students. Stress that pressure in a liquid depends on the density and depth of the liquid.

Activity 7-2

In this activity the student observes that the pressure in a container of water increases as the depth increases. The amount of pressure is indicated by the level of the colored water in the U-tube. You may choose to do this activity as a demonstration. Have each student draw a diagram of the apparatus and record the problem, procedure, and observations in a notebook.
Materials: U-tube, thistle tube, thin rubber sheet (may be cut from a balloon), rubber tubing (1 m), ink, ring stand/clamp, rubber band, beaker
Questions: The water level rises as the depth of the thistle tube increases, showing an increase in pressure.

Making Sure, page 132
4. Mercury. The greater density means a greater weight at 1 m than for water.
5. Air would compress when a force was exerted on the piston.
6. 1000 Pa, 0.1 N/cm²
7. 3600 Pa
8. The bottom of the dam must withstand a greater pressure because water pressure increases with depth.

7:3 Gas Laws

The gas laws describe changes in gases related to temperature, pressure, and volume. Boyle's law states that the volume of a dry gas varies inversely with the pressure exerted on it, provided the temperature remains constant. Charles' law states that if the pressure is constant, the volume of a dry gas is directly proportional to the kelvin temperature.

Inflate a balloon and demonstrate the effect of increased pressure. Ask students to hypothesize as to what changes would occur if the balloon were cooled. You can test the hypothesis by making a knot in the balloon and keeping it in a freezer for an hour.

Making Sure, page 134
9. (a) The pressure of a gas increases as its temperature increases.
(b) Water turns to steam when heated. If the volume remains constant, the pressure of the gas increases as it is heated.
(c) As a gas is heated its volume increases.
(d) When the temperature of a gas decreases, the volume (pressure) decreases.
(e) When the pressure of a gas decreases, its temperature decreases.
(f) The pressure of a gas increases when the volume decreases.

7:4 Buoyancy

Begin this section by asking your students to suggest hypotheses to explain why things float. Then do the two Activities 7-3 and 7-4 to test the hypotheses. Discuss the observations made in the activities in relation to accepting, rejecting, or amending the hypotheses. Then have students read the section and answer the guide questions and Making Sure questions. Go over the questions for review and reinforcement of density, buoyancy, and floating.

You may wish to introduce the term specific gravity. It is used frequently to compare the density of one substance with that of another. Specific gravity is defined as the ratio of the density of a solid or liquid to the density of water. Water is the standard chosen to compare the densities of all solids and liquids.

Activity 7-3

In this activity the student observes that the volume of a stone is equal to the volume of water it displaces when submerged.
Materials: graduated cylinder (25 mL), small rock or stone, string (30 cm), balance, water
Questions: The mass of the stone is greater than the mass of water it displaces because rock material has a greater density than water. The stone sinks because the buoyancy of the water is not great enough to support the weight of the stone.

Activity 7-4

In this activity the student observes that an aluminum saucer containing several coins will float on water. The saucer and coins will sink when the foil is folded tightly around the coins.
Materials: displacement can, beaker (100 mL), heavy aluminum foil, small coins or washers, balance, graduated cylinder (25 mL)
Questions: Because the weight of the displaced water is equal to the floating saucer, the water supports the saucer and the saucer floats. When the saucer is crushed around the coins, the weight of water displaced is less than the weight of the aluminum foil and coins. Therefore, the buoyant force is less than the weight of the aluminum foil and the coins.

Making Sure, page 137
10. Lower, because fresh water is less dense than ocean water. Thus, it exerts less buoyant force on the wood than saltwater.
11. Alcohol is less dense than water. The wood would float lower in the alcohol since it exerts less buoyant force than the water.
12. Mercury has 13 times the density of water; therefore, has greater buoyant force.

7:5 Bernoulli's Principle

Introduce this section by having students read and do the activity described in the first paragraph. If possible, obtain a Venturi tube and set up the apparatus to demonstrate how pressure within a streamline flow of a fluid is determined by the speed of the fluid. Conclude the section with Activity 7-5.

Activity 7-5

In this activity the student observes the effects of pressure within a streamline flow on an object.
Materials: funnel, Ping-Pong ball
Questions: If the funnel is held vertically, the Ping-Pong ball will not be blown out of the funnel because the pressure of the moving air under the ball is less than the pressure above the ball. The air above the ball pushes the ball into the funnel no matter how hard the student blows.

7:6 Lift

In this section Bernoulli's principle is used to explain the lift force produced on the wing of an airplane. Use Figure 7-15 to explain how the decreased air pressure produces the upward force on the wing that balances the downward force of gravity on the airplane. Have students offer explanations as to why a spinning baseball will follow a curved path. Use the figure below to assist your discussion.

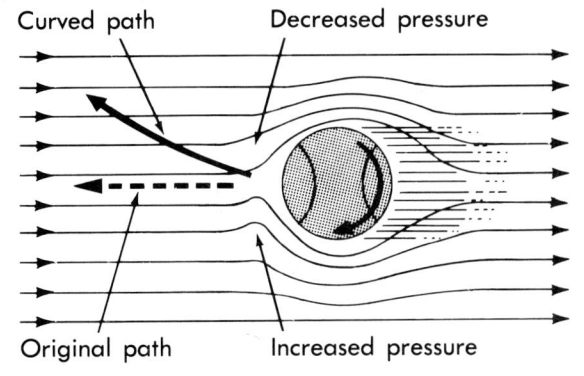

7:7 Flight

The flight of an airplane illustrates the effect of air pressure. Air pressure lifts the airplane into the air, keeps it up when in flight, and produces forces on the control surfaces of the wings and tail that are used to turn the airplane and increase or decrease altitude. One or more students may have model airplanes they can bring to class to demonstrate these actions. You can compare a submarine in water to an airplane. The buoyancy of the submarine prevents it from sinking, and the force of water against the moving submarine's rudder and other control surfaces are used to steer the vessel.

Making Sure, page 141
13. There is no air in space to create a force on the control surfaces. Explain that the space shuttle orbiter uses control surfaces to control its flight as it reenters the earth's atmosphere.
14. Both use escaping gas to produce force.
15. The streamlined shape of a submarine helps to reduce drag as it moves through the water.
16. Strong winds are sufficient to cause lift on the wings of small aircraft.
17. The amount of drag is directly proportional to the speed.

Perspectives

First Flight

Students may wish to set up a display of models of airplanes or boats. Retired flight attendants or pilot groups may provide speakers for the classroom. A field trip to a local airport or boatyard can provide students with useful information. A local museum or science center may have displays and exhibits of rockets or airplanes.

Study Questions

A. True or False
1. True **2.** False (Nitrogen and oxygen are plentiful on the earth.) **3.** True **4.** False (When water boils, its volume increases.) **5.** False (Surface tension is a property of liquids.) **6.** False (Fluids are often used to cool things.) **7.** True **8.** True **9.** True **10.** True

B. Multiple Choice
1. elevator **2.** Bernoulli's principle **3.** increase **4.** increases **5.** forces **6.** 1 N/m² **7.** pressure **8.** less **9.** buoyancy **10.** less than

C. Completion
1. decreases **2.** increases **3.** weight **4.** more **5.** higher **6.** weight **7.** less **8.** lift **9.** greater **10.** rudder

D. How and Why
1. A gas can be compressed by increasing pressure and decreasing volume.
2. Hydraulic lifts are based on the use of the pressure transmitted through liquids.
3. A ship floats because the weight of the water displaced by the ship is equal to the ship's weight.
4. An airplane's flight path is controlled by the control surfaces on the wings and tail.
5. Warm air is less dense, therefore, there is less lift. The extra speed gets more air molecules flowing over the wings. See Bernoulli's principle.

Unit 3 Earth Science

Overview

Unit 3 covers the major topics and basic principles of earth science. The unit opens with a discussion of the major features of the earth such as its size, rotation, and revolution. It then continues with geology in which the structure and composition of the earth are discussed. In Chapter 10, topics related to the geologic history of the earth are presented. Two chapters cover the properties of the atmosphere and factors in the atmosphere, such as air masses, pressure systems, and fronts that are responsible for weather and climate. The last chapter in the unit, Oceanography, is a survey of basic concepts in this field.

Teaching Tips

Obtain topographic maps for areas of your state that have interesting geological surface features. These maps may be obtained from the U. S. Geological Survey, Washington, DC, a county extension agent, or the office of the state geologist.

Purchase a collection of rocks from a scientific supply company and/or ask your students to assist in building a collection for the school science department. Donations of rocks may be solicited from local rock collectors.

Take your class on a field trip to a museum of natural history to observe the geology and paleontology collections. Obtain the required administrative and parental approval.

Obtain copies of weather maps for use in studying factors that affect weather and climate. Weather maps may be obtained from the National Weather Service, Washington, DC, or a local office of the National Weather Service.

If you live close to the seacoast, investigate the opportunity to take your class on a trip aboard a college or school district oceanographic boat. Obtain the required administrative and parental approval.

Students interested in aquariums can build a seawater aquarium under your guidance. Check with the local aquarium shop for assistance.

---

8 The Earth

Overview

Chapter 8 presents information concerning the shape and size of the earth, the earth's rotation on its axis, and its revolution around the sun. The season, caused by the earth's revolutions, are described and explained. Mapping the earth's surface, topographic maps, and time zones are three topics covered in detail. The chapter concludes with a discussion of Earth's magnetism.

Performance Objectives

Goal: Students will learn some of the features of Earth's surface and how maps are used to show them.

Objectives: Upon completion of the reading and activities and when asked to diagram, demonstrate, or respond either orally or on a written test, students will:

8:1	K	describe the shape of the earth.
	K/P	explain how the circumference and mass of the earth can be measured.
8:2	K/P	explain how a Foucault pendulum indicates the rotation of the earth.
	K	describe the rotation and revolution of the earth.
8:3	P/S	diagram the position of the earth in relation to the sun during the four seasons of the year.
	K/P	explain how the hours of daylight change during the four seasons.
	P/S	demonstrate with a globe or large ball and a flashlight the earth's rotation and revolution around the sun.
8:4	K/S	identify lines of latitude and longitude on a map.
	K/S	locate a point on a map when given the latitude and longitude.
8:5	K/S	locate a hill, valley, and stream on a contour map.
8:6	K	explain why a scale is used in map making and give two examples of such a scale.
	K	explain the use of time zones.

8:7	K	identify the locations of the magnetic north pole and magnetic south pole.
	K	explain how a compass is used to tell direction and how the angle of magnetic declination is used to locate a true direction.
	A/S/K	carry out one Challenge at the end of the chapter.
	A/K	read a science article or book and prepare a short report.

Equipment for Activities
flashlight
globe or large rubber ball
pebbles or small sticks
wooden stick, 1 m

8:1 Earth's Shape and Size

Review the observations that Eratosthenes used to determine the earth's circumference as illustrated in Figure 8–1. Use the chalkboard to explain how sunlight striking a verticle stick at an angle forms a shadow and when directly overhead, does not cast a shadow. The classic experiment done by von Jolly to determine Earth's mass is also discussed. Figure 8–2 represents a simplified explanation of the experiment. In the figure, the weight in the left pan represents a spherical flask of mercury with a mass of 5 kg. The mass of the lead sphere is 5775 kg. When the lead ball is positioned beneath the mercury, a tiny mass of 0.589 mg (right pan) is needed to restore the balance. The gravitational attraction of the lead was compared with the attraction of the earth using Earth's acceleration due to gravity (9.8 m/s^2).

8:2 Earth's Motions

Point out that a Foucault pendulum is very long compared to a clock pendulum. Demonstrate the rotation of the earth with a globe that spins and show the tilt of the earth at a 23½° angle with respect to Earth's plane of orbit. Review the concept of degrees as it is used in measuring angles. Relate the west to east rotation to the fact that the sun rises in the east and sets in the west. The position of the sun in the sky moves 15° per hour. Review the facts in Table 8–1. See Appendix A for a discussion of scientific notation.

Activity 8–1
In this activity the student observes how the changing position of the sun affects the shadows produced by a stick.
Materials: wooden stick (1 m), 3 pebbles or small sticks
Questions: The shadow is longest early in the morning and later in the afternoon. From morning until noon, the shadow gradually decreases in size. After noon, it gradually increases in size to its maximum length. Students should observe that the shadow will always appear on the north side of the stick in the northern hemisphere. The length of the stick at noon depends on the time of year and the latitude of your location.

8:3 Seasons

The seasons of the year are likely to be most obvious to students in the Midwest or Northeast sections of the United States and other regions where changes in the environment resulting from seasonal changes in climate are most apparent. Discuss these seasonal changes that occur in your region and relate these changes to the revolution of the earth around the sun. Using a globe and flashlight, demonstrate how the North and South Poles are illuminated 24 hours a day during the summer months.

Activity 8–2
In this activity the student demonstrates Earth's revolution and rotation. Have students compare the positions of their spheres with the positions of the earth as shown in Figure 8–6.
Materials: globe or large rubber ball, flashlight
Questions: The side of the sphere (Earth) facing the flashlight (sun) is illuminated, and the opposite side is dark. Rotation causes alternate light and dark (day and night) for a particular location on the sphere. As the sphere is moved around the flashlight, the light strikes more directly the surface of the sphere tilted toward the flashlight. The motion of the sphere around the flashlight is like the revolution of the earth around the sun.

8:4 Latitude and Longitude

The location of any point on the Earth's surface can be stated in terms of the geographic coordinates—latitude and longitude. The prime meridian (0°) was chosen arbitrarily as the meridian passing through the former location of the Royal Observatory near London, England.

The coordinates of a location may be given in degrees, minutes, and seconds, e.g., lat. 35° 12′31″ N. It is common, however, to state coordinates in decimal parts, e.g., long. 35.2086° W.

Activity 8-3

In this activity the student uses latitude and longitude to locate cities on a map. You may wish to assign additional cities that students can locate on a map or globe using geographic coordinates.
Materials: Figure 8–9, page 156
Questions: Boston—lat. 42.21° N, long. 71.04° W; Philadelphia—lat. 39.57° N, long. 75.07° W; Atlanta—lat. 33.45° N, long. 84.23° W; New Orleans—lat. 29.58° N, long. 90.07° W; Dallas—lat. 32.47° N, long. 96.48° W; Omaha—lat. 41.16° N, long. 95.57° W; St. Paul—lat. 44.58° N, long. 93.07° W; Phoenix—lat. 33.27° N, long. 112.05° W; San Francisco—lat. 37.48° N, long. 122.24° W; Helena—lat. 46.36° N, long. 112.01° W.

Making Sure, page 156
1. 45° N

8:5 Mapping the Earth's Surface

Explain the use of a scale in making maps and other drawings. For example, a plan for a house, model of a car, and floorplan of an apartment would be made according to scale rather than life size. Have students find examples of scales used on a variety of maps. Some students may be interested in displaying scale models of cars, aircraft, etc.

Distribute an assortment of topographic maps to the students and have them locate various landforms and cultural features.

Activity 8–4

In this activity the student determines map scales for their state and neighborhood.
Materials: notebook paper, pencil
Questions: The scale ratio for the state map would be small because the whole state must be reduced to the size of the page. For a neighborhood, the scale ratio would be much larger since a relatively smaller area than the state must be reduced to page size.

Activity 8–5

In this activity the student identifies surface features depicted on a topographic map.
Materials: Figure 8–13, page 158
Questions: (1) C (2) E (3) South, the elevation increases to the north. (4) 70 m (5) 80 m (6) Steeper between C and F because the contour lines are closer. (7) from point E eastward through the valley (8) Point G is lower than the land around it. (9) no difference in elevation

Making Sure, page 158
2. The inner contours of a hill represent higher elevations than the outer contours. The inner contours of a valley represent lower elevations than the outer contours.

8:6 Time Zones

Relate time zones to personal experiences in traveling east and west and the scheduling of events for television programs. Ask students what time a baseball World Series game in New York City should be scheduled so it can be seen nationwide on television. The three-hour time difference from east to west coast must be taken into account. If the game is scheduled before 8:00 P.M. EST (5:00 P.M. PST) most working people on the west coast would still be at work. Time zones in the United States were established when railroads were built to the west. Timetables for the arrival and departure of trains required a systematic scheduling of time to avoid the confusion of time differences between places that were only 50 to 100 miles apart. Be certain students understand the International Date Line and how the days change traveling east to west and west to east.

8:7 The Earth as a Magnet

Earth may be thought of as a large sphere containing a bar magnet oriented somewhat along the axis of rotation. Use a bar magnet and a small compass to demonstrate the effect of a magnetic field on the orientation of a compass needle. Referring to Figure 8–16, stress the fact that magnetic and geographic poles represent different points on Earth's surface.

Making Sure, page 163
3. A compass helps to correctly orient the features on the map with respect to the true layout of the earth's surface.
4. The magnetic declination is determined by the angle made between the direction of geographic and magnetic poles and depends on the longitude of your location.

Perspectives

Reading Maps

A student may not give attention to maps unless you stress their importance and make certain that the student understands them. Make certain that the student also understands how the maps relate to the material in the text.

1. Answers will vary.
2. Answers will vary.
3. towards the top of the map; north; Answers will vary.
4. about 1350 km; Answers will vary.
5. Pacific
6. The United States occupies part of the North American continent.
7. A stream is blue; contour lines are brown or black.
8. 20 meters
9. Point E is almost at sea level. Sea level would be indicated by a zero (0) elevation contour line.
10. 4; 6:00 A.M.
11. prime meridian
12. 6; 3:00 A.M., local time
13. northern and southern; They show the curved surface of the earth.
14. Answers will vary between 20° east declination to about 15° west declination.
15. magnetic north pole

Study Questions

A. True or False

1. True 2. True 3. True 4. False (One revolution of Earth takes about a year.) 5. False (Sunlight strikes Earth most directly at the equator.) 6. True 7. True 8. False (In the United States, the degrees of longitude are west of the prime meridian.) 9. False (The equator is 0° latitude.) 10. True

B. Multiple Choice

1. degrees 2. 0° 3. 1 cm = 500 m 4. close together 5. the same 6. contour 7. time zones 8. backward 9. west 10. 1 day

C. Completion

1. International Date Line 2. magnetic 3. magnetometer 4. declination 5. shadow 6. circumference 7. mass 8. changes 9. east, west 10. 3

D. How and Why

1. The Foucault pendulum consists of a heavy weight suspended from a wire that is free to rotate; thus, swinging back and forth. Each hour the path is marked and is found to change, even though the pendulum is swinging in the same direction. The path changed during the hour because the earth moved slowly beneath the pendulum.
2. Two factors cause the seasons. The first factor is the change in the number of daylight hours, and the second is the change in the angle at which the sun's energy strikes Earth's surface. Both factors are a result of the 23½° tilt of the earth's axis with respect to its plane of orbit.
3. If there were no time zones, every city and town might have a different time which could be quite confusing. The width of a time zone is based on Earth's rotational speed of 15° of longitude per hour.
4. A scale is chosen to show the ratio between the distance on the earth and distance on the map. A compass is used to orient the features on the map with respect to the true direction.
5. The magnetic declination will be 0° when your location is aligned with geographic north and magnetic north along the same meridian.

9 Geology

Overview

Chapter 9 covers the basic principles of geology from the structure of the earth through disturbances such as earthquakes and volcanic activity. Emphasis is given to the plate tectonics theory and its use in explaining major geologic features. Section 9:3 discusses different kinds of rocks and the minerals of which they are made. Features used in the identification of rocks are presented and two activities involve rock identification and testing minerals for hardness. A section is devoted to each of the three main groups of rocks—igneous, metamorphic, and sedimentary. Changes in the earth such as weathering, erosion, folding, and faulting are explained.

Performance Objectives

Goal: Students will gain knowledge about the properties of rocks and minerals and the changes that occur in the earth's surface.

Objectives: Upon completion of the reading and activities and when asked to diagram, demonstrate, or respond either orally or on a written test, students will:

9:1	K	describe the main features of the core, mantle, and crust.
9:2	K	state the plate tectonics theory.
	K/P	describe evidence that supports the plate tectonics theory.
9:3	K	describe the features of igneous, metamorphic, and sedimentary rocks.
	K/P	identify common rocks found in the local area.
	P/S	test the hardness of mineral samples and use scales of mineral hardness to rate the hardness of the samples.
9:4	K	define weathering and erosion.
	K/P	explain how weathering and erosion affect the surface of the earth.
9:5	K	explain how igneous rock is formed.
	K	describe the differences between extrusive and intrusive igneous rock.
	K/P	classify igneous rocks as extrusive and intrusive
9:6	K	explain how sedimentary rock is formed.
	K	name four kinds of sedimentary rocks.
	P/S	use dilute acid to identify rocks that contain calcium carbonate.
	P/S	use a hand lens and millimeter-lined graph paper to analyze a sample of sand to determine the fractions of different particle size.
9:7	K	explain how metamorphic rocks are formed.
	K	name three kinds of metamorphic rock.
9:8	K	explain how pressure within the earth produces anticlines, synclines, and faults.
	K	describe the characteristics of an anticline, syncline, and fault.
	P/S	make clay models of an anticline and syncline and determine the relative age of the rock beds in the formation.
9:9	K	explain how an earthquake is caused by a fault.
	K/P	explain how seismographs are used to detect earthquakes and locate their epicenters.
9:10	K	describe how a volcano forms.
	K	describe the main characteristics of lava.
	A/S/K	carry out one Challenge at the end of the chapter.
	A/K	read a science article or book and prepare a short report.

Equipment for Activities

cardboard or heavy construction paper
egg carton
glass plate
graph paper, mm-lined
hand lens
hydrochloric acid, dilute
knife, pocket
milk carton, empty
mineral samples: talc, gypsum, calcite, fluorite, orthoclase, quartz, topaz, corundum
modeling clay, 4 different colors
penny, copper
rock samples: igneous, metamorphic, sedimentary
sand, coarse
world map

9:1 Structure of the Earth

Study of the interior of Earth is an example of the use of indirect evidence from research to develop a scientific theory. Much of what is known about Earth's inner composition has been determined by studying shock waves sent through the earth from earthquakes.

Seismic waves travel at different speeds depending on the nature of the layer in which they travel and, therefore, indicate the position of each layer as well as giving clues to the composition.

An important zone was discovered within the upper mantle, called the asthenosphere, located between a depth of 70 km and 700 km. Findings indicate that this zone consists of partly melted rock (approximately 10%). The discovery of the asthenosphere contributed to the development of the plate tectonics theory.

Making Sure, page 171
1. The heat and temperature are too great to support human life.

9:2 Plate Tectonics

Plate tectonics is the newest theory to explain changes in the earth's crust caused by forces within the earth. The concept of plate tectonics states that the crust and upper mantle comprise the lithosphere which consists of several individual plates. The lithospheric plates, in turn, lie upon the pliable asthenosphere. The plates move in relation to one another due to a type of convective flow within the asthenosphere. Each plate moves as a distinctive unit and, therefore, all major changes occur along plate boundaries. Be certain students understand the three types of movement associated with plate boundaries.

Activity 9–1

In this activity the student observes how the edges of continents may have once fit together to form one large continent.
Materials: cardboard or heavy construction paper, world map or atlas
Questions: For the most part, the edge of eastern North and South America and the edge of western Europe and Africa tend to fit together. According to the plate tectonics theory, these continents were once joined together. They separated and moved away from each other.

9:3 Rocks and Minerals

Introduce the section with a discussion of the rock cycle (Figure 9–6). The rock cycle helps students to understand the role of various geologic processes in producing a rock and in transforming one type of rock into another. Be certain students can associate the different rock types with the processes that produce them. For example, igneous rocks are produced by the crystallization and cooling of magma.

Stress that minerals are solids formed by inorganic processes and that each mineral has an orderly arrangement of atoms (crystals) and a definite chemical composition that gives it a unique set of physical properties. Go over the information in Table 9–3. Hardness and density are two important properties used to identify mineral samples.

Activity 9–2

In this activity the student identifies common rocks from the local area and develops a system for classifying them. Students may identify their rock samples using pictures or a collection of rocks that has been identified and labelled.
Materials: a collection of rocks samples from the local area, egg carton
Questions: The surface rocks in most locations are sedimentary. Some outcrops of metamorphic and igneous rocks may be found near mountain ranges.

Activity 9–3

In this activity the student determines the hardness of several mineral samples. Go over Tables 9–1 and 9–2 before students begin the activity. Have them compare their rating of mineral hardness with the values given in Tables 9–1 and 9–3.

You may wish to supplement this activity by having students determine additional properties such as streak, crystal shape, cleavage, and specific gravity. Instructions for identifying the properties may be found in books such as the *Guide to Mineral Identification: A Laboratory and Field Manual* by William F. Kohland, Allegheny Press, 1977.
Materials: mineral samples (talc, gypsum, calcite, fluorite, orthoclase, quartz, topaz, corundum), copper penny, pocket knife, glass plate
Questions: Quartz, topaz, and corundum will scratch a glass plate.

Making Sure, page 177

2. (a) 2.3 g/cm^3 (b) gypsum
3. The hardness ranges between 7 and 9.

9:4 Weathering and Erosion

Call attention to the difference between weathering, a process that turns rock into soil particles, and erosion, a process in which the particles are carried away. Refer to Figure 9–16 and explain that weathering involves chemical processes as well as mechanical processes. Introduce the term agent of erosion. Agents include wind, water, ice in motion, as well as gravity mass movements. If possible display photographs of various erosional features produced by each agent. Allow students to compare and contrast the different features.

Activity 9–4

In this activity the student observes the destructive force of freezing water on rocks.
Materials: empty milk carton, freezer
Questions: A carton filled with water and placed in a freezer will expand when the water freezes. If the container is made of glass, the glass will break as the water expands. Freezing water in the cracks within rocks pushes the rock fragments apart and breaks them up.

Making Sure, page 178
4. the faster the speed, the greater the rate of erosion
5. Wind and ice roughen the surface of a pebble. A pebble bouncing around in a stream of water is smoothed.

9:5 Igneous Rock

Be certain the students understand the difference between intrusive and extrusive rocks. Exhibit specimens of igneous rocks such as granite, basalt, and olivine, if available. Point out that the composition of igneous rocks varies because of origin.

Activity 9–5

In this activity the student examines the crystalline structure of igneous rock samples and classifies them according to crystal size. The students should conclude that fine-grained rocks such as basalt are extrusive, and coarse-grained rocks such as granite are intrusive.
Materials: igneous rock samples such as granite, rhyolite, andesite, basalt and peridotite

9:6 Sedimentary Rock

Sedimentary rocks may be classified into two major categories—clastics and nonclastics. Clastic rocks contain fragments of rocks or mineral grains. The sizes of the rock particles range from microscopic clay particles to large boulders. Clastics are classified primarily on the basis of the size of the rock and the mineral fragments that compose them. Nonclastic rocks are produced from deposition from solution or by organic processes. Precipitates form from chemical reactions that produce solids. Many limestone formations are precipitates. Evaporites form when water evaporates, leaving behind dissolved solids. Evaporites include chert, rock salt, and gypsum. Organic nonclastics are formed from shells or bones (calcite or silica) that collect in place. If possible, arrange a field trip to study various formations of sedimentary rocks in your area.

Activity 9–6

In this activity the student tests rock samples for the presence of carbonates using hydrochloric acid. Calcium carbonate or magnesium carbonate in a rock reacts with hydrochloric acid to produce bubbles of carbon dioxide. The test can be used to identify limestone, dolomite, and marble.
Materials: dilute hydrochloric acid, several rocks including at least one limestone rock

Activity 9–7

In this activity the student observes a variety of particles found in a sample of sand. Explain to students how to use the lines on the graph paper to measure the size of the particles.
Materials: coarse sand, millimeter-lined graph paper, hand lens
Questions: The particles range in size and vary according to the minerals from which they are composed. Most of the minerals present are quartz and feldspar.

Making Sure, page 183
6. Sedimentary rocks form from compaction of sediments or by cementation from materials such as calcite, silica, or iron oxide.
7. Sandstone contains larger quartz crystals while shale contains small particles of clay.
8. Sandstone is resistant to erosion and weathering and has attractive colors.

9:7 Metamorphic Rock

Metamorphic rocks form when the original rock material undergoes a rearrangement of mineral grains, an enlargement of crystals, or a change in chemical composition. Regional metamorphism occurs when large quantities of rock are subjected to intense heat and pressure associated with large-scale deformation such as mountain building. Contact metamorphism takes place when rock is in contact or in close proximity to a mass of magma. The changes in the rock are caused primarily by the high temperature of the molten material.

Making Sure, page 183
9. The acid in the lemon juice reacts with the carbonates in the marble to produce carbon dioxide.
10. Sandstone contains individual sand particles.

9:8 Rock Formations

Stresses within rocks caused by forces within the earth result in deformation of rock layers. Folded rocks generally result from compressive forces that are slowly applied. Anticlines and synclines are two types of folds. An overturned fold occurs when the rock layers are bent more than 90°.

If possible, arrange a field trip where students can observe road cuts or other outcrops exhibiting folded or faulted formations.

Activity 9–8

In this activity the student constructs a model of an anticline and a syncline.

Materials: 4 large pieces of modeling clay (different colors), knife

Questions: By cutting off the top of the anticline, the layers of clay are exposed as beds of rock in a weathered anticline. The oldest beds are located in the center of the anticline. In the model of the syncline, the inner layers of clay represent the younger layers.

Making Sure, page 185
11. The folded rock layers within an anticline or syncline will be more vertical than undisturbed sedimentary layers.

9:9 Earthquakes

A display of pictures showing damage caused by earthquakes is a good way to introduce this section. There may be students in class who have experienced an earthquake and can describe the experience. Most large earthquakes are related to the boundaries between the continental plates.

P-waves are compressional waves and can be demonstrated by laying a large spring (Slinky) on a table and plucking one end. S-waves are similar to waves in the ocean. These vertical waves can be demonstrated by tying the end of a clothesline to a doorknob and applying a quick up and down motion to the other end.

Making Sure, page 188
12. (a) 13 minutes after beginning of tape
    (b) P-, S-, and surface (d) about 14 minutes
    (c) surface waves   (e) about 5 minutes

9:10 Volcanoes

As a project, students can build a plaster model of a mountain building volcano based on information obtained through library research. A large colored diagram showing the cross-section of a volcano makes a good bulletin board display. The factors that determine the nature of volcanic eruptions are the chemical composition of the magma, its temperature, and the amount of dissolved gases present. The lower the percentage of silica, the greater the fluidity. Fluid magma flows freely as hot lava. Thicker (more viscous) magma may cause an explosion because gases in the magma build up under great pressure that is suddenly released in an explosion.

Making Sure, page 190
13. Volcanoes tend to be located along earthquake belts because they are related to the movements along crustal boundaries.

Study Questions

A. True or False
1. False (The crust of the earth consists entirely of solid rock.) 2. False (The thickness of the crust varies from 5 to 40 km.) or (The thickness of the mantle is about 2885 km.) 3. False (All intrusive igneous rocks have a coarse-grained texture.) 4. True 5. True 6. True 7. False (Intrusive igneous rocks are always formed beneath the earth's crust.) 8. True 9. True 10. True

B. Multiple Choice
1. loose rock particles 2. soil 3. limestone 4. slate 5. limestone 6. inner core 7. decreases 8. folding 9. rising 10. fault

C. Completion
1. plate tectonics 2. magma 3. seismograph 4. surface waves, P-waves, S-waves 5. boundaries 6. mineral 7. magma 8. soil 9. fault 10. weathering

D. How and Why
1. Rocks can be identified by their physical properties such as shape, color, hardness, and texture. The region in which a rock is found is also helpful. A mineral has a definite chemical composition and crystal form. Most can be identified by properties such as luster, color, hardness, and crystal form.
2. Sedimentary rocks are formed when particles of eroded rock are deposited together and compressed or cemented together.
3. The earth's crust constantly changes due to internal forces and surface processes. Tectonic forces as well as erosion and weathering produce mountains, valleys, caves, etc.
4. Finding the epicenter of an earthquake requires reports from three or more recording stations. A circle is drawn around each station denoting the distance from the epicenter. The point where the three circles intersect gives the exact location of the epicenter.
5. The Himalaya mountains resulted from the collision of the Indian and Asian plates. The Alps were formed when the plate that contained Italy collided with Southern Europe.

Challenges

2. Suggest students build a working model of a seismograph to detect and measure earth vibrations in your immediate surroundings.

10 Earth History

Overview

Chapter 10 opens with a discussion of the Grand Canyon, an area that is a rich source of data from which scientists can learn about Earth's geologic history. The chapter focuses on the kinds of questions scientists ask about the geologic history, and how scientists obtain information used to develop theories that answer these questions. Fossils, methods employed to date rocks, and several theories based on fossil evidence are discussed. The concept of geologic time is introduced in Section 10:5 with a table that shows the main characteristics of the periods of geologic history. In the sections that follow, the details of each geologic era and period are discussed. The last section considers changes in the earth that may occur in the future, based on what scientists have learned about Earth's past.

Performance Objectives

Goal: Students will learn how the earth and its life have changed during geologic time.

Objectives: Upon completion of the reading and activities and when asked to diagram, demonstrate, or respond either orally or on a written test, students will:

10:1	K	describe the major features of the Grand Canyon.
	K/P	explain how features of the Grand Canyon provide clues to Earth's past.
	K/P	state the law of uniform change and explain how it applies to the Earth's geologic history.
10:2	K/P	explain how radioactive dating is used to determine the age of a rock.
	K	explain why in a series of horizontal rock beds the oldest are at the bottom and the youngest are at the top.
10:3	K	define the terms fossil and paleontology.
	K	explain how fossils are formed and give examples of fossils.
	P/S	use plaster of paris to make an imprint of a leaf and explain how the process is like the formation of a fossil.
	P/S	use a map of the United States to estimate the area of the seafloor where fossils may now be forming off the coast of the United States.
10:4	K	state three facts based on fossil records.
	K/P	explain how index fossils are used to determine the age of sedimentary rocks.
10:5	K	list the eras and periods of geologic history.
	K/P	explain how geologists divide prehistoric time.
10:6	K	describe the rocks and fossils of the Precambrian Era.
10:7	K	name the six periods of the Paleozoic Era and describe the major features of each period.
10:8	K	name the three periods of the Mesozoic Era and describe the major features of each period.
10:9	K	name the two periods of the Cenozoic Era and describe the major features of each period.
10:10	K/P	explain how scientists predict future changes in the earth based on changes that occurred in the past.

Equipment for Activities

aluminum pie pan   petrified wood
balance   plaster of paris
leaf   wood block, oak
map of the United States

10:1 The Grand Canyon

Geologic evidence is used to piece together the prehistoric history of the earth. One rich source of such data is the Grand Canyon located on the Colorado plateau, a region of nearly horizontal rockbeds uplifted to an elevation of over 1500 meters. The Canyon has been cut into the plateau by the flowing waters of the Colorado River. The Grand Canyon illustrates weathering and erosion, sedimentary rock formations, fossil evidence of past life and environment, unconformities, and the law of uniform change. Discuss why it is thought that the plateau was formed under water. (The Canyon is made of sedimentary rock.)

***Making Sure,* page 198**
1. 30 million years

10:2   Age of Rocks

In this section the theories and methods for determining the age of rocks are presented. Relative ages of sedimentary and metamorphic rocks are determined by the principle of superposition, meaning the younger beds are laid down on top of older ones. Discuss the measurement of time by the rotation and revolution of the earth, an egg timer, and a watch or clock. Demonstration: Use a wristwatch with a second hand to measure the time it takes for sand to fall through an egg timer. Then relate the two time-measuring devices to radioactive dating. Go over the concept of half-life and how the fixed rate of decay of an element is the basis of radioactive dating.

10:3   Fossils

Ask students to bring fossils thay have at home to class for an exhibit. Point out known local sites for fossil collecting and describe the kinds of fossils found in these sites. Ask students to describe fossils they may have seen in a natural history museum. The collection, description, and analysis of fossils is an example of science process; that is, using scientific methods to develop knowledge of the prehistoric life and climate of the earth.

Activity 10-1

In this activity, the student compares the weight and density of petrified wood with non-petrified wood. In petrified wood, the organic matter has been replaced by minerals which have greater density. Other woods may be used if oak, which is a heavy hard wood, is not available.
Materials: piece of oak, piece of petrified wood, balance
Questions: Petrified wood has more mass because it contains minerals.

Activity 10-2

In this activity, the student makes an imprint of a leaf in plaster of paris that illustrates the formation of a fossil.

Materials: plaster of paris, aluminum pie pan (or saucer), water, leaf.
Questions: The imprint of the leaf that forms in the plaster is like the fossil imprint of the leaf found in sedimentary rock. Hardened plaster of paris is like sedimentary rock.

Activity 10-3

In this activity, the student uses a map to estimate the total area off the coasts of the United States where fossils are now forming.
Materials: map of the United States
Questions: Examples of animals that might be preserved as fossils are fish, starfish, jellyfish, clam, oyster, crab, seal, whale, and birds such as the seagull and pelican.

10:4   Fossil Records

Explain the difference between simple and complex as it pertains to fossils. Complex refers to the number of different kinds of parts in something. It is helpful to use analogies in developing this concept. For example, comparing the complexity of a television with a radio, a single propeller airplane with a Boeing 747, and a clock with a sundial. Relate these examples to differences in the complexity of animals such as a fish and a brachiopod fossil, and an oak tree and a prehistoric fern. Go over the material on page 203 so that students understand how index fossils are used to date rocks in different parts of the world.

10:5   Geologic Time

Students may need to be reminded that the descriptions of geologic time refer to prehistory and are based on indirect evidence. No people were alive in those ancient ages to record descriptions of the living creatures and climate present in the eras and periods. Introduce this section with a review of Table 10-1. Note the time the major life forms appeared, the time each era and period began.

***Making Sure,* page 205**
2. Approximately 6 250 000 years
3. Eras are longer times that are divided into periods.

10:6   Precambrian Era

Remind students that most of the geologic history of the earth, four billions years of time, is in the Precambrian Era. A major characteristic of this time is the very few fossils left in Precambrian rocks. A worthwhile teaching approach is to have students learn one or two major facts about each of the following eras and period that describe the life, climate, and other conditions of the time interval.

10:7 Paleozoic Era

In Cambrian rocks 60% of the animal fossils are trilobites, 30% are brachiopods, and 10% are other animals. By the late Ordovician Period most classes of invertebrates were established. Changes in species within these classes have occurred over time, but no new classes have been formed. Vertebrates have been found in all periods since Ordovician time. Of the vertebrates, only birds and mammals have not given rise to more advanced forms of animal life.

Making Sure, page 209
4. Presence of fossils of plants from the Carboniferous Period may indicate the presence of petroleum.

10:8 Mesozoic Era

Reptiles spread through all land areas during the Mesozoic and became the dominant life of this era. Birds and reptiles probably had the same ancestors. Only four groups of reptiles have survived from the Mesozoic Era to the present. These are the snakes, lizards, turtles, and crocodiles. Most marine reptiles were extinct by the end of the Mesozoic Era. Fossil grasses and cereal grain plants as well as trees similar to trees living today were characteristic of the seed-bearing plants that became dominant in the late Mesozoic Era. Like the Paleozoic Era, the Mesozoic Era ended with the extinction of many groups of animals, including the dinosaurs.

Making Sure, page 211
5. It had the skin and teeth of a reptile but the features and wings of a bird.

10:9 Cenozoic Era

The Cenozoic Era began as the dinosaurs disappeared. Rapid changes in the life of the Cenozoic followed and many new species developed. Continents were enlarged and uplifted from the seas, bringing about the present distribution of land and sea, mountains, and plains. During Cenozoic time, many land areas such as Australia and South America were separated from other continental land masses and thereby cut off from the general trends in the development of mammals. Several kinds of unusual marsupials still exist in Australia. A great ice age marked by the invasion of North America and Europe by huge glaciers was a major event in the late Quaternary Period.

Making Sure, page 214
6. The rock movements that began during the Tertiary Period are still occurring.
7. The environment became too cold.
8. It was much smaller and had fewer toes.
9. It is only a small fraction of geologic time.

10:10 Earth's Future

Scientific predictions are based on events believed to have occurred during the 4.5 billion years of Earth's history. However, not all scientists agree on what the future will be. For example, some scientists believe that the earth's climate is slowly cooling and others believe that the greenhouse effect, the increasing carbon dioxide in the atmosphere, will cause a gradual warming accompanied by melting and receding of the glaciers. Allow students an opportunity to speculate on the future of the earth, but remind them that the best predictions are likely to be those based on known facts. For sure, no person can predict the future with absolute certainty.

Making Sure, page 215
10. earthquakes, volcanoes, erosion
11. Carbon dioxide acts as a blanket to hold heat in and may cause the earth to warm up.
12. Answers will vary, any city on the coast.
13. produce the tides that slow rotation

Perspectives

Mike Hansen: Earth Scientist

Invite a state geologist to speak to your students. Encourage students to prepare questions for the geologist to answer after the presentation.

Study Questions

A. True or False
1. True 2. True 3. True 4. True 5. False (Geologic forces of the past are the same as those of today.) 6. False (Radioactive dating is based on the use of decay of radioactive elements.) 7. False (Mammals appeared on Earth more than 1 million years ago.) 8. True 9. True 10. False (Trilobites are extinct today.)

B. Multiple Choice
1. index 2. different 3. shallow sea 4. 4 5. Cenozoic 6. 4.5 7. few 8. many 9. Devonian 10. Devonian

C. Completion
1. carboniferous 2. plants 3. Mesozoic 4. Jurassic 5. Jurassic 6. 10 000 7. glaciers 8. Cenozoic 9. 6 h 10. mammals

D. How and Why
1. Thousands of feet of rock layers representing many geologic ages have been exposed by the erosive forces of the Colorado River on the Grand Canyon.
2. Ages of rocks are determined by radioactive dating or by index fossils. A radioactive element is a "clock" because its established rate of radioactive decay can be used to measure time. The geologic period during which the fossil was formed is said to be the age of the fossil.
3. Fossils are the remains of prehistoric animals and plants. They provide information concerning prehistoric environment. They also aid in determining the age of rocks.
4. The geologic time scale has been divided into four major eras: Precambrian, Paleozoic, Mesozoic, and Cenozoic. Each of these eras (except Precambrian) is divided into periods. See Table 10–1, page 204.
5. Precambrian rocks contain few fossils. Since fossils are the major clues to the past, little is known about this time.

11 The Atmosphere

Overview
Chapter 11 begins with a discussion of the major characteristics of the atmosphere such as composition and structure. The concept of air pressure is introduced, and operation of aneroid and mercury barometers is explained. The processes by which the atmosphere is heated and the major circulation patterns within the atmosphere are also covered. The final three sections of the chapter include a discussion of humidity and the processes by which clouds and precipitation are formed.

Performance Objectives
Goal: Students will learn the major features of the atmosphere and how winds, clouds, and precipitation are formed.

Objectives: Upon completion of the reading and activities and when asked to diagram, demonstrate, or respond either orally or on a written test, students will:

11:1	K	describe the composition of air.
11:2	K	describe the main characteristics of the troposphere, stratosphere, mesosphere, thermosphere, exosphere, and ionosphere.
11:3	K	explain how air pressure is exerted in the atmosphere.
	K/P	explain how a mercury barometer is used to measure air pressure.
	K/P	explain how an aneroid barometer is used to measure air pressure.
	P/S	do an activity that shows the effect of air pressure by crushing a plastic milk container filled with water.
	P/S	record the air pressure on five consecutive days and correlate changes in pressure with the daily weather.
11:4	K	explain how the sun warms the atmosphere.
	K/P	describe how connection currents are formed and how they transfer heat in the atmosphere.
	P/S	do an activity that demonstrates the collection of heat in a plastic covered cardboard box.
11:5	K	describe the major winds systems in the atmosphere.
	K	describe the location and movements of the jet streams.
	P/S	use a rotating globe to observe the path of a drop of water moving from the North Pole on the globe to the equator.
11:6	K	explain how a sea breeze and land breeze occur near an ocean.
	K	describe the local winds in mountain regions.
	K/P	describe the effect of compression in warming winds that blow down to lower altitudes.
	P/S	observe a wind vane for five consecutive days and record the direction of the wind each day.
11:7	K	define the dew point of air.
	K	define relative humidity.
	K/P	explain how a hygrometer works.

	P/S	measure the dew point of air.
	P/S	use a sling psychrometer to measure relative humidity.
11:8	K	describe the characteristics of the three basic cloud forms—cirrus, cumulus, and stratus.
	P/S	use a large jar to demonstrate how the water vapor in human breath can produce a cloud in the jar.
11:9	K	define precipitation.
	K	name and describe the kinds of precipitation that form in the atmosphere.
	A/S/K	carry out one Investigation at the end of the chapter.
	A/K	read a science article or book and prepare a short report.

Equipment for Activities

barometer
cardboard box
chalk dust
dropper
glass tubing, 10 cm
glass jar, 4 L (1 gallon)
globe
glycerol
ice cubes
plastic milk container
plastic sheet, clear
rag or towel
rubber stopper, 1-hole
rubber tubing, 1 m
sling psychrometer
stirring rod
thermometer
wind vane
wooden match

11:1 Air

Introduce this section by directing students to Table 11–1. Briefly identify each of the gases and point out that oxygen and nitrogen together constitute 99.03% of air. Have students make a list of the gases for their notebooks and identify some chemical and physical properties of each. Explain that the word trace means that the amount of gas present in the air is sufficient to be detected by analytical instruments. Stress that even though carbon dioxide, water vapor, and ozone make up a small percentage of the air, they are important in terms of weather and atmospheric processes. Some students may be interested in researching and reporting on the theories of origins of Earth's atmosphere.

11:2 Parts of the Atmosphere

Atmospheric structure as discussed in this section is based on the thermal properties of the atmosphere. Each layer is characterized by an increase or decrease in temperature. Be certain that students are aware that density of air decreases with height. Review the definition of ion before discussing the nature of the ionosphere, which actually involves three thermal layers. As a review, have students diagram and label correctly the layers of the atmosphere.

Making Sure, page 224
1. 43.5°C (The temperature in the troposphere decreases 6.5°C/km.)

11:3 Air Pressure

Before beginning this section, review the concept of pressure in fluids as discussed in Chapter 7. Remind students that the height of the fluid as well as density determines the pressure at any depth within the fluid. Unlike a liquid, air can be compressed and is therefore most dense near the earth's surface. The pressure of air decreases 50% within the first 6000 m above the surface. The average sea level pressure is about 101 kPa. Most meteorologists use millibars, however. (1 mb = 0.1 kPa) The standard sea level pressure is 1013.3 mb. Students may be interested in learning about the purpose and function of an altimeter used in most airplanes.

Activity 11–1
In this activity the student observes the force of air pressure crush a plastic milk container.
Materials: empty plastic milk container, 1-hole rubber stopper, glass tubing (10 cm), rubber tubing (1 m), stack of books
Questions: When the rubber tubing is lowered into the sink, the weight of the water causes the water to fall out of the tube initially, thereby reducing the pressure inside the tube. As a result, air pressure on the surface of the plastic container causes it to collapse and push the water out of the container. The container is full of water at the end of the activity because as it collapses, its volume decreases to the reduced volume of water left in the container.

Activity 11–2
In this activity the student observes daily changes in atmospheric pressure and associates these changes to changes in weather. Generally, a rising barometer is associated with fair weather and a falling barometer indicates approaching bad

weather. A discussion of high and low pressure systems is found in Chapter 12.

Making Sure, page 226
2. Earth's gravity holds most of the gases in the air close to the surface.
3. Air pressure is directly proportional to the density of the air.

11:4 Heating the Atmosphere

You may want to begin this section by comparing the earth's surface and atmosphere to a greenhouse. On the chalkboard, draw a greenhouse to show how incoming solar radiation is trapped within the greenhouse. Use the term "greenhouse effect" to explain the chief mechanism for controlling surface temperatures. Before discussing the concept of convection currents, review buoyancy as presented in Chapter 7.

Activity 11-3
In this activity the student demonstrates the principle of the greenhouse effect.
Materials: cardboard box, a sheet of clear plastic, thermometer
Questions: The temperature increases when the box is covered with clear plastic. Sunlight penetrates the plastic and is absorbed and changed into heat. The plastic prevents the heat from escaping.

11:5 Winds

Be certain students understand that convection is the primary cause of atmospheric circulation. The major wind systems correspond to what is called the "three-cell model." You may wish to discuss the movements and causes of each individual system or cell before explaining the interrelationships among wind systems throughout the entire global pattern.

Jet streams are located in the region called the tropopause. The tropopause is the name given to the boundary between the stratosphere and the troposphere. Be certain that students understand that the mid-latitude jet streams continuously change position and follow a meandering path around the earth.

Activity 11-4
In this activity the student observes the Coriolis effect. It is important to stress that the Coriolis force is an apparent force that results from the fact that the earth rotates. Some students may require more time than others to fully grasp the concept.
Materials: globe, chalk dust, dropper
Questions: The drop follows a curved path and is deflected to the right of its original path. Remind students that a drop moving from the equator toward the North pole will also be deflected to the right (east).

Making Sure, page 228
4. increase speed going toward the equator; decrease speed going from equator
5. The winds are deflected to the east or west.
6. moves north and south and changes altitude

11:6 Local Winds

Local winds refer to small-scale circulation patterns that disrupt the prevailing wind circulation. Local winds such as land and sea breezes as well as mountain and valley breezes are produced as a result of differential heating. Other types of wind may result from cold air draining into a valley or compressional warming as air descends a mountain slope. If there are local winds in your area such as a Santa Ana or Chinook, discuss the characteristics of these winds.

Activity 11-5
In this activity the student determines prevailing wind direction in the local area for five days. Direct students to wind vanes in your area that can be observed daily. If a wind vane is available, you may wish to position it on the roof of your school or any other location where it will not be disturbed by students. Have students keep a daily record of wind direction in their notebooks.
Questions: Wind direction will usually vary over a five day period. The prevailing direction will be determined by the overall weather pattern affecting your area as well as the local wind systems.

Making Sure, page 230
7. A sea breeze is caused by the uneven heating of the water and the land. See Figure 11–11.
8. Air moving downslope is compressed. Compressed air increases in temperature because air molecules are forced closer together. See the discussion of gas laws in Chapter 7.

11:7 Moisture in Air

Begin by reviewing the water cycle illustrated in Figure 11–13. Students should be aware that water vapor is the single most important component of the air in terms of weather and climate, and that water vapor is a gas and varies from place to place. Stress the relationship between saturation of air and dew point and the difference between humidity and relative humidity.

Activity 11-6

In this activity the student measures the dew point of air by observing the temperature at which water vapor condenses on a glass jar containing ice and water. As the water is stirred, the ice melts and cools the water to the temperature (dew point) where water vapor in the air begins to condense on the cooled surface of the jar.
Materials: glass jar (1 L), water, several ice cubes, thermometer, stirring rod
Questions: Dew forms on the ground when the temperature near the ground cools to dew point and the water vapor condenses on the blades of grass.

Activity 11-7

In this activity the student measures relative humidity using a sling psychrometer. Caution students when swinging the psychrometer to stand clear of objects and other students to avoid breakage of thermometers and personal injury or injury to others.
Materials: sling psychrometer, Table C-1 in Appendix C
Questions: The wet bulb temperature is usually lower than the dry bulb temperature because the evaporation from the wet bulb cools it and lowers the temperature. (Recall that evaporation is a physical change that requires energy.) The rate of evaporation varies inversely with the relative humidity.

Making Sure, page 233

9. 100%
10. Higher relative humidity makes you more uncomfortable.
11. sinks, plants, water heated on stove, fish tank, etc.

11:8 Clouds

Clouds are good indicators of weather conditions. Clouds form as moist layers or parcels of air that are cooled to the dew point temperature. Obtain a book that illustrates the ten basic cloud types. Slides of clouds are also useful. Students interested in meteorology may want to learn about the 27 states of sky that are used in weather observation and forecasting. Encourage students to observe clouds on a daily basis and note the weather associated with each type of cloud.

Activity 11-8

In this activity the student produces a cloud inside a bottle by exhaling hard into the bottle and releasing the pressure. The water vapor within the exhaled air condenses to form droplets of liquid water. Smoke from a burning match adds condensation nuclei to the air in the jug. Condensation nuclei are tiny particles such as salt and dust around which water vapor condenses.
Materials: glass jug (4 L), wooden match
Questions: When a person's breath is exhaled hard into the jar several times, water vapor enters the jar, the air pressure inside the jar increases, and there is adiabatic heating. Adiabatic refers to temperature change without adding or removing heat. Adiabatic heating occurs when molecules of air are squeezed together. When the pressure is suddenly released, adiabatic cooling occurs and the water vapor condenses around the condensation nuclei.

Making Sure, page 235

12. Cumulus clouds are puffy clouds resembling cotton. Stratus clouds are flat layers. Cumulus clouds are formed when rising columns of moist air are cooled. Stratus clouds result when layers of moist air are cooled.

11:9 Precipitation

Precipitation represents all forms of water that fall from the atmosphere and reach the ground. Raindrops form as a result of rapid condensation and grow by coalescence with other droplets. Drizzle is a type of rain composed of droplets less than 0.5 mm and is produced in low-lying nimbostratus clouds. Snow consists of ice in branched hexagonal crystals which may stick together to form larger snowflakes. Sleet or frozen rain is made up of small ice pellets formed as rain falls through a freezing layer of air. Hail is made up of concentric ice layers and is formed only in the more severe thunderstorms when strong updrafts are present.

Study Questions

A. True or False

1. False (Air is a mixture.)   2. False (There is no definite boundary at which the atmosphere

ends.) **3.** True **4.** False (The number of air particles per liter is less at 300 m altitude than at sea level.) **5.** True **6.** False (Temperature within the mesosphere decreases with height.) **7.** False (Rain and snow occur often in the troposphere.) **8.** True **9.** True **10.** False (Some clouds are made of water droplets.)

B. Multiple Choice
1. nitrogen **2.** pressure **3.** less than **4.** decrease **5.** air pressure **6.** cumulus **7.** ionosphere **8.** hail **9.** increases **10.** oxygen

C. Completion
1. high, low **2.** westerly **3.** convection currents, winds **4.** rotation **5.** condensation **6.** sleet **7.** hygrometer **8.** faster **9.** cloud **10.** rising

D. How and Why
1. Pressure of a gas varies directly with density. Air is less dense at high altitudes; therefore, air pressure is less at high altitudes than at sea level.
2. Warm air at the equator rises and moves out toward the poles. Cooler surface air flows toward the equator to replace the rising air. The eastward and westward movement of air is a result of the Coriolis effect.
3. The air near the ground is saturated when it is cooled to the dew point temperature. Condensation usually occurs when air is cooled to or below the dew point.
4. Warm air rises because of its lower density. The buoyant force of the denser air pushes the warm air upward.
5. Freezing rain occurs when raindrops fall through air above 0°C and freeze at the surface, which has a temperature of 0°C or slightly below. Freezing rain will usually occur along a frontal boundary when warm air rises above the frontal surface.

12 Weather and Climate

Overview

Chapter 12 continues the theme from Chapter 11 by developing the student's knowledge and understanding of the atmosphere and its effects. The chapter begins with a discusson of weather observations and proceeds to an explanation of the causes of the changes in weather. Air masses, fronts, thunderstorms, pressure systems, and the use of weather maps in forecasting are covered. A concluding section describes the major factors that affect the climate of a region and presents the concept of microclimates. Principles of meteorology introduced in Chapter 11 are used to explain and interpret the topics covered in this chapter.

Performance Objectives

Goal: Students will learn the factors that affect weather and how weather forecasts are made.

Objectives: Upon completion of the reading and activities and when asked to diagram, demonstrate, or respond either orally or on a written test, students will:

12:1	K	define weather and describe different kinds of weather.
	K	explain how cloud cover and cloud ceiling are related to the weather.
	P/S	observe and record weather observations such as temperature, pressure, relative humidity, clouds, wind speed, wind direction, and precipitation.
12:2	K	define and describe the characteristics of air masses.
	K/P	explain and give examples of how air masses are classified.
12:3	K	describe the characteristics of a front.
	K	explain how a warm front is different from a cold front.
	K/P	explain how fronts cause weather conditions such as storms and changes in temperature.
12:4	K	describe the factors that cause the development of a thunderstorm.
	K	explain how hail and lightning are formed in a thunderstorm.
12:5	K	describe the differences between high and low pressure areas.
	K	describe the major characteristics of a hurricane.
	K/P	describe the characteristics of a tornado and state a theory that explains how tornadoes are formed.
12:6	K	define meteorology.
	K/P	describe how weather maps are made.
	K/P	explain how weather maps are used to forecast weather.

	P/S	locate fronts on a weather map.
	P/S	read the cloud cover, temperature, relative humidity, air pressure, wind direction, and wind speed for a weather station on a weather map.
	P/S	locate an isobar on a weather map and read the air pressure in millibars.
12:7	K	define weather forecast and explain why forecasts are based on percentages.
	P/S	explain how weather forecasters use weather data to predict the weather.
12:8	K	define the term climate.
	K	list the major factors that affect the climate of a region.
	K	describe the differences between temperate, tropic, and polar climates.
	K	define the term microclimate and describe an example of a microclimate.
	A/S/K	carry out one Challenge at the end of the chapter.
	A/K	read a science article or book and prepare a short report.

Equipment for Activities

anemometer
barometer
can, empty
hygrometer or psychrometer
metric ruler
thermometer
weather map
wind vane

12:1 Observing the Weather

Section 12:1 opens the chapter by relating the content to everyday observations of the weather. Begin by discussing the day's weather, specifically the temperature, air pressure, relative humidity, wind speed, precipitation, and cloud cover. Compare the day's weather with yesterday's weather. Then ask students to predict the weather for tomorrow. Discussion of this question provides a transition into the science of weather forecasting covered in sections that follow.

Activity 12-1

In this activity the student observes the weather each day for one week and formulates conclusions based on the data collected. You may wish to assign the recording of the different weather elements to individual students or small groups. Each day the students should pool the data from the previous day. Data may be recorded on a classroom chart or in students' notebooks.

Materials: anemometer, barometer, empty can, hygrometer or psychrometer, wind vane, metric ruler

Questions and Conclusions: **1.** Highest temperatures are likely to be recorded during mid-afternoon due to maximum solar heating. Lowest temperatures occur just before sunrise when much of the heat absorbed by the surface during the day has been lost by radiational cooling. **2.** Cloud cover tends to be greater when the air pressure is low or the barometer is falling. (See Section 12:5 for explanation.) **3.** Amounts of precipitation will vary depending on the sites where it was collected and the type of weather activity affecting your local area. **4.** Continuous precipitation is usually associated with stratus-type clouds (nimbostratus). Intermittent showers or thunderstorms are associated with convective clouds (cumulonimbus).

Making Sure, page 243

1. (a) hot, dry, clear, and windy (b) summer

12:2 Air Masses

Study of the characteristics of air masses provides a background for the material on fronts that follows in the next section. Have students locate and identify air masses on a weather map. Air masses are usually associated with high pressure regions and are produced principally within the anticyclonic flow of the subtropical and polar high pressure belts. Changes in the weather result from air masses moving into a region. Stress that properties of an air mass will change or modify as the air mass moves out of the source region. A cold, dry air mass from Canada, for example, will become warmer as it moves to the south through the United States.

Making Sure, page 244

2. desert
3. cP—polar, land; mP—polar, ocean; cT—land, near equator; mT—ocean, near equator

12:3 Fronts

A front represents the boundary zone between two different air masses. Go over the details of each type of front as illustrated in Figures 12–3 and 12–4. Be certain students understand the structure of each type of front and how the fronts are related to clouds and precipitation. Explain that fronts are zones of surface convergence. Along convergent zones, air rises and becomes cooler usually producing clouds and precipitation. The passing of a front is usually accompanied by a change in the direction of the wind and a change in temperature.

12:4 Thunderstorms

Review the details of the formation of a cumulonimbus cloud and a thunderstorm. The most common type of thunderstorm, called an air-mass thunderstorm, is discussed in this section. Air-mass thunderstorms occur most often in temperate latitudes during the summer. The more severe thunderstorms are associated with frontal systems. Most tornadoes and hail-producing thunderstorms usually occur with fast-moving cold fronts or ahead of a front along a squall line. The interaction of low-level and upper-level jet streams plays an important role in determining the severity of thunderstorms along a front.

Making Sure, page 247

4. Cumulonimbus clouds form when warm, moist, unstable air rises swiftly. The cloud builds vertically as the rising currents are cooled and water vapor condenses to form droplets.
5. When rain falls from the cloud, the drag effect of the raindrops produces downdrafts that eventually stop the upward flow of air.

12:5 Highs and Lows

As discussed in Section 12:2, high pressure systems in the mid-latitude regions are associated with air masses. Surface high pressure systems are characterized by a divergent outflow of air along the surface and subsiding air aloft. Low pressure systems that form along air mass boundaries are migrating systems and represent surface convergence of air and ascending air aloft. The passage of alternating high and low pressure systems characterizes the changeable and often stormy weather patterns within the mid-latitude temperate zones.

Point out that mid-latitude cyclones and tropical cyclones are different. Explain that hurricanes and other tropical storms result from disturbances in easterly trade winds and are not associated with fronts.

Making Sure, page 250

6. A tremendous amount of energy is used in evaporating the water and producing the high winds. The sun is the ultimate source of this energy.
7. A tornado is smaller, moves faster, has less total energy, and has a greater wind force over a given area.

12:6 Weather Maps

Weather maps are essential in weather forecasting. They are used together with pictures obtained from weather satellites. Each National Weather Service Station makes hourly observations of the temperature, humidity, air pressure, cloud cover, and other weather data, which is reported and collated in making a weather map. An important aspect of a weather map is that it shows the movements of highs, lows, and fronts from one place to another. Readings from two stations in opposite sides of a front often show significant differences in temperature, humidity, and wind direction. As an enrichment activity, provide National Weather Service maps, both surface and upper air, for students to study.

Activity 12-2

In this activity the student locates and identifies fronts on a newspaper weather map. Also the student determines weather information given for selected cities. Direct students to cut out the weather map and forecast from the newspaper and bring both to class. Discuss possible reasons for the forecast. Follow up in the next day's class by noting whether or not the forecast was correct.
Materials: weather map from newspaper
Questions: Weather information will vary. Be certain that students are able to locate and identify air masses and pressure systems from the map.

Making Sure, page 252

8. There is very little difference between temperature across the frontal boundary. This lack of contrast is due to the fact that most frontal systems during the summer are very weak because thermal and moisture properties of adjacent air masses are similar.
9. Low pressure over southern California is producing rain in Los Angeles. The high pressure over Oklahoma results in clear skies in Oklahoma City.

10. Pressure ranges from 1024 mb high to 1012 mb low.
11. cold front

12:7 Weather Forecasting

Relate the material in this section to the study of weather maps in the previous section. Define the term probability and relate it to weather forecasts, pointing out that forecasters often forecast the occurrence of precipitation on a precentage chance basis. Students are familiar with weather forecasts reported on television. You may wish to ask students which television weather person or forecast they rate the highest. Then discuss the reasons for the choices.

Activity 12–3

In this activity the student watches and records an evening weather forecast as a homework assignment. Follow up the assignment the following day with a discussion of the accuracy of the forecast. You may also want to read and discuss the Perspectives article found on page 256 in this chapter. Many television weather forecasters use severe weather radar extensively.

Questions: Accuracy of the weather forecast depends on the complexities of the systems affecting your area. Factors that may cause incorrect forecasts include changing rate and direction of movement of fronts and air masses, a lack of sufficient moisture to produce clouds and/or precipitation, and disturbances aloft not reflected on surface maps.

Making Sure, page 253

12. Forecasts are made using weather maps and images made from satellites to keep track of movements of storms, fronts, and pressure systems. Changes in temperature, humidity, wind speed and direction, cloud cover, and pressure are observed. Weather radar is also used to forecast movement of severe weather.

12:8 Climate

To a great extent, the climate of a region is due to the amount of sunshine received, the angle at which solar radiation strikes the earth, and the geography of the region. Climates are also a result of the effects of the three-cell circulation pattern as discussed in Chapter 11. Using a climate map and a relief map of the world, allow students to offer reasons why a climate exists for a particular region. For example, the deserts in southwestern United States are a result of the rain shadow effect. Moist air from the Pacific Ocean that is forced to rise over the coastal ranges becomes cooler, resulting in precipitation along the western slopes. As the air descends the eastern slope, it becomes warmer and absorbs any available moisture. Also, tropical rain forests along the equator are a result of the warm, moist air that rises, producing large annual rates of precipitation.

Making Sure, page 255

13. mild—temperature about 21°C and humidity about 40%
14. warm and humid

Perspectives

Warning: Severe Weather

The local office of the National Weather Service is a good source for severe weather information. A field trip to this office could provide information about conventional severe weather alert systems. Often local television weather forecasters are available as classroom speakers. Students can do further research and present a report on Doppler radar systems.

Study Questions

A. True or False
1. True   2. True   3. False (Fair weather is usually found in high pressure areas.)   4. True
5. False (There is much difference in the weather on opposite sides of a front.)   6. False (Air masses travel from west to east across the United States.)   7. True   8. True   9. False (Clouds are usually found along fronts.)   10. True

B. Multiple Choice
1. a desert   2. mP   3. less rapidly than   4. overcast   5. tropical   6. warm   7. low   8. lightning   9. cumulonimbus   10. high

C. Completion
1. easterly   2. warm   3. thunder   4. broken   5. cumulonimbus   6. weather   7. climate   8. microclimate   9. 23½° north   10. ceiling

D. How and Why
1. Weather is the condition of the atmosphere at a particular time. Climate is the average weather over a period of time.
2. A weather map contains information such as temperature, pressure, fronts, etc, for a partic-

ular day and time. Maps are used to record and forecast weather.
3. temperature, humidity, cloud types and amounts, wind speed and direction, percipitation type and amount, and visibility
4. The first step in constructing a forecast is to record current observations. The next step is to compare observations each hour to determine trends such as rising or falling barometer and temperature readings, wind shifts, and increases or decreases in cloud cover. The third step is to note the positions and movements of weather systems on daily weather maps. A forecast should be made using all available weather data.
5. Tornadoes are associated with severe thunderstorms. The most severe weather occurs when very moist and warm air is forced to rise rapidly along a swiftly advancing wedge of cold, dry air. This interaction occurs only when highly contrasting air masses are next to each other.

Challenges

Additional topics for research and discussion related to this chapter include (1) the development of severe winter storms and their effects on people, (2) weather modification, (3) tornado, lightning, and hurricane safety (if applicable to your area), and (4) the effects of changing weather patterns on human behavior.

13 Oceanography

Overview

Chapter 13 is an introduction to the study of oceanography. The chapter begins with a discussion of seawater and its characteristics such as salinity, temperature, and density. Features of the sea floor, sea life, ocean currents, waves, and tides are presented. Practical applications of oceanography such as using the energy in tides to generate electricity are discussed. A section on ocean resources and a Perspectives page focuses on obtaining valuable minerals, biological materials, food, and fossil fuels from the oceans.

Performance Objectives

Goal: Students will learn the major features of the oceans, ocean currents, tides, and sea life.
Objectives: Upon completion of the reading and activities and when asked to diagram, demonstrate, or respond either orally or on a written test, students will:

13:1	K	define the term oceanography.
	K	describe the composition of seawater.
	P/S	prepare a brine solution similar in composition to seawater.
13:2	K	explain how ocean temperatures affect marine life.
	K	state the average density of seawater and the relationship between water temperature and density.
	P/S	use a laboratory balance to find the density of seawater.
13:3	K/P	explain how the seafloor is mapped.
	K	define the terms continental margin, continental shelf, and ocean basin floor.
	K	describe the major features of the ocean floors.
13:4	K	define the term plankton and explain why plankton are important to ocean fish.
	K	explain how the oxygen supply of the oceans is maintained.
13:5	K	define the term ocean current.
	K/P	explain how convection currents are produced.
	K/P	describe how oceanographers chart the paths of ocean currents.
	P/S	demonstrate the formation of convection currents in a large pan of water containing ice cubes.
13:6	K	explain how ocean waves are produced.
	K	describe the amplitude, wavelength, and period of an ocean wave.
	K/P	calculate the period of an ocean wave when given the number of waves that pass a point in a given time.
	P/S	demonstrate the formation of waves in a pan of water.
	P/S	use a clear plastic container, water, wooden dowel, metal washers, and a glass plate to demonstrate the effect of shallow depth on the wavelength and refraction of ocean waves.

13:7	K	describe changes tides cause in ocean levels along a seacoast.
	P/S	use records of daily tides to form conclusions regarding the time intervals between tides.
13:8	K	define the terms seismic sea wave and surge.
	K	explain why seismic sea waves and surges cause property damage and loss of life.
13:9	K	list the major ocean resources.

Equipment for Activities
balance
beaker or jar (iL)
burner
cork
container, clear plastic (rectangular)
daily tide listings
egg
glass plate, 10 cm^2
ice cubes, 2 dozen
ink or food coloring
litmus paper or pH paper
matches
metal washers, 40
pan, large
saline solution
salt compounds: sodium chloride (23 g), magnesium chloride (5 g), sodium sulfate (4 g), calcium chloride (1 g), potassium chloride (0.7 g)
spoon
thermometer
tripod for heating
wooden dowel, 2 cm diameter and 15 cm long

13:1 Seawater

Introduce this section by displaying a globe or map of the world and call attention to the world's major oceans and seas. Then introduce the term oceanography and explain what oceanographers do when they study the ocean. If you are near an ocean you may be able to obtain samples of seawater for display and use in class. You may also be able to obtain a product from a local aquarium store which mixes with water to simulate seawater. The main difference between seawater and freshwater is the relatively large amount of salt dissolved in seawater. Seawater contains dissolved oxygen and carbon dioxide.

Activity 13–1
In this activity the student prepares a solution that is similar to seawater in composition. Review the use of the balance, and caution students to measure the chemicals accurately without waste.
Materials: beaker or jar (1 L), 1 L freshwater, litmus or pH paper, salts: NaCl (34 g), MgCl$_2$ (5 g), Na$_2$SO$_4$ (4 g), CaCl$_2$ (1 g), KCl (0.7 g).
Question: The solution produced will test neutral.

Making Sure, page 264
1. sodium chloride
2. sodium, chlorine, magnesium, calcium, potassium, sulfur, oxygen, hydrogen
3. sodium

13:2 Temperature and Density

Heat in the oceans comes mostly from the sun. Heat may also be added from volcanic action and radioactive rocks in the ocean floor. Temperature is the most influential factor in determining the amount and kind of life in any part of the ocean. Sunlight is the second most influential factor. Dissolved salts lower the freezing point of seawater below 0°C. Differences in temperatures in different parts of the ocean cause differences in the density of seawater.

Activity 13–2
In this activity the student determines the density of the saline solution prepared in the previous activity. Density in g/cm^3 is calculated by measuring the mass of 1 L of the solution and dividing by 1000.
Materials: saline solution from previous activity, balance, thermometer, tapwater, egg.
Questions: The density of the saline solution is 1.034 g/cm^3 which is more dense than tapwater. Yes, the egg floats in brine and seawater.

Making Sure, page 266
4. It is cooler.
5. sun
6. does not receive as much solar energy
7. Increased salinity increases the density.

13:3 The Seafloor

The first scientific attempt to map the ocean bottom was undertaken by the four-year *Challenger* expedition that set out from England in 1872. Since this time many oceanographic expeditions have crisscrossed the oceans. Sonar is used to map ocean floors, where sound waves pass from the surface to the ocean bottom. In a given length of time, the wave is reflected back to the surface. The velocity of sound in water is about 830 km/s. Depth is computed by the equation $D = \frac{1}{2}t \times V$. D represents the time elapsed between sending and receiving vibrations and V represents the velocity of the sound waves.

Making Sure, page 268
8. Soundings may be made by lowering a weight on a cable and measuring the length of the cable. Sonar is also used.
9. Ocean ridges are like mountains, and the trenches are like valleys.

13:4 Sea Life
Life in the sea is a carefully balanced web of ecological relationships. All fish in the oceans are part of food chains that reach back to tiny animal and plant plankton. Sea life provides food for people and offers recreational opportunities such as fishing, scuba diving, and whale watching during whale migrations. The oceans are a great reservoir of oxygen, which is released into the atmosphere by diffusion of dissolved oxygen into the water. About 90% of the oxygen in air is produced by photosynthesis within the oceans.

13:5 Ocean Currents
Section 13:5 begins with a definition of ocean currents and a description of the Gulf Stream, a major current in the Atlantic Ocean. Rotation of the earth causes the deflection of currents in the ocean just as it causes deflection of winds in the atmosphere. However, the effect on ocean currents is greater. Movement of the surface currents is almost due west in both the Northern and Southern Hemispheres in the zone of the trade winds.

Activity 13–3
In this activity, students add ice cubes to one edge of a pan of water and heat another edge, which causes the formation of water currents. The addition of ink or food coloring on the side that is heated serves to detect the path of the currents. Water cooled on the surface of the water increases in density and sinks to the bottom of the pan. Water that is heated decreases in density and rises to the surface.
Materials: ice cubes, water, pan, tripod, burner, ink or food coloring, pencil, paper, matches
Questions: The ice cubes cool the water at the surface and increase its density. Heating decreases the density at the bottom.

Making Sure, pages 272 and 273
10. Both are caused by differences in density due to solar heating.
11. It is warmer than would be expected if there were no Gulf Stream.

13:6 Waves
Introduce this section with Activity 13–4. If an overhead projector is available, you may wish to do this activity in a clear plastic pan and project the waves onto a screen. Then discuss the formation of waves in the ocean. Draw a wave diagram on the chalkboard and label the amplitude, wavelength, and wave period. Define these terms as they apply to water waves. Explain how the wave period is calculated using the equation given on page 274.

Activity 13–4
In this activity, students observe wave motion and its effect on a floating cork. The cork is moved up and down by the action of the waves. However, it is not carried forward by the wave motion.
Materials: cork, pan, water, spoon.
Questions: The wave makes the cork go up and down.

Activity 13–5
In this activity, students observe wave motion along a shoreline, simulated in a clear plastic container.
Materials: clear plastic container, water, thin flat plate of metal or glass, metal washers, wooden dowel.
Questions: Waves generated by motion of the wooden dowel slow in speed as they pass over the glass, and the wavelength of the waves decreases. When waves approach the glass at an angle, they are reflected or they bend toward the plate.

Making Sure, page 277
12. 4 s
13. 5 s

13:7 Tides
Students who have been to the seashore are likely to be familiar with effects of the tides. Sailings of ocean vessels often coincide with high tide to be certain there is adequate depth for navigation out of a harbor. Relate tides to the worldwide need for energy and the possibilities of harnessing the tides to produce electric power.

Activity 13-6

In this activity the student studies the times for tides for successive days and observes that tides occur at regular intervals, and that each successive tide occurs several minutes later than the previous tide.

Materials: records of daily tides for one week or the tide listings in the activity.

Questions: The time between high tides on successive days varies. Answers will vary depending on the data used. The time between high and low tides is not exactly the same, but tends to be equal.

13:8 Seismic Sea Waves and Surges

Seismic refers to earthquake activity. The shock of an earthquake in the sea floor produces a huge wave that travels for many miles and is often called a tidal wave because when the wave reaches a seacoast it raises the water level as would a tide. Discuss the difference between a seismic sea wave and a surge. A surge is accompanied by a storm and a seismic sea wave is not storm related. A surge may be produced by a hurricane and other large cyclonic storms over ocean waters.

Making Sure, page 280

14. Seismic sea waves travel low in the water and are faster than surface waves.

13:9 Ocean Resources

Major ocean resources currently used by people and prospects for the future use of ocean resources are discussed in this section. Have students make a list of ocean resources to impress upon them the great value of the oceans to all people. You may wish to follow with a discussion of the pollution of oceans caused by human activities and the need for conservation, meaning the wise and careful use of ocean waters.

Making Sure, page 281

15. Waves have kinetic energy, tides have kinetic energy, the ocean is a huge source of heat energy.
16. A drilling rig must be on a platform over the ocean. Waves can damage equipment; more protection is needed against oil spills.

Perspectives

Commercial Diving

Procure a film on commercial diving for the students to view. Invite an instructor from a local diving school to discuss the training needed to become a professional diver. Have students do some career research to see what other opportunities are available to professional divers.

Study Questions

A. True and False

1. True   2. True   3. False (More than one half of the earth is covered by water.)   4. False (The salinity of most ocean water is 3.3–3.7.)   5. False (When seawater is cooled, its density increases.)   6. True   7. False (Water within the Arctic Circle is more dense than surface water in the tropics.)   8. False (Two high tides and two low tides occur each day.)   9. False (Water near the ocean surface is much warmer then deep ocean water.)   10. True

B. Multiple Choice

1. 4000   2. the Mariana Islands   3. sound   4. continental margins   5. 200   6. rip currents   7. plants and animals   8. undertow   9. surge   10. gills

C. Completion

1. decreases   2. manganese   3. earthquakes   4. less   5. plankton   6. sunlight   7. tropics   8. tides   9. density   10. waves

D. How and Why

1. As the temperature decreases, the density increases; when the temperature increases, the seawater becomes less dense.
2. The rotation of Earth acts on the flow of the ocean currents as well as the wind.
3. As a wave approaches the shore, it slows and begins to break.
4. Life is plentiful where there is warmth and sunlight with which the plant life can make food.
5. Surface currents are caused by wind sweeping across broad expanses of water.

Side Roads

The Sandy Seashore

Have students create a blackboard mural with the land forms and life forms that make up the subtidal zone, intertidal zone, pioneer zone, fixed dune zone, sea grapes, and scrub-woodland zone. Pictures or drawings can be used. Be sure to label all additions to the mural. Contributing students should also be able to share some information about the life form they add to the mural.

Unit 4 Living Things

Overview

Unit 4 presents basic topics in the study of living organisms. Characteristics of different representative species, life activities, and ecological relationships between organisms are described and explained. The unit begins with the features that are characteristic of all living organisms and introduces the modern five-kingdom classification system. The principles explained early in the unit are used throughout the unit as strands that tie the content together and make it more meaningful. Features of each phylum in the classification system are given along with descriptions of specific organisms within each phylum that exemplify the organisms that make up the phylum. Details of the structure and function of plant and animal organisms are described. Emphasis throughout the unit is given to interesting examples and illustrations having meaning for students and to practical applications of the study of living organisms.

Teaching Tips

Collect and exhibit specimens of as many different species of living things as possible. Use these specimens for study and to motivate interest in the study of life activities and classification.

Collect pictures of a variety of different species and use them for a bulletin board display. Wildlife magazines and old calendars having animal and plant pictures are good resources for this material.

Obtain potted plants for the classroom and obtain the assistance of your students in maintaining these plants. A local nursery or garden shop may provide information regarding plant species for classroom use and the requirements for care of each species.

Plan a field trip to a zoo for your students. Be certain to secure the necessary administrative and parental approval.

Check with the local Audubon Society for a guest speaker on birds or for student participation in an early morning bird watching walk.

14 Life and the Cell

Overview

Chapter 14 details and explains the major features of living organisms. The structure and activities of a cell as the basic unit of living organisms is presented. Using this information as background, the classification of living organisms is explained, and the rules for the scientific naming of organisms are given. Characteristics of the monera, fungi, and protists are introduced and used to illustrate basic activities that keep an organism alive and enable it to reproduce. A section on viruses describes their living and nonliving aspects and relationships to other organisms.

Performance Objectives

Goal: Students will learn that all living organisms are alike in certain ways. However, every kind (species) of organism is unique and different from all the other species.

Objectives: Upon completion of the reading and activities and when asked to diagram, demonstrate, or respond either orally or on a written test, students will:

14:1	K	describe the features of living organisms.
14:2	K	list the parts of a cell.
	P/S	draw a diagram of a cell that includes the main parts of the cell.
14:3	K/P	explain how diffusion and osmosis occur in a cell and the importance of these activities.
	K	define respiration and explain why respiration is necessary for the life of a cell.
	P/S	use a carrot and corn syrup to demonstrate diffusion.
14:4	K/P	name the five kingdoms and one example from each kingdom.
	K/P	explain how plants and animals are given scientific names and list examples of scientific names.
14:5	K	describe the main characteristics of bacteria.
	K	explain how bacteria reproduce.

	P/S	use a microscope to study prepared slides of bacteria.
14:6	K	describe the main characteristics of fungi.
	K	describe fermentation and explain why it is important to yeasts.
	K	describe the main characteristics of molds.
	K/P	explain how molds reproduce.
	P/S	do an activity that shows the growth of yeast and the production of carbon dioxide by yeast.
14:7	K	describe the main characteristics of an amoeba, its life activities, and how it reproduces.
	K	describe the main characteristics of a paramecium, its life activities, and how it reproduces.
	P/S	prepare a microscope slide of an amoeba culture and study it under a microscope.
	P/S	make a hay infusion culture and observe the changes in organisms that occur in the culture over time.
14:8	K	list three examples of flagellates and state one fact about each.
14:9	K	list the main characteristics of sporozoans.
14:10	K	describe the main characteristics of viruses.

Equipment for Activities

aluminum pie pan
amoeba culture
bacteria slides, stained
beaker, 1000 mL, or glass jar
beaker, 400 mL
Bunsen burner
candle
carrot
corn syrup
cover slip
dropper
glass bottles, 3 identical
glass slide
glass tube, 10 cm
glycerol
hay or dried grass
jar
limewater
match
microscope
petroleum jelly
pond water
rag or towel
rice, boiled
3 rubber balloons, identical
rubber tube, 30 cm
sealing wax
stopper, one-hole
test tube
sugar, 80 g
yeast

14:1 Features of Living Things

Begin by asking students to name their favorite plant and animal. Then discuss the special features of these organisms that make them different from other living organisms. Follow with a discussion of how all living things are alike. Point out that reproduction is not necessary for the life of an individual organism, but it is necessary for the survival of the species.

Activity 14–1

In this activity, the student observes a candle and relates the observations to the features of living things described in Section 14:1. The purpose of this activity is to review basic life processes and increase the student's understanding of these processes.
Materials: candle, dish or pan, match
Questions: Of course, a burning candle is not alive. However, it possesses certain features that are similar to those present in plants and animals. It does not use food. It does not reproduce.

Making Sure, page 293

1. (a) human—thinks, eats (b) grasshopper—eats, hops (c) spider—spins webs, moves (d) robin—flies, eats worms (e) cactus—has spines, uses water (f) frog—hops, eats insects (g) mushroom—produces spores, grows (h) snake—has scales, crawls (i) octopus—lives in water, moves
2. Answers will vary. Possible examples may include sound of a bell or flash of a bright light.
3. Plants move slower and usually cannot move from place to place.

14:2 The Cell

The cell is the basic unit of structure and function in living organisms. You may wish to have students learn the parts of a cell well enough so they can draw and/or label a diagram of a cell. Prepared microscope slides of different kinds of cells, if available, to illustrate the fact that cells differ somewhat in their structure and appearance. Specialized cells have features different from the idealized or generalized cell students learn to diagram. Point out that the life code in DNA regulates the chemistry of the cell and carries a code for the inherited traits that are passed from one generation to another.

14:3 Cell Activities

It is the life activities inside a cell that keep the cell alive and enable it to perform the functions needed to keep a whole organism alive. Diffusion occurs in gases and liquids such as water. Osmosis is diffusion of water through a membrane and the process by which a cell gains or loses water. The concentration of dissolved materials inside a cell affects the rate at which water is lost or gained. A cell placed in salt will lose water quickly because the water diffuses from inside the cell where it is most concentrated to the salt outside the cell.

Activity 14-2

In this activity, the student observes the osmosis of water into a carrot. The carrot may be hollowed out with a knife or a cork borer. Place the beaker, carrot, and tube on the ring stand and support the tube with a test tube clamp. Heat sealing wax in a small pan and pour it around the area where the stopper contacts the carrot. After several hours, the liquid (syrup and water mixture) will rise in the tube due to the pressure of the water moving from the carrot into the corn syrup.

Materials: carrot, corn syrup, one-hole stopper, glass tube (100 mm), glycerol, rag or towel, sealing wax, beaker (1000 mL) or glass jar

Questions: The liquid in the tube is the syrup and water mixture. Water enters the carrot by osmosis.

Making Sure, page 297
4. decrease.
5. Water is more concentrated inside the cell.
6. Digestion breaks large food molecules into smaller molecules. Respiration combines food with oxygen to yield energy.
7. The cell would be poisoned and die.

14:4 Classification

Review the principle of classification as the grouping of things based on their similarities and differences. Classification is used by scientists to help develop an understanding of the things they study and is therefore one of the methods of science. Students may be familiar with a three kingdom system of classification. The modern five kingdom system (Monera, Protist, Fungi, Plant, Animal) used in this text takes into account the special characteristics of bacteria that warrant placing it in a separate kingdom (Monera) and, similarly, the special characteristics of yeasts, molds, and mushrooms that indicate logical grouping of these organisms in a single phylum (Fungi).

Making Sure, page 299
8. They are classified into groups based on their similarities and differences.
9. See Table 14-2 for kingdoms and examples. Also see Appendix E.

14:5 Bacteria

Although bacteria cannot be seen without a microscope, there are many examples of their presence. Go over examples such as disease, souring of milk, decay of food, nitrogen fixing bacteria in the roots of legumes, and the making of sauerkraut, yogurt, and butter. Point out the importance of cleanliness, refrigeration, and cooking in preventing the growth of bacteria. Although some bacteria are harmful, most are useful or have no harmful effect on people.

Activity 14-3

If students are unfamiliar with the use of a microscope, use the material in Appendix D to teach the proper procedures in the use of a microscope before students do this activity. In this activity the student observes different kinds of stained bacteria. Use slides that show the three main types of bacteria: coccus, bacillus, and spirilla.

Making Sure, page 302
10. Bodies contain food, warmth, and moisture.
11. They produce endospores.
12. fission (cell division)
13. Lack of food, and wastes produced are poisonous.
14. rods, spheres, spirals

14:6 Fungi

To illustrate the growth of mold, place a piece of bread that does not contain a preservative in a jar together with a few drops of water and seal it tight with a cap. Allow to stand in a warm place for several days. Keeping bread in a refrigerator aids in preservation of the bread because the cool temperature retards the growth of bread mold. You may wish to make some bread dough from a simple package mix and place it in a container. Have students observe and discuss the increase in the volume of the dough caused by the production of carbon dioxide by yeast in the dough.

Activity 14-4

In this activity, the student prepares mixtures containing yeast and observes the growth of yeast. Remind students to use warm water and not hot water in making the mixtures. The sugar provides food for the growth of yeast. Carbon dioxide gas produced by the fermentation of sugar inflates the balloon attached to the container of sugar, water, and yeast.
Materials: beaker (400 mL), 3 identical glass bottles, glycerol, limewater, 3 identical rubber balloons, rubber tube (30 cm), 1-hole stopper, sugar (80 g), yeast
Questions: Yes, the carbon dioxide turns the limewater milky due to the formation of calcium carbonate. $Ca(OH)_2 + CO_2 \longrightarrow CaCO_3 + H_2O$
Yeast cells have an oval shape.

Making Sure, page 304

15. no chlorophyll; live on other organisms or dead organic matter.
16. Yeast reproduces by budding.
17. fermentation
18. only one parent
19. Wind blows them; they are carried on bodies of people and animals.

14:7 Amoebas and Paramecia

Amoebas and paramecia are good illustrations of one-celled organisms and the life activities that keep these organisms alive. Remind students that a one-celled organism is a complete living thing and performs all the necessary life activities needed to stay alive. A very large amoeba species such as *Chaos chaos* can be seen without using a microscope. Amoebas belong to the phylum Sarcodina. Ciliates such as paramecia are found in pond water near decaying vegetation. Ciliates are more complex than other one-celled organisms.

Activity 14-5

In this activity students observe amoeba with a microscope. Because an amoeba is transparent, it is more likely to be observed in dim light.
Materials: glass microscope slide, test tube, petroleum jelly, cover slip, dropper, amoeba culture, microscope
Question: The culture is sealed with petroleum jelly to prevent the loss of water by evaporation.

Activity 14-6

In this activity, the student makes a hay infusion culture and observes the culture with a microscope. Water from a bird bath or any stagnant water may be used in place of pond water. Paramecia and other ciliates will be observed in the culture. The numbers of different kinds of organisms observed will change over time due to the ecological succession of species. Hay is dried grass or legumes.
Materials: hay or dried grass, jar, boiled rice, pond water, microscope, glass slide and cover slip, Bunsen burner, matches
Question: Answers will vary. See examples of the various organisms in Section 14:7.

Making Sure, page 308

20. It would poison itself and die.
21. 4096
22. Both take in food through a food vacuole. An amoeba uses pseudopods. Paramecia take in food through the oral groove.
23. reproduction
24. forms a cyst

14:8 Flagellates

A *Euglena* is probably the best known flagellate. Because it contains chlorophyll, a *Euglena* is sometimes classified as a plant. Flagellates are classified on the basis of how they move; therefore, organisms with flagella are grouped into one phylum. Colonial flagellates are one-celled organisms. However, the cells associate together somewhat like cells in a multicellular organism.

Making Sure, page 309

25. paramecia—cilia, amoebas—pseudopods, euglena—flagella
26. has chlorophyll

14:9 Sporozoans

All sporozoans are parasites, meaning they obtain food from living organisms in whose bodies they live. Review the meaning of parasite. A sporozoan has a life cycle in which spores are produced for reproduction. Perhaps the best known sporozoan is the one that causes malaria and is carried by the *Anopheles* mosquito.

Making Sure, page 309

27. cannot move to obtain food
28. obtain food from other organisms

14:10 Viruses

Wendall Stanley isolated the first virus from a tobacco plant. Scientists are now searching for a vaccine to immunize people against the common cold, which is believed to be caused by many different viruses. Cancer may be caused by a virus or viruses. Diseases known to be caused by viruses include chicken pox, smallpox, rabies, and polio. Review the nature of DNA and RNA (14:2, p. 295) and protein, a substance used in the growth and repair of animals and plants.

Making Sure, page 311
29. reproduction
30. can be crystallized and stored for many years as with any nonliving substance.
31. Present-day viruses can only grow inside other organisms.

Perspectives

Natalie Jones: A Dedicated Microbiologist
Have students research the job description of a microbiologist working in a hospital, a private laboratory, the food industry, and an environmental agency.

Study Questions

A. True or False
1. True 2. False (The number of plant species is less than the number of animal species.) 3. True 4. False (One genus is *Canis*.) 5. True 6. False (Yeasts and molds do not contain chlorophyll.) 7. False (Amoebas move by using fingerlike projections.) 8. True 9. False (One type of flagellate causes sleeping sickness.) 10. False (Most viruses are smaller than bacteria.)

B. Multiple Choice
1. cell membrane 2. species 3. fungi 4. osmosis 5. kingdom 6. monera 7. cilia 8. fermentation 9. sporozoan 10. spores

C. Completion
1. species 2. high, low 3. two 4. organism 5. wall 6. electron 7. cells 8. mushrooms 9. *Volvox* 10. fission

D. How and Why
1. Most living things have the following features: definite size and shape, definite life span, reproduction, metabolism, movement, response to the environment.
2. See Figure 14–3, page 294.
3. Scientific names are the same throughout the world; the system is convenient; the system shows the relationships among organisms.
4. Reproduction is necessary to replace members of a species that die.
5. Bacteria can form endospores, which are structures containing a protective cell wall.

15 Plants

Overview

Plants are living organisms that contain chlorophyll and make their own food by photosynthesis. In this sense, plants are "free-living" organisms. Chapter 15 presents four major groups of plants: algae, mosses and liverworts, ferns, and seed plants. The characteristics of each group are described and contrasted. The structure of various parts of seed plants, such as roots, stems, leaves, flowers, and seeds are covered and the importance of these parts to the life of a seed plant is discussed. Photosynthesis is explained as a chemical change involving reactants (carbon dioxide and oxygen), a catalyst (chlorophyll), and end products (food and oxygen).

Performance Objectives

Goal: Students will learn the main features of different kinds of plants and their life activities.
Objectives: Upon completion of the reading and activities and when asked to diagram, demonstrate, or respond either orally or on a written test, students will:

15:1	K	describe the importance of plants to animals.
	K	explain why most algae live in water.
	K	describe the main features of *Protococcus*.
15:2	K	list the main features of mosses and liverworts.
15:3	K	list the main features of ferns.
15:4	K	list the main features of seed plants.
	P/S	explain how gymnosperms are different from angiosperms.

	K	define tissue and organ.
15:5	K	list the functions of plant roots.
	K	explain how gravity and water affect the growth of plant roots.
	P/S	grow roots from radish seeds and study the roots with a hand lens.
	P/S	do an activity that investigates the effect of water on the growth of roots.
15:6	K	list the functions of plant stems.
	K	describe the difference between herbaceous and woody stems.
	K	identify the parts of a stem and describe the function of each part.
	K/P	explain how parts of a stem such as a tuber or bulb may be used to reproduce the plant.
15:7	K	describe the structure and function of a leaf.
	K	define photosynthesis and transpiration.
15:8	K/P	describe the chemical change that occurs in photosynthesis.
	P/S	write an equation for the photosynthesis reaction.
	P/S	do an experiment with geranium plants to determine if sunlight and chlorophyll are needed for photosynthesis.
	P/S	do an experiment to determine if carbon dioxide is needed for photosynthesis.
15:9	K	describe the difference between complete and incomplete flowers.
	K/P	explain how pollination and fertilization occur in flowers.
	K/P	explain how a fruit and seed are formed.
15:10	K	list the parts of a seed and state the function of each part.
	P/S	do an experiment to study the factors affecting the germination of seeds.
15:11	K	define seed dispersal and explain how it aids plant reproduction.
	A/S/K	carry out one Challenge at the end of the chapter.
	A/K	read a science article or book and prepare a short report.

Equipment for Activities

alcohol
baby food jar
beakers, 2
bean seeds, 74
bell jar or large, clear plastic bag
blotters, small, 20
celery stalk
coleus leaf
construction paper, black or aluminum foil
cord for tying bundles
electric lamp, if no direct sunlight available
geranium plants, potted, 2
glass plate to cover terrarium
gravel
hot plate or ring stand and laboratory burner
humus
iodine solution
jar
leaf, magnolia if possible
Marchantia
marking pencil
metric ruler
microscope or hand lens
oat seeds, 1 pkg
paper towels, 3
petri dish
petroleum jelly
plastic bags, 1 small and 1 large
plastic sheets
pond water containing infusorians
radish seeds, 1 pkg
red ink
refrigerator
rubber gloves
sodium hydroxide pellets
sphagnum moss
spray bottle
string or rubber bands
sweet potato
terrarium (glass container)
test tube holder
test tubes/stoppers, 18
thermometer
toothpicks, 3
topsoil
tray or pan
tree trunk, cross-section

15:1 Algae

Algae are different from fungi in that most algae contain chlorophyll and can therefore make their own food. Algae include freshwater species such as *Spirogyra*, *Protococcus* that live on the bark of trees, single-celled marine algae living in the oceans, and seaweed that grows to great lengths. Water-living algae obtain nutrients and carbon dioxide from the water in which they live.

Making Sure, page 318
1. They dry out and die when not moist.
2. fission

15:2 Mosses and Liverworts

All mosses and liverworts are many-celled organisms, in contrast to many species of algae that are single-celled organisms. The absence of true circulation tissue within mosses and liverworts prevents the development of tall-growing species because food and water cannot be moved as it can be in plants having food- and water-conducting tissues.

Activity 15–1

In this activity the student makes a terrarium in a glass container for the purpose of studying the growth of a liverwort. First, a layer of gravel should be placed in the bottom of the container. Then, add a layer of topsoil mixed with humus. Plant the *Marchantia* in the topsoil mixture. A few other small plants may be planted in the terrarium, too. Adjust the cover of the terrarium so that air may enter, but keep it covered enough to maintain a high level of moisture within. Keep the terrarium in a place where it does not receive direct sunlight.
Materials: terrarium tank or large wide-mouth glass jar, *Marchantia* and other plants, gravel, topsoil, humus, glass plate to cover the terrarium
Question: The humidity is kept high so that the plants do not dry out.

Making Sure, page 319

3. They do not have a system to carry water and food very far.
4. They do not have roots to obtain water. They dry out and die if not kept moist.
5. Similar: contain chlorophyll, need moisture, made of cells. Different: Most algae are one-celled organisms. Mosses and liverworts are many-celled.

15:3 Ferns

Ferns are the simplest vascular plants. Tubelike structures within the roots, stems, and leaves of a fern facilitate the transport of water and food throughout the plant. You may be able to obtain fern specimens from a local florist. Ferns may be grown in a terrarium or in pots in the classroom, also. Examine the underside of fern leaves, called fronds, to look for spore cases. Review the asexual (spore) and sexual (egg and sperm) stages in the life cycle of a fern.

Making Sure, page 320

6. Ferns have a system to transport water and food. Mosses do not.
7. The sperm cells need water to move to the egg cells for fertilization.

15:4 Seed Plants

Seed plants are the most complex and advanced plants. The existence of modern civilization and the development of agriculture is connected to the production of food from seed plants. Seed catalogs contain good examples of many different kinds of domesticated seed plants. Flowers or vegetable plants may be grown from seeds in the classroom so that students can observe the plant development that occurs.

15:5 Roots

Many plants have more root structure below ground than stems above ground. Sometimes tall trees fall over after heavy rainstorms because the ground becomes too soft to hold the roots. Pressure from growing roots often lifts up pieces of concrete sidewalk. Have students list the functions of roots and discuss each function.

Activity 15–2

In this activity the student germinates radish seeds and studies the young roots of radish plants. Soaking the seeds softens the seed coat and aids in germination. The petri dish must be kept covered or the radishes will dry out and die. If a binocular microscope is available, use it to study the roots. Observe the root hairs attached to each root. Be certain students identify root hairs.
Materials: radish seeds, petri dish, moist paper towel, thermometer, binocular microscope or hand lens

Activity 15–3

In this activity the student plants seeds within a sphagnum moss ball to observe root growth. The sphagnum ball must be sprayed daily to keep the seeds moist at all times. The ball can be placed in a sealed plastic bag and taken home for watering on weekends. Observe in which directions the roots grow.
Materials: sphagnum moss, cord, radish or oat seeds, water, spray bottle
Question: Moisture is a greater stimulus to root growth than gravity.

Making Sure, page 324

8. Water increases the growth rate of roots.
9. Food is available for use by the plant as needed.
10. Roots anchor plants, take in minerals, store food, absorb, store and transport water.

15:6 Stems

Prepared, stained slides showing the cross-sections of woody and herbaceous stems may be obtained from a biological supply house. Review the diagrams showing the arrangement of tissues in woody and herbaceous stems. Relate these stem parts to the definition of a tissue as a group of cells working together to perform a specific function. Review the definitions of annual and perennial and the examples of each kind of plant.

Activity 15-4

In this activity the student observes the movement of red ink upward through xylem tissue in a celery stalk. The liquid moves into the cells of the stalk by diffusion. Water is lost by transpiration from the leaves at the top of the stalk. The ink is then drawn into and upward in the tubelike structures within the stalk.
Materials: celery stalk, bottle of red ink
Question: The ink moves up the celery stalk causing the stalk to turn red. As the leaves on the stalk release water by transpiration, ink is drawn into and through the stalk.

Activity 15-5

In this activity the student observes annual rings in a cross-section of a piece of a tree trunk. Each ring includes a layer of spring wood and a layer of summer wood. The number of rings is the age in years. You may wish to make copies of a cross-section on a photocopy machine and have each student tape a copy in their science notebook. Have students count and write the number of annual rings.
Materials: cross-section (slice) of a tree trunk, hand lens (optional)

Activity 15-6

In this activity the student grows a sweet potato plant from a sweet potato. Be certain to keep the water level constant in the jar. Growth will begin within a week.
Materials: fresh sweet potato, jar, 3 toothpicks, water
Question: Leaves and roots develop from buds on the sweet potato.

15:7 Leaves

Students may help make a collection of leaves for a bulletin board display or scrapbook. Leaves can be temporarily preserved for display by placing them on stiff poster board and affixing them with clear contact paper. Call students' attention to the characteristic leaf shape of each plant species. Point out that the flat, thin structure of leaves maximizes the absorption of light. A wet mount of the epidermis of a leaf may be made on a microscope slide. Strip off a layer of lower epidermis with tweezers and place it in a drop of water on a slide. Then, place a coverslip on top of the tissue. Observe the stomata in the leaf. Use a microscope or microprojector for viewing.

Activity 15-7

In this activity the student studies transpiration. by covering a geranium plant with a plastic bag. Water vapor given off by the plant condenses on the inside of the plastic bags.
Materials: potted geranium plant, small plastic bag, large (clear) plastic bag, string or rubber bands
Question: Water vapor is given off by the plant in a process called transpiration. The water vapor condenses on the inside of the plastic bags.

Activity 15-8

In this activity the student prepares a skeleton of a leaf by soaking a leaf in pond water containing infusorians. The infusorians and bacteria feed on the leaf, stripping away the soft parts of the leaf and leaving the tougher skeleton.
Materials: leaf (magnolia if possible), tray or pan, pond water containing infusorians, water faucet, blotters or paper towels, 2 sheets of plastic
Questions: Infusorians (and some bacteria) are microorganisms that cause decay. The infusorians grow faster in warm temperatures. Sunlight kills infusorians. The microorganisms caused the leaf to decay.

15:8 Photosynthesis

Light is necessary for the formation of chlorophyll and for photosynthesis. There are at least six different kinds of chlorophyll. Four kinds are found in plants and two are present in certain species of bacteria that carry on photosynthesis. Glucose (sugar) is the product of photosynthesis and these molecules are bonded together to form starch molecules. The equation given for photosynthesis is an overall reaction showing the net effect of a complex series of reactions.

Activity 15-9

In this activity the student determines whether sunlight and chlorophyll are needed for photo-

synthesis. For safety, do not allow any burners or matches in the vicinity of the alcohol. Keeping the geranium in darkness for a few days prevents photosynthesis and rids the leaves of starch. When the leaves are soaked in alcohol, the chlorophyll is removed from the leaves. Since starch turns iodine a blue-black color, the appearance of this color when iodine is added to a leaf shows the presence of starch and indicates that photosynthesis has occurred.
Materials: alcohol, baby food jar, beaker, coleus leaf, construction paper (black) or aluminum foil, electric lamp (optional), geranium plant, hot plate or ring stand and burner, iodine solution
Questions and Conclusions: **1.** to start food production in the leaves **2.** photosynthesis **3.** starch

Activity 15–10
In this activity the student places a "starved" geranium plant in air free of carbon dioxide to show that the plant needs carbon dioxide for photosynthesis. Sodium hydroxide pellets absorb carbon dioxide and remove it from the air inside the jar. **CAUTION:** Do not touch sodium hydroxide pellets. Severe burns could result. A control plant is used to show that with carbon dioxide, photosynthesis occurs and starch is produced.
Materials: 2 geranium plants that have been kept in darkness for a few days, bell jar or large, clear plastic bag, sodium hydroxide pellets, rubber gloves, petroleum jelly, iodine solution
Question: When the leaves of the plant in the bell jar are tested with iodine, no color change is observed showing that photosynthesis has not occurred and starch is not present. The leaf from the plant outside the jar (control) turns blue-black when iodine is added showing that starch is present and photosynthesis has taken place.

15:9 The Flower
Introduce this section by showing flowers or pictures of flowers to the class. A local florist may be a good source of these materials. You may wish to have students dissect one or more flowers to observe the parts after they have learned the structure and function of the parts. Have students label cross-section diagrams.

Making Sure, page 333
11. (a) f   (b) s   (c) s   (d) f   (e) f
12. apple, peach, lemon, pear, beans

15:10 Seeds
Many of the foods people eat are seeds. Some examples are peas, beans, corn, wheat, oats, and rice. Seeds are rich in stored food that is available to the developing embryo plant. Sunlight is not necessary for the germination of many kinds of seeds, but it is necessary for continued growth of a new seedling after the stored food in the seed is used up.

Activity 15–11
In this activity, students observe and compare the effects of moisture, light, and temperature on seed germination.
Materials: 54 bean seeds, 18 small blotters, 18 test tubes, 9 test tube stoppers, marking pencil, lamp, refrigerator
Question: The seeds that are warm and moist will germinate first (B,F), indicating these are the best conditions for germination. The dry, cool seeds do not germinate (C). Moist, cool seeds take the longest to germinate (D). Warmth and moisture are the best conditions for germination.

Activity 15–12
In this activity the student studies the seedlings produced by the germination of bean seeds. Fill the tray or wooden flat half-full with a flat layer of topsoil. Plant seeds at a depth of about 2 cm. Keep the flat in sunlight or under the light of a lamp and water daily but do not overwater. Have students make a sketch of the seeds they dig up.
Materials: small tray or wooden flat, metric ruler, lamp (optional), 20 bean seeds, topsoil

15:11 Seed Dispersal
Stress the fact that seed dispersal is not necessary for the survival of an individual plant. It increases the survival of a plant species. Most seeds in the wild never grow into new plants. For a seed to germinate, it must be in soil suited to its species, have sufficient warmth and moisture, and be able to compete with other nearby plants for its space, water, and mineral nutrient needs. Go over the features of plants and fruits that disperse their seeds and give examples of these features. Most seeds are dispersed by wind, water, and animals.

Perspectives
A Marine Botanist
Have students read more about *halimeda* or other types of algae. Address the topic of algae and

algae products. Have students find out about products people use that have algae in them. Set up a classroom display of algae products.

Study Questions

A. *True or False*
1. False (Most of the land plants are seed plants.)
2. True 3. True 4. False (Some stems grow below the ground.) 5. False (A moss has no true roots, stems, and leaves.) 6. True 7. True 8. False (Transpiration occurs most rapidly when the stomata of a plant are open.) 9. False (Ferns reproduce through flowers.) 10. True

B. *Multiple Choice*
1. root hair 2. white potato 3. osmosis 4. xylem 5. tomato 6. cambium 7. flower 8. wind 9. transpiration 10. glucose

C. *Completion*
1. solar (light) 2. incomplete 3. food 4. chloroplasts 5. oxygen 6. chlorophyll 7. petals 8. ovule 9. fertilization 10. moisture, oxygen, warmth

D. *How and Why*
1. Desert plants have small leaves or no leaves at all in order to conserve water.
2. The seed is moved to a favorable environment where it can grow and produce more of the species.
3. Bees go into flowers to obtain nectar. They pick up pollen on their legs and transfer it from flower to flower as they forage.
4. See Figure 15–2 (b), page 318.
5. Photosynthesis provides the energy and basic organic compounds a green plant needs for growth and maintenance.

16 Animals

Overview

Chapter 16 begins with a discussion of animal tissues, organs, and systems. The scientific principles that explain the nature of animal life and the special features of animals that enable them to carry on their life activities are emphasized. A discussion of some invertebrate members of the animal kingdom is included. (The vertebrates are discussed in Chapter 17.) The special characteristics of each invertebrate phylum and examples of animals classified in each phylum are described. A section on arthropods, the phylum with the largest number of species, focuses on the structure and functions of the grasshopper as an example of insect life.

Performance Objectives

Goal: Students will learn the characteristics and life activities of invertebrate animals.

Objectives: Upon completion of the reading and activities and when asked to diagram, demonstrate, or respond either orally or on a written test, students will:

16:1	K	define tissue, organ, and system.
	K	define specialized cell and name three examples.
16:2	K	list the major features of a sponge.
16:3	K	list the major features of jellyfish and their relatives.
16:4	K	list the features of a flatworm.
	K	list the major features of a roundworm.
	K/P	explain how trichinosis is caused and how it can be prevented.
16:5	K	list the major features of an earthworm.
	K/P	explain how an earthworm moves, digests food, and circulates material through its body.
	KP	explain how an earthworm obtains oxygen and excretes wastes.
	K/P	explain the process of reproduction in earthworms.
	K/P/S	dissect a preserved earthworm, and identify its major organs.
16:6	K	list the major features of spiny-skinned animals.
	K	define radial symmetry.
	K/P	explain how a starfish obtains food, circulates materials in its body, and moves about.
16:7	K	list the major features of the mollusks.
	K	name four kinds of mollusks.
	K/P	describe how squids and octopuses are different from oysters and clams.

16:8	K	list the features of arthropods.
	K	name six different arthropods.
	K/P	describe the stages in the complete metamorphosis of an insect.
	K/P	explain how a grasshopper obtains and digests food.
	K/P	explain how a grasshopper obtains oxygen and circulates it through its body.
	K/P	describe the major parts of a grasshopper's nervous system and state the function of each.
	K/P	explain the process of reproduction in the grasshopper.

Equipment for Activities

aquarium water, 10 mL
cardboard, thick or Styrofoam
cotton
dissecting needle
dissecting pan or soft wooden board
dropper
earthworm, preserved
fingernail polish remover
hand lens
hydra culture
insect catching net (optional)
insect field guide
jar and lid, small
marking pen
straight pins
scalpel or sharp knife
watch glass or small jar

16:1 Tissues, Organs, and Systems

This section introduces basic concepts as background for the study of the characteristics of specific animals in the sections that follow. Specialized cells in an animal may be compared to workers in a factory where each worker does a special job that contributes to the overall production of the factory. Prepared slides of different specialized animal cells may be purchased from a scientific supplier and studied by students with microscopes.

16:2 Sponges

Discuss the ways in which animals are different from plants. A sponge is similar to a plant in some ways, but is classified as an animal because its cells do not have cell walls and it does not contain chlorophyll. Discuss some basic needs of animals—food and oxygen and excretion of wastes. The sponge obtains food and oxygen from the water that flows through its body and excretes wastes, such as carbon dioxide, into this water.

16:3 Jellyfish and Their Relatives

This section focuses on the coelenterates and discusses the jellyfish as an example of this phylum. Have students compare the structure of a jellyfish with a sponge, noting that the jellyfish is more complex, has tentacles to capture prey, and has a central cavity in its body where food is digested.

Activity 16-1

In this activity the student studies the hydra to observes its major features. Hydra can be obtained from a scientific supply company or a freshwater pond.
Materials: watch glass or small jar, aquarium water (10 ml), dropper, hydra culture, hand lens, dissecting needle
Questions: The tentacles of a hydra are located in a circle around the entrance to its body cavity. Movement of the tentacles moves water and food into the body cavity; the hydra will move away when touched by sharp objects such as the point of the dissecting needle. A hydra pulls food into its mouth with its tentacles.

Making Sure, page 344
1. has a central cavity, can swim
2. tentacles and poison threads

16:4 Flatworms and Roundworms

Many species of flatworms and roundworms are parasites. Review the definition of a parasite and discuss a tapeworm living in the digestive tract of an animal as an example of a parasite. Students may be familiar with the term "worms" referring to parasitic roundworms as pests. Discuss the importance of cleanliness, sanitation, and food inspection in the prevention of worms. Planaria are free-living flatworms that are not parasites. Planaria in pond and stream water are attracted to bits of meat. Collect some planaria for class study.

Making Sure, page 345
3. A parasite lives off the host.
4. A tapeworm needs a host animal to live.
5. A tapeworm is a parasite. A planarian is free-living.
6. can absorb food through its body wall; special structures on the head to attach to host

16:5 Earthworms and Their Relatives

Segmented worms, such as an earthworm, are more complex than a roundworm or flatworm. The study of the body parts and functions in this section provides students with a specific example of an invertebrate and how it carries on life activities, including reproduction. Review the concepts of asexual and sexual reproduction, pointing out that for reproduction the earthworm must have a sexual partner.

Activity 16–2

In this activity the student dissects an earthworm and identifies its body parts. Go over the steps in the procedure with students before they begin. Then provide individual assistance as needed during the dissection.
Materials: earthworm, dissecting needle, dissecting pan or board of soft wood, scalpel or sharp knife, 20 straight pins
Questions and Conclusions **1.** mouth, pharynx, esophagus, crop, gizzard, intestine, anus **2.** It allows all the digestive processes to be completed. **3.** Digested food diffuses through these blood vessels **4.** These vessels are muscular organs that pump blood through the earthworm. **5.** male—testes, female—ovaries **6.** nerves

Making Sure, page 350

7. sexual reproduction—Two earthworms produce eggs and sperm that are exchanged. Fertilized eggs develop into young earthworms.
8. for exchange of gases (O_2 and CO_2) through the skin

16:6 Spiny-Skinned Animals

Spiny-skinned animals are echinoderms. Students may be able to bring specimens of dried sand dollars or starfish from home for an exhibit of spiny-skinned animals. Discuss the concept of radial symmetry in a starfish and the bilateral symmetry of a person. Have students give examples of each type of symmetry. A spoked-wheel is an example of radial symmetry and a water tower or tall skyscraper building could be examples of bilateral symmetry.

Making Sure, page 351

9. pulls itself with its arms
10. radial symmetry, tough outer skin with coarse spines

16:7 Mollusks

Snails can be maintained in a classroom aquarium tank and observed by students. Land snails can be found on garden plants or in ponds. Species of mollusks can be identified by the size, shape, and color of the shell. Invite students to bring seashell collections to class for exhibit. Have interested students check the school or local library for a natural history guide they can use to identify the shells.

16:8 Arthropods

This is the animal phylum with the most species and it includes the insects, which number over 700 000 different species. Have students distinguish between spiders, which have two body segments (head and thorax) and four pairs of legs, and insects, which have three body segments (head, thorax, and abdomen) and six legs. Contrast the exoskeleton of an arthropod with the internal skeleton of a mammal.

Making Sure, page 354

11. to protect soft internal body parts
12. external skeleton, segmented bodies, jointed legs

16:9 Insects

Have students learn the names of the main body parts of an insect and the features (exoskeleton, segmented body, and jointed legs) used to classify insects as arthropods. Discuss why insects are the most plentiful and widespread of animals. There are many different species of insects that are adapted to living conditions from the arctic to the equator. There is a short time between generations and each pair of insects produces many offspring. Use the grasshopper as an example of insect life and have students learn the parts and functions of a grasshopper's body. Compare the grasshopper to the earthworm.

Activity 16–3

In this activity the student collects and mounts insects, identifying each insect collected. Fingernail polish remover contains acetone. **CAUTION:** The fumes are poisonous and it is flammable. Put only enough fingernail polish remover on the cotton to moisten it. Do not use an excess amount. Caution students to keep the bottle capped when not in use. Handle the dead insects with tweezers.

Obtain an insect field guide for identification and naming of the insect species.
Materials: net (optional), small jar with lid, cotton, fingernail polish remover, cardboard or Styrofoam, straight pins, insect field guide, marking pen

Making Sure, page 358
13. Differences: spider—two body segments, four pairs of legs; insects—three body parts, three pairs of legs. Similarities: all have exoskeleton, segmented bodies, jointed legs
14. A caterpillar emerges from a butterfly egg. A caterpillar is the larva stage of a butterfly. The pupa stage occurs when a caterpillar spins a cocoon around itself. The caterpillar changes in the cocoon, and an adult butterfly hatches out of the cocoon.
15. A grasshopper has all the body parts of a typical insect. See Figure 16-19, page 356.
16. Earthworm—heart consists of pairs of large blood vessels. Blood is in closed vessels and contains hemoglobin dissolved in plasma. Grasshopper—blood fills body cavity; no blood vessels. Blood does not contain hemoglobin and does not carry oxygen.
17. The male deposits sperm inside the female. Fertilization takes place in the body of the female. A female grasshopper lays eggs in a hole in the ground. These eggs hatch in the spring.

Study Questions

A. True or False
1. False (A blood cell is a specialized cell that carries oxygen to all parts of the body.) 2. True 3. False (One example of an arthropod is a grasshopper.) 4. True 5. False (An earthworm is free-living.) 6. True 7. False (Food is digested in an earthworm's digestive system.) 8. True 9. True 10. True

B. Multiple Choice
1. clam 2. arthropod 3. Arthropods 4. Spiracles 5. sexual 6. pupa 7. cavity 8. arthropods 9. Blood 10. sexual

C. Completion
1. organ 2. Trichinosis 3. arthropod 4. Insects 5. Larva 6. eyes 7. Insects 8. male and female 9. animal 10. organs

D. How and Why
1. They would die also because they live off the host.
2. Answers will vary. Circulatory—moves blood through the animal; digestive—breaks down food into simpler substances that cells can use; respiratory—breathing, getting oxygen; excretory—removes wastes; nervous—sends messages to all parts of the body, coordinates movements; reproductive—producing offspring
3. See Figure 16-19, page 356.
4. egg, larva, pupa, adults
5. Answers will vary. They lay many eggs; they are adaptable to adverse conditions.

17 Animals with Backbones

Overview
Chapter 17 continues the study of animals that was begun in Chapter 16. In the first section, vertebrates are defined as animals with internal skeletons. Fish, amphibians, reptiles, birds, and mammals are all vertebrates that make up the Phylum Chordata. The features of each of these groups of animals is described and examples of each group are given. Emphasis is placed on the increasing complexity of animal life among the vertebrates and to the body structures and functions that keep each kind of vertebrate animal alive. One section is devoted to the frog as a representative amphibian and includes an activity on the dissection of a frog. The last section in the chapter covers the mammals, the most advanced animals, and explains why people are classified as mammals.

Performance Objectives
Goal: Students will learn the characteristics of the species classified in the chordate animal phylum.
Objectives: Upon completion of the reading and activities and when asked to diagram, demonstrate, or respond either orally or on a written test, students will:

17:1	K	explain how a vertebrate is different from an invertebrate.
	K	list the functions of an internal skeleton.
17:2	K	describe the main features of a fish.

	K	explain how the skeleton of a shark is different from a bony fish skeleton.
	K/P	explain how a fish obtains oxygen, digests food, and circulates material through its body.
17:3	K	describe the main features of an amphibian.
17:4	K	explain how a frog obtains oxygen and circulates materials through its body.
	K/P	list the parts of a frog's digestive tract and state the function of each part.
	K	describe the nervous system of a frog.
17:5	K	describe the major features of reptiles.
	K	explain how a reptile is different from an amphibian.
17:6	K	describe the major features of birds.
17:7	K	describe the major features of mammals.
	K	list examples of different kinds of mammals.
	K/P	explain how mammals are more complex than other species of animals.
	K	explain the process of reproduction in mammals.

Equipment for Activities

air pump and filter
aquarium tank
aquarium with variety of fish
beaker
chicken, internal body parts removed when the chicken is "cleaned"
cotton
dissecting needle
dissecting pan or soft wooden board
dropper
frog, preserved
goldfish
lettuce, chopped spinach, or cooked egg yolk
microscope
microscope slide
pan, flat
paper towels
petri dish
rubber band
scalpel or small, sharp knife
scissors, pointed—small and sharp
straight pins
tadpoles
tweezers
water

17:1 Backbones and Skeletons

Vertebrates are defined and examples are given as background for the study of specific animals in the sections that follow. Animals studied in this chapter are classified as chordates. A field trip to a zoo offers an excellent study of this phylum because most animals maintained in a zoo are vertebrates.

Making Sure, page 364
1. vertebrates—backbone, internal skeleton
   invertebrates—no backbone, soft bodies, exoskeleton
2. fish—backbone; jellyfish—no backbone

17:2 Fish

A classroom aquarium tank is a good resource for the study of fish. Students with experience in setting up and maintaining an aquarium at home may volunteer for this assignment in the science classroom. Complete aquarium kits may be purchased locally. Ideally, the tank should be 40 to 60 liters in size. Keep a glass or plastic cover on the tank to reduce evaporation, but leave a space at each end for air to enter. Remind students that porpoises, dolphins, and whales are mammals and not fish.

Activity 17-1

In this activity the student observes the circulation of blood in the tail of a fish. The water the fish is placed in should be "aged," that is allowed to stand exposed to the air for 2-3 days before use. During the activity, instruct students to be certain to keep the goldfish moist. You may prefer to do the activity as a demonstration.
Materials: goldfish, beaker, cotton, dropper, petri dish, microscope, microscope slide, aged tap water (or aquarium water), rubber band (2)
Questions: Yes, the blood flows in blood vessels (in this case, capillaries). Yes, the blood always flows in the same direction.

Activity 17-2

In this activity the student studies a fish in an aquarium and makes a labeled sketch of the fish.

This activity may be done as an out of class assignment for students who have an aquarium at home.
Materials: aquarium with a variety of fish
Questions: The drawings of fish may differ in the size and shape of the fish and the size, shape, and location of body parts such as eyes and fins. The colors of the fish may also be different.

Making Sure, page 368
3. fins, gills, scales, internal skeleton
4. from water through gills
5. Blood carries food and oxygen to the cells.
6. fins

17:3 Amphibians

You may wish to set up a vivarium to house amphibians in the classroom. Use an old aquarium tank or a very large, wide mouth jar. Cover the bottom with a layer of pebbles and a little charcoal. Add a layer of soil rich in humus (potting soil). Place a pan of water at one end and put sand around it to make a beach. Plant ferns and mosses in the soil and place a small rock in the water. Do not keep the vivarium in direct sunlight. Water newts are the easiest salamanders to raise. Feed the newts daphnia, tubifex, or earthworms. Remove food that is not eaten within an hour.

17:4 The Frog

Study of the frog in this section is intended to illustrate a specific example of the body parts and life activities of an amphibian. Further, many of a frog's organs are similar to a human's organs. Therefore, by studying frogs students may learn more about the human body.

Activity 17-3
In this activity the student dissects a frog and studies the external and internal parts. You may wish to prepare a diagram of the internal organs of a frog and have students label the diagram as they do the dissection.
Materials: dissecting needle, dissecting pan or soft wooden board, frog (preserved), paper towel, pointed scissors (small and sharp), scalpel or other small, sharp knife, 6 straight pins, tweezers
Questions and Conclusions: **1.** skeleton, muscles, skin **2.** Answers will vary. **3.** Organs in the digestive tract are the mouth, tongue, pharynx, stomach, small intestine, large intestine or cloaca, and anus. The liver, gall bladder, and pancreas are part of the digestive system since digestive juices move from these organs into the digestive tract.

Activity 17-4
In this activity the student studies the development of tadpoles into adult frogs. Do this activity in the spring when tadpoles may be obtained from pond water. Keep the tadpoles in an aquarium. You may want to have student volunteers do the activity as a project.
Materials: tadpoles, aquarium tank, air pump and filter, chopped lettuce or spinach or cooked egg yolk

Making Sure, page 373
7. One life stage has gills and lives in water.
8. Frog—female deposits eggs in water which are fertilized by sperm from males. Fertilized egg develops into tadpole.
9. They hibernate in mud at the bottom of a pond.
10. Body temperature changes as the temperature around it changes.
11. hemoglobin

17:5 Reptiles

If you take your class on a field trip to a zoo, plan to visit the reptile house. Advance arrangements may be made for the reptile curator to talk to the class about the reptile collection. Arrangements may also be made for the students to observe at feeding time. However, be aware that some students may be sensitive about watching the feeding. Make participation optional. Be certain students understand the differences between reptiles and amphibians.

Making Sure, page 375
12. A reptile has scales and does not have a stage that lives in water.
13. Predators eat most of the eggs and offspring.
14. The dry scales on the skin resist water loss and protect the animal from rough surfaces.
15. Eggs develop into young inside the female.

17:6 Birds

Birds are warm-blooded in contrast to reptiles, amphibians, and fish, which are cold-blooded. Birds rank high in student interest and some students may be able to discuss experiences in raising canaries, finches, or parakeets, or experiences with backyard bird feeders. You may want to place a bird feeder outside a classroom window to attract birds for observation. Pigeons, sparrows, and starlings are common birds in urban areas that students can observe to study the behavior of birds. Binoculars are an important piece of equipment for bird watching.

Activity 17–5
In this activity the student examines the internal body parts of a chicken. Have students work in small groups. You may wish to have them label each part with a pin.
Materials: Internal body parts of a chicken removed when it is "cleaned" by a butcher, a dissecting pan or flat pan, cold water
Question: The gizzard is called the "hen's teeth" because it contains small stones and grinds up the chicken's food.

Making Sure, page 377
16. They can react quickly to escape predators and to catch food. They are colorful and many species fly in the air where they can be seen.
17. This structure decreases density and makes flight possible.
18. Bright color of males attracts females for mating.
19. The call enables one member of a species to locate another member of the same species.

17:7 Mammals
Discuss the features of mammals that make them the most advanced animal life. Mammals have the most developed brains and exhibit the most complex patterns of behavior. The fertilization and initial development of mammals occurs inside the female, and mammals provide the most care for their offspring. Remind students that people are mammals. People have the traits common to mammals as a group: hair on the body, feed young on milk, internal fertilization, warm-blooded.

Making Sure, page 379
20. They have the largest brains and the most complex bodies.

Perspectives
Why Rattlesnakes Have Rattles
Pose a similar query for students to hypothesize about. For example, why do some roses have thorns? You may choose to have students devise their own questions. Hypotheses will vary and be based on available resource material, student observations, and personal experience. Accept all reasonable hypotheses.

Study Questions
A. True or False
1. False (The fertilization of eggs in a reptile occurs inside the female's body.) 2. True 3. False (A fish's heart has two chambers.) 4. True 5. True 6. False (Tadpoles have gills.) 7. True 8. False (Reptiles are cold-blooded animals.) 9. False (Most reptiles have four legs.) 10. True

B. Multiple Choice
1. crayfish 2. lungs 3. Capillaries 4. amphibian 5. intestine 6. chordate 7. four 8. Mammals 9. Mammals 10. inside

C. Completion
1. bone 2. Muscles 3. Capillaries 4. mammals 5. brain 6. reptile 7. Feathers 8. stars 9. chicken (Answers will vary.) 10. predator

D. How and Why
1. vertebrates—backbone, internal skeleton; invertebrates—no backbone, soft bodies, exoskeleton
2. fish—breathe through gills, live in water, fins for swimming, two-chambered heart, scales; amphibians—live part of life in water, part on land, adult breathes through lungs, moist skin, three-chambered heart
3. Answers may include digestive—breaks down food into substances cells can use; circulatory—movement of blood; nervous—sends messages throughout body; respiratory—getting oxygen, expelling CO_2; excretory—getting rid of cell wastes; reproductive—producing offspring
4. Birds are identified by their markings, sometimes distinctive shapes or calls.
5. Answers will vary. All mammals are warm-blooded vertebrates, have hair, feed their young with milk, provide more care for their young than other animals, and are the most advanced of all animals.

Side Roads
Reading the Environment
Lead a discussion in which students give examples of how they use their senses to receive information from their environment.

Unit 5 Environment and Heredity

Overview
Unit 5 covers the basic concepts of ecology and genetics. In Chapter 18, the concept of a natural community is defined and explained. The cooperative and competitive roles of the populations within communities are contrasted. In Chapter 19, the roles of producers, consumers, and decomposers in communities are explained. Energy pyramids are used to illustrate the concepts of energy flow in the environment. Chapter 20 introduces heredity, explaining cell division (mitosis) and reduction division (meiosis) as the mechanisms of inheritance. Use of genetic knowledge for selective breeding is explored. Chapter 21 discusses scientific research concerning the origin of living things and the concepts of mutation and natural selection.

Teaching Tips
Collect old magazines with pictures of landscapes, seascapes, plants, and animals for bulletin boards and activities throughout the unit. Ask for student volunteers to make bulletin boards for use with each chapter. Suggestions are: Chapter 18—communities, populations; Chapter 19—food chains, energy pyramids; Chapter 20—animal families, animal and plant breeds; Chapter 21—theories about the origin of living things.

Contact people in careers related to ecology or genetics and invite them to speak to your class. Or let students choose careers of interest and arrange to interview someone employed in that career and report to the class.

---

18 Populations and Communities

Overview
Chapter 18 begins with a definition of a community and goes on to a discussion of populations within a community. Factors that affect the size of a population and changes in the kinds of populations in a community are discussed. Elements of competition and cooperation in populations are detailed. The last section in the chapter defines ecology, gives examples of ecological principles, and explains the importance of this subject.

Performance Objectives:
Goal: Students will learn about the interrelationships among living things and their environments.

Objectives: Upon completion of the reading and activities and when asked to diagram, demonstrate, or respond either orally or on a written test, students will:

18:1	K		define community and describe examples of communities.
18:2	K		define population and describe examples of populations.
	K		explain the meaning of the term dominant species.
	K		define habitat and give examples of habitats for different species.
	K		define environment.
	P/S		study life in a community.
18:3	K		list the factors that affect the size of a population.
	K/P		explain how the size of a population changes over time.
18:4	K		explain how organisms within a community compete for survival.
	P/S		do an activity to study the effects of competition on the size of a population.
18:5	K		explain how succession occurs in a community.
	K		define climax community.
	P/S		prepare a hay infusion culture and study succession of populations within the culture.
18:6	K		define the term boundary community.
18:7	K		explain how species of animals cooperate for survival.
	K		define the term society and describe examples of societies.
18:8	K		define the term ecology.
	K/P		explain the changes that occur in the oxygen-carbon dioxide cycle.
	K/P		explain the changes that occur in the nitrogen cycle.

Equipment for Activities

aquarium with variety of populations
beaker, 50 mL
Bunsen burner
cover slip
dishpan, plastic
dropper
hand lens
hay or dried grass
hot plate or lab burner and ring stand
jar
meter stick
microscope
microscope slide
pan, 2 L
pond water
rice, boiled
shovel
spoon
string
sugar cube
terrarium with variety of populations
yeast, dry
water
wooden stakes, 4

18:1 Communities

Introduce this discussion with a description of the local community of people in which your school is located. Then identify living things, other than people, within the community; for example, species of trees, pigeons, sparrows, rats, mice, and cockroaches. Compare this community to a forest community and the species of living things that live there.

Making Sure, page 390
1. A pond is smaller in size and the water is not moving as in a river or stream. A pond will have different species of fish.
2. Where organisms go, what they eat, how they move, and how they relate to other organisms can be learned by studying organisms in their communities.

18:2 Populations

Introduce this section by writing on the chalkboard the four main vocabulary words in this section: population, dominant species, habitat, and environment. Have students copy these terms in their notebooks, look up the definitions in the text, and write the definitions in their notebooks. If there is an aquarium and/or terrarium in the classroom, have students do Activity 18–1, working in small groups.

Activity 18–1
In this activity the student examines the organisms that live in a sample of soil.
Materials: hand lens, meter stick, plastic dishpan, shovel, string, 4 wooden stakes
Questions and Conclusions: **1.** Answers will vary. **2.** Answers will vary. **3.** microorganisms **4.** break down organic material into compounds plants can use **5.** provide food **6.** sun

Activity 18–2
In this activity the student compares the populations living in two different communities.
Materials: aquarium, terrarium (each with a variety of populations)
Questions: Answers will vary. The aquarium contains gravel, water, plants, fish, snails, algae. (A terrarium contains rocks, soil, insects, plants, microorganisms.) The aquarium has water and not air; it contains fish and other aquatic life. Both the room and the aquarium contain living things. (A terrarium has soil and many populations that the room does not have.)

Making Sure, page 394
3. cockroaches—dark places near sinks and food cabinets in homes, bakeries, and restaurants; rats—walls of buildings, basements, attics, restaurants, bakeries, markets; pigeons—ledges and under eaves on buildings, rooftops; All live near people and their food supplies. Cockroaches and rats live in buildings. Cockroaches live in rooms that people inhabit. Pigeons live outside buildings.

18:3 Population Change

Go over the factors that affect the size of a population. Have students distinguish between immigration and emigration. Point out that cyclical increases in the population of species, such as foxes and rabbits, are natural. Have students review the photos and captions in this section and relate these to the factors that affect population size.

Making Sure, page 397
4. amount of food, poisoning by people, presence of predators such as cats, disease
5. food supply, disease
6. Decrease in mice reduces the food supply for the natural enemies.
7. It causes the population to increase and decrease. For example, more insects are present in the summer than in winter.

18:4 Competition

Competition among living organisms results in survival of the fittest. Discuss this idea with your students and point out that the fittest are those organisms best adapted to their environment. The fit are not necessarily the biggest or strongest. Adapted to the environment means the needs of the organism are met by that which is available in the organism's habitat.

Activity 18-3

In this activity the student prepares a yeast culture and observes the effects of competition within the culture. Yeast need water and sugar to survive. Adding sugar to a mixure of yeast and water stimulates the growth of yeast. As the yeast grow, there is competition for food. Crowding, decrease in the amount of food, and wastes produced by the yeast have a negative effect and slow the rate of growth until it eventually stops.
Materials: beaker (50 mL), dropper, dry yeast, microscope, slide and coverslip, spoon, sugar cube, water
Questions and Conclusions: **1.** sugar, water **2.** increases over time **3.** Answers will vary. **4.** food—increases; crowding and waste products—decreases

Making Sure, page 400
8. It reduces the competition for resources in an area.
9. decreases the number

18:5 The Climax Community

You may wish to have students draw a sequence of pictures in their notebooks that show succession. Suggest that the first picture be a pond, lawn, or forest after a fire, and then draw the stages in succession of the area to a climax community.

Activity 18-4

In this activity the student prepares a hay infusion culture and observes the succession changes that occur in the culture. Any stagnant water may be used instead of pond water. Demonstrate how to remove water from the culture with a dropper, and how to prepare a slide with a drop of the water.
Materials: hay or dried grass, dropper, jar, boiled rice, pond water, microscope, slide and coverslip, Bunsen burner
Question: Different species of protozoa will appear in the culture.

Making Sure, page 401
10. The climax community is suited to the climate.
11. When the climate changes, the community changes.
12. Dominant animal species change during succession.
 grass—shrubs and weeds—trees

18:6 Boundary Communities

A beach, the edge of a forest, and the shore of a lake are examples of boundary communities. Have students make a bulletin board display using pictures of animals unique to each community and common to the boundary community for one of these examples.

18:7 Cooperation in Populations

Compare cooperation in a society to the division of labor and cooperation of workers in a factory that makes shoes, cars, or other products. You may wish to give students a library assignment on life in a bee hive or ant farm and have students do a short report. Stress the idea that by contributing to the survival of its society, the individual aids its own survival.

Making Sure, page 403
13. queen—no new bees produced
 drones—queen's eggs not fertilized
 worker—food (honey) not produced

18:8 Ecology

This section summarizes the ideas of the chapter and explains why the study of ecological relationships is important. Remind students that people are living organisms whose lives, like the lives of all living organisms, depend on their environment. Ecology is the scientific study of the environment and an ecologist is a scientist who specializes in ecology. The oxygen-carbon dioxide cycle and the nitrogen cycle described in this section are essential to the life of people and other organisms.

Making Sure, page 405
14. The bodies of dead plants and animals would litter the environment. The minerals and nutrients stored in the bodies of the organisms would not be naturally recycled and might eventually be depleted.
15. Carbon, in the form of carbon dioxide, is used by plants to make food. Animals use the plants

for food, thus taking the carbon into their bodies. Some of the carbon is given off from the animal's body as carbon dioxide.

Perspectives

Devotion to Wildlife

Locate a professional in your area whose work is in wildlife preservation. Invite the individual to visit your class for a presentation. If it can be arranged, you may prefer to take students to the zoo or wildlife preserve for a guided tour and presentation. Students should come away with career information and a better understanding of the necessity for preserving wildlife.

Study Questions

A. True or False
1. True   2. False (The environment of an organism includes the living and nonliving things around it.)   3. True   4. True   5. False (Food supply is only one factor that affects population size.)   6. True   7. False (The population of a species tends to increase when there is a large food supply.)   8. False (Succession usually occurs over a long period of time.)   9. True   10. True

B. Multiple Choice
1. population   2. birth and immigration   3. succession   4. decreases   5. community   6. a decrease   7. succession   8. decrease   9. 30% to 90%   10. boundary

C. Completion
1. population   2. society   3. foxes   4. forest   5. succession   6. carbon dioxide   7. lack of food, predators, emigration, disease   8. succession   9. climax   10. ecologists

D. How and Why
1. A decrease in the population of one species could cause a decrease in the population of another dependent species, such as a predator population. A decrease could also cause an increase in the population of a competing species.
2. Through continuous succession, there is a change in dominant species. Eventually, the species most adapted to the environment survives in the largest numbers and becomes dominant, forming a community that does not change much through time.
3. The numbers of a population increase to a maximum, then decrease to a minimum then increase to a maximum again.
4. Answers will vary. pond—algae, aquatic plants, fish, frogs, turtles; forest—trees, grass, birds, deer, fox.
5. Answers will vary. Competition may reduce population size. An example is trees competing for sunlight.

19   Food and Energy

Overview

Chapter 19 describes the food relationships between living organisms and the energy needs of organisms. The chapter begins with a discussion of the nature of food chains and the importance of green plants in food chains is explained. Characteristics of producers, consumers, decomposers, scavengers, and predators are described and examples of each are given. The energy flow in a food chain is discussed, and the representation of this flow by an energy pyramid is explained. A section on predators discusses the ecological significance of predation in the natural culling of weak and sick animals. One section is devoted to parasitism and the last section in the chapter presents orchids, decoy fish, and lichens as examples of symbiosis.

Performance Objectives:

Goal: Students will learn the food and energy relationships among organisms that exist in food chains and food webs.

Objectives: Upon completion of the reading and activities and when asked to diagram, demonstrate, or respond either orally or on a written test, students will:

19:1	K	describe how organisms use energy.
	K/P	explain what a food chain is and describe the links in one food chain.
19:2	K	define producer, consumer, decomposer, scavenger.
	P/S	study producers and consumers in pond water.
19:3	K	explain how energy is transferred through a food chain.

	P/S	draw a diagram of an energy pyramid and explain how it represents the energy lost in a food chain.
	P/S	construct a cardboard model of an energy pyramid.
19:4	K	define the term predator and describe examples of predation.
	K	describe the effect of predators on their prey species.
19:5	K	define the term parasite and describe examples of parasites.
19:6	K	define the term symbiosis.
	P/S	prepare a microscope slide of a lichen and observe it with a microscope.

Equipment for Activities

cardboard
coverslip
dissecting needles
dropper
jars, widemouth, with lids, 4
lichen
metric ruler
microscope
microscope slide
old magazines
paste
pen
pond water, 4 samples (surface, under the surface, bottom, near shore)
scissors
slide
tape
water

19:1 Food Chains

Old magazine pictures of organisms can be used for a bulletin board display of a food web. Connect the pictures together with yarn to show the food chains that make up the web. As an enrichment activity, have students make a food chain mobile. Paste old magazine pictures of organisms on 3″ × 5″ cards and attach the cards together with string so they hang vertically in order of their position in the food chain.

Making Sure, page 414
1. Food web consists of more than one food chain.
2. Green plants use the sun's energy to make food. Animals depend on plants for food.
3. Plants make the food on which animals depend.
4. corn-chicken-person. Answers will vary.

19:2 Producers and Consumers

Introduce the definitions of producer, consumer, decomposer, and scavenger as they apply to a food chain. Then have students read the section and answer the guide questions and Making Sure questions.

Activity 19-1
In this activity the student prepares microscope slides and studies producers and consumers in pond water. Demonstrate the procedure for making a slide of pond water. Producers tend to be found in the surface water and decomposers are most plentiful near the bottom.
Materials: dropper, 4 jars (large, widemouth with lids), microscope, 4 samples of pond water (surface, under the surface, bottom, near shore), microscope slide and coverslip
Questions and Conclusions: **1.** producers—Reasons will vary. **2.** Producers make their own food. Consumers eat other organisms. **3.** surface **4.** under surface **5.** eat consumers **6.** bottom **7.** Sketches will vary depending on the organisms observed.

Making Sure, page 416
5. (a) P (b) C (c) P (d) D (e) D (f) C
6. They help clean up the environment and recycle nutrients.

19:3 Energy Pyramids

You may wish to introduce this section by displaying a cardboard model of a pyramid similar to the one described on page 418. Discuss the pyramid model, and relate it to the decreasing energy present in a food chain from the bottom to the top. Remind students that the stored energy is chemical energy stored in food molecules. Organisms in the food chain use this energy for life activities and much of the energy is lost as heat.

Activity 19-2
In this activity the student makes a cardboard model of an energy pyramid for a food chain. Request that students bring to class a used cardboard garment box large enough to cut out the four triangles, 30 cm on each side. Supply old magazines or colored pens and pencils.
Materials: cardboard, ruler, scissors, tape, old magazines, paste.
Questions: The number of producers and consumers will vary depending on diet of students. Producers make the food that contains the energy for all the organisms in the pyramid.

Making Sure, page 418
7. Energy is lost as it is used for life activities.
8. (a) There is no loss of energy in the minnows. All of the energy in the plankton goes to the bass.
9. Green plants trap the sun's energy.

19:4 Predators

Some students may view predation as "cruel." Develop the view that predation is part of nature and is a natural aspect of food webs. Within food webs animals eat and are eaten. These processes are essential to the movement of food and energy through a food chain. Stress the fact that predators hold down the numbers of prey species and thereby prevent the inevitable starvation that would occur if the numbers grew so large they outstripped their food supply.

Making Sure, page 419
10. increase in size
11. decrease in number

19:5 Parasites

Bacteria and other organisms that cause disease are parasites because they live on a host organism in which they produce disease. Review the concept of a life cycle and have students diagram the life cycle of the beef tapeworm described in this section.

Making Sure, page 421
12. It loses its food and habitat.
13. Cook beef thoroughly.
14. A mosquito obtains its food from living organisms.

19:6 Symbiosis

Point out that symbiosis is defined as two organisms living together. Therefore, the predator-prey relationship is an example of symbiosis. Point out the additional symbiotic relationships discussed in the section and emphasize how they are similar and different from each other.

Activity 19-3

In this activity the student prepares a slide containing a dissected piece of lichen. Review the preparation of a wet mount and the use of a microscope before students begin work. Students will see that the algae cells are colored and the fungi lack color. The fungi cells are threadlike.

Materials: lichen, microscope slide, coverslip, dissecting needles, dropper, water
Question: Algae are green.

Making Sure, page 423
15. Symbiosis is two organisms living together. Parasitism is one organism living off another.
16. Alga makes food. Fungus supplies water and minerals.
17. Alga—producer; fungus—consumer

Perspectives

A World Without Sunlight
Pursue the topic of the deep sea communities that exist without sunlight. Provide students with reading materials and a film about the expedition if possible. Emphasize the fact that scientists actually observed heretofore unnamed species.

Study Questions

A. True or False
1. False (Food energy is lost at each level of an energy pyramid.) 2. True 3. True 4. True
5. False (Both organisms sometimes benefit from their relationship in symbiosis.) 6. False (Few parasites kill their hosts.) 7. True 8. False (A parasite is not a predator.) 9. True 10. True

B. Multiple Choice
1. decomposers 2. increases 3. producers 4. energy pyramid 5. predator 6. parasite 7. Producers 8. algae on fungi 9. second 10. human intestine

C. Completion
1. Decomposers 2. symbiosis 3. energy 4. predators 5. scavenger 6. food chain 7. producer 8. decomposer 9. producer 10. scavengers

D. How and Why
1. A food web is a complex feeding system containing more than one chain.
2. Both feed on dead organisms. Scavengers are animals that feed on other animals. Decomposers are microbes that decay both animals and plants.
3. Answers will vary. grain-mouse-cat
4. Answers will vary. algae on fungi sea anemone-decoy fish
5. They live off dead organisms and change them into compounds that return to the air and soil.

20 Heredity

Overview

Chapter 20 covers the basic principles of heredity and their application to plant and animal breeding. The concept of heredity is introduced and the history of genetics is discussed. Mendel's laws are explained. Chromosomes and genes are described as the physical basis of inherited traits, and their effects are described. The mechanisms of reduction division and fertilization and their role in maintaining the species number of chromosomes is explained. Determination of sex, mutations, and the use of selection and crossing in developing breeds of plants and animals are also presented in the chapter.

Performance Objectives:

Goal: Students will learn how inherited traits are passed from parents to offspring.

Objectives: Upon completion of the reading and activities and when asked to diagram, demonstrate, or respond either orally or on a written test, students will:

20:1	K/P	define heredity and describe how scientists study the inheritance of traits.
20:2	K	describe Mendel's study of heredity in pea plants.
	K/P	explain how Mendel developed the theory of dominance.
	P/S	observe and record the traits of fruit flies in an experiment.
20:3	K/P	explain how traits may skip a generation.
	P/S	use coin tossing to explain probability or chance in the inheritance of traits.
20:4	K	define blending as incomplete dominance and describe examples of blending.
	P/S	grow hybrid soybean seeds to investigate blending in soybeans.
20:5	K	describe the features of chromosomes and tell where they are located.
	K/P	explain how genes control the inheritance of traits.
	K	describe the relationship between genes and DNA.
20:6	K	explain how a new organism obtains its genes.
	K	describe reduction division and fertilization.
	P/S	observe and identify cell division and reduction division in a prepared slide of onion root tip, whitefish eggs, and/or animal testis.
20:7	K	describe the sex chromosomes of a male and female animal.
	K	explain why sex is determined by the sperm cell in animals.
20:8	K	explain how a zygote is formed.
	K	describe the early stages in the development of an embryo.
	K/P	describe the difference between identical and fraternal twins.
20:9	K	define mutation and explain how a mutation is produced.
	K	list three causes of mutations.
	P/S	do an activity in which mutant tobacco seeds are grown and the percent of albinos is calculated.
20:10	K	define selection, breed, purebred, crossbred, and hybrid vigor.
	K/P	explain how new varieties and breeds of plants and animals are developed.
	K	describe how mutations have been used to develop new breeds of plants.

Equipment for Activities

baby food jars with lids, 2	hand lens
cloth, dark color	masking tape
coins, 2	microscope
container of water (small)	nail
Drosophila fruit fly culture	paper towels
flower pot, large, or growing flat	prepared slides of onion root tip, whitefish eggs, animal testis
glass baking dish	soybean seeds, hybrid, 20
hammer	straight pin
	tobacco seeds, 60
	topsoil or sand

20:1 Inherited Traits

Use discretion in references to the inheritance of human traits. Most human traits, including eye color, are polygenetic, meaning they are influenced by two or more pairs of genes. Personal references to inherited traits in students should be made cautiously since some children may not live with their natural parents. In addition to *Drosophila,* much has been learned about the biochemistry of genetics through the study of heredity in mold and other microorganisms.

Making Sure, page 430
1. They are expensive to raise and have a long time (12 years) between generations.
2. They are cheap to raise and have a short time (10 days) between generations.

20:2 Law of Dominance
You may wish to assign your students a library report on the life of Gregor Mendel and his study of pea plants. Use a Punnett square and letters to represent dominant tall (TT or Ts) and recessive short (ss). Show students how to diagram a cross between pure tall (TT) and pure short (ss) pea plants.

Activity 20-1
In this activity the student crossbreeds fruit flies and observes the traits in the parents and offspring. Review the materials and procedure carefully before students begin work. Then, direct students to copy the Observations and Data tables into their notebooks.
Materials: 2 baby food jars/lids, dark cloth, *Drosophila* fruit fly culture, hammer, hand lens, masking tape, nail, straight pin
Questions and Conclusions: 1. eggs from the parents 2. Answers will vary depending on the cultures used. 3. cheap to raise and a short time between generations

20:3 Crossing Hybrids
Introduce this section by going over the diagrams of crosses provided. Draw a Punnett square on the chalkboard and diagram the cross between hybrid tall pea plants that is shown on Figure 20–4. Do the same for the cross between hybrid yellow seed plants (Figure 20–5). Point out to students that the ratios predicted for traits in offspring are a function of the number of offspring.

Activity 20-2
In this activity the student investigates the probability of a coin landing heads or tails. Relate the concept of probability in coin tossing to the probability of traits appearing in offspring.
Materials: two coins
Questions: 1 chance out of two for either head or tail. one out of 4, 2 out of 4, 1 out of 4; Actual results agree more closely with the probability ratios. The head-tail combination represents the dominant-recessive gene combination.

Making Sure, page 435
3. Two recessive genes must be present for the recessive trait to show up.
4. Both parents are hybrid black and contain a gene for white.

20:4 Blending
Use a Punnett square to diagram on the chalkboard a cross between four o'clocks which shows the production of red, pink, and white flowers. (Cross two hybrid reds (Rr).)

Activity 20-3
In this activity the student plants soybean seeds hybrid for leaf color, observes the leaf colors of the new plants, and calculates the percent of plants having each leaf color.
Materials: soybean seeds hybrid for leaf color, topsoil or sand, large flower pot or garden flat
Question: About 25% of the new plants have green leaves, about 25% have yellow leaves, and about 50% have yellow-green leaves.

Making Sure, page 436
5. Cross the plant with another plant that is pure recessive for the trait. If the trait appears in the offspring, it is dominant. If the recessive trait appears in all the offspring, then the trait is recessive. If the cross results in three traits appearing in the offspring, the trait is caused by blending.

20:5 Chromosomes and Genes
Genetics is the science that is concerned with the study of genes and their function in heredity. Genes (DNA) control the production of enzymes within a cell. These enzymes help regulate the activities of a cell, including the production of cellular materials. You may wish to assign a library report on the latest developments in gene splicing (recombinant DNA). Transplanting genes from one species to another is used for the production of useful products, such as insulin for diabetics.

Making Sure, page 438
6. It contains the genes that control inherited traits.

20:6 Reduction Division and Fertilization

Diagram the steps in reduction division in a male and female fruit fly (8 chromosomes) on the chalkboard. Then show how a sperm and egg unite togther and restore the chromosome number for the species. Remind students that in animals the production of sex cells, including reduction division, occurs in the male testes and female ovaries. In seed plants it occurs inside the pollen and ovules formed in flowers.

Activity 20–4

In this activity the student studies prepared microscope slides of plant and/or animal tissues in which body cell division and reduction division can be observed. The onion root tip is a rapidly growing tissue in which cell division (mitosis) can be observed. Reduction division (meiosis) can be observed in animal testis tissue.
Materials: microscope, prepared slides of onion root tip, whitefish eggs, and/or animal testis

Making Sure, page 441
7. (a) ten in each
   (b) Chromosomes in the pollen and egg nuclei are combined together in the egg.

20:7 Sex Determination

Review the explanation of sex determination in animals having XX and XY sex chromosomes. Use chalkboard diagrams to show how an X-bearing sperm always results in a female being produced at fertilization and Y-bearing sperm results in conception of a male.

Making Sure, page 442
8. XY—males; XX—females
9. 1:2, 1:2
10. sperm cell

20:8 Reproduction

Section 20:8 continues the development of the concept of genetic continuity introduced in the earlier sections of this chapter. The new organism receives its species number of chromosomes at fertilization and this number of chromosomes is replicated over and over again in each cell division that forms the embryo.

Making Sure, page 443
11. identical—Identical twins are alike in inheritance. Differences are due to the effects of environment.

20:9 Mutations

Remind students that genes are made of DNA and that DNA is a chemical compound. A change in the chemistry of DNA causes it to have different properties and different effects on cell activities. Carcinogens such as X rays, nuclear radiation, and certain chemicals cause changes in the chemistry of the DNA in genes. Remind students that most mutations are harmful because they make an organism less adapted to its environment.

Activity 20–5

In this activity students germinate irradiated tobacco seeds and observe the resulting percent of albino plants. Be certain to keep the paper towels moist and the room temperature fairly warm.
Materials: 60 irradiated tobacco seeds, paper towels, glass baking dish
Questions: Answers will vary. No, they do not have chlorophyll needed to make food.

Making Sure, page 445
12. A white animal would stand out and not be camouflaged. Thus, it might be easy prey.

20:10 Plant and Animal Breeding

Introduce this section by showing the wide variety among the breeds of animals such as dogs, cats, horses, or cattle. Charts showing the different breeds are available from animal feed and veterinary supply companies. Purebred and crossbred are presented and related to hybrid vigor. Various plants are produced from corms, bulbs, runners, and other asexual methods to insure predictable offspring rather than by sexual reproduction where the offspring's traits are less carefully controlled.

Study Questions

A. True or False
1. False (Tallness in pea plants is an example of dominance. Answers will vary.) 2. False (The law

of dominance explains the inheritance of many traits.) **3.** True **4.** True **5.** True **6.** False (The body cells and sex cells of a species have a different number of chromosomes.) **7.** True **8.** True **9.** True **10.** False (A purebred animal or plant has less hybrid vigor than a crossbred animal or plant.)

B. Multiple Choice
1. crossbred **2.** Blending **3.** Purebred animals **4.** Hybrid corn **5.** Fruit flies **6.** tall **7.** zygote **8.** 1 of 4 **9.** remains the same **10.** half

C. Completion
1. dominance **2.** blending **3.** female **4.** genes **5.** identical **6.** chromosomes **7.** mutation **8.** selection **9.** recessive **10.** seedless grape (Answers will vary.)

D. How and Why
1. Genes carrying recessive traits may be passed from one generation to another without appearing because they are masked by dominant traits. The recessive trait sometimes appears in a later generation.
2. Identical twins come from the same fertilized egg. Fraternal twins come from two fertilized eggs from the same mother. The genetic composition of different egg cells varies.
3. The hybrid organism is usually hardier and more disease resistant.
4. Answers will vary. For example, consider developing an apple without seeds: (1) Search for a seedless mutant. (2) Graft branch on which it grew to an apple tree stock. (3) Continue to propagate more seedless trees by grafting.
5. The male rabbit has two different sex chromosomes, an X and a Y. The female rabbit has only one kind of sex chromosome, an X. If the egg is fertilized by an X-carrying sperm, it will be female. A Y-carrying sperm will produce a male.

21 Descent and Change

Overview
Chapter 21 discusses the changes in organisms that fossil evidence indicates has occurred over many millions of years of time. The first section reviews a theory which postulates a scientific explanation for the origin of life on earth. The work of Charles Darwin and Alfred Wallace, which led to a theory of evolution, is described. Evidence for Darwin's theory is presented and related to the concept of natural selection. In Section 21:6, the effect of mutations on the survival of organisms and species is discussed.

Performance Objectives:
Goal: Students will study the theories that scientists have proposed to explain the origin of life and the changes in species.

Objectives: Upon completion of the reading and activities and when asked to diagram, demonstrate, or respond either orally or on a written test, students will:

21:1	K	state a scientific theory which explains how life may have begun on earth.
	K/P	describe an experiment which supports the scientific theory.
21:2	K/P	explain the methods Darwin used to obtain evidence for his theory of evolution.
	K	state Darwin's theory of evolution by natural selection.
	P/S	study the number of seeds produced by a pine tree or a milkweed pod and relate the data to the survival of the species.
21:3	K	describe examples of fossil evidence that indicate how some species have changed over time.
21:4	K	explain why natural selection of organisms occurs.
	K	describe evidence that supports the concept of natural selection.
21:5	K	explain the meaning of descent and change in species.
	K	list the main points of Darwin's theory.
21:6	K	explain how mutations affect the survival of organisms.
	K	describe the causes of mutations.
	K/P	explain how mutations can cause changes in species.

Equipment for Activities
milkweed plant with pods
pine tree with cones

21:1 Origin of Living Things

This section presents some of the theories scientists have proposed to explain the origin of life on earth. The possible formation of organic compounds from inorganic material is discussed and evidence for this idea is presented. You may wish to point out to your students that science is concerned with developing and testing theories on *how* life began and developed on earth. Science cannot answer the question of *why* life began and developed or whether or not a divine being or influence was responsible for the origin and development of life.

Making Sure, page 452
1. They were made from gases when an electric discharge passed through the gases.
2. Land was bare rock and not suited to life. Seawater contained nutrients needed to support life.

21:2 Darwin's Theory

This section provides a good example of the process of science. It shows how a scientist, Charles Darwin, gathered evidence over many years and used it to develop a scientific theory. Darwin was educated in theology and planned on becoming a preacher. He started out as an amateur naturalist whose skills and abilities were of such high quality that they were recognized by professional scientists who recommended him to the British Admiralty for the position of naturalist on the *Beagle*. Recommended Reading: Stone, Irving. *The Origin.* Garden City, N.Y.: Doubleday, 1980.

Activity 21-1

In this activity the student estimates the number of seeds produced by a pine tree and estimates the number of seeds produced by a milkweed plant. Any other plant may be substituted for the pine tree or milkweed plant provided the total number of seeds produced by the plant can be estimated. Relate the observations to the general overproduction of offspring by all species and the survival of those offspring that are most fit; that is, suited to the environment.
Materials: pine tree with cones, milkweed plant with pods
Question: At least a few seeds will land where they can grow.

Making Sure, page 455
3. long hair, keep animals warm
4. Artificial selection is used to develop new varieties. According to Darwin's theory natural selection causes species to change.

21:3 Fossil Evidence

Introduce this section by reviewing the definition of fossils and how fossils are formed. Exhibit and discuss fossil specimens from a school collection or those brought to school by students. Then explain how the fossil evidence is used by scientists to support the theory of evolution. You may wish to have students do library research on the fossil evidence showing changes in the human species over the past million years.

21:4 Natural Selection

Have students read the descriptions of the hay-pasture experiment and write up the problem, hypothesis, procedure, observations, and conclusion for the experiment. Then discuss the other examples of natural selection in this section. Remind students that the organisms with the greatest chance of survival are not necessarily the biggest or strongest, but those whose traits make them most adapted to their environment. Discuss characteristics of organisms that are most likely to survive if the environment changes. Consider protective coloration, for example.

Making Sure, page 459
5. A rat is suited to the environment in a large city. It eats garbage and reproduces at a rapid rate. There are many secluded places to hide from enemies.
6. A housefly is adapted to conditions under which many people live.

21:5 Changes in Species

A species is a group of closely related organisms that can produce fertile offspring. A species does not naturally mate with another species and produce offspring. For a new species to develop, the changes in organisms must be great enough to produce a distinct group (species) that does not mate with another species. Review the main points in Darwin's theory listed on page 461 and have students give examples of each point.

Making Sure, page 462
7. More light-colored moths appear. As pollution decreases the trees become lighter in

color and the light-colored moths are less visible against the bark.
8. Through many generations it has increased in size and lost all but one toe on each foot.
9. Climate changes and the species is not suited to the new climate.
10. whether or not it agrees with known facts, whether or not the theory can be used to predict events

21:6 Mutations and Change

Introduce this section by reviewing the definition of mutation and examples of mutations. Darwin and other scientists of his time were not aware of mutations, therefore they could not explain the cause of the changes in species that were theorized. Mutations create the changes that make organisms suited or unsuited to their environment and provide the traits that may make organisms suited to a changed environment.

Making Sure, page 463
11. Mutations cause changes that produce the traits of new species.

Advances in Science
Evolutionary Clues

Review with students the connection between mutations and natural selection with evolution. Evidence that evolution occurs is further supported by other lines of scientific inquiry. The evidence becomes even more meaningful considering that there are so many different lines of inquiry and that the lines tend to support one another.

Perspectives
Improving a Theory

Set up a fossil display in your classroom. Students may have samples of their own to contribute. If possible, take students on a field trip to a museum where they can study and compare fossils.

Study Questions
A. True or False
1. False (Charles Darwin developed his theory of evolution 20 years after his explorations.) 2. False (Darwin's theory of natural selection was published.) 3. True 4. True 5. True 6. True 7. False (Changes in living things over the years have generally been slow and gradual.) 8. False (The early atmosphere of earth is believed to have been different from what it is today.) 9. False (Scientists have found evidence of natural selection.) 10. True

B. Multiple Choice
1. survive 2. changing 3. Amino acids 4. birds 5. sometimes 6. oxygen 7. greater than 8. larger 9. theory 10. changes in genes

C. Completion
1. oceans 2. changed 3. fossils 4. Fossil 5. climate 6. nonliving 7. natural selection, survival of the fittest 8. tall 9. oversupply 10. extinct

D. How and Why
1. Darwin collected and studied various plants, animals, and rocks found along the coast of South America. He became interested in how the many different species of living things originated. Upon returning to England, he studied the breeding of domestic animals and plants. Twenty years later he proposed his theory of natural selection.
2. Living things overproduce offspring. Some living things are more adapted to the environment than others. Those most adapted to the environment survive and reproduce; they pass their characteristics to a new generation of living things.
3. Natural selection has taken place. Bacteria with a resistance to antibiotics have survived and reproduced; bacteria not resistant to antibiotics have died.
4. Polar bear: white protective coloring; rabbit: protective coloration, speed, burrow
5. It could not fly rapidly to escape its enemies.

Side Roads
Biomes

Students can do research and prepare an exhibit or bulletin board showing how climate, plants, and animals have an effect on the economy, types of work, and lifestyles available to the humans living within each biome. Water biomes have not been discussed; however, students may wish to do research on them and prepare a report or bulletin board. Students can create "mini" biomes using terrariums and aquariums.

Unit 6   Conservation

Overview

Unit 6 presents the major issues involved in the conservation of natural resources. Chapter 22 covers soil, water, and air conservation. Chapter 23 covers forest and wildlife conservation. Natural resources are defined, examples are given, and conservation methods are described. The connection between soil conservation and water resources is explained. A section on air pollution discusses the nature of pollution, causes of pollution, and pollution detection. Positive efforts to conserve forests and wildlife through permanent wilderness areas and wildlife refuges, and efforts to restore forests and wildlife are examined.

Teaching Tips

Take a field trip to a local water purification plant or sewage treatment plant. Be certain to obtain the necessary parental and administrator approvals.

Obtain information from the local air pollution control agency pertaining to air pollution control in your area. Request a guest speaker on this subject.

Have students do a library search for magazines that contain conservation topics. Each student should prepare a list of the names of the magazines and a short report on one of them.

---

22   Soil, Water, and Air Conservation

Overview

Chapter 22 begins with a discussion of renewable and nonrenewable resources and the need for conservation of resources. The composition of soil is described and methods for conserving soil are explained. Soil conservation is related to water conservation. Public waste supplies and sewage treatment are discussed. Air pollution is also discussed in terms of pollutants, causes, and methods of control. Conservation of natural resources is presented as a way of insuring a healthy and quality world today and in the future.

Performance Objectives

Goal: Students will learn the methods used to conserve the natural resources of soil, water, and air.

Objectives: Upon completion of the reading and activities and when asked to diagram, demonstrate, or respond either orally or on a written test, students will:

22:1	K	define the terms natural resource and conservation.
	A/K	explain why conservation is important to people.
22:2	K	explain how topsoil is formed and describe the conservation of topsoil.
22:3	K	describe the factors that cause erosion of topsoil.
	P/S	do an activity that compares the water holding capacity of sand, clay, and loam.
22:4	A/K	explain why soil conservation is important to both city and rural people.
	K	list and describe soil conservation methods.
22:5	A/K	explain why water resources are important.
22:6	K	list the three main goals of water conservation.
	K	describe how water becomes polluted.
22:7	K	explain how ocean water may be used as a source of fresh water.
22:8	K	list the major air pollutants.
	K	explain how air pollution can be reduced.
	P/S	carry out an activity to show how particulate air pollution can be detected.

Equipment for Activities

baking dish or pie pan	microscope or hand lens (optional)
beakers	
bean seeds	nail
clay soil	plaster of paris
cotton	sand, 1 L
flowerpots	sprinkling can or bottle
funnels, same sizes	
glass jar and screw cap	subsoil, 1 L
graduated cylinder, 100 mL	tape, cellophane or masking
hammer	
loam, soil	topsoil, 1 L
marking pen or pencil	water
metric ruler	wood block

22:1 Conservation of Natural Resources

A modern industrial nation makes extensive use of resources of all kinds. As the supply of a resource decreases, it may become scarce and its price may increase rapidly. These changes tend to have a negative effect on the development of industries and the increase in jobs needed for the increasing number of people in the world. Proper conservation methods, including recycling, can insure a supply of adequate resources for the future.

22:2 Soil

Soil is the basic resource essential to the life of plants and animals that live on land. Point out that people depend on soil indirectly for many things. For example, most of the foods we eat are grown in soil or derived from animals that eat plants that grow in soil. Wool and cotton clothing, paper, and wooden furniture are other examples of products we use that depend on soil for their development.

Activity 22-1

In this activity the student grows beans in topsoil, sand, and subsoil and observes the growth of the bean plants. Topsoil or potting soil is available in most stores that carry garden supplies. Subsoil can be obtained from construction sites or along highways where a road cut has been made and the subsoil is exposed.
Materials: 12 bean seeds, 3 flowerpots, marking pen or pencil, subsoil, sand, topsoil, sprinkling can and water, sunny location
Questions: topsoil—most, sand—least; Topsoil contains mineral elements available to plants. Also, it retains more moisture than subsoil and sand.

22:3 Soil Erosion and Mineral Loss

Review the definition of soil erosion as the wearing away of topsoil. Point out that the amount of erosion caused by rainwater depends on the amount of water and the speed with which the water runs off the land. In areas in which the soil is frozen during the winter, there is less leaching of minerals than in areas in which the soil is continually penetrated by water throughout the year.

Activity 22-2

In this activity the student observes the amount of water that drains through sand, clay soil, and loam. If clay soil is not available, then use only sand and potting soil. Water passes most readily through the sand and drains most poorly through the clay. Clay soils, such as adobe soil, tend to be too wet during the spring planting season, and sandy soils tend to dry out too fast for favorable plant growth.
Materials: 3 funnels, 3 beakers, cotton, sand, clay soil, loam
Questions: sand; sand

22:4 Soil Conservation

Introduce this section with a discussion of why soil conservation is important to people living in cities and suburbs. You might want to arrange for a speaker from the cooperative extension service to conduct this discussion. Have students make a chart listing soil conservation methods with a short description of each.

Activity 22-3

In this activity the student makes a model of a contoured field with plaster in a pan. You may wish to have some students make a few models the day before class. Then have students in your class take turns using the models to do this activity.
Materials: plaster of paris, small baking dish or pie pan, sprinkling can or bottle, water, metric ruler
Questions: runs off; water stays in; Water trapped in furrows soaks into the ground.

22:5 Water Resources

Have students report on the way in which the local community obtains its water supply and treats its drinking water. As part of the report have students find out how local people are billed for water, either by meter or at a flat rate, and the cost of the water.

22:6 Water Conservation

Relate water conservation methods such as planting grass and trees to soil conservation methods. Have students list and discuss water conservation methods people can use to save water and decrease their water bills.

22:7 Water Sources for the Future

Many fertile soils are in arid regions where there is not sufficient rainfall for agriculture. Irrigation from wells in some areas can provide this water. You may wish to have students report on methods used to obtain fresh water from ocean water.

22:8 Air Pollution

Review the substances that cause air pollution and discuss the properties of each. Review the properties of acids and relate these to acid rain. You may wish to place a piece of concrete, marble, or limestone in a dilute solution of acid and have students observe the effect of the acid. Also, place a hard-boiled egg with the shell removed in a dilute solution of nitric acid and have students observe the effect of nitric acid on the egg-white protein.

Activity 22–4

In this activity the student makes and sets up a simple device used to detect particulate pollution. Remind students to place the jar in a spot where it will not be disturbed. You might suggest labeling the jar with a description of its purpose and the student's name. The place on the jar that has the greatest density of pollution particulates indicates the direction from which most of the pollution arrives. If a microscope is available, the tape can be removed from the jar and placed on a slide for microscopic examination of the particles. A hand lens may be used.

Materials: glass jar and screw cap, block of wood, cellophane or masking tape, nail, hammer

Question: The darkest side of the tape points to the direction.

Making Sure, page 485

1. The two major air pollutants in the United States are carbon monoxide, which is produced by automobile emissions, and sulfur dioxide, which results from the burning of coal.

Perspectives

Using Graphs

Visual presentations of information enhance understanding and improve retention. Stress the importance of referring to graphs, tables, and illustrations as students read their textbooks. Discuss methods of interpretation and the advantages of visual aids. Also, consider the kinds of information best presented with each type of graph.

Bar graphs:
1. homes—about 8500 ha
2. schools and hospitals
3. about 20 500 ha

Line graphs:
1. coincides with heaviest automobile, bus, and truck activity together with buildup of industrial emissions
2. use of cars, buses, trains by people going to work
3. Lowest levels of pollution would probably occur from midnight to 6:00 A.M., since there are very few emissions from automobiles and industry.

Circle graphs:
1. toilet and bath
2. 52 L
3. Flush toilet fewer times; take a shower instead of a bath; wait until washer is fully loaded before washing dishes/clothes, etc.

Study Questions

A. True or False

1. False (Natural resources such as minerals and forests have economic value.) **2.** True **3.** False (It takes about 200-400 years to form 1 cm of topsoil) **4.** False (Topsoil is rich in humus.) **5.** True **6.** True **7.** True **8.** True **9.** False (Most carbon monoxide is produced by automobile engines.) **10.** False (Decreasing the runoff of rainwater on farm fields is good for water conservation.)

B. Multiple Choice

1. Fertilizing **2.** corn **3.** A tree **4.** Subsoil **5.** grass **6.** contour plowing **7.** fertilizing **8.** row crops **9.** smog **10.** natural gas

C. Completion

1. polluted **2.** reservoirs, lakes, rivers **3.** conservation **4.** lime **5.** clover (Answers will vary). **6.** contour **7.** sulfur dioxide **8.** Dams, reservoirs, planting grass and trees, contour plowing, terracing **9.** Distillation **10.** carbon dioxide, sulfur dioxide, monoxide

D. How and Why

1. People who live in cities are dependent on food and other products developed from plants grown in watered topsoil.
2. Running water, wind, and ice cause soil erosion.
3. The rate of erosion is directly proportional to the amount and speed of water runoff. Contour plowing and strip-cropping slow down the speed and amount of water runoff.
4. Rotating crops from year to year allows depleted soil minerals to be replaced by natural processes.
5. Flush the toilet fewer times; take a shower instead of a bath; wait until the washer is fully loaded before washing dishes/clothes, etc.

Sideroads

Using Fires to Grow Forests

Have students prepare a bulletin board showing nature's step by step recovery process from a fire or other natural disaster. You may also want to discuss plant succession and make a bulletin board display about the subject as it relates to natural disasters.

23 Forest and Wildlife Conservation

Overview

Chapter 23 begins with a delineation of the products obtained from forests, and the effects of forest fires on forests. Forest conservation practices are described and explained. The need for conservation of wildlife, and the extinction of species is discussed. Different kinds of wildlife resources are identified, and the methods used in the conservation of wildlife are covered. Specific wildlife conservation methods discussed include wildlife refuges, conservation of migratory birds, and permanent wilderness areas.

Performance Objectives

Goals: Students will learn the methods used in the conservation of forest and wildlife resources.
Objectives: Upon completion of the reading and activities and when asked to diagram, demonstrate, or respond either orally or on a written test, students will:

23:1	K	list products obtained from forests.
	K	define the term watershed.
	P/S	make a model of a watershed and investigate the flow of water in a watershed.
23:2	A/K	explain why forest conservation is important.
23:3	K	define improvement cutting, selective cutting and block cutting and explain how each is related to forest conservation.
23:4	K	define the terms extinct and endangered species.
	K/P	explain how scientists think the passenger pigeon became extinct.
23:5	A/K	describe how wildlife resources benefit people.
	K	explain how game laws and the raising game birds and fish may aid conservation of wildlife.
	P/S	construct and maintain a bird feeder and observe the number and species of birds that visit the feeder.
23:6	K	describe how changes in wildlife habitats and unwise hunting practices reduce wildlife.
23:7	K	define the term wildlife refuge and explain how wildlife refuges aid the conservation of wildlife.
	K	explain how a permanent wilderness area aids wildlife conservation.

Equipment for Activities

aquarium tank
materials to make a bird feeder
metric ruler
plants, small
putty
rock pieces, flat (about 5 cm in length)
sand
sod or moss
sprinkling can
wild bird seed

23:1 Forest Resources

You may wish to have students obtain a tree identification guide and have them learn to identify ten local tree species. Discuss students' experiences in visiting forests and their attitudes towards planting and caring for trees. If there is a nearby watershed that drains into a river or lake, call attention to the location and geography of the watershed.

Activity 23-1

In this activity the student makes a model of a watershed inside an aquarium tank. You may wish to have a team of students make this model for the class and have all students make a diagram of the model in their notebooks. Plastic plants and a piece of synthetic outdoor grasslike carpet may be used in place of real plants and moss. Punch holes through the backing of the carpet to allow water to drain through it.

Materials: aquarium tank, metric ruler, pieces of flat rock about 5 cm in length, small plants, putty, sand, sod or moss, sprinkling can

Questions and Conclusions: **1.** The water drained slowly into the sand. **2.** The water drained less slowly or not at all. **3** Streams may be formed by overflow from a lake. **4.** Lakes may be formed when water remains in a depression which is saturated with water. **5.** The water level will decrease and the lake may dry up. **6.** The area of the watershed is all the area within the aquarium. **7.** The watershed of the model drains into the lake.

23:2 Forest Conservation

Information on forest conservation in your state may be obtained from the state forestry department. Information pertaining to national forests may be obtained from the U.S. Forest Service, Department of Agriculture, Washington, DC. You may wish to hold an Arbor Day celebration and plant a tree in the schoolyard. A local nursery or other business may be willing to donate a young tree suited to the schoolyard environment. Consult a local nursery for tree planting instructions. Let the class plan the tree planting ceremony.

Activity 23-2

In this activity the student works with a group of students to make a list of wood products. The list may include all sorts of things made of wood and products of wood such as turpentine and paper.
Materials: pencil and paper
Questions: Answers will vary.

23:3 Forest Conservation Practices

Point out to students that the overall aim of forest conservation is to increase the number and quality of trees grown in forests and to reduce losses caused by forest fires and unwise lumbering practices. Review and explain each forest conservation practice described in this section. You may wish to have students prepare a forest conservation bulletin board by having students make drawings that picture different forest conservation practices. Have students check with the school or local library for information to be used in preparing the bulletin board.

23:4 Vanishing Wildlife

Many species of animals and plants are endangered and are threatened with extinction. Have each student prepare a report on an endangered species that includes a description of the species, its habitat, reasons why the species is endangered, and what can be done to increase the numbers of the species. A list of threatened and endangered species can be obtained from the National Wildlife Federation or the U.S. Dept. of the Interior, Fish and Wildlife division. Students can also report on the success of efforts to increase the numbers of the American alligator and other endangered species. If there is a wildlife refuge in your area, you may wish to plan a field trip or, if possible, invite a game warden to speak to your class. Remind students that the conservation of wildlife involves maintaining and increasing the kind of habitat that is suited to wildlife, including an adequate food supply.

23:5 Wildlife Resources

Have students read this section, then discuss the ways wildlife benefits people. Ask students if they have a favorite wildflower or wild animal and the reasons for their choices. Point out that wildlife is part of food webs that keep populations of mice, rats, and harmful insects in a natural, healthy balance.

23:6 Wildlife Conservation

Make a list of things that can be done to conserve wildlife. Then ask students to identify ways they can aid in wildlife conservation. Joining organizations such as the Audubon Society, Sierra Club, and other conservation organizations may be suggested and writing letters in support of wildlife conservation to state and federal legislators may also be included. A copy of your state's game laws may be obtained and reviewed by students in relation to the effects of these laws on wildlife conservation.

Activity 23-3

In this activity the student constructs a bird feeder and observes and records the kinds and number of birds that visit the feeder. It is a good idea to drill a few tiny holes in the bottom of the feeder to allow rainwater to drain. Wild bird seed, bread, or other food scraps may be used in the feeders. It is essential that a bird feeder set up during the autumn or winter be continually supplied with food throughout the winter. Birds will become dependent on the feeder as a source of food and may die without it during winter when other foods may be in short supply.

Materials: wood and other materials for making bird feeder, wild bird seed

Question: There will probably be a difference in numbers and kinds of birds depending on the type of feed. For example, cardinals like sunflower seeds.

Making Sure, page 502
1. reforestation, allow brush to grow where possible, supply feed when necessary, reduce use of harmful pesticides, game laws, stocking fields, lakes, and streams

23:7 Wildlife Refuges

Poll students to ascertain whether any have visited national parks such as Yellowstone and ask them to describe their experiences. Have students prepare library reports on wildlife refuges and permanent wilderness areas. If there is a migratory bird flyway in your area, you many be able to arrange with the local chapter of the Audubon Society for a class fieldtrip to observe the seasonal migration of birds. Students interested in stamps may wish to start a collection of wildlife stamps. Have them check with the post office or a local stamp dealer for information.

Perspectives

Grizzly Bears and Humans

Have students gather additional information about grizzly bears. Perhaps students can prepare and present team reports based on their research. Encourage students to use visual aids with their presentations.

Study Questions

A. True or False
1. True 2. False (Loss of ground cover during a forest fire increases erosion.) 3. True 4. False (Reforestation renews forests.) 5. True 6. True 7. False (Clear cutting is an unwise forest conservation practice.) 8. True 9. True 10. False (Block cutting is helpful to forest conservation.)

B. Multiple Choice
1. 90 2. American Chestnut 3. passenger pigeon 4. whooping crane 5. 20 6. 1905 7. extinct 8. watershed 9. Selective cutting 10. bears

C. Completion
1. western 2. game 3. watershed 4. Game laws, stocking fields and streams 5. U.S. Forest Service 6. improvement cutting 7. refuges 8. Answers will vary. 9. migratory 10. extinct

D. How and Why
1. Lumber, plywood, timber, and veneer are major direct wood products. Wood pulp and particle board are wood byproducts.
2. preventing forest fires, improvement cutting, selective cutting, block cutting, reforestation, fighting harmful insects and disease
3. Wildlife is preserved by protecting its natural habitat and protecting it from unnatural enemies. Soil, forest, and water conservation aid wildlife conservation.
4. Refuges are areas in which the environment is suitable for survival of wildlife species and wildlife is protected from hunting.
5. No. The crow is a natural enemy of insects, and the hawk is a natural enemy of rats and mice. Excessive killing of crows and hawks results in an increase of insects, rats, and mice. Overhunting of a species may upset the balance of nature.
Note: Answer may be yes if killing controls the population of animals that, if allowed to overpopulate, could cause extensive damage to grain crops. In general, game laws are subject to change when observations reveal that the number of any species is imbalanced.

Teacher Questionnaire

One of the best ways to ensure that effective educational materials are produced is to let authors and publishers know how you feel about the materials you are using. Please help us in planning our revisions and new programs by completing and returning the questionnaire found on the next page. Additional comments are also welcome. After removing the questionnaire from the book, fold and staple it so that the address label shows. We will certainly appreciate hearing from you.

Principles of Science Program

Circle the number which corresponds most nearly to your opinion of each of the following items of the **Principles of Science** program. Please also star (*) three factors which most influence your evaluation or choice of a text.

	Excellent	Very Good	Satisfactory	Fair	Poor	Comments
Student Text						
1. Format	5	4	3	2	1	_____
2. Readability	5	4	3	2	1	_____
3. Approach	5	4	3	2	1	_____
4. Organization	5	4	3	2	1	_____
5. Factual accuracy	5	4	3	2	1	_____
6. Coverage of science principles	5	4	3	2	1	_____
7. Concept development	5	4	3	2	1	_____
8. Visual impact	5	4	3	2	1	_____
9. Margin notes	5	4	3	2	1	_____
10. Treatment of new terms	5	4	3	2	1	_____
11. Activities	5	4	3	2	1	_____
12. Problems	5	4	3	2	1	_____
13. Tables	5	4	3	2	1	_____
14. Illustrations	5	4	3	2	1	_____
15. Chapter-end materials	5	4	3	2	1	_____
16. Perspectives	5	4	3	2	1	_____
17. "Side Road" features	5	4	3	2	1	_____
18. Appendices	5	4	3	2	1	_____
19. Glossary	5	4	3	2	1	_____
20. Skills pages	5	4	3	2	1	_____
Teacher's Annotated Edition						
1. Teachability	5	4	3	2	1	_____
2. Planning Guide	5	4	3	2	1	_____
3. Effective program use	5	4	3	2	1	_____
4. Demonstrations	5	4	3	2	1	_____
5. Performance Objectives	5	4	3	2	1	_____
6. Teaching aids	5	4	3	2	1	_____
7. Chapter-by-chapter materials	5	4	3	2	1	_____
8. Annotations	5	4	3	2	1	_____
9. Tint block answers	5	4	3	2	1	_____
Supplements						
1. Evaluation Program	5	4	3	2	1	_____
2. Activity-Centered Program, Teacher's Guide	5	4	3	2	1	_____
3. Teacher Resource Book	5	4	3	2	1	_____

School Information

1. Grade level of students   6   7   8   9   10
2. Total number of students enrolled in that grade   1-50   51-100   101-200   200+

3. Total number of students enrolled in general science	1-50	51-100		101-200	200+
4. Average class size	25 or less	26-30	31-35	36-40	41 or more
5. Total school enrollment	1-200	201-500		501-1000	1000+
6. Locale of school	rural	small town		suburban	large city
7. Ability level of class	below average		average		above average
8. Appropriateness of text for your class	easy		about right		difficult
9. Number of years text used	1	2	3	4	5
10. May we quote you?			yes		no

Please feel free to attach an additional sheet to make further comments.

Name _____ Date _____

School _____ City _____ State _____ Zip _____

Fold

NO POSTAGE
NECESSARY
IF MAILED
IN THE
UNITED STATES

BUSINESS REPLY MAIL
FIRST CLASS   PERMIT NO. 284   COLUMBUS, OHIO

POSTAGE WILL BE PAID BY ADDRESSEE

Managing Editor, Elhi Science

CHARLES E. MERRILL PUBLISHING CO.

A BELL & HOWELL COMPANY

**1300 ALUM CREEK DRIVE
COLUMBUS, OHIO 43216**

principles of SCIENCE Book One

Author
Charles H. Heimler is a Professor of Science Education and Director of Teacher Education at California State University, Northridge, CA. He received his B.S. degree from Cornell University and his M.A. and Ed.D. degrees from Columbia University and New York University. He has 32 years of teaching experience at the junior high, high school, and university levels. Dr. Heimler is a member of the American Association for the Advancement of Science, National Science Teachers Association, and the National Association of Biology Teachers. He is currently a consultant in science education to several California school districts. Dr. Heimler is author of the Merrill *Focus on Life Science* program and co-author of the *Focus on Physical Science* program.

Consultant
Charles D. Neal is Professor Emeritus of Education at Southern Illinois University, Carbondale, IL. Dr. Neal has taught at the junior high school level and has lectured extensively on the teaching of science in the elementary and junior high school. He received his B.S. degree from Indiana University, his M.A. degree from the University of Illinois, and his M.S. and Ed.D. degrees from Indiana University. He is a member of various professional organizations and is the author of numerous science books for children and adults.

Content Consultants
Earth Science: Dr. Jeanne Bishop, Parkside Junior High School, Westlake, OH
Life Science: Lucy Daniel, Rutherfordton-Spindale High School, Rutherfordton, NC
Physical Science: Dr. Richard H. Moyer, University of Michigan-Dearborn

Reading Skills Consultant
Dr. David L. Shepherd, Hofstra University

Charles E. Merrill Publishing Co.
A Bell & Howell Company
Columbus, Ohio
Toronto, London, Sydney

A Merrill Science Program

Principles of Science, Book One Program
   Principles of Science, Book One and Teacher's Annotated Edition
   Principles of Science, Book One, Teacher Resource Book
   Principles of Science, Book One, Activity-Centered Program, Teacher's Guide
   Principles of Science, Book One, Evaluation Program, Spirit Duplicating Masters
Principles of Science, Book Two Program
(This program contains components similar to those in the Book One Program.)

Reviewers

Donald Baumann, Science Teacher: Franklin Junior High, Nutley, NJ
David Graham, Science Teacher: Squaw Peak School, Phoenix, AZ
Jerry Hayes, Science Coordinator: Chicago Board of Education, Chicago IL
John P. Newton, Principal/Science Teacher: Immokalee Day School, Fort Myers Beach, FL
Betty Rivinius, Science Teacher: Merrill Junior High, Denver, CO
Greg Rottengen, Science Teacher: Brooklawn Junior High, Parsippany, NJ
Donna Seligman, former Science Department Chairperson: Riley Middle School, Livonia, MI

Special Features Consultant

Douglas E. Wynn, Science Teacher: Westerville North High School, Westerville, OH

Cover Photograph: Earth is the environment for all living creatures. A seashore is home to the fairy tern, a bird nicknamed by naturalists as the "sea swallow." Terns lay their eggs on almost any surface, as they do not build nests. A pile of driftwood on a beach can be one of those unusual egg-laying sites. Interaction of living things with their environments is only one of the concepts you will study in *Principles of Science, Book One*. Changes in matter and energy, the basic laws of physics, and fundamentals of earth science are other topics you will study.
Photograph by *Frans Lanting*

Series Editor: Terry B. Flohr; *Project Editor:* Francis R. Alessi, Jr.; *Editors:* Janet Helenthal, Peter R. Apostoluk; *Book Design:* Patricia Cohan; *Project Artist:* Dick Smith; *Illustrators:* Don Robison, Jim Shough, Bill Robison; *Photo Editor:* Lindsay Gerard; *Production Editor:* Joy E. Dickerson

ISBN 0-675-07080-5

Published by
Charles E. Merrill Publishing Company
A Bell & Howell Company
Columbus, Ohio 43216

Copyright 1986, 1983, 1979, 1975, 1971, 1966 © by Bell & Howell

All rights reserved. No part of this book may be reproduced in any form, electronic or mechanical, including photocopy, recording, or any information storage or retrieval system without written permission from the publisher.
Printed in the United States of America

Preface

Principles of Science, Book One is a modern general science program. The interrelationships among the life, earth, and physical sciences are emphasized. Basic science principles of matter and energy and their interaction in understanding natural phenomena are stressed. Interesting everyday examples using science concepts and their practical applications are presented throughout the text to stimulate student interest and motivation. Science activities are included where appropriate to teach the methods of scientific inquiry through direct student involvement. Science is presented as an ever expanding body of knowledge that is useful in everyday life and in science-related careers.

Principles of Science, Book One is organized to provide the maximum flexibility and adaptation to middle, junior high, or high school programs. Emphasis is given to the organization of the material for improved teachability and maximum student learning. The six units provide comprehensive coverage of important topics in the life, earth, and physical sciences. The first unit introduces the basic principles of metric measurement, matter, and energy. The following units discuss mechanics, earth science, animals and plants, ecology and heredity, and conservation. Science content has been carefully selected to provide a unified, basic general science program that furnishes a proper background for the student's future study of science. Concrete examples of abstract concepts are provided. Relevant topics such as cancer, solar energy, nuclear energy, pollution, and conservation are included.

Chapters are subdivided into numbered sections that form logical teaching blocks. The chapter organization provides students with a basic outline for the chapter. Reading level is carefully controlled by monitoring sentence length and the introduction of new terms. Margin notes, most in the form of questions, are printed in the margins of each chapter. These notes serve as guides for learning and reviewing important concepts, thereby increasing student comprehension.

Each chapter contains attractive, scientifically accurate diagrams, photographs, and tables that are related to the text material both in content and placement. Important new terms are emphasized by boldfaced type and are defined where they first appear. Many new terms are spelled phonetically to assist the student in learning the correct pronunciations. Science activities that provide for student participation are included in all chapters. These activities are especially designed to increase student motivation and strengthen skills in problem-solving methods. *Making Sure* questions at the end of some sections provide for immediate review and reinforcement.

Chapter-end material contains study aids designed to enhance student achievement. The *Main Ideas* section summarizes the content of the chapter. Important new terms introduced in the chapter are listed in *Vocabulary*. The *Study Questions* section provides for a thorough review of the chapter. This section provides four types of questions including both recall and application questions. The *Challenges* section contains ideas for projects and special assignments. Selected reading references are listed under *Interesting Reading*.

Many chapters contain magazinelike pages entitled *Perspectives*. These pages contain articles and photographs describing interesting careers, current developments in research and technology, and skills. The pages discussing skills are designed to assist students in developing reading and other communication skills important to the study of general science. *Side Roads* are special feature pages that present interesting additional information related to the unit. There is one *Side Roads* in each unit of **Principles of Science, Book One.** The goals of **Principles of Science, Book One** are to develop the student's understanding of the environment and the interrelationships of science and technology. The textbook will be valuable in preparing each student for future studies in science and in developing knowledge necessary for making decisions regarding the choice of a science-related career.

To the Student

Science is interesting and exciting. Why? Science helps you understand the world in which you live. For example, science explains how airplanes fly and how birds find their way when they travel long distances. In addition, people use science to make discoveries that have practical value. One of these discoveries is the use of light to carry telephone messages through a glass wire. Another is the lengthening of human life through the use of heart pacemakers and other mechanical devices.

In your future, there will be an endless number of new scientific discoveries. These discoveries will affect your career and your daily life. Television, computers, and space shuttles are part of today's world. Who can imagine what new, yet to be discovered developments lie ahead? In the future, scientists may discover how to predict earthquakes and how to produce an endless supply of energy. Someday you may live and work in a space station in orbit around the earth. Scientists will continue to make discoveries that will change the world in which you live.

A science class provides a great opportunity to increase your understanding of the world around you, to study your environment, and to learn about ways in which you can play an active role in the future that is science. You may find you like science so much that you may plan to become a scientist. There are many interesting careers related to science such as nursing, engineering, and electronics.

This textbook contains many study aids to help you achieve your goals. For example, pictures and diagrams are included to increase your understanding. Margin questions are designed to help you get more out of your reading. Questions at the end of many sections as well as the chapter review questions will help you review and sum up what you have learned. Many interesting hands-on activities allow you to experience some of the methods scientists use.

As with any opportunity, much depends on you. Your success in the study of science will be related to the time and effort you give. Learning science is not any more difficult than learning other subjects. However, it does take time and practice, just like sports or anything else worth doing. Success in the study of science will be richly rewarding. It will give you knowledge and skills you will use throughout your life.

See Teacher's Guide.
Textbook Inventory

A student's textbook is his/her primary learning tool. Therefore, an inventory is an excellent way to introduce a new textbook. An inventory introduces the general structure of the book. It also points out the various features of the textbook as well as highlighting the different kinds of study aids.

Use your textbook to answer the following questions.

A. Introduction
1. What is the name of your textbook?
2. Who is the author?
3. Read To the Student on page v. Why should science be important to you?
4. Look at the Table of Contents. How many chapters are in this book? What section in Chapter 9 would you learn about sedimentary rocks?

B. Graphic Aids
5. Where does Unit 1 begin? How is the photograph at the beginning of Unit 1 used to introduce the unit?
6. Look at the beginning of Chapter 5. Why was the photograph chosen to begin Chapter 5?
7. What is the title of Table 11-1 on page 222? What are the two main types of information given in the table?
8. How could the information in Figure 6-5 on page 110 be shown in a table? In what ways is the picture better than a table?

C. Study Aids
9. What should be your goal as you read Chapter 7?
10. In Chapter 7, what is discussed in Section 7:2?
11. How can you use the blue margin notes on page 135 to help you study?
12. What is purpose of the example problem on page 95? How are example problems helpful?
13. Why are Making Sure questions included at the end of many sections?
14. List the important terms you should know when you read Section 8:4 on page 155.

D. Activities

15. How is the activity on page 330 different from the activity on page 331?
16. What is the purpose of the activity on page 330?
17. What two types of plants are used in the activity on page 330? What is the purpose of the iodine solution in the activity?
18. What may happen if you touch the sodium hydroxide in activity on page 331?
19. In the activity on page 399, what is to be done with the data once it has been collected and organized in table form?

E. Chapter Review

20. Look at the end of Chapter 6 that begins on page 123. What section should you review if you did not understand how centripetal force causes an object to move in a curved path?
21. How many new terms were introduced in Chapter 6?
22. How many kinds of study questions are included in the Chapter Review?
23. Where can you find activity ideas to extend your knowledge of the chapter subjects?
24. Where can you find sources of additional information about the subjects in the chapter?

F. Other Features

25. What is the purpose of the Perspectives and Side Roads pages?
26. What six learning skills are covered in the Perspectives features?
27. What information is provided in Appendix B?
28. In what section are all vocabulary terms defined?
29. Turn to the Index. On what pages is the term mineral discussed? What page shows a photograph of a cougar?

Table of Contents

unit 1
Matter and Energy — 2

1 Science and Measurement — 4
- 1:1 What is Science? — 5
- 1:2 Careers in Science — 6
- 1:3 Using Science Skills — 8
- 1:4 Experiments — 9
- 1:5 Metric Measurement — 12
- 1:6 Area and Volume — 14
- 1:7 Mass — 17
- Perspectives
  - Reading for Meaning — 18

2 Matter — 22
- 2:1 Properties — 23
- 2:2 States of Matter — 25
- 2:3 Density — 26
- 2:4 Elements — 29
- 2:5 Atoms — 30
- 2:6 Compounds — 32
- 2:7 Formulas — 34
- 2:8 Mixtures — 35
- Perspectives
  - Using Tables — 38
- Side Roads
  - Superconductors — 42

3 Energy and Changes in Matter — 44
- 3:1 Energy — 45
- 3:2 Heat and Temperature — 47
- 3:3 Freezing and Melting — 50
- 3:4 Evaporation and Boiling — 51
- 3:5 Chemical Change — 52
- 3:6 Chemical Equations — 54
- 3:7 Nuclear Change — 56
- 3:8 Sources of Energy — 56
- Perspectives
  - Fluidized Bed Combustion — 60

unit 2
Mechanics — 64

4 Work and Energy — 66
- 4:1 Force — 67
- 4:2 Weight — 68
- 4:3 Friction — 70
- 4:4 Work — 73
- 4:5 Work and Energy — 75
- 4:6 Power — 75
- 4:7 Engines — 77

5 Machines — 82
- 5:1 Machines — 83
- 5:2 Levers — 84
- 5:3 Mechanical Advantage — 86
- 5:4 Pulleys — 89
- 5:5 Wheel and Axle — 91
- 5:6 Inclined Plane — 91
- 5:7 Wedge — 93
- 5:8 Screw — 93
- 5:9 Efficiency — 94
- 5:10 Compound Machines — 98
- Perspectives
  - Organizing Notes in Outline Form — 100
- Side Roads
  - The World of Robots — 104

6 Motion — 106
- 6:1 Inertia — 107
- 6:2 Speed — 109
- 6:3 Change in Speed — 111
- 6:4 Action and Reaction Forces — 114
- 6:5 Falling Objects — 115
- 6:6 Circular Motion — 117
- 6:7 Momentum — 120
- Perspectives
  - Understanding Science Words — 122

7 Fluids and Pressure — 126

- 7:1 Fluids — 127
- 7:2 Pressure in Liquids — 129
- 7:3 Gas Laws — 133
- 7:4 Buoyancy — 134
- 7:5 Bernoulli's Principle — 138
- 7:6 Lift — 139
- 7:7 Flight — 140
- Perspectives
  - First Flight — 142

unit 3
Earth Science — 146

8 The Earth — 148

- 8:1 Earth's Shape and Size — 149
- 8:2 Earth's Motions — 151
- 8:3 Seasons — 152
- 8:4 Latitude and Longitude — 155
- 8:5 Mapping the Earth's Surface — 156
- 8:6 Time Zones — 159
- 8:7 The Earth as a Magnet — 161
- Perspectives
  - Reading Maps — 164

9 Geology — 168

- 9:1 Structure of the Earth — 169
- 9:2 Plate Tectonics — 171
- 9:3 Rocks and Minerals — 174
- 9:4 Weathering and Erosion — 177
- 9:5 Igneous Rock — 179
- 9:6 Sedimentary Rock — 180
- 9:7 Metamorphic Rock — 183
- 9:8 Rock Formations — 184
- 9:9 Earthquakes — 186
- 9:10 Volcanoes — 189

10 Earth History — 194

- 10:1 The Grand Canyon — 195
- 10:2 Age of Rocks — 198
- 10:3 Fossils — 200
- 10:4 Fossil Records — 202
- 10:5 Geologic Time — 204
- 10:6 Precambrian Era — 205
- 10:7 Paleozoic Era — 206
- 10:8 Mesozoic Era — 210
- 10:9 Cenozoic Era — 212
- 10:10 Earth's Future — 214
- Perspectives
  - Mike Hansen: Earth Scientist — 216

11 The Atmosphere — 220

- 11:1 Air — 221
- 11:2 Parts of the Atmosphere — 222
- 11:3 Air Pressure — 224
- 11:4 Heating the Atmosphere — 226
- 11:5 Winds — 227
- 11:6 Local Winds — 229
- 11:7 Moisture in Air — 231
- 11:8 Clouds — 234
- 11:9 Precipitation — 236

12 Weather and Climate — 240

- 12:1 Observing the Weather — 241
- 12:2 Air Masses — 243
- 12:3 Fronts — 245
- 12:4 Thunderstorms — 246
- 12:5 Highs and Lows — 247
- 12:6 Weather Maps — 251
- 12:7 Weather Forecasting — 252
- 12:8 Climate — 254
- Perspectives
  - Weather Watching — 256

13 Oceanography — 260

- 13:1 Seawater — 261
- 13:2 Temperature and Density — 264
- 13:3 The Seafloor — 266
- 13:4 Sea Life — 269
- 13:5 Ocean Currents — 271
- 13:6 Waves — 273
- 13:7 Tides — 277
- 13:8 Seismic Sea Waves and Surges — 279
- 13:9 Ocean Resources — 280
- Perspectives
  - Commercial Diving — 282
- Side Roads
  - The Sandy Seashore — 286

unit 4
Living Things 288

14 Life and the Cell 290
- 14:1 Features of Living Things 291
- 14:2 The Cell 293
- 14:3 Cell Activities 295
- 14:4 Classification 297
- 14:5 Bacteria 300
- 14:6 Fungi 302
- 14:7 Amoebas and Paramecia 305
- 14:8 Flagellates 308
- 14:9 Sporozoans 309
- 14:10 Viruses 310
- Perspectives
  - Natalie Jones: A Dedicated Microbiologist 312

15 Plants 316
- 15:1 Algae 317
- 15:2 Mosses and Liverworts 318
- 15:3 Ferns 320
- 15:4 Seed Plants 321
- 15:5 Roots 322
- 15:6 Stems 324
- 15:7 Leaves 327
- 15:8 Photosynthesis 329
- 15:9 The Flower 331
- 15:10 Seeds 333
- 15:11 Seed Dispersal 335
- Perspectives
  - A Marine Botanist 336

16 Animals 340
- 16:1 Tissues, Organs, and Systems 341
- 16:2 Sponges 342
- 16:3 Jellyfish and Their Relatives 343
- 16:4 Flatworms and Roundworms 344
- 16;5 Earthworms and Their Relatives 346
- 16:6 Spiny-skinned Animals 350
- 16:7 Mollusks 351
- 16:8 Arthropods 353
- 16:9 Insects 354

17 Animals with Backbones 362
- 17:1 Backbones and Skeletons 363
- 17:2 Fish 364
- 17:3 Amphibians 368
- 17:4 The Frog 370
- 17:5 Reptiles 373
- 17:6 Birds 375
- 17:7 Mammals 378
- Perspectives
  - Why Rattlesnakes Have Rattles 380
- Side Roads
  - Reading the Environment 384

unit 5
Environment and Heredity 386

18 Populations and Communities 388
- 18:1 Communities 389
- 18:2 Populations 390
- 18:3 Population Change 394
- 18:4 Competition 397
- 18:5 The Climax Community 400
- 18:6 The Boundary Communities 402
- 18:7 Cooperation in Populations 403
- 18:8 Ecology 404
- Perspectives
  - Devotion to Wildlife 406

19 Food and Energy 410
- 19:1 Food Chains 411
- 19:2 Producers and Consumers 414
- 19;3 Energy Pyramids 416
- 19:4 Predators 418
- 19:5 Parasites 420
- 19:6 Symbiosis 421
- Perspectives
  - A World Without Sunlight 424

20 Heredity — 428

20:1	Inherited Traits	429
20:2	Law of Dominance	430
20:3	Crossing Hybrids	433
20:4	Blending	435
20:5	Chromosomes and Genes	436
20:6	Reduction Division and Fertilization	439
20:7	Sex Determination	441
20:8	Reproduction	442
20:9	Mutations	444
20:10	Plant and Animal Breeding	445

21 Descent and Change — 450

21:1	Origin of Living Things	451
21:2	Darwin's Theory	453
21:3	Fossil Evidence	456
21:4	Natural Selection	457
21:5	Changes in Species	459
21:6	Mutations and Change	462
	Advances in Science	
	Evolutionary Clues	464
	Perspectives	
	Improving a Theory	466
	Side Roads	
	Biomes	470

unit 6 Conservation — 472

22 Soil, Water, and Air Conservation — 474

22:1	Conservation of Natural Resources	475
22:2	Soil	476
22:3	Soil Erosion and Mineral Loss	478
22:4	Soil Conservation	479
22:5	Water Resources	481
22:6	Water Conservation	482
22:7	Water Sources for the Future	483
22:8	Air Pollution	484
	Perspectives	
	Using Graphs	486
	Side Roads	
	Using Fires to Grow Forests	490

23 Forest and Wildlife Conservation — 492

23:1	Forest Resources	493
23:2	Forest Conservation	496
23:3	Forest Conservation Practices	497
23:4	Vanishing Wildlife	499
23:5	Wildlife Resources	500
23:6	Wildlife Conservation	501
23:7	Wildlife Refuges	502
	Perspectives	
	Grizzly Bears and Humans	504

Appendices

A	Scientific Notation	508
B	Science Classroom Safety and First Aid	509
C	Weather Tables	510
D	The Microscope and its Use	511
E	Classification of Living Organisms	512
F	Measuring with the International System (SI)	513

Glossary — 514

Index — 523

Matter and energy are all around you. Your clothes, food, and even your body are matter. There are different kinds of matter. How is matter classified? Of what is matter made? How are matter and energy related? How is the energy of the water affecting the people in the raft? How does energy affect you?

Appalachian Wildwaters, Inc.

Matter and Energy
unit 1

At an outdoor education center, students collect data to learn more about the ocean. There are still many unanswered questions about the ocean. Using the methods of science, a student like yourself or a scientist may find solutions to these questions. What is the scientific method? Why is measurement important in science?

Science and Measurement   chapter 1

Introducing the chapter: Display a meter stick, graduated cylinder, and balance. Identify and explain briefly the use of each. Point out that these instruments are some of the tools of science. Ask students to name some other tools scientists might use. Be sure students also understand that skill, luck, trial and error, and intelligent guessing are part of science.

1:1 What Is Science?

Why is tap water safer to drink than pond water? Why do some birds fly south for the winter? How can solar energy be collected and used? You may already know the answer to the first question. Scientists are searching for answers to the second and third questions.

Answers to these questions can be useful in many ways. New methods for cleaning water can be used to increase our water supply. Knowing why birds migrate may help to protect them. Sunlight may someday become a common source of energy for home, work, and travel. In this chapter you will learn how scientists search for answers.

Science can be thought of as having two parts. These two parts are the *process* and *product* of science. The process of science is the way in which scientists go about discovering facts and finding answers. It is the methods scientists use to study things and to solve problems. Measuring and doing experiments are two kinds of methods used by scientists.

The product of science consists of the facts and ideas that have been discovered by scientists. For example, flu and some other diseases are caused by viruses. Nine planets move around the sun. These facts represent the product of science.

GOAL: You will gain an understanding of scientific methods and the use of SI units.

Begin this chapter by assigning Section 1:1 and going over the guide question in the margin. After students do the reading, review the question and answer with the class.

What are the two parts of science?

Science is not just a body of facts. It is the process scientists use to make discoveries.

FIGURE 1–1. The process of science is the way in which scientists discover facts and search for answers.

Review briefly the major concepts in Main Ideas at end of each chapter prior to intensive study of each chapter. Stress the use of the study-guide questions in the margin to guide students' reading and learning.

How are the product and process of science related?

Check with your school librarian before assigning this activity. A local public library is another source for students. Examples of science magazines are: *Science Digest, Natural History, Scientific American*

Name two kinds of scientists.

Scientists use both the product and process of science. They read books and journals to obtain facts and ideas discovered by other scientists. Scientists make careful observations of events and changes in the environment. They may also do experiments to test their ideas. New facts that result from this research become part of the product of science. Due to the process of science, scientific knowledge is always changing.

activity

READING ABOUT SCIENCE   Activity 1-1

Go to a library and locate the periodical literature section. Make a list of the titles of five magazines that deal with science. Choose one magazine and write a brief report that describes its contents.

See Teacher's Guide.

making sure

1. Why is reading an important activity in science?
By reading, a person can learn what other scientists have discovered.

1:2   Careers in Science

Have you ever thought about your future job or career? There are many jobs in science and related fields. Being a scientist is one kind of career. Scientists usually specialize in one field of science. For example, a geologist (jee AHL uh just) studies rocks and changes in the earth such as earthquakes. An entomologist (ent uh MAHL uh just) is a scientist

6   **Science and Measurement**

who studies insects. Some scientists may teach science and do research at colleges and universities. Many scientists work in various kinds of research laboratories. Not all scientists are found in laboratories, however. Some scientists do their work in outdoor areas.

Many jobs require a background in science. For example, science is used in nursing, agriculture, and electronics. Nursing students take science courses concerning the human body. They also learn about different medicines and how they affect a person's health. Agricultural students study science related to soils, climate, plants, and animals. A person who plans a career in electronics studies science courses related to electricity.

Relating the science class to careers increases student interest and motivation.

FIGURE 1-2. X-ray technicians work in hospitals, clinics, and doctors' offices (a). Entomology students work outdoors locating and identifying many types of insects (b).

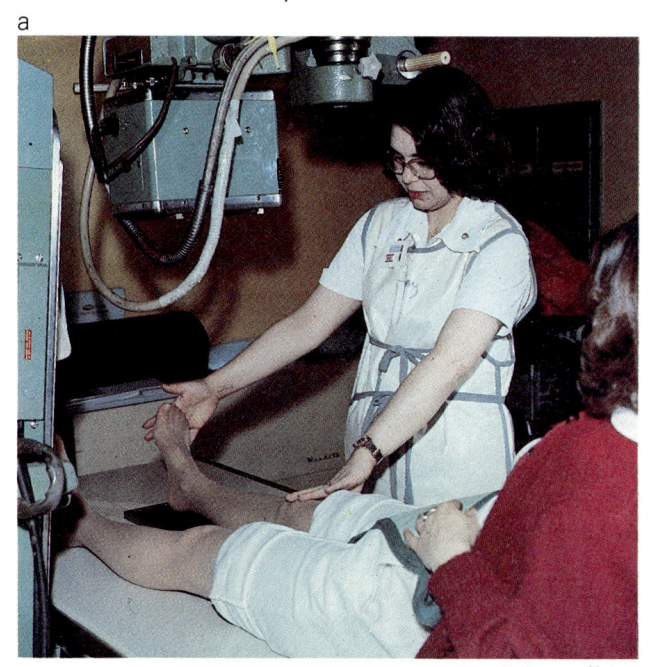

Table 1-1.
Careers Related to Science

Electrician	Optician	Surveyor
Engineer	Pharmacist	Veterinarian
Forester	Physical therapist	Welder
Nurse	Science teacher	X-ray technician

Name and describe five jobs related to science.

1:2 Careers in Science 7

Check with your school guidance department for information related to this activity. It may be possible to invite a guest speaker on this topic from the local government employment office.

activity

SCIENCE CAREERS

Activity 1-2

Choose a career or job related to science. You may select a career listed in Table 1–1. Obtain information about this career or job from a library or school guidance office. Find out how a person prepares for a position in the field you have selected.

See Teacher's Guide.

FIGURE 1–3. The microscope is a tool used by scientists to observe nature.

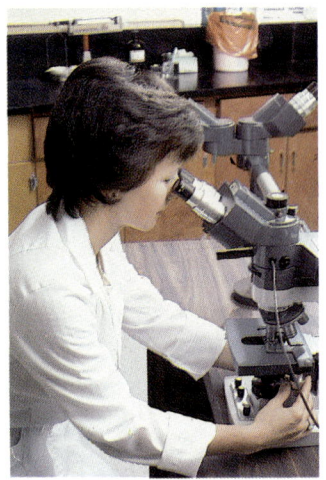

Latent Image

Stress the fact that scientific skills can be applied to many kinds of activities. For example, bicycle repair, sewing, gardening, health care, etc.

What steps are used in scientific problem solving?

1:3 Using Science Skills

Scientists learn to observe nature with great skill. They record their observations by keeping notes, taking photographs, and making tape recordings. Another important science skill is organizing information so it can be understood. Scientists make graphs, charts, and tables in which information is organized and classified.

In your science class you will learn many skills used by scientists. Some of the skills have practical value and can be applied in everyday life. For example, you will learn the methods scientists use to solve problems. Then you can use these methods to solve your own problems.

What methods do scientists use to solve problems? Here are the steps in scientific problem solving.

(1) *Make a clear statement of the problem.* Try to understand the problem for which you seek an answer.

(2) *Collect information that relates to the problem.* Find as much information as you need from books and other written records.

(3) *Form a hypothesis.* Form your own hypothesis (hi PAHTH uh sus), or "best prediction based on information." A **hypothesis** is a possible solution for a problem. A hypothesis is formed after studying the facts and ideas relating to the problem.

(4) *Test the hypothesis.* Design and perform some experiments to see if your hypothesis is acceptable. An **experiment** is a method of testing a hypothesis. You can test many ideas by trying them.

(5) *Accept or reject the hypothesis.* To accept or reject your hypothesis, study the results obtained in experiments. In some cases, you may change your hypothesis as a result of testing. This new hypothesis may then be tested by another experiment.

(6) *Report the results.* You must inform others of your work so they can test your results.

These steps might not be followed in the order listed here. The work of successful scientists shows that the process is not always a step-by-step process. Scientists are creative people and they approach their work in many ways. Skill, luck, trial and error, and intelligent guessing are all involved in their success.

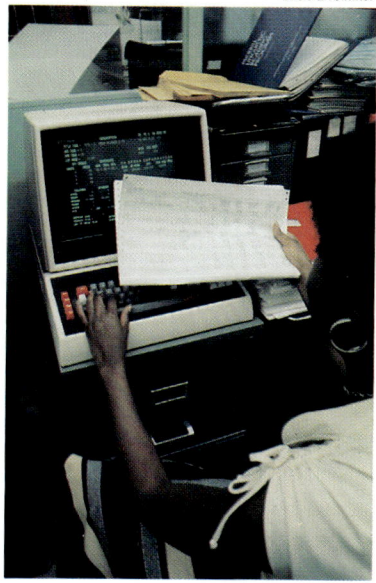

FIGURE 1–4. Computers are important for collecting data and are frequently used by scientists to test their hypotheses.

making sure See Teacher's Guide at the front of this book.
2. How could scientific methods be used to solve each problem?
   (a) Repairing a car engine that is not working properly.
   (b) Choosing a new typewriter for your personal use.

1:4 Experiments

An experiment is used to seek an answer to a question or to test a hypothesis. In an experiment, you perform tests on an object or group of objects and then observe what happens. Often you are able to see something happen, or hear it, or smell it, or taste it, or touch it. The observations made in an experiment should be carefully recorded. You must have accurate information to accept or reject your hypothesis.

In an experiment, a person manipulates things in the environment and observes what happens. A good experiment is done under carefully controlled conditions.

1:4 Experiments 9

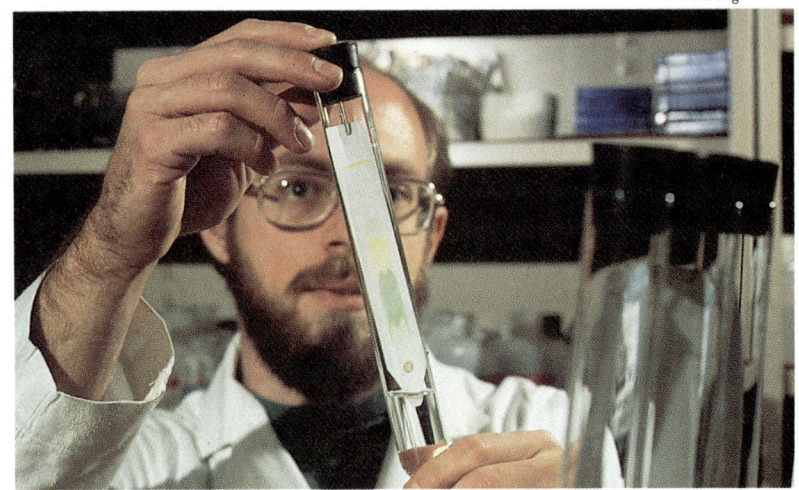

FIGURE 1–5. Experiments must be carefully controlled and observations must be accurately recorded.

Review the science classroom safety rules (Appendix B) with students.

CAUTION: Be certain that students secure hair and loose clothing while using a burner. Remind them that if the flame rises from the burner or appears to "blow out" after lighting, reduce the supply of gas.

Throughout your course in science, you will do many experiments. In this book, experiments and other activities are under the heading ACTIVITY. Keep a record of the activities in your notebook. The following activity is an example of the type of activity found in this book. Read the activity carefully. *When performing the activity, be sure safety procedures are observed at all times.*

activity
See Teacher's Guide.

HEATING AIR   Activity 1-3

(1) Insert a 10-cm piece of glass tubing into a one-hole rubber stopper. **CAUTION:** Moisten the end of the glass tubing with a lubricant such as glycerol. Hold the tubing with a rag or towel as you slide it into the stopper. (2) Fit the stopper into the mouth of a flask. (3) Place a rubber balloon over the end of the tubing and tie it tightly with a piece of string. (4) Then warm the flask for a short time with a laboratory burner. What happened to the balloon? *inflates* Explain your observations. What would happen if the flask did not contain any air?   *no change in the balloon*

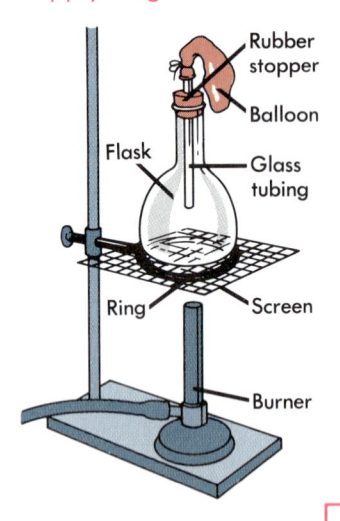

FIGURE 1–6.

Heated air expanded and increased the size of the balloon.

Each activity should be organized into five parts: Title, Objective, Procedure, Observations, and Conclusion. The previous activity would be organized as follows:

10   Science and Measurement

(1) **Title**—Heating Air
(2) **Objective**—Purpose of the activity. The question you wish to answer: What happens to air when it is heated?
(3) **Procedure**—Your explanation of how you did the activity.
(4) **Observations**—Your answer to the question, "What happened to the balloon?"
(5) **Conclusion**—Your answer to the question, "What happens to air when it is heated?" Base your answer on the observation you made.

An important part of many experiments is a control. The **control** is a part of the experiment which is held constant. It is a standard of comparison for the results. The control for the preceding activity would be a flask and balloon left to stand without heating.

If a scientist believes a certain drug will cure a disease, an experiment is done to test this idea. In the experiment, a control should be used. The scientist selects two groups of animals that have the disease. One group, called the experimental group, is given the new drug. The other group, called the control group, does not get the drug. Both groups have exactly the same living conditions, except for the drug. The groups have the same diet, housing, and activities. Only one aspect is different—the condition being tested.

How can an experiment be organized into five parts?

You may wish to teach students this format for recording each activity in their notebooks.

What is the control in an experiment?

When subjects such as animals are used, they may be divided into a control group and an experimental group.

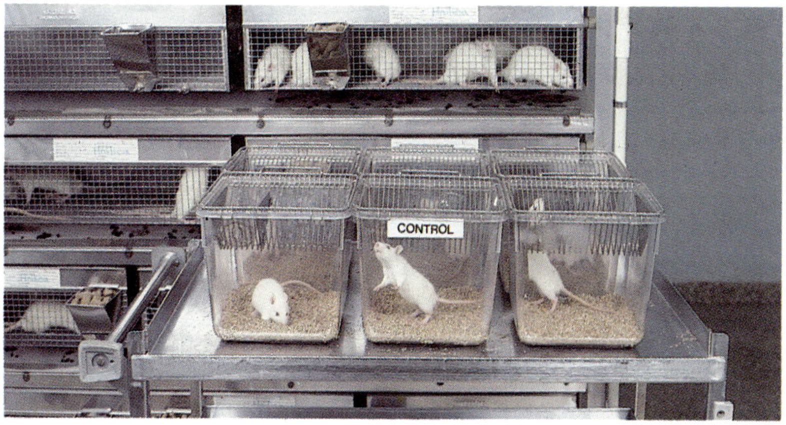

FIGURE 1–7. The use of a control group is part of the procedure in many experiments. The rats in the control group are the standard for comparsion.

Latent Image

1:4 Experiments 11

> You may wish to begin by calling attention to road signs and containers listing metric units. At many service stations, gasoline is sold by the liter.

> SI is an abbreviation for the French Système International d'Unités.
>
> What are SI units?

1:5 Metric Measurement

Over 90 percent of the world's people use the metric system of measurement. Scientists in all countries use metric units. The metric system of measurement is a decimal system, just like the United States' money system. Metric units are based on ten and multiples of ten. For this reason, it is easy to change from one unit to another. **The modern form of the metric system is called the International System of Units or SI.** Table 1–2 lists some metric units. Note the abbreviation for each unit.

Table 1–2.
Frequently Used Metric Units

Length
1 millimeter (mm) = 1/1000 meter (m)
1 centimeter (cm) = 1/100 meter
1 decimeter (dm) = 1/10 meter
1 dekameter (dam) = 10 meters
1 hectometer (hm) = 100 meters
1 kilometer (km) = 1000 meters

Mass
1 milligram (mg) = 1/1000 gram
1 gram (g) = 1/1000 kilogram (kg)

Time
1 millisecond (ms) = 1/1000 second(s)

Volume
1 liter (L) = 1000 milliters (mL)
1 cubic meter (m^3) = 1 000 000 cubic centimeters (cm^3)

> What units are used to measure length?

A **meter** (m) is the basic unit of length in SI. A meter is about the same as the width of most doors in your home. Other units of length that are smaller than a meter include the millimeter (mm) and the centimeter (cm). A metric ruler is marked in centimeters and millimeters. The smallest divisions are millimeters. A kilometer (km) is a unit of length that is larger than a meter. A kilometer is about the length of five city blocks.

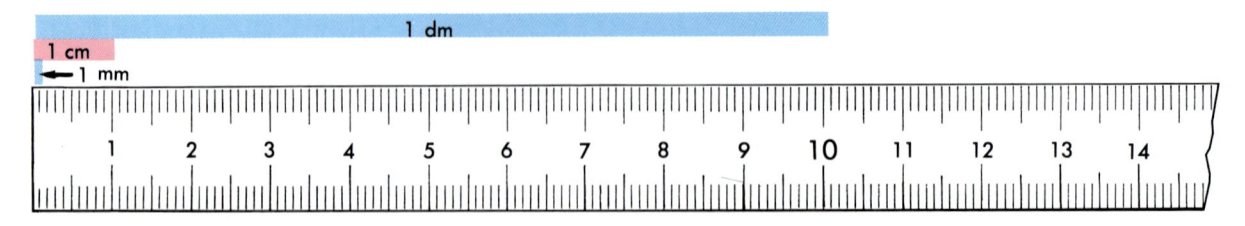

FIGURE 1–8. On a meter stick, the smallest divisions are millimeters; the largest are decimeters.

12 Science and Measurement

Most SI units contain a prefix which gives a clue to the size of the unit. The prefix *milli-* means one-thousandth (1/1000). *Centi-* means one-hundredth (1/100). *Kilo-* means one thousand (1000). These prefixes indicate how to change from one unit to another. Changing from one unit to another is easy because you multiply or divide by multiples of ten. For example, to change millimeters to meters you divide by 1000. To change kilometers to meters you multiply by 1000.

Example
Change 5000 mm to centimeters.
Solution
Step 1: Determine the number of millimeters in a centimeter.

$$10 \text{ mm} = 1 \text{ cm}$$

Step 2: You are changing from a smaller unit to a larger unit. Divide by 10 to find the answer.

$$\frac{5000 \text{ mm}}{10 \text{ mm/cm}} = 500 \text{ cm}$$

$$5000 \text{ mm} = 500 \text{ cm}$$

Another method for changing units is through the use of a scale (Figure 1–9). The scale tells you whether to multiply or divide. It also indicates the number with which you are to multiply or divide. For example, to change 8000 m to kilometers you must move to the left three places from meter to kilometer on the scale. The number of places moved indicates the power of ten with which you will divide or multiply (10 × 10 × 10 = 1000). Since you are moving to the left, you divide. Thus,

$$\frac{8000 \text{ m}}{1000 \text{ m/km}} = 8 \text{ km}$$

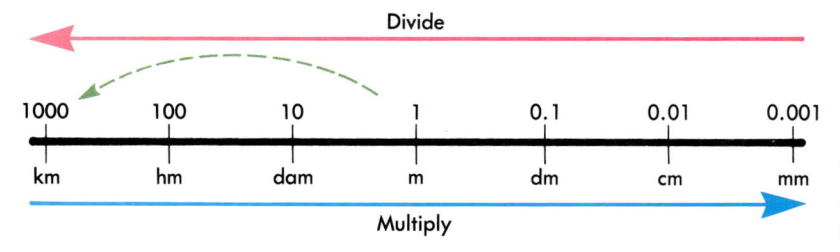

Review these prefixes carefully to ensure student understanding.

State the meaning of *milli-*, *centi-*, and *kilo-*.

In the SI system a space is used instead of a comma when writing or printing numerals. For example: 10 000 and 1 000 000.

Work examples step by step on the chalkboard. Stress the use of factor labels when working with measurements.

Make scales demonstrating the other base units.

Divide ←

| 1000 | 100 | 10 | 1 | 0.1 | 0.01 | 0.001 |
| km | hm | dam | m | dm | cm | mm |

→ Multiply

FIGURE 1–9. This scale may be used when changing from one metric unit of length to another.

1:5 Metric Measurement

To estimate the area of the leaf shown in Figure 1–11 first count the number of totally filled squares. Count the number of partially filled squares and divide by two. Add this number to the number of totally filled squares. The sum represents the approximate area of the leaf in square centimeters.

Diagram a square, rectangle, triangle, and circle on the chalkboard and review the concepts of area using these geometric figures.

What units are used to express area?

FIGURE 1–10. There are 100 square millimeters in 1 square centimeter.

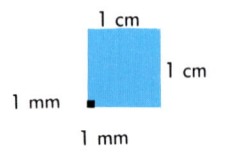

FIGURE 1–11. The area of an irregular shape can be estimated by counting the number of square centimeters covered by the shape.

Example
Use Figure 1–9 to change 24.5 m to centimeters.
Solution
Step 1: Determine the direction and the number of places you move from m to cm on the scale. Since you are moving two places to the right, you multiply by 100 (10 × 10).
Step 2: Multiply to find the answer.
24.5 m × 100 cm/m = 2450 cm
24.5 m = 2450 cm

making sure

3. How many centimeters are in 1 m? 100
4. Change 3 m to centimeters. 3 m × 100 cm/m = 300 cm

1:6   Area and Volume

Area is the number of square units required to cover a surface. Finding area is one kind of measurement in science. For example, the size of a forest is determined by the area of land on which the trees grow.

Two ways of expressing area are square centimeters (cm^2) and square millimeters (mm^2). A square centimeter is a square that is 1 cm long on each side. The area of the leaf in Figure 1–11 can be given in square centimeters. The surface area of a blood vessel may be expressed more simply in square millimeters. A square millimeter is a square that is 1 mm long on each side.

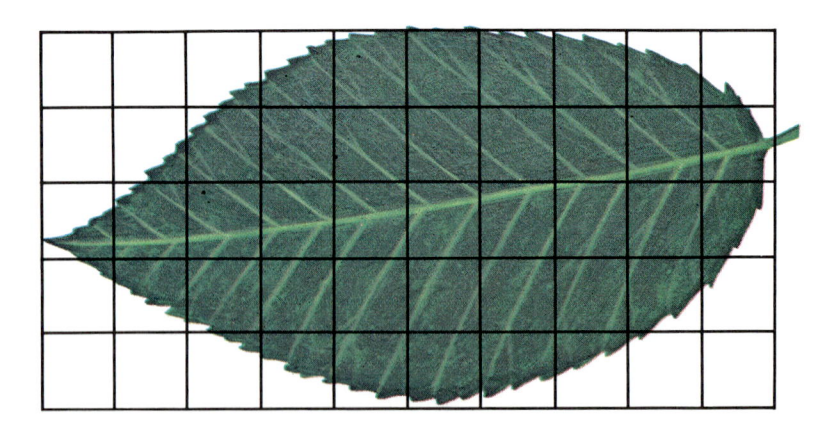

14   *Science and Measurement*

The **volume** of an object is the space occupied or filled by the object. One important use of volume is measuring the amount of liquid used in an experiment. The units for volume are derived from units for length. One unit for volume is the **cubic meter** (m³). A wooden tub 1 m in length on each side would hold a cubic meter of water. A much smaller unit of volume is the cubic centimeter (cm³). The **liter** (L) is a unit of volume often used for measuring liquids. A milliliter (mL) is 1/1000 of a liter and is the same size as a cubic centimeter.

To measure the volume of a liquid, a graduated cylinder, or graduate, can be used. The correct way of reading the volume of water in a graduate is shown in Figure 1-12. A graduate has unit markings printed on its side.

SI units of volume are used to mark other kinds of lab equipment. Three types of equipment used for liquids are test tubes, beakers, and flasks. These containers are made in a number of sizes and shapes (Figure 1-13). They are often made of a heat-resistant glass to avoid breaking when heated.

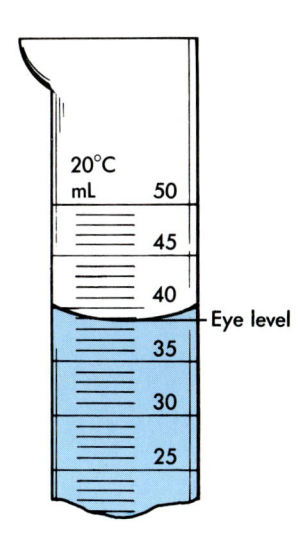

FIGURE 1-12. For accurate volume measurement, read the level of liquid at the lowest point on its curved surface.

What units are used for volume?

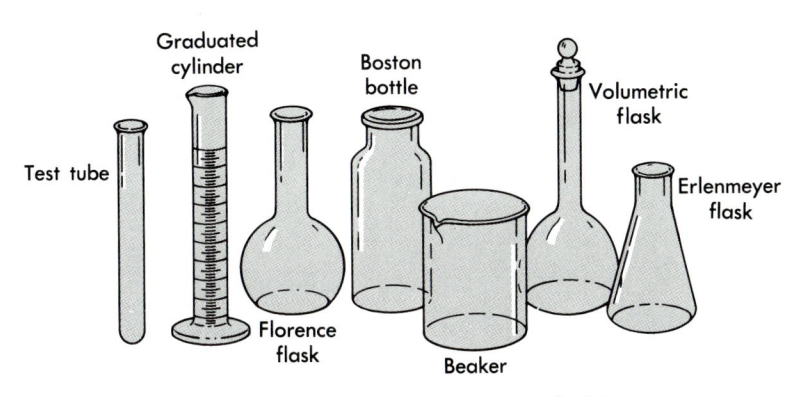

FIGURE 1-13. These are examples of laboratory equipment used to hold and measure liquids.

activity   See Teacher's Guide.
FINDING AREA   Activity 1-4

Find the length and width of the two rectangles in Figure 1-14. Record the measurements in millimeters and centimeters. What is the area of each rectangle in square millimeters and square centimeters? How many square millimeters are in a square centimeter?   100

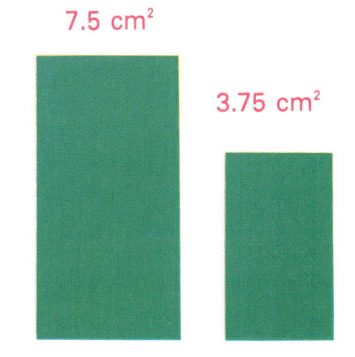

7.5 cm²

3.75 cm²

FIGURE 1-14.

1:6 Area and Volume 15

activity

MEASURING LENGTH, AREA, AND VOLUME

Activity 1-5

Objective: To practice measuring using SI units

The format of an expanded activity such as this lists the materials required for the activity. Review the objective and materials list with students before they begin work.

Materials *See Teacher's Guide.*

2 boxes, different sizes
water glass
dropper
graduated cylinder, 50 mL
meter stick
teacup

Observations and Data

Box	Estimated volume (cm³)	Actual volume (cm³)	Volume (mL)
A	2000	1570	1570
B	4000	6400	6400

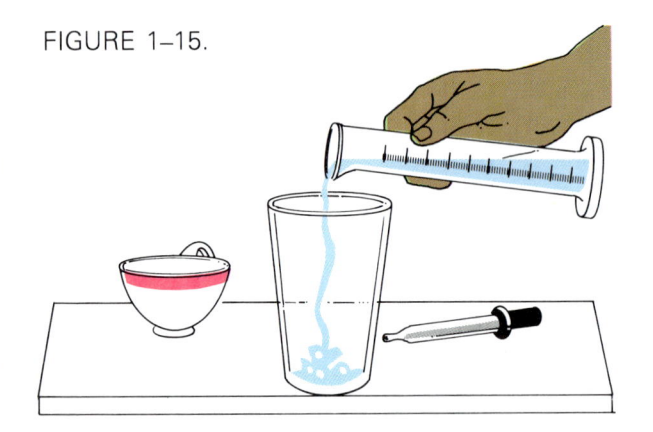

FIGURE 1–15.

Procedure *Do not write in this book.*

Part A

1. Measure the water needed to fill the water glass, dropper, and teacup.
2. Determine the number of drops from the dropper that equals 1 mL. Calculate the number of drops equal to 50 mL.
3. Using the meter stick, find your height in cm.
4. Using the meter stick, measure the length and width of your desk or table top. Find the area in cm².

Caution students to be precise in their measurements.

Part B

1. Label the boxes A and B.
2. Before measuring the boxes, estimate the volume in cm³ of each box. Record your estimated values in the data table.
3. Measure in cm the length, width, and height of boxes A and B.
4. Determine the volume in cm³ of boxes A and B by multiplying length × width × height. Record the volumes in a table.
5. Change the volume of each box in cm³ to mL. Record the volumes.

Questions and Conclusions

Answers will vary.

1. What is your height in meters?
2. What is the area of your desk in square meters? $A = l \times w$
3. How well did you estimate the volumes of boxes A and B? How close to the correct volumes were your estimations? *Answers will vary.*
4. If you were to fill your largest box with water using a 10-mL graduate, how many times would you have to fill the graduate? *number of graduates = $\frac{\text{volume in cm}^3}{10}$*
5. What procedure would you use to find the volume of your classroom in liters? *Use the formula $V = l \times w \times h$ to find the volume in cm³. Divide by 1000 cm³/L to calculate the liters.*

16 *Science and Measurement*

making sure

5. How much water may be added to a dish that is 20 cm long, 10 cm wide, and 5 cm high? Remember, a milliliter is the same as a cubic centimeter. 1000 mL or 1 L

6. Calculate the area of your notebook cover. Answers will vary.

7. Calculate the area of this page in square centimeters. 23 cm × 19.5 cm = 348.5 cm²

1:7 Mass
Demonstrate the use of a balance to measure the mass of several objects.

The **mass** of an object is the amount of matter in the object. For example, there is more matter in a truck than in a bicycle. Thus, a truck has more mass than a bicycle. A sledge hammer has more mass than a tack hammer.

The **kilogram** (kg) is the basic unit of mass in SI. A **gram** (g) is 1/1000 of a kilogram. A paper clip has a mass of about 1 gram. A stack of 10 "D" size flashlight batteries is about 1 kilogram.

A balance, such as the one in Figure 1-16, can be used to measure mass. An object is placed on the left pan, and objects of known masses or standard masses are placed on the right pan. The pans will balance when the mass of the standard masses on the right pan equals the mass of the object on the left pan.

What units are used for mass?

Hickson-Bender Photography

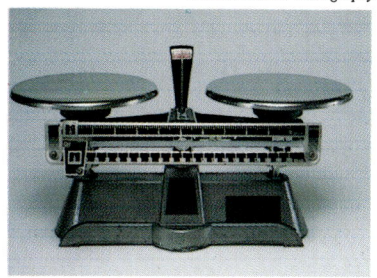

FIGURE 1-16. With a balance, an unknown mass can be compared to a known mass.

making sure

8. What procedure would you use to determine the mass of water in a beaker?

9. An object has a mass of 40 580 g. What is the object's mass in kilograms?

Jet Propulsion Labs

FIGURE 1-17. Io and Europa photographed against the background of Jupiter by Voyager I on February 13, 1979. The planet Jupiter in the background has 21 000 times more mass than its moon Io (on the left), and 40 000 times more mass than the moon Europa (on the right).

See Teacher's Guide.
PERSPECTIVES
skills

Reading for Meaning

An important reason for reading your science textbook is to get ideas and learn information. The text is designed to help you get information quickly and easily. However, there are some methods that can make learning easier. Two of these methods include (1) how to know what information you should look for as you read and (2) how to determine the meaning of new scientific words when they are first introduced.

We concentrate better when we know the type of information we are to learn. You will note that each chapter in your textbook has section titles in heavy black print. For example, look at Section 1:1 on page 5. The title asks a question "What Is Science?" The paragraph that follows will tell you about the two basic parts of science. Most of the section titles are not questions, however. Look at Section 1:3, Using Science Skills, page 8. To learn the information in Section 1:3 the title can be made into a question. A logical question would be "What are science skills?" You may also ask yourself, "How do we use science skills?" Again, the information that follows will answer the questions.

We have used the words *what* and *how* to ask questions about the information in two sections. In other sections we may also use the words *when*, *where*, *who*, and *why*. Now, try to turn the rest of the section titles in Chapter 1 into questions.

The second method is to be alert to definitions when they are given. New words or terms are usually highlighted in boldface type when each is first used. At this time the definition will be given. The definition may be given in the same sentence in which the word appears or in the same paragraph. You can note the definition by the verbs that link the word with its meaning. Look for verbs such as *are*, *is*, *mean*, or *means*. There may also be verbs such as *explains* and *states*. The important thing to remember is that whenever you see a word in boldface print, the definition is nearby.

Look at Section 1:3. See if you can determine the meaning of *hypothesis*. What verb signals its meaning? What is the subject of the sentence? You should conclude that the word being defined and the subject of the sentence are often the same.

All boldface terms used in each chapter are listed in "Vocabulary" at the end of the chapter. The words from Chapter 1 are listed below. Practice your skills by (1) finding the definition of each word, (2) noting the verb used to signal the definition, and (3) stating the section where the definition is found.

area	kilogram
control	liter
cubic meter	mass
experiment	meter
hypothesis	volume
International System of Units (SI)	

You may wish to check each of your textbook definitions with either the glossary in your book or a dictionary. The dictionary definitions may contain more information about the words, or more than one meaning may be given for some words. The meanings that you find in your glossary will apply to those which have been discussed in the chapter.

> Chapter end material provides a thorough summary of the chapter. The Study Questions give excellent reinforcement prior to a test on the information in the chapter.

Chapter Review

main ideas

1. Science can be thought of as a product and a process.   1:1
2. Scientific methods are used to solve problems and discover new knowledge.   1:3
3. A hypothesis is a "proposed answer" to an unsolved scientific problem.   1:3
4. An experiment is used to test a hypothesis. Some experiments contain a control.   1:4
5. The five parts of an experiment are the title, objective, procedure, observations, and conclusion.   1:4
6. Measurements of area and volume are derived from units for length.   1:6
7. The amount of matter in an object is its mass.   1:7

vocabulary

> You may choose to have a weekly vocabulary quiz using five to ten words selected from the Vocabulary lists.

> Suggestion: You may wish to put these words on the chalkboard as you begin this chapter. After the students have studied the text, check for understanding by having them use the words in sentences or paragraphs.

Define each of the following words or terms.

area	International System of Units (SI)	meter
control		volume
cubic meter	kilogram	
experiment	liter	
hypothesis	mass	

study questions

DO NOT WRITE IN THIS BOOK.

A. True or False

Determine whether each of the following sentences is true or false. If the sentence is false, rewrite it to make it true.

T   1. Scientific methods are used to solve problems and discover answers to questions.

T   2. Scientific facts and ideas are the product of science.

T   3. A hypothesis may be tested by an experiment.

F   4. Scientists always report their discoveries ~~before~~ they do an experiment. after

F   5. A control group is the ~~same as an experimental group~~. group that is held constant.

F   6. There are ~~100~~ g in a centigram. $^1/_{100}$

F   7. A cubic centimeter is ~~larger~~ than a cubic meter. smaller

Science and Measurement   19

T 8. The prefix *milli-* means 1/1000.
F 9. A paper clip has a mass of about ~~1 kg.~~ 1 g
T 10. A liter is equal to 1000 mL.

B. Multiple Choice

Choose the word or phrase that completes correctly each of the following sentences.

1. The way in which an experiment is done is called the (*problem, procedure, observations, conclusion*).
2. The first step in an experiment is to state clearly the (*problem, procedure, observation, conclusion*).
3. A (*meter, dollar, pint, pound*) is an SI unit.
4. To change centimeters to meters you must (*divide by, multiply by, add, subtract*) 100.
5. The abbreviation mm³ stands for (*millimeter, square millimeter, cubic millimeter*).
6. A (*gram, liter, cubic meter*) is a unit of mass.
7. A beam balance is used to measure (*volume, weight, mass, length*).
8. When a dry sponge gets wet its mass (*increases, decreases, stays the same*).
9. The large divisions on a metric ruler are (*millimeters, decimeters, kilometers, centimeters*).
10. A (*m, m², m³, kg*) is a unit of mass.

C. Completion

Complete each of the following sentences with a word or phrase that will make the sentence correct.

1. The _____ is a base unit of mass in SI. kilogram
2. The prefix *centi-* means _____. $1/100$
3. A centimeter is equal to _____ m. $1/100$
4. There are __2500__ grams in 2.5 kg.
5. A cubic centimeter is equal to __1__ mL.
6. Milliliters of liquid may be measured with a(n) __graduated cylinder__.
7. A __balance__ is used in the laboratory to measure mass.
8. A 1-L flask would be filled by the water in __50__ 20-mL test tubes.

9. A square 10 cm on each side has an area of 100 cm² .
10. It is easy to change from one SI unit to another because SI units are multiples of 10 .

D. How and Why
1. Why is a control group used in some experiments?
2. What scientific knowledge would be useful to a welder?
3. How might skill, luck, trial and error, and intelligent guessing be involved in a scientist's success?
4. Explain how a laboratory balance works.
5. Calculate the volume of a dish 10 cm long, 5 cm wide, and 2 cm high. How many 25-mL graduates of water would be needed to fill this dish?

challenges

1. Arrange for a scientist to talk to your class. If this is not possible, have a member of the class visit a scientist and act as a reporter. Find out why the scientist chose science as a career, what kind of work is required, and what makes the work rewarding.
2. Prepare a "great scientist" report. Through library research obtain information about why a certain scientist is considered important. Make a report to your class.
3. Use a notebook to keep a log of current events in science reported on TV and in newspapers. Make a short summary report of the log to your classmates.

interesting reading

Harpur, Patricia (ed.), *The Timetable of Technology.* New York: Hearst, 1982.

O'Connor, Karen, *Maybe You Belong in a Zoo! Zoo and Aquarium Careers.* New York: Dodd, Mead, and Co., 1982.

Keller, Mollie, *Marie Curie.* New York: F. Watts, Inc., 1982.

Dank, Milton, *Albert Einstein.* New York: F. Watts, Inc., 1983.

Many different types of matter are found on the earth. This matter has many different forms and properties. Balloons, for example, are constructed of matter that is lightweight and flexible. Air fills the balloon and is heated by a burning gas. What are the characteristics of matter? How is an element different from a compound?

Alan Benoit

Matter

chapter 2

Introducing the chapter: Write the following formulas on the chalkboard: H_2O, O_2, Fe_2O_3, $C_{12}H_{22}O_{11}$, NaCl, and SiO_2. List the common names of these substances: water, oxygen, rust, sugar, salt, and sand. Point out that the formula shows the composition and amount of each substance. Lead students, through discussion, to conclude that each substance listed can be identified by its physical properties.

2:1 Properties

How are snowflakes, hailstones, and ice cubes alike? All three are forms of water that are frozen solid. Each will melt into water again when warmed. Water and ice are two kinds of matter. Air, plastic, gasoline, and soil are other examples of matter. Anything that has mass and takes up space is matter. All matter can be identified by its physical properties. Color, odor, hardness, and melting point are some physical properties of matter. The physical properties of a substance are the features that make it different from other substances. For example, glass is brittle but when heated properly, can be formed easily into many shapes. Can you name some properties of wood that make it different from glass?

The use of a substance depends upon its properties. You look for a solid to sit on. You drink liquids such as water and milk. Wool is used in winter clothes because it prevents body heat from escaping rapidly. Airplanes are made of aluminum metal because it is light in weight. Copper, a reddish-brown metal, is used to make electric wires. It is a good electrical conductor and can be stretched to form a wire.

GOAL: You will learn the properties and composition of different kinds of matter.

Define the terms matter and property.

Wood is softer, does not shatter when broken; you cannot see through it, etc.

How are properties related to the use of matter?

Properties such as density, color, hardness, conductivity, and malleability are physical properties. Other physical properties include crystalline shape, melting point, and boiling point.

Lyndon B. Johnson Space Center *Larry Hamill*

FIGURE 2–1. The properties of the special tiles enable space shuttle orbiters to withstand very high temperatures (a). Steel's properties are demonstrated in the construction of the sculpture (b).

FIGURE 2–2.

activity   See Teacher's Guide.

A PROPERTY OF MATTER   Activity 2-1

Crumple a soft tissue and fit it tightly into the bottom of a paper cup. Push the cup, top down, into a container of water until the whole cup is underwater. Then pull the cup, top still down, out of the water. Is the tissue wet? no Punch a small hole in the bottom of the cup. Repeat the procedure. Explain your observations. What property of matter did you observe? Matter takes up space.

making sure   See Teacher's Guide at the front of this book.

1. What properties allow you to tell the difference between each of the following pairs?
    (a) rubber and aluminum
    (b) chalk and snow
    (c) milk and water
    (d) salt and sugar
    (e) wood and plastic
2. Name one use for each substance listed in Question 1

24 *Matter*

2:2 States of Matter

All matter can be classified by its physical state. These states are solid, liquid, gas, and plasma. A **solid** has a definite volume and a definite shape. Solids include such things as your desk, this book, and most of the objects around you. Some solid substances are used to build things such as furniture, houses, and airplanes.

Define "definite" as meaning not changing.

Water, milk, and oil are examples of matter normally in liquid form. A **liquid** has a definite volume, but it does not have a definite shape. Milk in a bottle will have the same volume after it is poured into a pitcher. But the milk in the pitcher will have a different shape. The shape of a liquid can be changed easily. What shape will the milk take if you spill it on the floor? What will its volume be?

Name four states of matter.

Milk will spread out to form a thin layer on the floor. Its volume will remain the same.

FIGURE 2–3. A liquid takes the shape of its container.

Describe the change in the shape and size of an icicle as it melts.

Latent Image

A **gas** has mass and occupies space, but it does not have a definite shape or volume. A gas fills its container, regardless of the shape or size of the container. The balloon in Figure 2–4 occupies more space than it did before it was blown up. Gas added to the balloon takes up space. When you blow up a balloon, air fills the whole balloon and takes the shape of the balloon. Air is a gas. Hydrogen, helium, oxygen, and carbon dioxide are also gases.

Demonstration: Obtain a package containing balloons of different shapes. Blow up 5 balloons to illustrate that a gas takes the shape of its container.

FIGURE 2–4. A gas takes up space but has no definite shape.

Cameron Balloons U.S.

2:2 States of Matter 25

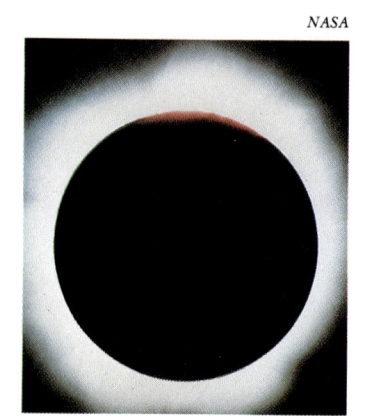

FIGURE 2–5. A region of plasma surrounds the sun.

How is plasma different from a gas?

Another state of matter, plasma, is formed only when matter is heated to unusually high temperatures. **Plasma** is a state of matter that is extremely hot and composed of electrical particles. Its properties are like those of a gas. The hot surface of our sun is made up of plasma. Also, the intense heat from a lightning bolt can change matter into the plasma state.

Table 2–1.
Properties of Matter

State	Properties	Examples
Solid	Definite shape Definite volume	Wood, lead, sugar, sand
Liquid	No definite shape Definite volume	Alcohol, water, milk, molasses
Gas	No definite shape No definite volume	Air, hydrogen, oxygen, nitrogen
Plasma	Similar to a gas Composed of electrical particles at high temperatures	The sun and other stars

making sure

3. Under normal room conditions, which of the substances below are gases? Which are liquids? Which are solids? Use a dictionary to look up words you do not know.
   (a) carbon dioxide g   (c) helium g   (e) neon g
   (b) gasoline l   (d) mercury l   (f) steel s
4. How is a plasma different from a gas?
   Plasma has electrically charged particles, gas does not.

2:3 Density

An important measurable property of matter is density (DEN suht ee). **Density** is mass per unit of volume. To find density, you must be able to measure mass and volume. You can measure mass with a balance. To measure volume, you can use a meter stick for solids and a graduated cylinder for liquids. Density can be expressed in grams per

Density is expressed in what units?

milliliter (g/mL). Since 1 mL = 1 cm³, density is also expressed in grams per cubic centimeter (g/cm³).

Table 2–2.
Densities of Some Common Substances

Substance	Density (g/cm³)
Air	0.0013
Gasoline	0.68
Water	1.0
Iron	7.9
Lead	11.3
Gold	19.3

Which would you rather carry up a flight of stairs—a suitcase filled with lead or the same suitcase filled with feathers? The suitcase filled with feathers would be easier to carry. Why? The density of lead is greater than the density of feathers. The lead has more mass per volume than the feathers.

To find the density of an object you must first determine its mass and volume. You must then use the following equation:

$$\text{density} = \frac{\text{mass}}{\text{volume}} \qquad D = \frac{m}{V}$$

Example

A plastic box has a mass of 12.9 g and a volume of 15 cm³. Determine the density of the box.

Solution

Step 1: Write the equation for density.

$$D = \frac{m}{V}$$

Step 2: Substitute the values for mass and volume given in the problem.

$$D = \frac{12.9 \text{ g}}{15 \text{ cm}^3}$$

Step 3: Divide to find the answer.

$$\frac{12.9 \text{ g}}{15 \text{ cm}^3} = 0.86 \text{ g/cm}^3 \qquad D = 0.86 \text{ g/cm}^3$$

Review gram, milliliter, cubic centimeter, and the abbreviations for these units.

Ask students to compare the density of styrofoam with a solid plastic and to explain the difference.

FIGURE 2–6. The density of a bowling ball is greater than the density of a volley ball.

Latent Image

How do you calculate density?

Enrichment: Find the mass of a 25 mL graduate with a balance. Fill the graduate to the 25 mL mark with water. Then find the combined mass and calculate the density of the water in g/mL.

2:3 *Density* 27

activity

See Teacher's Guide.

Activity 2-2
FINDING THE DENSITY OF WOOD

Find the density of two types of wood: oak and pine. Obtain a block of each wood. Measure the length, width, and height of each block in centimeters. Find the volume of each block. Use a balance to find the mass of each block in grams. Calculate the density. Which type of wood is more dense? oak

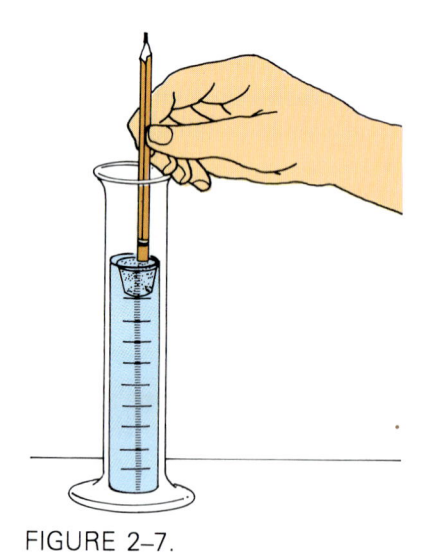

activity

Activity 2-3
FINDING DENSITY USING A GRADUATED CYLINDER

(1) Obtain a rubber stopper and a cork about the same size. (2) Find and record the mass of each with a balance. (3) Fill a graduated cylinder to the 15-mL mark with water. (4) Drop the cork into the water and push it down gently until it is just below the surface. How many milliliters does the water rise? The change in water level is the volume of the cork. Record the volume. (5) Remove the cork and make sure the volume of the water is at the 15-mL mark. (6) Find and record the volume of the rubber stopper. Use the same procedure as Step 4. Calculate the density of the stopper and cork. How do they compare? The rubber stopper is more dense than the cork.

FIGURE 2-7.

making sure

See Teacher's Guide.

5. Why is aluminum used to make airplanes?
6. What volume of air would have the same mass as one cubic centimeter of water?

1.0 g/cm^3 ÷ 0.0013 g = 769.2 cm^3

7. A metal block has a mass of 418.7 g and a volume of 53 cm^3. Use Table 2-2 to identify the metal of which the block is made. density = 7.9 g/cm^3; iron

block = 60 cm^3
density = 16.7 g/cm^3

8. A block of "gold" has a length of 5 cm, a width of 4 cm, and a height of 3 cm. Its mass is 1 kg. Determine if the block is pure gold. (*Hint:* You will need to use Table 2-2 to determine the answer). not pure gold because density is too low

28 Matter

2:4 Elements

All matter consists of simple substances called elements. An **element** is a substance that cannot be broken down into simpler substances by ordinary chemical means.

Gold, carbon, lead, and oxygen are elements. At least 109 elements have been discovered. At room temperature, most of these elements are solids and some are gases. Two, mercury and bromine, are liquids.

Of the 109 known elements, 90 occur naturally on the earth. The other elements have been made in experiments. Scientists are still searching for new elements. Thus, the total number of known elements may increase.

A **symbol** is a shorthand method or abbreviation for writing the name of an element. Each element has a different symbol containing one or two letters. H is the symbol for hydrogen. Fe is the symbol for iron. Note in Table 2–3 that a symbol has a capital letter. When there are two letters in a symbol, the second letter is a small letter.

Table 2–3.
Some Elements and Their Symbols

Element	Symbol	Element	Symbol
Aluminum	Al	Oxygen	O
Calcium	Ca	Silicon	Si
Carbon	C	Sodium	Na
Gold	Au	Sulfur	S
Lead	Pb	Zinc	Zn
Magnesium	Mg	Californium *	Cf
Mercury	Hg	Einsteinium *	Es
Nitrogen	N	Plutonium *	Pu

*synthetic elements

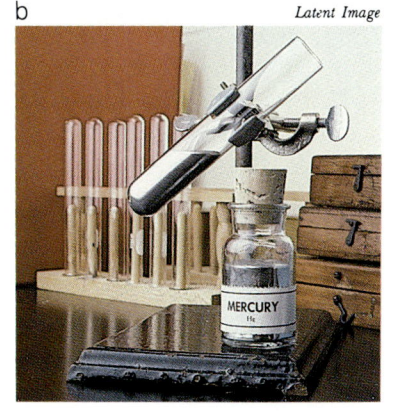

FIGURE 2–8. Copper is an example of an element that is a solid (a). Mercury is a liquid element (b). Chlorine is a gas (c).

making sure

9. Write the symbols for the following elements: aluminum, gold, carbon, lead, oxygen. Al, Au, C, Pb, O

10. List one property of each element in Question 9.
Al—silver color, Au—yellow color, C—black color, Pb—high density, O—colorless gas

Stress the fact that atoms are very tiny and cannot be seen with ordinary microscopes. Some Scanning Transmission Electron Microscopes can pick up the image of individual atoms.

Name three kinds of particles in an atom.

Point out the meaning of prefix *sub-* in subatomic.

Draw diagrams of different atoms on the chalkboard showing the protons, neutrons, and electrons they contain.

2:5 Atoms

Every element is made of tiny particles called **atoms.** Atoms are so small you cannot see them with a microscope. One part of an atom is the central core, called the **nucleus.** Around the nucleus is a sphere or cloud of negative charges. Inside the nucleus are two types of tiny subatomic particles. A subatomic particle is a particle within an atom. One type of subatomic particle is the **proton** which has a positive charge. Another kind of subatomic particle in the nucleus is the **neutron.** It has the same mass as a proton. A neutron, however, does not have an electrical charge. Both the proton and neutron have a very high density. For this reason, almost all the mass of an atom is in the nucleus.

Outside the nucleus are particles called **electrons**. The number of electrons in an atom is equal to the number of protons in the nucleus. An electron has a negative electric charge and its mass is almost zero. Electrons move about the nucleus at very high speeds. The rapid movement of the electrons creates the cloud of negative electricity that is part of the atom. Locate the parts of an atom in Figure 2–9.

Electrons in an atom are arranged in different energy levels. The first energy level, which is closest to the nucleus, can have no more than two electrons. The second energy level can contain a maximum number of eight electrons. The energy levels for atoms and the number of electrons they can hold are shown in Table 2–4.

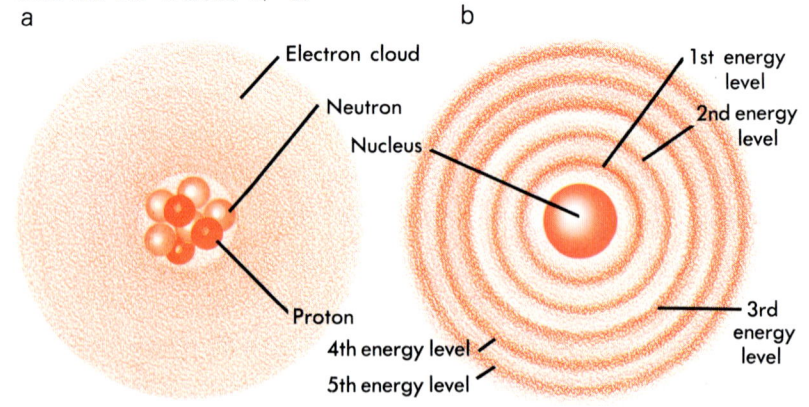

FIGURE 2–9. Protons and neutrons are found in an atom's nucleus. An electron cloud surrounds the nucleus (a). Electrons are arranged in energy levels (b).

30 Matter

Table 2–4.
Electrons in Energy Levels

Level	Maximum Number of Electrons in Each Level	Maximum Number of Electrons in the Atom
1	2	2
2	8	10
3	18	28
4	32	60
5	50	110

Atoms of different elements have different numbers of protons. For example, a hydrogen atom has one proton in its nucleus. An oxygen atom has eight protons. There are 26 protons in an iron atom. The properties of elements are determined by the number of particles in their atoms.

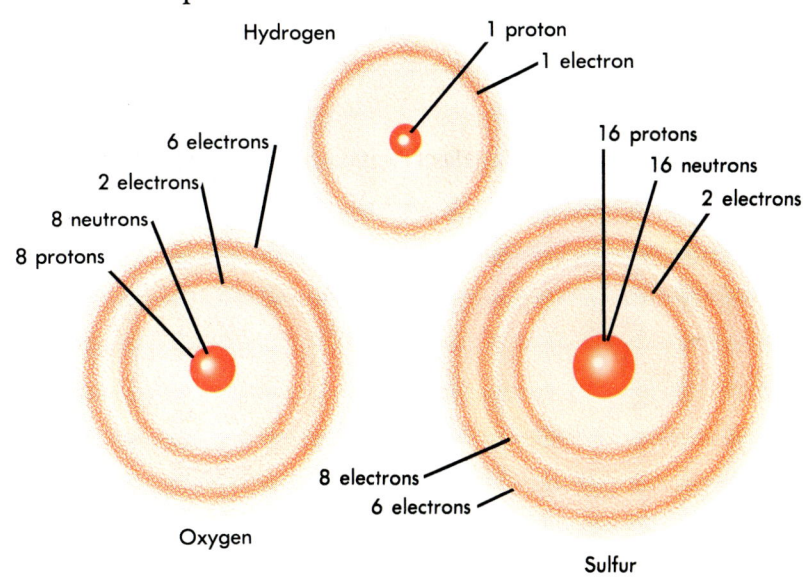

How is an oxygen atom different from a hydrogen atom?

Physical and chemical properties are determined by the electron configuration of the atoms. Nuclear reactions such as fission and fusion depend on the number of protons and neutrons in the nuclei of the elements.

FIGURE 2–10. The properties of hydrogen, oxygen, and sulfur are determined by the number of particles in their atoms.

making sure

11. What is the relationship between protons and electrons in an atom? *The number of electrons in an atom is always equal to the number of protons in the nucleus.*
12. How many electrons are in a hydrogen atom? Oxygen atom? Sulfur atom? *1, 8, 16*
13. How many energy levels are in an oxygen atom? *2*

How is a compound different from an element?

2:6 Compounds

Elements combine to form substances called compounds. **A compound** contains two or more elements joined by a chemical bond. The bond makes it difficult to separate the elements. Water, sugar, salt, soap, and rust are examples of compounds.

Tim Courlas

FIGURE 2-11. When atoms of copper (a), sulfur (b), and oxygen are joined by chemical bonds, copper sulfate (c) is formed.

How do atoms bond?

Use diagrams on a chalkboard to show how atoms bond in a covalent bond and an ionic bond.

When elements combine to form compounds their atoms gain, lose, or share electrons. The electrons involved in bonding are those in the outermost energy level. For example, sodium combines with chlorine to produce sodium chloride. Table salt is the common name for sodium chloride. In the process of bonding, the sodium atom loses one electron and the chlorine atom gains one electron. An atom that loses one or more electrons gains a positive charge. Why? The atom now has more

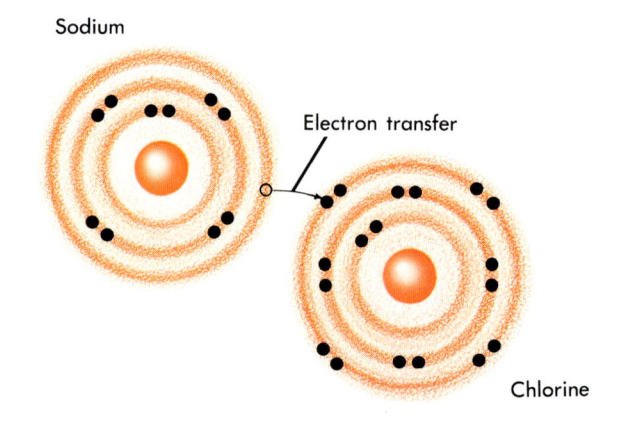

FIGURE 2-12. Sodium chloride (table salt) is formed when one electron from the sodium atom is transferred to an atom of chlorine.

32 Matter

protons than electrons. An atom that gains electrons becomes negatively charged because it has more electrons than protons.

An atom with an electric charge is called an **ion.** Sodium chloride consists of sodium and chlorine ions held together by chemical bonds. These bonds are called ionic bonds. An **ionic bond** forms when electrons are transferred from one atom to another.

A **covalent** (koh VAY lunt) **bond** is formed when atoms share electrons. For example, oxygen and hydrogen may combine to form water. Water consists of molecules. **Molecules** are particles that form when atoms are joined by covalent bonds. Each molecule of water contains two hydrogen atoms and one oxygen atom. See Figure 2–13.

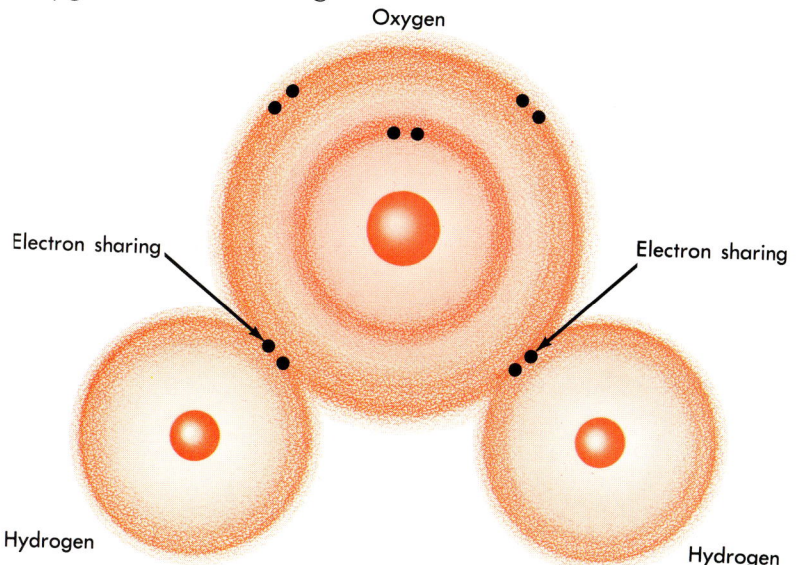

What is an ion?
A substance formed when atoms are joined by ionic bonds is called an ionic compound. Salt is one example of an ionic compound.

What is the difference between an ionic bond and a covalent bond?

FIGURE 2-13. A molecule of water contains two covalent bonds. The oxygen atom shares two electrons with each of the hydrogen atoms.

Table 2–5.
Some Compounds and Their Elements

Compound	Elements
Ammonia	Nitrogen, hydrogen
Carbon dioxide	Carbon, oxygen
Rust	Iron, oxygen
Sand	Silicon, oxygen
Sugar	Carbon, hydrogen, oxygen
Sulfuric acid	Hydrogen, sulfur, oxygen

FIGURE 2–14. A molecular model shows the kinds and number of atoms in a molecule.

How is a subscript used in a formula?

A symbol may be used to represent an atom of an element and a formula may be used to represent a molecule of a compound.

carbon, oxygen

Stress that subscripts must be precise. For example, Fe_3O_4 is a different compound from Fe_2O_3.

Write three formulas.

You may wish to begin this activity by making a model of an ammonia molecule. Point out that the carbon atoms in sugar are bonded in a chain.

2:7 Formulas

Just as a symbol is an abbreviation for an element, a **formula** is used to show a compound. A formula may have one or more symbols. For example, NaCl is the formula for sodium chloride. This formula contains the symbol for sodium and the symbol for chlorine. H_2O is the formula for water. What symbols does it contain?

Formulas are also used to show substances like hydrogen gas. The formula H_2 shows that a hydrogen molecule has two atoms of hydrogen. O_2 is the formula for oxygen gas, which has two atoms in its molecule. The small numbers in a formula, called **subscripts,** show how many atoms are in a molecule of the substance.

Subscripts also show the ratio between elements in a compound. The subscript 2 in H_2O means two parts of hydrogen to one part of oxygen, a ratio of 2:1. CO_2 is the formula for carbon dioxide. What elements are in carbon dioxide? What is the ratio between the elements in this compound? 2 oxygen to 1 carbon

Table 2–6.
Some Compounds and Their Formulas

Compound	Formula
Ammonia	NH_3
Carbon dioxide	CO_2
Rust	Fe_2O_3
Sand	SiO_2
Sugar	$C_{12}H_{22}O_{11}$
Sulfuric acid	H_2SO_4

activity

MOLECULAR MODELS Activity 2-4

Models of molecules may be made with clay spheres and toothpicks. Different colors of clay are used to represent different kinds of atoms. Obtain some modeling clay and make models of three of the molecules listed in Table 2–6. Are your models exactly like real molecules? Explain. See Teacher's Guide.

making sure

14. Each formula below represents one unit of a compound. How many and what kinds of atoms are in each unit?
(a) N_2
(b) Cl_2
(c) HCl
(d) H_2S
(e) CCl_4
(f) N_2O_5
(g) Al_2O_3
(h) CH_4

(a) 2 nitrogen
(b) 2 chlorine
(c) 1 hydrogen
  1 chlorine
(d) 2 hydrogen
  1 sulfur
(e) 1 carbon
  4 chlorine
(f) 2 nitrogen
  5 oxygen
(g) 2 aluminum
  3 oxygen
(h) 1 carbon
  4 hydrogen

15. What is the ratio between the elements in Questions 14c–14h?
(c) 1:1
(d) 2:1
(e) 1:4
(f) 2:5
(g) 2:3
(h) 1:4

2:8 Mixtures

Salt water is a mixture. **A mixture contains two or more elements or compounds that are mixed together but not chemically joined.** The parts of a mixture can be separated by physical means such as sifting, pouring, or evaporating. For example, salt water can be separated into salt and water through the process of evaporation.

How is a mixture different from a compound?

Air and soil are also mixtures. Air is made up of oxygen, nitrogen, and other gases. Soil is made up of rock, sand particles, and decayed plant and animal matter. Describe one way to separate soil into its parts. Soil can be separated into its component parts by sifting it through screens of different size.

FIGURE 2–15. A screen may be used to separate the large and small particles in soil.

One kind of mixture is a solution. In a solution one or more substances are dissolved in another substance. The particles in the solution are so small they cannot be seen with the naked eye. Water solutions are the most common because so many substances dissolve in water. Milk, soft drinks, and soapy water are examples of water solutions.

Another kind of solution occurs when a substance is dissolved in a gas. Air is a solution containing oxygen, nitrogen, carbon dioxide, and other gases.

In a solid solution, solids, liquids, or gases are dissolved in solids. For example, many crystals contain water molecules mixed together with the atoms in the crystal. Most alloys (AL oyz) of metals are also solid solutions. An **alloy** is a mixture of two or more metals. Stainless steel is an alloy of chromium metal in iron. Yellow gold is a solution of gold, silver, and copper.

Name three kinds of solutions.

FIGURE 2–16. Bubbles form when gas that was dissolved in the water escapes (a). An opal consists of crystals that contain water molecules (b).

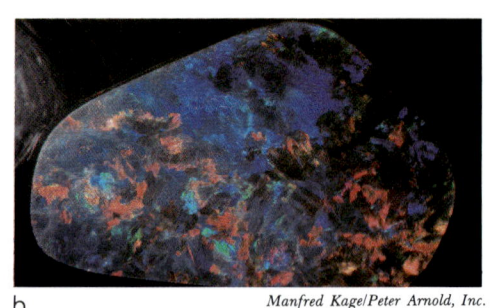

a *Hickson-Bender Photography*   b   *Manfred Kage/Peter Arnold, Inc.*

activity

Activity 2-5

SEPARATING A MIXTURE See Teacher's Guide.

Steps 1 and 2 may be done by students. However, you may wish to do steps 3 through 8 as a teacher demonstration.

(1) Make a mixture by stirring iron filings and sulfur together. (2) Pass a magnet over the mixture. What happened to the iron filings (Figure 2–17)? The filings separate. (3) Now place a mixture of iron filings and sulfur in a test tube. (4) Heat the mixture until it glows (Figure 2–17). **CAUTION:** Do the heating under an exhaust hood or other place where there is good ventilation. (5) Let the test tube cool. (6) Wrap the cooled test tube in a paper towel and place it on a table. (7) Crack the test tube inside the towel by striking it with a hammer. (8) Open the towel and try to remove the iron filings with a magnet. What happened? Why? The filings do not separate. The iron and sulfur formed the compound iron sulfide.

FIGURE 2–17.

activity

MAKING SOLUTIONS

Activity 2-6

Objective: To observe properties of different solutions   See Teacher's Guide.

FIGURE 2–18.

Materials

alcohol
bromothymol blue
dropper
lime (calcium oxide)
mineral oil
phenol red
soda straw
soil
5 test tubes/stoppers
test tube rack

Procedure

1. Label the test tubes A, B, C, D, and E. Set them in the test tube rack.
2. Fill test tubes A and C one-half full of water. Add 5 drops of mineral oil to test tube A and 3 g of soil to test tube C.
3. Fill test tube B one-half full of alcohol. Add 5 drops of mineral oil to the alcohol.
4. Fill test tube D one-half full of phenol red. Add 3 g of lime to the test tube.
5. Place the stoppers in test tubes and mix by carefully shaking each test tube. Record any changes that occur.
6. Fill test tube E one-half full of bromothymol blue. Place a straw in the test tube and carefully bubble your breath through the liquid for a few minutes. Record any changes that occur.

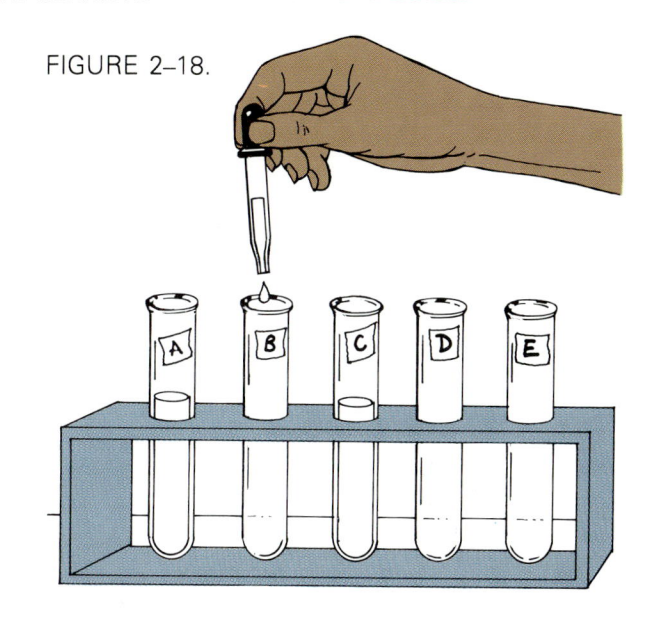

Observations and Data

Test Tube	Observations
A	Oil and water did not mix
B	Oil dissolved in alcohol
C	Soil did not dissolve
D	Changed colors
E	Changed colors

Questions and Conclusions

1. Which test tube(s) did not contain a solution? A, C
2. Which test tube(s) contained a gas dissolved in a liquid?   E
3. How could you determine if some substances were dissolved in test tube C?   Possible answers may include using phenol red, using bromothymol blue, and filtering then evaporating the water.

2:8 Mixtures   37

PERSPECTIVES
skills

Using Tables

Being able to use tables quickly and accurately is important in the study of science. They organize and classify large amounts of scientific information. In this book, tables have the following characteristics: (1) The information is condensed in as few words as possible. (2) Each table is about one kind of information. (3) The information is arranged in columns. (4) Each table is related to the information next to it in the text. The following activities will help you become better at using tables.

Use Table 2–1 to answer the following questions.

1. What type of information does this table present?
2. What two types of information are given for each "state"?
3. Using the table, can you answer Question 4 of Making Sure?

Use Table 2–2 to answer the following questions.

4. What is the density of water?
5. In what order are the substances listed? Does this order seem logical?

Use Table 2–3 to answer the following questions.

6. What type of information is given? Why do you suppose four columns were used?
7. Which would be easier to use in locating the information—Table 2–3 as it is shown or each item of information in a separate sentence? Why?

Use Table 2–4 to answer the following questions.

8. Why is it not possible to answer Questions 11 and 12 of Making Sure from the table alone?
9. In order to make use of Table 2–4, what other information do you need? Where do you find the additional information?
10. At what energy level would there be a maximum number of 32 electrons?
11. How could the maximum number of electrons in each atom be 60 when the energy level can only contain 32 electrons?
12. Can you easily see the relationship between the two columns showing the maximum number of electrons in each level and in the atom? Explain.
13. Is the relationship you get from the table made clear in the text? Explain.

Use Tables 2–5 and 2–6 to answer the following questions.

14. What relationship does the table show you between the compounds and elements?
15. What is the relationship btween Table 2–5 and Table 2–6?
16. Sugar is a compound and consists of three elements. What element occurs in the greatest amount? Where are you told about how to find this information?

After completing the activities you have probably concluded the following ideas about the use of tables:
(1) Information that is listed in column form is easier and quicker to locate.
(2) Columns of related information makes it easier to see the organization and classification of the information.
(3) Many times the information in the text helps to explain information in the table. The opposite is also true. The text and the table used together make the information clearer.

main ideas

Chapter Review

1. Properties are used to identify a substance.	2:1
2. Matter exists in four states—solid, liquid, gas, and plasma.	2:2
3. Mass and volume are needed to determine density.	2:3
4. There are 109 known elements. Of these elements, 90 occur naturally on Earth.	2:4
5. A symbol is used to represent an element. A formula is used to represent a compound.	2:4
6. An atom is the smallest part of an element.	2:5
7. All matter is composed of atoms.	2:5
8. Protons and neutrons are in the nucleus of an atom. Electrons are outside the nucleus.	2:5
9. A molecule contains two or more atoms which are chemically joined.	2:6
10. A formula may contain one or more symbols.	2:7

vocabulary

Define each of the following words or terms.

alloy	gas	nucleus
compound	ion	plasma
covalent bond	ionic bond	proton
density	liquid	solid
electron	molecule	subscript
element	mixture	symbol
formula	neutron	

study questions

DO NOT WRITE IN THIS BOOK.

A. True or False

Determine whether each of the following sentences is true or false. If the sentence is false, rewrite it to make it true.

F   1. The properties of solids and liquids are ~~the same.~~ different

T   2. Liquids and gases take the shape of their container.

T   3. Plasma is a state of matter.

T   4. Density is the mass per unit of volume of a substance.

Matter 39

F 5. An ~~element~~ is a substance in which two or more compound ~~compounds~~ are chemically joined. elements

T 6. The compound NH₃ contains nitrogen and hydrogen.

T 7. Matter is made of small particles called atoms.

T 8. The mass of a proton is greater than the mass of an electron.

F 9. ~~Both~~ a proton ~~and an electron have~~ a positive charge. has

T 10. Atoms join together to form molecules.

B. Multiple Choice

Choose the word or phrase that completes correctly each of the following sentences.

1. (*Liquids, Gases, Solids*) have a definite volume and a definite shape.
2. The number of electrons in an atom is (*equal to, less than, more than*) the number of protons.
3. (*Fe, O, O_2, Fe_2O_3*) is the formula for rust.
4. (*Water, Carbon dioxide, Hydrogen, Air*) is a mixture.
5. Salt water is a(n) (*element, compound, mixture*).
6. (H_2, N_2, O_2, H_2O) is the formula for oxygen gas.
7. In a water molecule (H_2O) there are (*3, 2, 6, 4*) atoms.
8. In a molecule of table sugar ($C_{12}H_{22}O_{11}$) there are (*12, 22, 11, 45*) atoms.
9. (*Protons, Neutrons, Electrons*) move around the nucleus of an atom.
10. The (*proton, neutron, electron*) has a positive electric charge.

C. Completion

Complete each of the following sentences with a word or phrase that will make the sentence correct.

1. The __mass__ of an atom is almost entirely in its nucleus.
2. There are always the same number of __protons__ and __electrons__ in an atom.
3. A compound contains two or more __elements__.
4. __H_2__ is the formula for hydrogen gas.
5. The parts of a __mixture__ can be separated by physical means.
6. The symbol Fe stands for the element __iron__.
7. The formula NH₃ represents the compound __ammonia__.

8. Ammonia contains the elements <u>nitrogen</u> and <u>hydrogen</u>.
9. Au is the symbol for <u>gold</u>.
10. Carbon dioxide contains <u>two</u> oxygen atoms.

D. How and Why See Teacher's Guide at the front of this book.
1. Tell whether each of the following substances is an element, compound, or mixture: oxygen, soil, nitrogen, water, carbon dioxide, mercury.
2. How is a gas different from a liquid? How is a liquid different from a solid?
3. What elements are present in $NaCl$, H_2O_2, CaO, $ZnSO_4$?
4. A fluorine atom contains 9 protons, 10 neutrons, and 9 electrons. How many of these particles are in the nucleus?
5. Explain how a molecule is different from an atom.

challenges

1. Obtain small samples of five elements and five compounds. Tape each to a large piece of cardboard. Print the name of each element, its symbol, and three of its properties. List the name, formula, and three properties for each compound.
2. Make a list of five elements you know. Obtain information from a library on how each element was discovered.

interesting reading

Lightman, Alan P., "The Loss of the Proton." *Science 82,* September, 1982, pp. 22-24.

Preuss, Paul, "The Shape of Things to Come." *Science 83,* December, 1983, pp. 80-87.

SIDE ROADS
Super Conductors

Latent Image *Doug Martin*

One important property of a substance is the ability to conduct electricity. Substances that allow an electric current to flow through are called conductors. Metals such as gold, silver, copper, and aluminum are good conductors, because they have many electrons that are free to move. Substances that slow down electron flow are called resistors, because they resist the electric current. Substances through which electricity does not flow are called insulators, because electron flow is stopped. Glass, rubber, and many plastics are good insulators.

About 70 years ago, scientists discovered that some metals and alloys lose resistance to electric currents when cooled to very low temperatures. The metals are called superconductors, because electricity passes through them very rapidly. Superconductors are very important in science and technology, because of their ability to allow easy electron flow. Recently, chemists discovered that some groups of chemicals can also be considered superconductors at low temperatures. Some scientists feel that chemical superconductors at normal temperatures will be discovered in the near future.

The process of superconduction is being used today in powerful magnets to conduct electricity. These superconducting magnets allow generators to operate more efficiently, thus reducing the cost of electricity.

Millions of dollars are being spent in the research of superconductors for use in computers. Computers generate heat as current flows through the circuits. Therefore, circuits must be spaced far enough apart to prevent the system from overheating. If a computer could operate at lower temperatures, the electrical circuits could be packed more closely together. A large number of circuits enables the computer to work faster and to retain more information. The results of this research remain questionable, because other types of circuits may be developed that do not require low temperatures to function.

Doug Martin *Image Workshop*

Despite the low temperature requirement of superconductors, other applications are being studied. It is predicted that superconductors will be used in large generating plants, electric motors, electric storage systems, and wires. Scientists have considered building trains with powerful magnets that would elevate the train above the tracks. As research continues, it appears that superconductors will play a major role in modern technology.

Doug Martin

43

Energy is used in many ways. In the photograph, the welder is using energy to heat and melt metal. The welder uses the chemical energy of a gas to produce heat. Light is also produced. Heat and light are two of many forms of energy. What is energy? How is energy related to changes in matter? What changes in the picture are the result of energy?

Joe DiChello, Jr.

Energy and Changes in Matter

chapter 3

Introducing the chapter: Place an ice cube in water in a beaker. Heat the water with a Bunsen burner until the ice melts and the water begins to boil. Discuss the use of heat energy to melt ice and boil water. Have students make a list of different examples of energy such as heat, light, electricity, and the kinetic energy of running water. Discuss the examples with the class and use this introduction as a springboard to the material in section 3:1.

3:1 Energy

Everything you do requires energy. Reading, running, throwing, climbing, and eating are all activities that use energy. Energy is also needed for travel and industry and to warm our homes, schools, and factories. Energy is used in refrigeration, lighting, and cooking. **Energy is the ability to do work.** In science, work means to move something.

Table 3-1.
Major Uses of Energy in the United States

Use	Percent
Industry	36
Transportation	27
Homes	21
Business	16
Total	100

Energy can be divided into two kinds—potential and kinetic. **Potential energy is energy of position, or stored energy. Kinetic energy is energy of motion.** For example, moving water in a river has

GOAL: You will learn the sources and forms of energy and how energy is related to changes in matter.

What is energy?

Go over the scientific definition of energy and explain the meaning of work in a scientific sense.

Ask students where the energy they used yesterday has gone.

How is potential different from kinetic energy?

45

FIGURE 3–1. Much energy is needed for transportation.

Name five forms of energy.
Explain that the energy people use becomes heat in the atmosphere.

Introduce the connection between matter and energy. For example, electric energy to start a car comes from a change in the matter in a storage battery.

FIGURE 3–2. The potential energy of the water behind the dam changes to kinetic energy as the water falls over the dam.

kinetic energy. This kinetic energy becomes potential energy as the water reaches a dam. When the water falls over the dam the potential energy changes back to kinetic energy. Another example of energy change is found in a wind-up toy. The wound spring in the toy has potential energy. The potential energy changes to kinetic energy as the spring unwinds.

Energy has many forms. Light and heat are two forms of energy. Some others are electric, chemical, nuclear, and mechanical energy. Energy can be changed from one form to another. In a light bulb, electric energy is changed to light and heat. In a battery, chemical energy is changed to electric energy. In a solar water heater, light energy is changed to heat. All forms of energy can be changed to heat.

Remember that one kind of energy can be changed to another kind of energy. For instance, potential energy can be changed to kinetic energy. Kinetic energy can be changed to potential energy.

Energy can be released from matter and stored in matter. When a match burns, energy is released as heat and light. When water is heated to boiling, steam is formed. Energy is stored in the steam as potential energy. Energy is released when the steam

46 *Energy and Changes in Matter*

cools and changes back to liquid water. Energy is gained or lost as matter changes from one form to another.

activity   See Teacher's Guide.

KINETIC ENERGY AND HEAT   Activity 3-1

Obtain two paper cups. Fill one cup one-third full of dry sand. Place the end of a thermometer in the sand and record the temperature. Remove the thermometer and invert the second cup over the first cup. Seal the cups with a piece of tape. Shake the sand back and forth inside the cup. Insert the thermometer. Did you observe a change in temperature? Explain. Yes, shaking the sand increases the kinetic energy of the atoms and molecules within the sand particles.

FIGURE 3–3.

making sure

1. Do the following have potential energy or kinetic energy?
   - (a) wound-up spring
   - (b) ocean wave
   - (c) stretched rubber band
   - (d) rotating windmill

   (a) PE (c) PE
   (b) KE (d) KE

3:2 Heat and Temperature

If you hammer a nail into a block of wood, the nail becomes warm because it gains heat energy. When your hands are cold, you may try to warm them by rubbing your hands together. In the winter, people keep comfortable indoors by heating their homes. Heat for homes is often produced by burning fuels such as oil or natural gas.

Heat is the kinetic and potential energy of molecules within a substance. At the same temperature, which has more heat, a glass of milk or a pitcher of milk? The amount of heat in an object depends on both its temperature and mass. Since the pitcher contains a greater mass of milk, it contains more heat. If you have a pitcher of cold milk and one of warm milk, which has the most heat? Both the warm and cold milk have the same mass. Therefore, the pitcher with the higher temperature has the most heat.

Define the terms heat and temperature.

Temperature is an indicator of heat. **Temperature** is the average kinetic energy of the molecules of a substance. It tells how hot an object is and the direction heat will travel. Your body, for example, loses heat to the air because your body has a higher temperature than the air. If you heat a kettle of water on a gas stove, the water gains heat. The heat comes from the fire beneath the kettle. Heat travels from regions of higher temperatures to regions of lower temperatures.

Temperatures are measured using the Celsius temperature scale. On this scale, water freezes at zero degrees (0°) and boils at 100 degrees (100°). There are 100 degrees between the freezing and boiling points. Some common Celsius temperatures are listed in Table 3–2.

FIGURE 3–4. The frozen pond (a) contains more heat than the boiling water (b) since the pond contains a greater mass of water.

Table 3–2.
Some Celsius Temperatures

Boiling point of water	100°C
Human body (average)	37°C
Pleasant room temperature	21°C
Freezing point of water	0°C

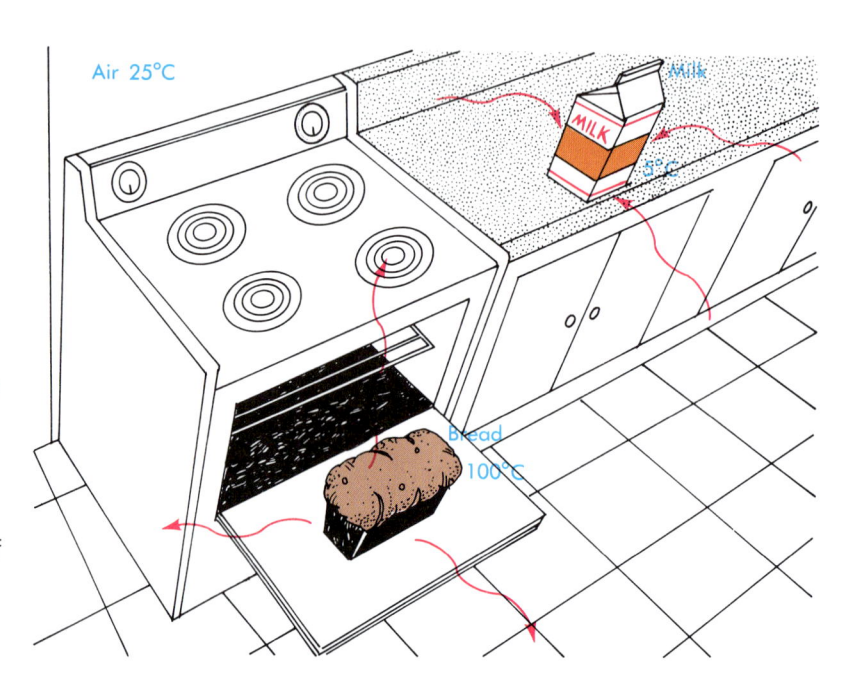

How is temperature measured?

The Celsius temperature scale is used by scientists throughout the world and has everyday use in many countries. Point out the increasing use of the Celsius scale in the United States.

FIGURE 3–5. Temperature determines the direction of heat travel. Eventually, the loaf of bread, the air, and the milk will have the same temperature.

48 *Energy and Changes in Matter*

activity

See Teacher's Guide.

GAINING AND LOSING HEAT

Activity 3-2

Objective: To determine the movement of heat by measuring temperature

Materials
balance
beaker, 250 mL
ice cube
stirring rod
stopwatch or watch with second hand
thermometer
tongs

Procedure

1. Determine the mass of the empty beaker.
2. Fill the beaker with 125 mL of water. Find the total mass of the beaker and the water. Determine the mass of the water by subtracting the mass of the empty beaker from the total mass of the beaker and water. Record the mass of the water.
3. Carefully place the thermometer into the water and record the temperature of the water. Note: Be certain that the thermometer does not touch the sides or bottom of the beaker.
4. Add the ice cube to the water and stir for 1 min. **CAUTION:** Use the stirring rod and not the thermometer to stir the water. After 1 min, measure and record the temperature of the water.
5. Continue stirring the water and ice and measuring the temperature for 5 min. Record the temperature after each minute.
6. After 5 min, remove any ice that is left. Stir the water gently again and measure the temperature each minute until the water returns to room temperature.
7. Find and record the mass of water in the beaker.

Observations and Data

Mass of the water before ice was added: __125__ g

Mass of the water after the ice melted: __175__ g

Time (min)	Temperature (°C)
0	10°
1	8°
2	5°
3	

See Teacher's Guide.

Questions and Conclusions

1. What was the lowest temperature to which the water cooled in the beaker?
2. How was the temperature of the water changed by the loss and gain of heat?
3. Where did the heat go when the water was cooled and warmed?
4. How did the rate in which the water cooled compare to the time it took for the water to return to room temperature?
5. How did the change in the mass of water affect the rate in which the water returned to room temperature?

3:2 Heat and Temperature 49

3:3 Freezing and Melting

When a liquid freezes, it changes to a solid. Water inside a freezer changes to ice. Just the opposite happens when something melts. During melting, a solid changes to a liquid. Freezing and melting are two kinds of physical changes.

Boiling, crushing, tearing, and grinding are also examples of physical change. A **physical change** occurs when a substance changes from one form to another without a change in its composition. Both ice and liquid water are composed of water molecules. Ice molecules are slightly farther apart and in a rigid solid position. The molecules in water can slide over and around each other. Energy may be gained or lost by a substance during a physical change.

At very high altitudes, ice particles in clouds are formed directly from water vapor. The water vapor forms solid ice without going through a liquid state. When carbon dioxide gas is cooled to a low temperature or placed under high pressure, it turns into a solid called dry ice. Dry ice is used to cool food and other things. When warmed, dry ice turns back into a gas without forming a liquid. The physical change in which a substance changes directly from a solid to a gas or gas to a solid is called **sublimation** (sub luh MAY shun).

FIGURE 3–6. A metal is melted to a liquid so it can be easily formed into different shapes.

What is a physical change?

Give two examples of physical change.

Stress that composition does not change in these physical changes, but there is always a gain or loss of energy.

FIGURE 3–7. Through the process of sublimation, solid iodine at the bottom of the container, changes to a gas. The gas returns to a solid at the top of the container.

50 *Energy and Changes in Matter*

making sure

2. Does a substance gain or lose heat when it condenses? **lose**
3. How is freezing different from sublimation?
4. Why is boiling an example of a physical change?
5. Frost consists of small ice crystals that form on windows and blades of grass. Explain the process by which frost forms during cold mornings.

3. freezing—liquid to solid; sublimation—gas to solid
4. change in state, but not in composition
5. By sublimation, water vapor in air forms ice crystals on cold surfaces.

3:4 Evaporation and Boiling

How do wet clothes dry when they are hung on a washline? The liquid water changes into water vapor, which enters the air. This change from a liquid into a gas is called **evaporation.** Heat energy is needed to evaporate a liquid. This explains why wet clothes dry faster on warm days than on cold days.

Condensation is the opposite of evaporation. In **condensation,** a gas changes to a liquid. You have seen the effect of condensation many times. Clouds contain tiny drops of water formed when water vapor changes to a liquid. Water forming on kitchen windows during the winter is a result of condensation. Gases lose heat energy as they become liquids. You can see the condensation of water vapor in your breath if you exhale on a mirror.

When a liquid is heated to its boiling point, the change from a liquid to a gas occurs rapidly. Just as in evaporation, heat is gained by a substance when it boils. If the heat source, such as a burner flame, is turned off, the boiling will stop. To keep a liquid boiling, heat must continuously be applied.

Latent Image

What happens to a liquid as it evaporates?

How is condensation different from evaporation?

The boiling point of a substance increases as the pressure increases. Boiling point is a property of a substance.

How is boiling different from evaporation?

FIGURE 3–8. Heat is gained by water as its temperature rises to the boiling point. The water vapor loses heat when it condenses.

FIGURE 3–9.

CAUTION: Ventilation may be inadequate if activity is done by the entire class. It may be preferable to have only one team per class perform the activity.

activity

SEPARATING PURE WATER FROM SALT WATER.   Activity 3-3

Soak a piece of sponge with salt water and set it in a small glass inside a glass jar. Place a plastic bag in the jar as shown in Figure 3–9. Set the jar in sunlight or under a lamp. After a few hours, note the changes. Where does evaporation occur inside the jar? Where does condensation occur? How is the water changed?   See Teacher's Guide.

activity

THE RATE OF EVAPORATION   Activity 3-4

(1) Label three beakers, *A*, *B*, and *C*. (2) Fill beaker *A* with 100 mL of water. Place the beaker outdoors where it will not be disturbed. (3) Fill beaker *B* with 100 mL of water and place it inside the classroom. (4) Fill beaker *C* with 100 mL of rubbing alcohol and set it next to beaker *B*. (5) Observe the beakers. Determine and record the amount of liquid that evaporates each day. (6) Draw a graph showing how much liquid evaporated from each beaker each day. In which beaker was evaporation fastest? The slowest? Explain why there was a difference. What factors affect the rate at which a liquid evaporates?

See Teacher's Guide.

making sure

6. Does water vapor gain or lose heat when it condenses on a cool window? lose
7. Why does alcohol placed on the skin feel cool? As the alcohol evaporates, it absorbs heat from the skin.

3:5 Chemical Change

What happens in a chemical change?

Distinguish between chemical change and physical change.

In a **chemical change,** new substances with different properties are produced. Elements and compounds may combine to form the new substance. A substance may break down into two or more elements or compounds. For example, rust is produced when iron combines with oxygen. A type of acid is formed when milk sours. Gases and ashes

Latent Image

are produced when wood burns. The new substances formed are no longer iron, milk, or wood. A chemical change has taken place in each case.

The bright light from a photo flashbulb is the result of a chemical change. Try to look at a flashbulb before and after the flash. You will see that the fine metal wire in the bulb changes to a white powder. In most flashbulbs, the fine wire is made of magnesium and the bulb is filled with oxygen. When the flashbulb is used, the magnesium (Mg) combines with the oxygen (O_2) and forms a white powder. The white powder is magnesium oxide (MgO) that forms during a chemical change.

Burning also involves a chemical change. When a substance burns in air, atoms of the substance combine with atoms of oxygen. Heat is always given off during burning. Thus, burning is a means of obtaining energy.

FIGURE 3-10. Magnesium oxide is formed during a chemical change when the flashbulb is used.

Name two examples of a chemical change.

activity

CHANGES IN A BURNING CANDLE Activity 3-5

Place a small candle on a watch glass and light it. Invert a glass jar over the candle and watch glass (Figure 3–11). What happened to the candle? Explain. Note the drops of water and the black substance on the inner surface of the jar. How can you explain these? Explain why the candle could not burn with the gas left in the jar. See Teacher's Guide.

FIGURE 3-11.

making sure

8. Which of the following chemical changes release energy?
   (a) burning of wood
   (b) decaying of cut grass
   (c) discharging a battery
   (d) charging a battery

(a) Heat and light are released.
(b) Heat is released.
(c) Electric energy is released.
(d) Energy is not released.

3:6 Chemical Equations

When a chemical change occurs, new substances are always formed. That is, there is a change in the arrangement of atoms. In some cases, atoms join to

Stress that substances are not used up in a chemical change.

3:5 Chemical Change

How is a chemical equation used to show a chemical change?

form molecules. In other cases, molecules break apart into atoms. These atoms may join together to form different molecules. However, atoms are never created or destroyed in a chemical change. There is the same number of each kind of atom at the end as there was at the start.

A chemical change may be shown by a word equation. A word equation shows both the substances that react and those that are formed. For instance, when hydrogen burns in air it combines with oxygen to form water. Below is the word equation for the change.

$$\text{hydrogen} + \text{oxygen} \rightarrow \text{water}$$

The symbol (+) means plus and the arrow ($\rightarrow$) means yield or produce. This word equation is read, "Hydrogen plus oxygen yield water."

A chemical change may also be shown by a balanced chemical equation. In a balanced chemical equation, symbols and formulas are used in place of words. Below is the equation for the formation of water.

What is a balanced chemical equation?

$$2H_2 + O_2 \rightarrow 2H_2O$$

In this reaction hydrogen combines with oxygen to form water. The number in front of a symbol or formula is called a **coefficient.** Coefficients are used to balance an equation. In a balanced chemical equation, the number of each kind of atom is the same on both sides of the arrow.

In the equation above, two molecules of hydrogen (H_2) combine with one molecule of oxygen (O_2). These substances are written to the left of the arrow. They are called reactants.

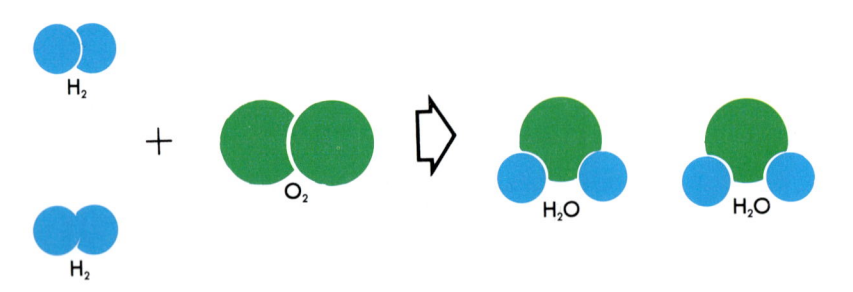

Present examples of word equations and chemical equations, such as:

hydrogen peroxide $\longrightarrow$ water + oxygen
$2H_2O_2 \longrightarrow 2H_2O + O_2$
carbon + oxygen $\longrightarrow$ carbon dioxide
$C + O_2 \longrightarrow CO_2$
sodium + chlorine $\longrightarrow$ sodium chloride
$2Na + Cl_2 \longrightarrow 2NaCl$

FIGURE 3–12. Two molecules of water are formed when four atoms of hydrogen react with two atoms of oxygen.

To the right of the arrow, symbols and formulas stand for the new substances that are formed. They are called products of the reaction. The $2H_2O$ in the equation shows that two molecules of water are formed. Water is the product of the reaction. A number in front of a symbol or formula shows the number of atoms or particles in the chemical change.

The word equation for the rusting of iron is

$$\text{iron} + \text{oxygen} \longrightarrow \text{iron oxide (rust)}$$

This can be read, "Iron plus oxygen yield iron oxide." Rust (iron oxide) is a compound of iron and oxygen.

The balanced chemical equation for rusting is

$$4Fe + 3O_2 \longrightarrow 2Fe_2O_3$$

This equation shows that four atoms of iron (4Fe) combine with three molecules of oxygen ($3O_2$) to form two molecules of rust ($2Fe_2O_3$). When iron rusts, millions of atoms of iron combine with millions of molecules of oxygen. The equation shows the ratio of atoms of iron, molecules of oxygen, and molecules of rust.

To write a balanced chemical equation:
1. Write the "word" equation.
2. Write the equation using symbols and formulas.
3. Add coefficients to balance the equation.

FIGURE 3-13. Rust forms in the presence of moisture.

Don Parsisson

making sure

9. Write a chemical equation for each reaction:
   (a) carbon + oxygen $\longrightarrow$ carbon dioxide $C + O_2 \longrightarrow CO_2$
   (b) water $\longrightarrow$ hydrogen + oxygen $2H_2O \longrightarrow 2H_2 + O_2$
   (c) iron + sulfur $\longrightarrow$ iron sulfide $Fe + S \longrightarrow FeS$

The equations need not be balanced.

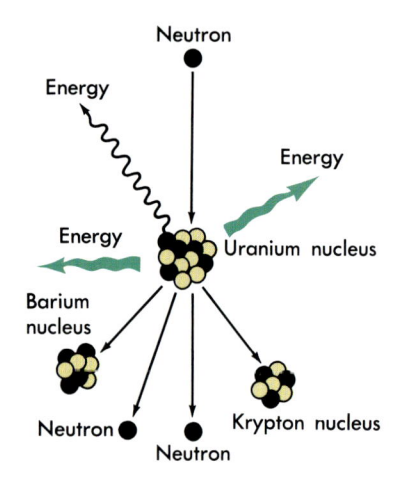

FIGURE 3–14. In nuclear fisson, a large nucleus splits to form two smaller nuclei.

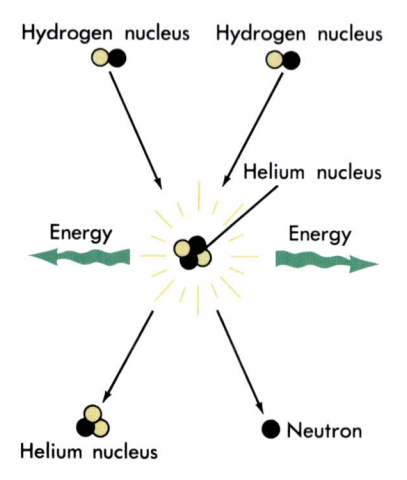

FIGURE 3–15. In nuclear fusion, small nuclei join together to form a larger nucleus.

How is nuclear fusion different from fission?

List five sources of energy.

What are fossil fuels?

3:7 Nuclear Change

A nuclear change differs from a chemical or a physical change. In a nuclear change, an element is changed into a different element. That is, the nucleus of an atom is changed. Energy is always released during a nuclear change.

There are two types of nuclear change—fission and fusion. Fission means to divide. In **nuclear fission,** a single, large nucleus splits into smaller nuclei. A uranium nucleus, for instance, may split into two nuclei—barium and krypton. Much energy is released during this change. A piece of uranium about the size of a golf ball can provide as much energy as one million kilograms of coal.

In nuclear fission, one or more neutrons are released along with the smaller nuclei. These neutrons can cause other atoms to split, thus producing a nuclear chain reaction. Since a chain reaction provides a steady supply of energy, it can serve as the energy source for an electric power plant.

Fusion is the opposite of fission. To fuse means to join. In **nuclear fusion,** two or more nuclei join to form one larger nucleus. For instance, hydrogen nuclei may fuse to form one helium nucleus. One gram of hydrogen can provide as much energy as 8 thousand kilograms of coal. Nuclear fusion takes place in the sun and stars.

3:8 Sources of Energy

Many people in modern countries depend upon a plentiful supply of energy at a reasonable cost. Most of the energy we use comes from fossil fuels, hydroelectric dams, nuclear power plants, and the sun. Other sources of energy being developed include geothermal (jee oh THUR mul) and wind power.

Coal, oil, gasoline, and natural gas are fossil fuels. **Fossil fuels** were formed from plants and animals

56 *Energy and Change in Matter*

FIGURE 3–16. Fossil fuels may be burned to produce electric energy.

that died many, many years ago. Fossils are the preserved remains of once living things.

When a fossil fuel such as coal is burned, light and heat are released. The burning of fossil fuels is an example of chemical change. Energy stored in the fuel is released as light and heat. When fossil fuels burn, carbon dioxide and water are formed.

Hydroelectric power is produced by moving water. Water from a river is trapped in a lake behind a dam. The water from the lake is allowed to fall through a shaft in the dam to a stream below. As the water falls, it flows through a turbine. The kinetic energy of the falling water is transferred to the moving turbine blades. As the turbine rotates, it turns an electric generator (JEN uh ray tur) which produces the electricity.

Geothermal energy refers to heat energy within the earth. In some areas, wells can be drilled into water heated by hot rock deep below the surface. Steam or very hot water is forced up the well and is used to generate electricity.

A nuclear power plant has a nuclear reactor (ree AK tur) that contains a radioactive fuel element such

Have students make a short outline of this section in which they list the sources of energy and important facts about each source. Review the outlines with your class and then give a short quiz on the material.

What is hydroelectric power?

How is geothermal energy obtained?

3:8 Sources of Energy 57

FIGURE 3-17. In a nuclear power plant, heat energy released during nuclear fission is used to generate electricity.

as uranium. Fission reactions in the nuclear fuel produce heat. This heat is carried out of the nuclear reactor by a water cooling system and is used to produce steam. The steam is used to produce electricity.

No way has yet been found to control nuclear fusion. Many problems must be solved before nuclear fusion can provide us with a steady supply of energy. For example, scientists must learn how to create a temperature of several million degrees. Nuclear fusion may someday provide an almost endless supply of energy for us on Earth.

Solar energy is energy that comes from the sun. It is produced by the nuclear fusion reactions of the hot gases that make up the sun. Solar energy may someday become a major energy resource. It is present in an almost endless supply, and it does not cause pollution. It presents little danger from accidents. One problem in the use of solar energy is collecting a large amount of it. One way is to use large curved mirrors that focus the sun's energy on a small area. Another way is through the use of a large surface area for collecting the energy.

FIGURE 3-18. Some homes use solar panels on the roof to collect the sun's energy for heating.

How is solar energy stored?

Solar energy is absent at night and greatly reduced when the sky is cloudy. Thus, there must be a means of storing solar energy. One way to store solar energy is by heating water or rocks. Another way

58 *Energy and Changes in Matter*

is to use solar cells that convert light into electric energy which is stored in batteries.

Solar energy is now being used in some places for cooking, heating, and cooling. Heating water for homes is an important practical use. Some buildings are being built with panels for collecting solar energy. Often the panels are placed on the roof.

See Teacher's Guide. Caution students to set the bottles where they will not be knocked over and broken.

activity

HEAT FROM THE SUN   Activity 3-6

Obtain two glass bottles. The bottles should be the same size and have a narrow neck. One should be clear glass and the other dark, colored glass. Attach a balloon over the neck of each bottle. Place the bottles in direct sunlight. What changes did you observe? What caused these changes? The balloon on the dark bottle inflated more than the balloon on the clear bottle. The dark bottle absorbed more energy than the clear bottle. As a result, the air in the dark bottle was heated more and expanded more than the air in the clear bottle.

activity

See Teacher's Guide.

COLLECTING SOLAR ENERGY   Activity 3-7

Tie a piece of thread to a thumbtack. Push the tack into the bottom of a cork that will fit tightly in the mouth of a clear glass bottle. Put the cork in the top of the bottle so that the thread hangs inside. Use a magnifying glass to focus the sunlight onto the thread. What happened? Will you get the same results if you set the bottle in the sun and do not use the magnifying glass? Explain. The thread began to burn. No, without the lens, the energy is not focused enough to ignite the thread.

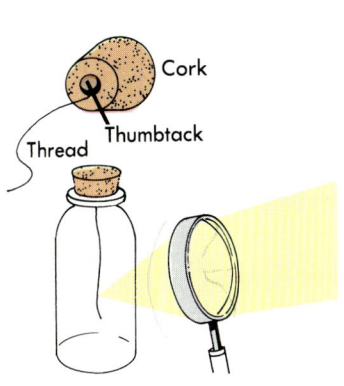

FIGURE 3–19.

making sure

10. Name three sources of energy you used today. Answers will vary. Some examples are sun, natural gas, gasoline, oil, coal.
11. What sources of energy are used most for transportation? gasoline and diesel fuel
12. Why are solar and hydroelectric power called "clean" sources of power? They do not produce pollution.
13. What can people do to reduce the waste of energy resources? Answers will vary. See Teacher's Guide.

PERSPECTIVES

frontiers

Fluidized Bed Combustion

Much of our energy is obtained from burning fossil fuels such as coal, natural gas, and petroleum. For years, burning coal was criticized as a major source of air pollution and cause of acid rain. Power companies switched to other fuel sources, such as nuclear fission reactions. But recently, coal has been regaining popularity as a fuel source, for a variety of reasons.

The economics and safety of nuclear power have been questioned for several years. Construction costs of nuclear reactors have soared and several "accidents" have released radioactivity into the environment. Petroleum reserves are expected to be depleted before coal, and will not provide a stable source of fuel for the future. On the other hand, coal is one of the most abundant resources, inexpensive to mine, and safe to handle.

A recent technological advance has also made coal more promising as a fuel, by cutting down on air pollution. The process is called fluidized bed combustion, and requires coal and limestone to be ground into small pieces and blown into a combustion chamber. Air currents are then pumped into the combustion chamber through holes in the bottom. Constant air flow keeps the coal mixture aloft, where it burns easily. The chemical bonds in the coal are broken, and heat and light energy are given off. Limestone retains the heat, helps the coal burn better, and captures most polluting gases. The energy produced is used to heat water into steam. The steam then turns generators to make electricity.

Conventional coal burning does not use air currents. Coal lays on a hot plate, and does not always burn completely. This method leaves many chemical bonds unbroken, and gives off less energy than fluidized bed combustion.

Fluidized bed combustion also has a number of advantages as a method of generating electricity. Besides coal, other fuel sources include wood, municipal solid wastes, paper mill wastes, and sewage sludge. Another advantage is that the fuel burns at a constant rate, resulting in more efficient power generation.

The cost of constructing a fluidized bed system is relatively low when compared to other methods of producing energy. Pollution from a fluidized bed is less than any conventional method of burning coal. By adjusting the amount of limestone pumped into the chamber, different grades of coal can be used. This would allow the burning of high sulfur coal that is most abundant, but also the worst polluting type of coal.

Fluidized bed combustion currently is being used in a variety of situations. Some problems must be overcome before it can be successfully applied to large scale operations. Much work is being done in the Ohio and Tennessee Valleys, where high sulfur coal is abundant and power plants are numerous. Other methods of generating electricity may work better in some situations, but the future for fluidized bed combustion appears promising.

Gary Milburn/Tom Stack & Associates

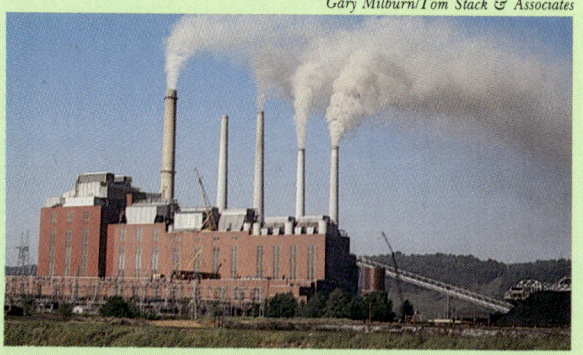

main ideas — Chapter Review

1. Energy exists in several forms and can be converted from one form to another.   3:1
2. Potential energy is energy of position. Kinetic energy is energy of motion.   3:1
3. Energy is stored in matter or released from matter in a physical, chemical, or nuclear change.   3:1, 3:5, 3:7
4. Temperature is an indicator of heat.   3:2
5. A physical change is a change in the properties of a substance without a change in its composition.   3:3; 3:4
6. During a chemical change, new substances with different properties are formed.   3:5
7. A chemical change may be shown by an equation.   3:6
8. Nuclear fission and nuclear fusion are two kinds of nuclear change.   3:7, 3:8

vocabulary

Define each of the following words or terms.

chemical change
coefficient
condensation
energy
evaporation
fossil fuel
geothermal
hydroelectric
kinetic energy
nuclear fission
nuclear fusion
physical change
potential energy
solar energy
sublimation
temperature

study questions

DO NOT WRITE IN THIS BOOK.

A. True or False

Determine whether each of the following sentences is true or false. If the sentence is false, rewrite it to make it true.

T   1. Potential energy can be changed to kinetic energy.

F   2. In a ~~physical~~ change, new and different substances are formed.   chemical

T   3. Water and ice both have the same chemical formula.

T   4. A chemical change takes place when a substance combines with oxygen.

T   5. Electric energy can be changed to heat energy.
F   6. ~~Potential~~ energy is energy of motion. *Kinetic*
T   7. Water behind a dam has potential energy.
T   8. A liquid absorbs heat when it evaporates.
F   9. Water boils at ~~37°C~~. *100°C*
F   10. When iron rusts, it combines with ~~nitrogen~~. *oxygen*

B. Multiple Choice

Choose the word or phrase which completes correctly each of the following sentences.

1. When a substance burns in air, it combines with (nitrogen, <u>oxygen</u>, carbon, hydrogen).
2. In a chemical equation the symbol → means (<u>yields</u>, plus, balanced).
3. In a (<u>chemical</u>, physical, nuclear) change, hydrogen combines with oxygen to form water.
4. (Coal, Oil, Natural gas, <u>Uranium</u>) releases nuclear energy.
5. Nuclear (fission, burning, <u>fusion</u>) occurs in the sun.
6. An element is changed into another element during a (chemical, physical, <u>nuclear</u>) change.
7. (Potential, Chemical, <u>Solar</u>) energy may be collected with large curved mirrors or through the use of large surface areas.
8. In a battery, chemical energy is changed to (<u>electrical</u>, solar, chemical) energy.
9. One example of a fossil fuel is (hydrogen, oxygen, uranium, <u>coal</u>).
10. (<u>Hydroelectric</u>, Nuclear, Geothermal) energy is produced by falling water.

C. Completion

Complete each of the following sentences with a word or phrase which will make the sentence correct.

1. Rusting is a(n) *chemical* change.
2. A(n) *physical* change takes place when a mixture is separated into its parts.
3. Freezing is caused by the __*loss*__ of heat.
4. In a balanced chemical equation, *symbols* and *formulas* are used to represent the reactants and products.

5. Burning is an example of a(n) __chemical__ change.
6. A(n) _____ reaction can provide a steady supply of energy.   __nuclear chain__
7. Hydroelectric power is produced by falling __water__.
8. _____ is an example of a fossil fuel.   __Coal, oil, natural gas__
9. Geothermal energy comes from within the __earth__.
10. Falling water has __kinetic__ energy.

D. How and Why See Teacher's Guide at the front of this book.
1. Explain the differences in a chemical change, a physical change, and a nuclear change.
2. Metal (iron) lawn furniture left outside becomes rusty. What causes the formation of rust?
3. How is hydroelectric power produced?
4. What is the difference between potential and kinetic energy? Give one example of each.
5. Why must we find new sources of energy to replace fossil fuels?

challenges

1. Collect free educational materials from power and utility companies in your area. Use these materials to make an energy conservation bulletin board display.
2. The use of nuclear energy is a controversial issue. Do library research on this topic and prepare a report for your class. List the pros and cons of nuclear power plants.
3. Fossil fuels are limited in supply. How much coal, petroleum, and natural gas are believed to be in the earth? How long will these fuels last? Through library research, prepare a report on this problem.

interesting reading

Coombs, Charles, *Coal in the Energy Crises*. New York, NY: Morrow, 1980.

Davis, George, *Your Career in Energy-Related Occupations*. New York, NY: Arco, 1980.

Hoke, John. *Solar Energy*. rev. ed., New York, NY: F. Watts, 1978.

Energy and Changes in Matter

In the days of sailing ships, the wind provided the energy to move the ships. People provided the energy to move the cargo. Today, large machines are used to power the ships as well as load and unload cargo. From where does the energy for these machines come? What forces enable the large steel ships to float? How can machines help us move?

John Youger

Mechanics
unit 2

A gondola is one way to transport people and materials over a rugged landscape. This gondola is being pulled up the face of a steep cliff. Work is being done in moving the gondola up the cliff. What is work? How is energy being used to do this work? What is necessary for work to be done? How are work and energy related?

Tom Stack & Assoc.

Work and Energy

chapter 4

Introducing the chapter: Have two students demonstrate the sharpening of two pencils. Have one student carefully sharpen a pencil with a knife and the other sharpen a pencil with a hand operated pencil sharpener. Explain that energy is needed and work is done in each case. Compare the use of the two tools: the knife and the pencil sharpener. Demonstrate the use of an electric powered pencil sharpener. Introduce the concept of force and relate it to the forces needed to sharpen each of the pencils. Follow with a reading and discussion of forces in Section 4:1.

4:1 Force

Every time you walk, jump, or pedal a bicycle, you use a force. The force you use makes your body move. When you push on the pedals of a bicycle, the force sets you and the bicycle in motion. The harder you push, the greater the force, and the faster you go.

A **force** is any push or pull on an object. The pull of a car on a trailer being towed is a force. So is the push of your hand against a ball when you throw it. When two people arm wrestle, they clasp each other's hand and push hard. The person who pushes with the most force wins.

You can measure force with a spring scale. Attach the spring scale to an object. Then pull the object up or sideways. The amount of force used will be indicated in newtons on the scale. The **newton** (N) is the unit of force. Forces may range from very small to very large.

GOAL: You will gain an understanding of forces, work, and the use of energy to do work.

Students are familiar with pounds. Explain that the newton is the SI unit of force.

List five examples of a force.

FIGURE 4–1. A large force is needed to lift a heavy object.

Ohio Dept. of National Resources

FIGURE 4–2.

Define the terms weight and gravity.

activity

MEASURING FORCES   Activity 4-1

(1) Obtain a spring scale and a piece of cord about 1 m in length. (2) Tie one end of the cord around a brick, wooden block, or other object of similar mass. (3) Tie a small loop in the other end of the string. Attach the spring scale to the loop. (4) Drag the object across your desk top. Note and record the readings on the spring scale when the object starts to move and after it is moving. (5) Using the spring scale lift the object off the desk. Again note and record the reading on the scale. Compare the amount of force needed to start, drag, and lift the object. See Teacher's Guide.

making sure

1. Make a list of forces. Which force do you think is largest? Smallest? Answers will vary. See Teacher's Guide.

4:2 Weight

The force of gravity is a force that every object exerts on another object. The force of gravity is always an attractive force. **Weight** is the measure of the force of gravity that the earth exerts on every object on or near its surface. Since weight is a force, it is measured in newtons like other forces. You can use a spring scale to measure the weight of an object.

68   *Work and Energy*

The force of gravity is not the same all over the world. It is less near the equator than at the North and South Poles. The force of gravity is different because the earth's diameter at the equator is greater than at the poles. The force of gravity is also less at higher elevations than at the earth's surface. If you move a sack of potatoes, for example, from the North Pole to the equator, it loses weight. Likewise, the sack weighs less at the top of a mountain than at sea level.

Although the weight of an object changes with location, its mass does not change. ==Weight is a force and mass is the amount of matter in an object.== Suppose you were to weigh a 1-kg sack of potatoes with a scale. The mass of the potatoes is 1 kg. The weight of the potatoes is 9.8 N. A size "D" battery has a mass of about 102 g. It weighs about 1 N. How much force would you use to lift a size "D" battery?   *a force of 1 N or greater*

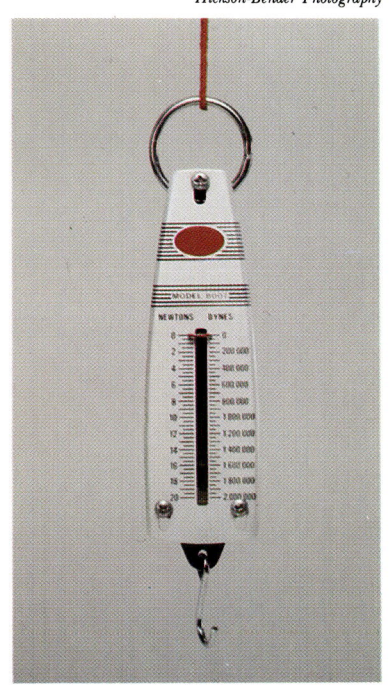

FIGURE 4–3. This spring scale measures the weight of an object in newtons.

How is mass different from weight?

FIGURE 4–4. Although the astronaut's mass remains the same, weight decreases as the force of gravity decreases.

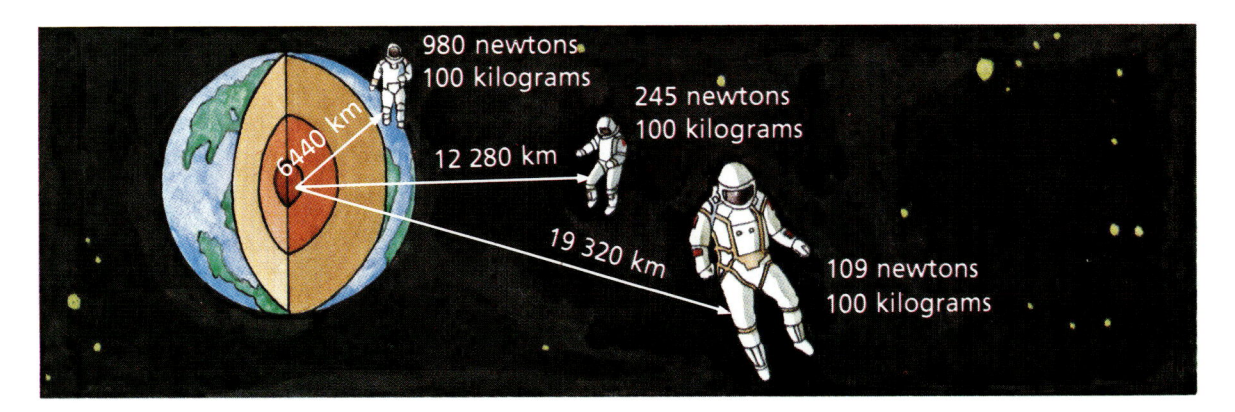

Table 4–1.
Some Weights in Newtons

Boeing 747 jet	3.3×10^6 N
Elephant	6.9×10^4 N
Average person	735 N
Desk telephone	20 N
Slice of bread	0.25 N
Postage stamp	2.0×10^{-4} N

4:2   Weight

Make certain students understand that grams and kilograms measure mass, not weight. Pass a "D" size 1½ volt battery around the class so that students may get the feel of a 1 N force.

activity

See Teacher's Guide.

MEASURING WEIGHT   Activity 4-2

Obtain five different objects such as a book, baseball, block of wood, shoe, and chalkboard eraser. Estimate the weight of each object in newtons and record it in your notebook. Obtain a spring scale and a long piece of string. Weigh each object and record the weight. Which object weighed the most? The least? Find your approximate weight in newtons using 4.5 N = 1 lb. **Remind students to estimate the weights in newtons.**

When the force of gravity is balanced by an equal and opposite force an object is weightless.

making sure

2. How is weight different from mass? **Weight is a force and mass is the amount of matter.**
3. How can an object be weightless but not massless? **See Teacher's Guide.**
4. According to *The Guinness Book of World Records*, R. E. Hughes reached a record weight of 4763 N. What was his mass in kilograms? **486 kg.**

4:3 Friction

Forces are used to move objects. Cars, roller skates, bicycles, and airplanes all depend on forces for their motion. As an object is pushed by a force, it must overcome friction. **Friction** is a force that slows down or prevents motion. The three types of friction are sliding, rolling, and fluid. Sliding friction exists when solid objects slide over each other. Rolling friction occurs when an object such as a ball or wheel rolls across a surface. Fluid friction occurs when things move across or through a fluid, such as water, oil, or air.

What are the three types of friction?

When a brick slides across a surface, a friction force opposes its motion. A spring scale can be used to measure the force needed to slide the brick. This force is equal to the friction force. The amount of friction force depends on the weight of the object and the type of surfaces that rub together.

The amount of sliding friction is a function of weight and the smoothness of the surfaces. A brick pulled across a table top has the same friction force when pulled on its narrow surface as it does when pulled on its wide surface.

In general, the smoother the surface between two objects, the less the friction. When you walk on a

Work and Energy

waxed floor, there is little friction between the floor and your feet. If there is too little friction, your feet may slide from under you. Accidents sometimes occur when one car slides into another on icy streets. When a driver applies the brakes, there may not be enough friction between the tires and the icy surface to stop the car.

A rolling object has less friction than a sliding object. A ball bearing and a roller bearing are examples of this principle. The bearing between a wheel and an axle is used to reduce friction.

Ask students how friction can be increased on icy surfaces—snow tires, chains, sand.

FIGURE 4–5. Examples of sliding friction (a), rolling friction (b), and fluid friction (c).

Friction causes moving surfaces that are in contact with each other to wear. Friction gradually strips away tiny particles from the surfaces. As a result, the moving parts of machines may wear down and break. Friction between moving parts also produces heat. You can find this out for yourself if you rub your hands together very rapidly. You will feel the heat in your palms. Heat from friction also causes the moving parts of machines to break down.

Lubricants (LEW brih kunts), such as oil and grease, are used to reduce friction between moving parts of machines. For example, the bearings of bicycles and roller skates are oiled and greased. Lubrication reduces wear and makes roller skating and bicycle riding easier. Proper lubrication extends the life of machinery and reduces the need for costly repairs.

How can friction be reduced?

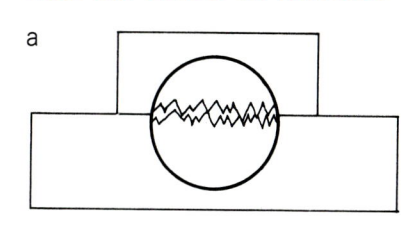

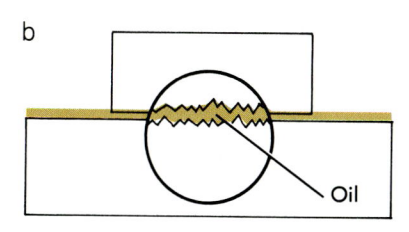

FIGURE 4–6. Friction between two surfaces (a) may be reduced by using a lubricant (b).

activity

FORCE OF FRICTION

Activity 4-3

Objective: To compare sliding and rolling friction   See Teacher's Guide.

Materials
board, (smooth) 1 m × 20 cm
brick
cord
spring scale
6 wooden dowels, 10 cm

Procedure

1. Tie the cord around the brick. Attach the spring scale to the other end of the cord.
2. Pull the brick slowly across a rough concrete surface such as a sidewalk. Determine the force needed to keep the brick moving (Figure 4–7). Record the force in the data table.
3. Place the brick on the pieces of dowel. Pull the brick with the spring scale so that the dowels roll underneath it (Figure 4–8). In the data table, record the force needed to keep the brick moving.
4. Measure the amount of force needed to keep the brick moving across the smooth board. Record the force.
5. Repeat Step 4 with the brick on top of the dowels. Record the force.

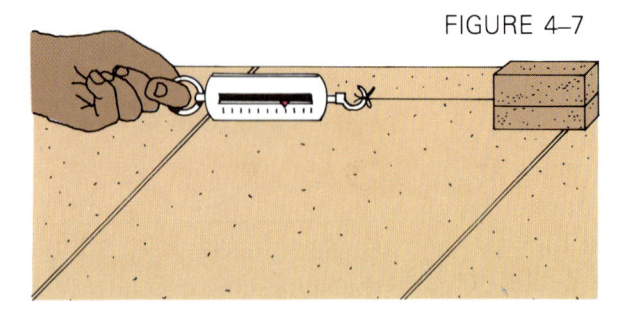

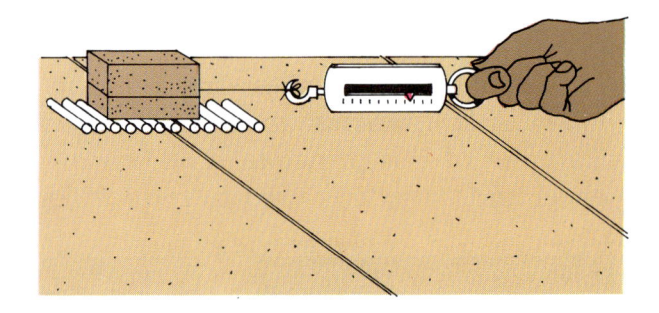

FIGURE 4–7

FIGURE 4–8.

Observations and Data

Type of surface	Type of motion	Amount of force
Rough concrete	Sliding	25 N
	Rolling	10 N
Smooth board	Sliding	15 N
	Rolling	5 N

Questions and Conclusions

1. Compare the forces needed to keep the brick sliding over the smooth and rough surfaces. Which required more force? Why? See Teacher's Guide.
2. How much more force was needed to keep the brick sliding than to keep it rolling over the rough surface? Answers will vary.
3. Would more force be required to keep the brick rolling using two wooden dowels rather than using the six dowels? Test your hypothesis and explain your results. no

72 **Work and Energy**

making sure

5. List the type of friction in the following situations.
   (a) A submarine moves through the ocean.
   (b) A bowling ball rolls down a bowling alley.
   (c) The Space Shuttle reenters the earth's atmosphere.
   (d) A person skates on a sidewalk.
   (e) A skier waxes skis and then glides down a slope.

   (a) fluid
   (b) rolling
   (c) fluid
   (d) rolling
   (e) sliding

6. How does sandpaper smooth rough wood?
   The sand particles exert a friction force that rubs off the surface particles of wood.

4:4 Work

If a force causes an object to move, work is done. In science, **work** is defined as a force acting through a distance. Work is done whenever something is moved by a force.

The girl in Figure 4–9 is lifting a box. To lift the box, she must exert a force to overcome the pull of gravity on the box. As she lifts the box upward, work is done. If she had not been able to lift the box, no work would have been done.

Note that the definition of work includes distance. When forces do not move objects, they do not produce work. Your body exerts a force on a chair when you are sitting, but no work is done. The chair does not move through a distance. When you push down on a desk top with your hand, you apply a force. Again, no work is done. The force does not move through a distance. Work is done only when a force is applied through a distance.

FIGURE 4–9. Work is done when force is exerted through a distance.

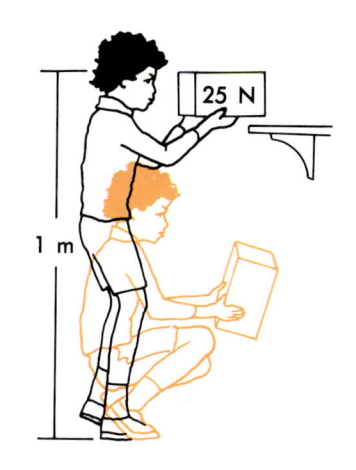

FIGURE 4–10. Although a force is exerted, no work is done on the car (a). Work is done only when the force causes the car to move (b).

a

b

Young/Hoffhines

How is work calculated?

Work is force multiplied by the distance through which the force is applied. To calculate work, use the equation

Work = Force × distance $W = F \times d$

The force of 1 N exerted through a distance of 1 m is a newton-meter (N·m) The **joule** (J) is equivalent to 1 N·m. The joule is a unit of work.

What is a joule?

Example

The girl in Figure 4–9 uses a force of 25 N to lift the box 1 m off the floor. How much work is done?

Solution

Step 1:   Write the equation for work.

$$W = F \times d$$

Work always involves motion. The same work can be done by a small force through a large distance as a large force through a small distance.

Step 2:   Substitute the values for force and distance given in the problem.

$$W = 25 \text{ N} \times 1 \text{ m}$$

Do the example on a chalkboard. Repeat it using a force of 10 N and a distance of 2 m. Change the values to 20 N and 5 m and let your students solve.

Step 3:   Multiply to find the answer.

$$25 \text{ N} \times 1 \text{ m} = 25 \text{ N·m} \qquad W = 25 \text{ J}$$

making sure   Note: work is done only when something is moved.

7. In which of the situations is work done? (b), (d), (e)
   (a) a ball hitting a brick wall
   (b) a breeze pushing a kite into the air
   (c) a small child trying to push a car
   (d) a hand turning the handle of an eggbeater
   (e) a power shovel lifting a load of soil
   (f) a hand holding a 10-N object

8. A boy was helping his parents move some stones to another part of the backyard. He hauled them in a wagon. The boy moved the stones 20 m and had to apply a force of 65 N to move the wagon. How much work did he do? 1300 J

9. A girl was collecting newspapers for a recycling project. She hauled some newspapers from a garage to a truck. A force of 30 N was required to move the newspapers 10 m. How much work was done? 300 J

4:5 Work and Energy

To do work, energy is needed. Energy is the ability to do work. It is needed to move a car, to lift a skier up a slope, or to push a roller coaster up a hill.

When work is done, energy changes from one form to another. The potential energy of a skier at the top of a slope changes into kinetic energy as the skier moves down the slope. The kinetic energy of a roller coaster changes into potential energy as the roller coaster moves up a hill.

The **law of conservation of energy** states that energy can change forms, but energy cannot be created or destroyed under ordinary conditions. It is considered a law because it has been tested many times and has always been true. Though energy can be changed from one form to another, the total amount of energy always stays the same. When work is done, energy is used, but the energy is never destroyed or used up.

FIGURE 4–11. As a roller coaster moves down a slope, potential energy changes to kinetic energy.

King's Island Amusement Park

State the law of conservation of energy.

making sure

10. You do work when you rub your hands together briskly. What energy change occurs? *Kinetic energy is changed to heat.*
11. Why does a bulldozer use more energy than a garden tractor when work is done? *The bulldozer does more work.*
12. How does the law of conservation of energy apply to the burning of coal? *Chemical energy is changed to light and heat when the coal burns.*

4:6 Power

Power is the rate at which work is done. It is the amount of work per unit of time. A large engine is more powerful than a small one. It can do more work per hour than the small one. This means it can move larger forces through greater distances in less time. The **watt** (W) is the unit of power. One watt is equivalent to 1 joule per second (J/s).

Define the terms power and watt.

FIGURE 4-12. The power shovel is more powerful than the person because it can do more work in less time than the person.

How is power calculated?

The following equation is used to find power:

$$\text{power} = \frac{\text{work}}{\text{time}} \qquad P = \frac{W}{t}$$

Example

A 450-N person walked up a flight of stairs 10 m high in 30 s. What power was used?

Solution

Step 1: Determine the amount of work done by the person climbing the stairs.

$$W = F \times d$$

$$W = 450 \text{ N} \times 10 \text{ m} = 4500 \text{ N·m} = 4500 \text{ J}$$

Step 2: Write the equation for power.

$$P = \frac{W}{t}$$

Step 3: Substitute the values for work and time given in the problem.

$$P = \frac{4500 \text{ J}}{30 \text{ s}}$$

Step 4: Divide to find the answer.

$$\frac{4500 \text{ J}}{30 \text{ s}} = 150 \text{ J/s} \qquad P = 150 \text{ W}$$

activity

Caution students not to run on the stairs.

USING POWER TO CLIMB STAIRS

See Teacher's Guide. Activity 4-4

Find the height of a flight of stairs. To do this, measure the height of one step and multiply by the total number of steps. Have a classmate clock the time it takes you to walk up the steps. Determine your weight using 1 lb. = 4.5 N. Find the power (in watts) used to climb the stairs.

FIGURE 4–13.

making sure

13. In 2 s, a person lifts a box that weighs 500 N to a height of 4 m. Find the power in watts. **1000 W**

4:7 Engines

At one time much of the world's work was done by horses, mules, and oxen. In the United States today, engines have replaced most of these work animals. **Engines** are machines that produce mechanical power. One example of an engine is the gasoline engine. It burns gasoline and thereby uses the energy in gasoline to do work. Gasoline engines are used to run cars, trucks, boats, lawn mowers, and other machines.

How does a gasoline engine produce power? A gasoline engine has one or more round tubes called cylinders. In each cylinder is a piston that moves back and forth. Each piston is attached to a rod. The rod is a piece of metal that connects to the crankshaft. When gasoline is burned inside a cylinder an explosion occurs. The explosion creates the force that moves the piston. As the pistons move back and forth, they cause the crankshaft to rotate and produce power. In a vehicle such as a car, the crankshaft is connected to the driveshaft. The driveshaft turns the car's wheels. A driveshaft can also be used to supply power for machinery such as a chain saw, a winch, or a snow blower.

A gasoline engine burns gasoline and produces waste gases. Gasoline is fed into the cylinders from a device called a carburetor. Inside the carburetor, liquid gasoline pumped from the fuel tank is

What is an engine?

A gasoline engine is called an internal combustion engine because a fuel is burned (combustion) inside the engine.

How does a gasoline engine operate?

An old lawn mower engine may be dismantled and used to show the parts of an engine.

4:7 Engines 77

changed to a vapor. Gasoline vapor is then mixed with air drawn in through an air filter. The gasoline-air mixture enters the cylinder through an opening called an intake valve. A sparkplug in the cylinder creates an electric spark that ignites the gasoline. Waste gases produced in the burning are forced out of the cylinder through the exhaust valve.

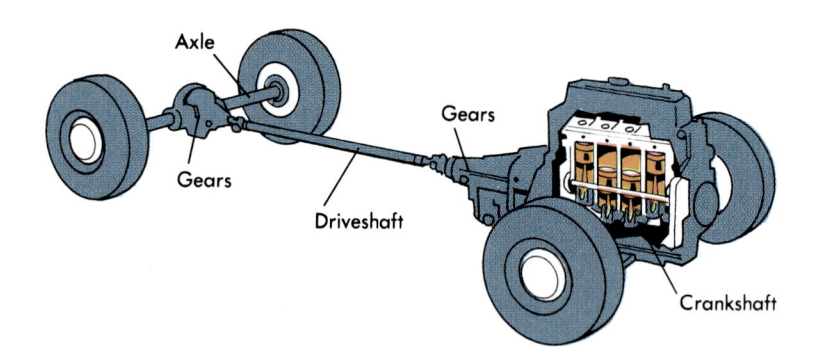

FIGURE 4–14. Automobiles are run by gasoline or diesel engines which produce mechanical power.

Estimates indicate that one in five cars on American roads will be diesel powered by 1990.

A diesel engine is similar to a gasoline engine. It is different in that it does not have spark plugs and burns a fuel called diesel oil. The fuel-air mixture is ignited by the heat produced when the mixture is compressed by the motion of a piston inside a cylinder. Both diesel and gasoline engines waste most of the fuel they use. Only about 12 percent of the energy in the fuel a gasoline engine burns is used for power. Most is lost as waste heat. However, a diesel engine is more efficient in that it uses a greater proportion of the energy in fuel to produce power. There is less wasted energy.

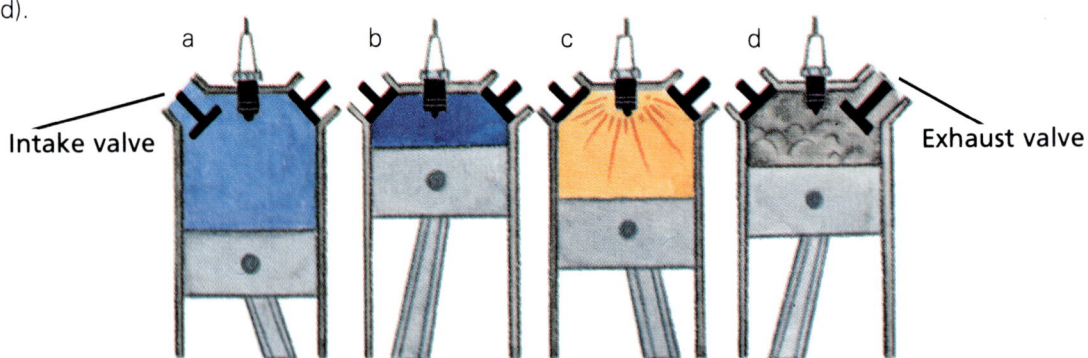

FIGURE 4–15. Operation of most gasoline engines involves four steps. Fuel mixture enters cylinder (a). The mixture is compressed (b). Sparkplug ignites mixture (c). Waste gases are forced out of cylinder (d).

78 *Work and Energy*

main ideas
Chapter Review

1. A spring scale is used to measure forces in newtons.   4:1
2. Every object exerts a gravitational force.   4:2
3. The weight of an object decreases as its distance from Earth's center increases.   4:2
4. Friction may be classified as sliding, rolling, or fluid.   4:3
5. A lubricant is used to reduce friction between moving surfaces.   4:3
6. Work is done when an object is moved by a force.   4:4
7. Energy can be changed from one form to another.   4:5
8. A large engine is more powerful than a smaller one because it can do more work per hour.   4:6
9. Power may be measured in watts.   4:6
10. An engine uses the energy obtained in burning a fuel to do work.   4:7

vocabulary

Define each of the following words or terms.

engine	law of conservation	watt
force	of energy	weight
friction	newton	work
joule	power	

study questions

DO NOT WRITE IN THIS BOOK.
A. True or False

Determine whether each of the following sentences is true or false. If the sentence is false, rewrite it to make it true.

F   1. A person does work by pushing against a solid brick wall.   no
T   2. The newton is a unit of force.
T   3. The joule is a unit of work.
F   4. If a 5-N force moves a box 10 m, ~~15~~ J of work is done.   50
F   5. Weight and mass are the same.   not
T   6. You can measure your weight in newtons.
F   7. A rolling object has ~~more~~ friction than a sliding object.   less
T   8. When you lift a chair, you do work.

~~F~~ 9. Adding oil to an engine ~~increases~~ the friction between the moving parts.   decreases

~~F~~ 10. ~~Work~~ is the amount of ~~power~~ per unit of time.   Power work

B. Multiple Choice

Choose the word or phrase that completes correctly each of the following sentences.

1. The definition of work includes (*force, distance, force and distance*).
2. (*Friction, Power, Weight*) is the force that opposes the motion of a sliding object.
3. An anchor being pulled up through water is an example of (*rolling, sliding, fluid*) friction.
4. Power is equal to (*work, force, distance*) per unit of time.
5. When a person lifts a 200-N sack of feed 2 m, (*100, 4, 400, 40*) J of work are done.
6. A metric spring scale measures weight in (*grams, newtons, ounces*).
7. Your (*weight, mass, action, work*) is the amount of force that gravity exerts on your body.
8. As gasoline burns inside a gasoline engine the (*spark plugs, cylinders, pistons*) move back and forth.
9. When work is done, energy is (*created, changed, destroyed*).
10. (*Rolling, Fluid, Sliding*) friction occurs when an object moves across water.

C. Completion

Complete each of the following sentences with a word or phrase that will make the sentence correct.

1. _____ can be measured with a spring scale.   Weight, force
2. Lubrication _____ friction force.   decreases
3. The smoother the surface between two objects, the less the _____.   friction
4. Ball bearings are used to reduce _____.   friction
5. The carburetor of a gasoline engine supplies the mixture of _____ and _____.   gasoline air
6. Work is the application of a(n) _____ through a distance.   force
7. The equation for work is _____. $W = F \times d$

8. When __work__ is done, energy changes form.
9. A(n) __force__ is a push or pull on an object.
10. The law of _____ states that energy cannot be created or destroyed. __conservation of energy__

D. How and Why See Teacher's Guide at the front of this book.
1. How does work differ from force?
2. What is the difference between sliding, rolling, and fluid friction? Give one example of each.
3. Why do the wheel bearings of a car sometimes need to be checked and repacked with grease?
4. Make a list of all the forces produced when a person rides a bicycle.
5. What is the law of conservation of energy? Give one example.

challenges

1. Use a spring scale to measure friction force. Compare the friction force of sliding, rolling, and fluid friction for the same object.
2. Devise an experiment to show the advantage of ball bearings.
3. Models of engines can be purchased in hobby stores. Obtain a kit for a model engine and put the parts together. Display the model in your classroom.

interesting reading

Feldman, Anthony and Bill Gunston, *Technology at Work*. New York, NY: Facts on File, 1980.

Tierney, John, "Perpetual Motion," *Science '83*, May, 1983, pp. 30-39.

Vergara, William C., *Science in Everyday Life*. New York, NY: Harper and Row, 1980.

An industrial arts class is a busy place. Students in the class use a variety of common tools. Tools such as a saw, screwdriver, and drill consist of one or more machines. Machines help you do work. How do they make tasks easier? How is the efficiency of a machine measured? Why is efficiency important?

Machines

chapter 5

Introducing the chapter: Display a pencil sharpener, bottle opener, mechanical can opener, and a nut cracker. Ask students to explain how each is used and how each makes work easier. Then discuss other machines with which students are familiar such as a vacuum cleaner and typewriter. Point out that each machine uses energy to do work.

5:1 Machines

When you use a machine you may do work faster and easier. Mowing a large lawn is much faster and easier with a power mower than with a hand mower. Washing clothes in a washing machine is easier than washing them by hand. Cutting firewood with a chain saw goes faster than chopping it with an ax. A **machine** is a device that can make work easier by changing the direction or amount of a force. For example, a bicycle can move you from place to place faster than you can walk. You can use a pulley to lift a heavy object. The pulley changes the direction and amount of the force needed to lift the object. A wrench can be used to loosen a bolt. The wrench increases the amount of force applied to the bolt.

All machines can be classified as either simple or compound machines. Simple machines produce work with one movement. There are six simple machines. They are the lever, pulley, wheel and axle, inclined plane, screw, and wedge. As you will see, the pulley and the wheel and axle are forms of a lever. The screw and the wedge are forms of an inclined plane.

Compound machines are composed of two or more simple machines. To understand compound machines, you should first be familiar with simple machines.

GOAL: You will gain an understanding of the use of simple and compound machines to do work.

What is a machine?

Go over the meanings of speed, direction, and amount of force. Use examples to illustrate these ideas.

Name the six simple machines.

83

A meter stick and a wedge-shaped block may be used to demonstrate a lever and fulcrum. Place weights on each side of the fulcrum to show how they affect the lever when moved to different positions.

Define the terms lever, fulcrum, effort force, and resistance force.

Name the two arms in a lever.

5:2 Levers

A **lever** is a bar that is free to move about a fulcrum. The **fulcrum** is the point at which a lever rotates. A seesaw is one kind of lever. When people ride a seesaw it rotates around a fulcrum. The fulcrum supports the seesaw at the center.

A lever may be used to do work. When a lever is used, a force is applied through a distance. The force applied to a lever is called the **effort force.** The force that is overcome when work is done is called the **resistance force.** In a lever such as a crowbar, the effort force is applied at one end. The resistance force is overcome at the opposite end. Between the two forces is the fulcrum. The distance from the effort force to the fulcrum is called the effort arm. The distance from the resistance force to the fulcrum is the resistance arm. Every lever has two arms, the effort arm and the resistance arm. Identify the parts of a lever in Figure 5–1.

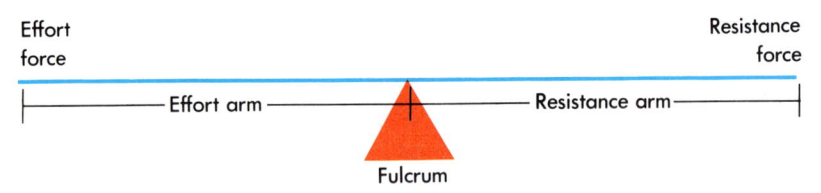

FIGURE 5–1. Diagram of a lever.

Levers are classified as first, second, or third class levers. The class of lever is based on the relative positions of the effort force, resistance force, and fulcrum. Examples of first, second, and third class levers are given in Figure 5–2.

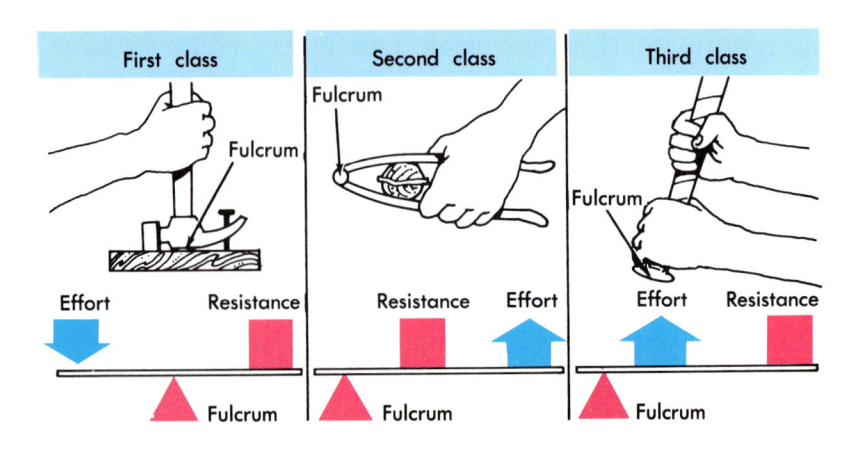

What are the three classes of levers?

Make certain students understand the differences between the classes of levers before going on to other material.

FIGURE 5–2. The three classes of levers.

84 Machines

Two things affect the action of a lever—the amount of force at both ends and the length of both arms. Remember, every lever has two forces, the effort force and resistance force. Every lever has two arms, the effort arm and resistance arm.

When the effort arm is longer than the resistance arm, the effort force is less than the resistance force. Thus, a lever can be used to multiply the effort force.

It is easier to pry open a can of paint with a long screwdriver than with a short one. Why? The long screwdriver has a longer effort arm. Thus, it takes less effort force to produce a large resistance force. The can is more easily opened because the effort force has been multiplied.

activity

USING A RULER AS A LEVER   Activity 5-1

Place a book on the edge of a table or desk. Insert about 5 cm of a wooden ruler under the edge of the book. The remainder of the ruler should extend out beyond the table. Pull up on the end of the ruler so it raises the book. Where are the fulcrum and the effort forces? *edge of table, hand pushing on end of ruler* Now put a pencil on the table in front of the book. Place one end of the ruler under the book. Place the center of the ruler over the pencil. Push down on the ruler with your hand. Where are the resistance force, effort force, and fulcrum? *resistance force—book; effort force—hand pushing; fulcrum—pencil*

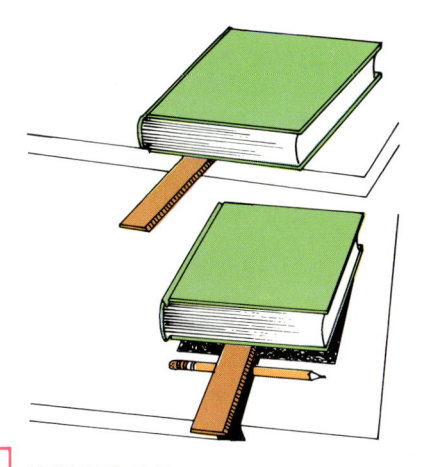

FIGURE 5–3.

making sure

1. Where is the fulcrum in each of these levers?
   (a) crowbar
   (b) seesaw
   (c) shovel
   (d) scissors
   (e) bottle opener

   (a) between the ends
   (b) between the ends
   (c) lower hand on handle
   (d) screw attaching the bars
   (e) where it touches the top of the cap

2. Name the lever(s) from Question 1 in which the resistance force is greater than the effort force. *crowbar, bottle opener, shovel*

3. Name the lever(s) from Question 1 in which the effort force moves through a greater distance than the resistance force. *crowbar, bottle opener, scissors, shovel*

4. What is the practical use of each lever mentioned in Question 3? *crowbar—separate things; bottle opener—remove bottle caps; shovel—move soil*

5:3 Mechanical Advantage

A lever is often used to multiply an effort force. The number of times that a machine multiplies the effort force is called the **actual mechanical advantage** (A.M.A.). The actual mechanical advantage is found by dividing the resistance force by the effort force.

What is the equation for A.M.A.?

$$A.M.A. = \frac{\text{resistance force}}{\text{effort force}}$$

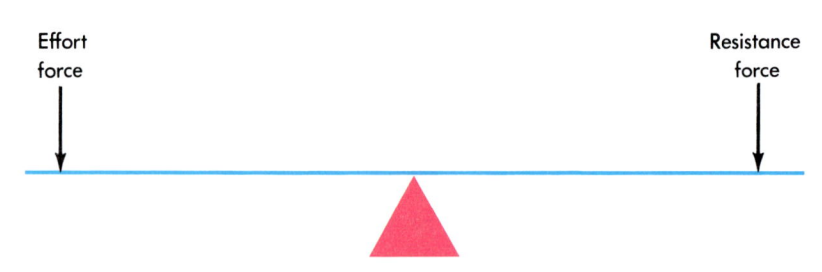

FIGURE 5–4. Divide resistance force by effort force to find the A.M.A. of a lever.

Go over the example on the chalkboard to ensure students grasp the concept.

Example

An effort force of 10 N is applied to a lever in lifting a 200-N load. What is the A.M.A.?

Solution

Step 1: Write the equation for A.M.A.

$$A.M.A. = \frac{\text{resistance force}}{\text{effort force}}$$

Step 2: Substitute the values for effort force and resistance force given in the problem.

$$A.M.A. = \frac{200 \text{ N}}{10 \text{ N}}$$

Step 3: Divide to find the answer.

$$\frac{200 \text{ N}}{10 \text{ N}} = 20 \qquad A.M.A. = 20$$

Note that the value for mechanical advantage has no units. The division of newtons by newtons cancels the units. The answer is the number 20. Thus, the effort force is multiplied 20 times. Each newton of effort moves 20 N of weight.

You have found the actual mechanical advantage of a lever. Sometimes it is useful to know the **ideal mechanical advantage** (I.M.A.) of a lever or other machines. Ideal means that friction and weight of the lever are not considered. In actual practice, friction reduces the mechanical advantage. Thus, the ideal mechanical advantage of a machine is greater than its actual mechanical advantage.

To find a lever's I.M.A., divide the effort arm length by the resistance arm length.

$$I.M.A. = \frac{\text{effort arm}}{\text{resistance arm}}$$

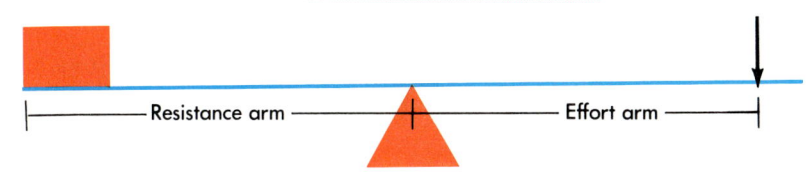

How is the A.M.A. of a machine different from the I.M.A.?

What is the equation for I.M.A.?

FIGURE 5–5. The I.M.A. of a lever is found by dividing effort arm length by resistance arm length.

Example
Find the I.M.A. for a lever that has a resistance arm 1 m long and an effort arm 2 m long.
Solution
Step 1: Write the equation for I.M.A.

$$I.M.A. = \frac{\text{effort arm}}{\text{resistance arm}}$$

Step 2: Substitute the values for effort arm and resistance arm given in the problem.

$$I.M.A. = \frac{2 \text{ m}}{1 \text{ m}}$$

Step 3: Divide to find the answer.

$$\frac{2 \text{ m}}{1 \text{ m}} = 2 \quad I.M.A. = 2$$

How long are the arms of a lever with an I.M.A. of 2? They can be any length. The effort arm of this lever, however, is two times longer than the resistance arm. What if the I.M.A. is 3? Then the effort arm is three times longer than the resistance arm. When the I.M.A. of a lever is 1, the effort force is equal to the resistance force. A lever with an I.M.A. of 1 does not change the amount of force.

The I.M.A. is also equal to the effort movement divided by the resistance movement.

$$I.M.A. = \frac{\text{effort movement}}{\text{resistance movement}}$$

Here the word movement refers to the distance traveled by a force acting on a rotating lever. For example, you may use the tip of a screwdriver to open a can of paint. You put the tip under the can's lid. Then, you push down on the handle. The distance your hand pushes the handle of the screwdriver is the effort movement. Resistance movement is the distance the tip of the screwdriver moves up as it pushes the lid open.

FIGURE 5-6. The effort movement of a lever divided by the resistance movement determines the I.M.A. of a lever.

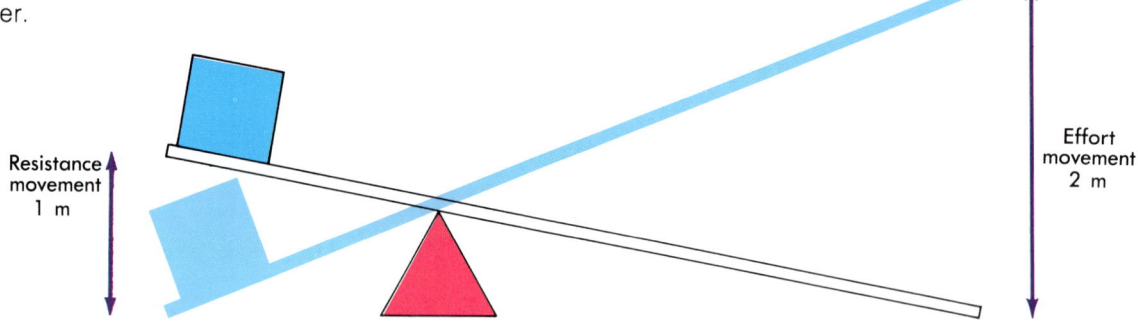

Example

In a certain lever, the effort force moves 4 m and the resistance force moves 2 m. Find the I.M.A. of this lever.

Solution

Step 1: Write the equation for I.M.A.

$$I.M.A. = \frac{\text{effort movement}}{\text{resistance movement}}$$

Step 2: Substitute the values for effort movement and resistance movement given in the problem.

$$I.M.A. = \frac{4 \text{ m}}{2 \text{ m}}$$

Step 3: Divide to find the answer.

$$\frac{4 \text{ m}}{2 \text{ m}} = 2 \qquad I.M.A. = 2$$

making sure

5. A person applies an effort force of 65 N to raise a 325-N stone. What is the A.M.A. of this lever? 5
6. A 150-N log is shifted with an effort force of 25 N. Find the A.M.A. of the lever. How many newtons of weight did each newton of effort move? 6 N of weight
7. What is the I.M.A. of a seesaw if a person 3 m from the fulcrum moves a person who is 1 m from the fulcrum? 3

5:4 Pulleys

A **pulley** is a wheel that turns on an axle. It is used with a rope, chain, or belt to change the amount or the direction of a force. A pulley can be either fixed or movable. There are single and double pulleys.

Figure 5–7a shows a rope and a single, fixed pulley. An effort force of 125 N is used to raise a 125-N metal block. The effort force equals the resistance force. Thus, the mechanical advantage is 1. What is the practical value of this pulley? The pulley changes the direction of force to make lifting easier. With a single, fixed pulley, the effort force pulls down and the weight goes up.

The pulley in Figure 5–7b has an I.M.A. of 2. In finding the I.M.A., the friction between the pulley and the rope is ignored. Two strands of a single rope support the 100-N weight. The total number of support strands of a movable pulley is always equal to the I.M.A. The number of support strands is determined by considering each part of the long rope as a separate strand. Then, count the number of strands that support the movable pulley. Since there are two support strands, the I.M.A. is 2.

What is a pulley?

FIGURE 5–7. The I.M.A. of the pulley system is 1 because only one rope strand supports the weight (a). The I.M.A. is 2 because two rope strands support the weight (b).

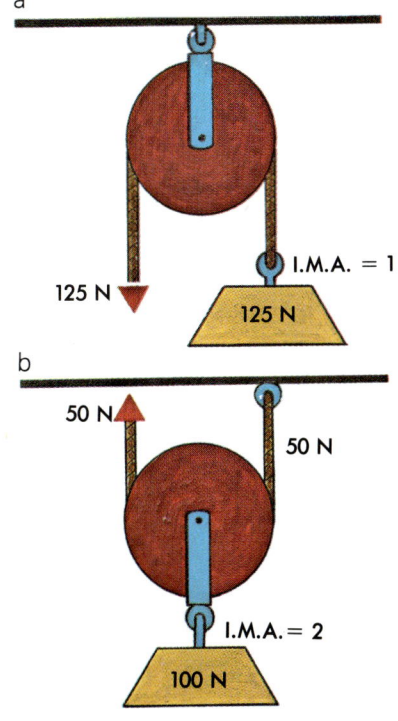

activity

USING PULLEYS

Activity 5-2

See Teacher's Guide.

Objective: To measure the I.M.A. of three pulley systems

Direct students to copy the data table in their notebooks and record their observations in the table.

Materials
double pulley
object (rock or large bolt)
ring stand
2 single pulleys
spring scale
string, 1 m

Procedure

1. Determine the weight of the object using the spring scale. Record the weight in the data table.

2. Set up the single pulley, the string, and the object to make a machine with an I.M.A. of 1. Use Figure 5–7 as a guide.

3. Attach the spring scale to the effort end of the string. Measure the force needed to raise the object. Record the force in the data table.

4. Set up the pulleys to make a machine with an I.M.A. of 2 (Figure 5–7). Measure the force needed to raise the object. Record the force in the data table.

5. Make a sketch and label the I.M.A. and effort force for each pulley system.

6. Using the double pulley, repeat Step 3 for a pulley system with an I.M.A. of 3 (Figure 5–8). Be sure to make a sketch of this system. Record the force in the data table.

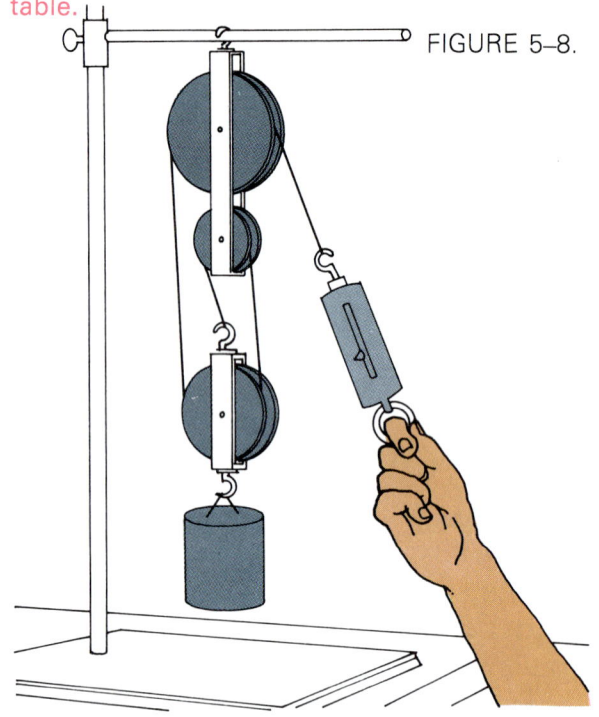

FIGURE 5–8.

Observations and Data

I.M.A.	Weight	Force
1	10 N	10 N
2	10 N	5 N
3	10 N	3 N

Questions and Conclusions

1. How did friction force affect the force needed to raise the weight? increased the effort force

2. Draw a diagram of a pulley system with an I.M.A. of 4. (*Hint:* Use 2 double pulleys.) See Teacher's Guide.

90 Machines

5:5 Wheel and Axle

Bikes, cars, trucks, and trains have two or more wheels and axles. A **wheel and axle** works on the same principle as the lever and pulley. The wheel always turns through a greater distance than the axle. The smaller force on the wheel moves through a greater distance than the larger force on the axle. A force applied to the wheel is increased at the axle. To find the I.M.A. of a wheel and axle, divide the diameter of the wheel by the diameter of the axle. Suppose a wheel has a diameter of 1 m. Its axle has a diameter of 10 cm. What is the I.M.A.? 10

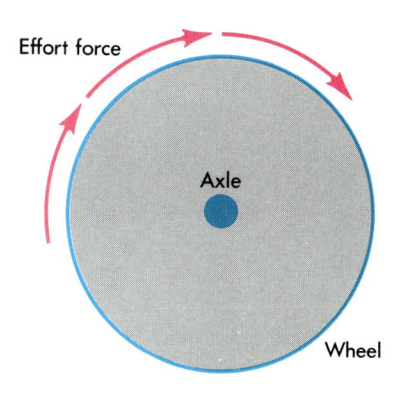

FIGURE 5–9. Effort force applied to a wheel is increased at the axle.

How is the I.M.A. of a wheel and axle calculated?

At the axle, a greater force moves through a small distance. At the wheel, a smaller force moves through a large distance.

making sure

8. When a bicycle wheel turns, is the force greater at the wheel or axle? Explain why.
9. What is the benefit of this change in force in a bicycle? The wheel turns faster than the axle making the bicycle move faster.

5:6 Inclined Plane

An **inclined plane** is a slanted surface which may be used for raising objects to higher places. A mountain road and a ramp are examples of inclined planes. Less effort force is needed with an inclined plane. This is because an object is moved through a longer distance on the slanted surface. When the same object is lifted straight up, a large effort force must be applied through a short distance.

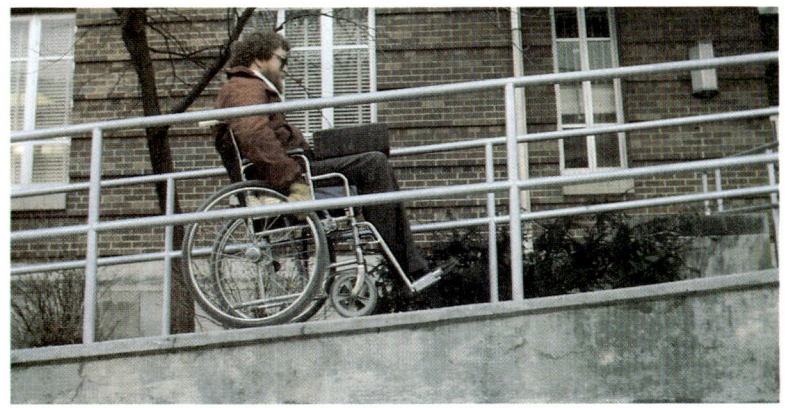

What is an inclined plane?

Doug Martin

FIGURE 5–10. A ramp allows less force to be used in going to different levels.

5:6 Inclined Plane 91

activity

USING INCLINED PLANES   Activity 5-3
See Teacher's Guide.

Objective: To determine the I.M.A. and A.M.A. of inclined planes

Materials
board, (smooth) 1 m × 20 cm
5 books
brick
meter stick
spring scale
string

Procedure

1. Tie the string to the brick. Attach the free end of the string to the spring scale and measure the weight of the brick. Record the weight in the table.
2. Place one end of the board on a book to form an inclined plane. Measure the length and height of the inclined plane. Record the measurements in your data table.
3. Pull the brick up the inclined plane at a constant speed. Do not jerk the brick. Read the spring scale while the brick is moving. Record the force needed to pull the brick up the inclined plane in your data table.
4. Change the angle of the inclined plane by adding a second book. Measure the length and height of the inclined plane. Record the measurements in your data table.
5. Repeat Step 3 and record the force used in the data table.
6. Change the height of the inclined plane by adding more books. Record the length and height of the inclined plane in your data table.
7. Repeat Step 3 and record the force used in the data table.

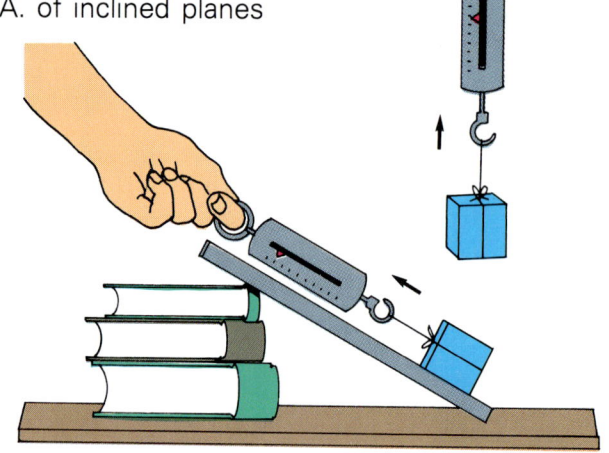

FIGURE 5-11.

Observations and Data See Teacher's Guide.
Weight of brick: _____ N

Number of books	Height of plane (cm)	Length of plane (cm)	Force (N)
1			
2			

Questions and Conclusions

1. Determine the I.M.A. for each inclined plane using the following equation. Answers will vary.

$$I.M.A. = \frac{\text{length of plane}}{\text{height of plane}}$$

2. Use your data and the following equation to calculate the A.M.A. for each inclined plane. Answers will vary.

$$A.M.A. = \frac{\text{weight of brick}}{\text{effort force}}$$

3. Compare the A.M.A. to the I.M.A. for each inclined plane. Why did they differ? The greater the angle, the closer the A.M.A. is to the I.M.A.

making sure

10. The I.M.A. of an inclined plane is its length divided by its height. Find the I.M.A. for the inclined planes in Figure 5–12. Then find the effort force needed to move the object up each of the inclined planes. Ignore the effect of friction when solving these problems.

5:7 Wedge

A **wedge** is a type of inclined plane. It consists of a piece of material, such as wood or metal, that is thicker at one edge than at the other. Examples of wedges are a knife, axe, needle, and can opener. The length of the wedge divided by its thickness determines its I.M.A. A long, narrow wedge requires a small effort force to overcome a great resistance force.

When you sharpen a knife, you are decreasing its thickness at the cutting edge. This increases its mechanical advantage. You need less force to cut with a sharp knife than to cut with a dull knife.

making sure

11. Determine the I.M.A. of two wedges that have the following dimensions.
    (a) thickness of 5 cm; length of 36 cm 7.2
    (b) thickness of 3 cm; length of 42 cm 14
12. Which wedge in Question 11 requires the least effort force to do work? Explain. b, it has a greater mechanical advantage

FIGURE 5–12.

FIGURE 5–13. Some common examples of wedges include a razor blade, chisel, and nail.

How is the I.M.A. of a wedge calculated?

5:8 Screw

A **screw** is a circular inclined plane. It looks like an inclined plane wound around a center to form a spiral. Turning a screw requires a small effort force to overcome a large resistance force. The small effort force turns through a large distance. It produces a large resistance which turns through a small distance.

See Figure 5–14.

List five examples of a screw.

5:8 *Screw* 93

FIGURE 5-14. These household objects show some common uses of screws.

(a) screw, lever
(b) wheel and axle
(c) lever, wedge
(d) lever
(e) inclined plane, wedge
(f) pulley
(g) wheel and axle
(h) inclined plane
(i) wedge
(j) lever, wheel and axle
(k) wedge, lever, wheel and axle
(l) lever

Define and explain the meaning of efficiency in terms of wasted energy. e.g.: efficient worker, energy efficient home. More efficiency means less waste.

What is the difference between work input and work output?

making sure

13. Which simple machine is present in each of the following items:
    (a) automobile jack
    (b) door knob
    (c) ax
    (d) pliers
    (e) snowplow
    (f) block and tackle
    (g) eggbeater
    (h) loading ramp
    (i) bow of a boat
    (j) faucet
    (k) screwdriver
    (l) broom

5:9 Efficiency

Do you think of machines as gadgets that do work? Actually, machines must use energy to do work. A machine can do work only if work is done on it. The work done on the machine is effort force times the distance the effort force moves. It is called the work input.

$$\text{work input} = \text{effort force} \times \text{effort distance}$$
$$W_i = F \times d$$

A machine also has a work output. The work output is the resistance force times the distance through which the resistance force moves.

$$\text{work output} = \text{resistance force} \times \text{resistance distance}$$
$$W_o = F \times d$$

94 **Machines**

Example

The effort force used with a certain set of pulleys is 50 N. When this effort force moves a distance of 3 m, 120 N of resistance moves 1 m. What is the work input for this machine? What is the work output for this machine?

Solution

Step 1: Write the equation for work input.
$$W_i = F \times d$$

Step 2: Substitute the values for effort force and effort distance given in the problem.
$$W_i = 50 \text{ N} \times 3 \text{ m}$$

Step 3: Multiply to find the answer.
$$50 \text{ N} \times 3 \text{ m} = 150 \text{ J} \quad W_i = 150 \text{ J}$$

Step 4: Write the equation for work output.
$$W_o = F \times d$$

Step 5: Substitute the values for resistance force and resistance distance given in the problem.
$$W_o = 120 \text{ N} \times 1 \text{ m}$$

Step 6: Multiply to find the answer.
$$120 \text{ N} \times 1 \text{ m} = 120 \text{ J} \quad W_o = 120 \text{ J}$$

Review the use of joules to measure work.

FIGURE 5–15. A gasoline engine provides the energy for this log-splitting machine to do work.

In an ideal machine, the work output would equal the work input. Since no machine is ideal, some force must always be used to overcome friction. Thus, the work output of a real machine is always less than the work input. The work output for the pulleys in the example is 120 J. The work input is 150 J.

Some machines are more efficient than others. The **efficiency** (ih FIHSH un see) of a machine is a comparison of the work output to the work input. Efficiency is measured in percent. An ideal machine, if it could exist, would have an efficiency of 100 percent. However, some work is always used to overcome friction. No machine can have an efficiency of 100 percent. You can never get as much work out of a machine as you put into it.

The efficiency of a machine is found by dividing the work output by the work input and then multiplying by 100%. The equation is

$$\text{efficiency} = \frac{\text{work output}}{\text{work input}} \times 100\%$$

$$E = \frac{W_o}{W_i} \times 100\%$$

Example

Find the efficiency of the pulleys whose work input is 150 J and work output is 120 J.

Solution

Step 1: Write the equation for efficiency.

$$E = \frac{W_o}{W_i} \times 100\%$$

Step 2: Substitute the values for work input and work output given in the problem.

$$E = \frac{120 \text{ J}}{150 \text{ J}} \times 100\%$$

Step 3: Divide, and then multiply to find the answer.

$$\frac{120 \text{ J}}{150 \text{ J}} \times 100\% = 80\% \qquad E = 80\%$$

FIGURE 5–16. Regular tuning of a car engine maintains its efficiency.

Efficient machines help conserve resources. They also help save money. A machine of low efficiency wastes energy. For example, a car that gets low gasoline mileage is not efficient. It wastes fuel.

making sure

14. Find the work input, work output, and efficiency of the machines in Table 5-1.

Table 5–1.
Input and Output for Some Machines

Machine	Effort force	Effort distance	Resistance force	Resistance distance	Efficiency
A	2 N	15 m	8 N	3 m	? %
B	100 N	20 m	200 N	5 m	? %
C	10 N	5 m	40 N	1 m	? %
D	125 N	4 m	200 N	1.5 m	? %
E	480 N	7 cm	840 N	3 cm	? %
F	80 N	5 m	53 N	4 m	? %

30 J, 24 J, 80%
2000 J, 1000 J, 50%
50 J, 40 J, 80%
500 J, 300 J, 60%
3360 J, 2520 J, 75%
400 J, 212 J, 53%

15. How does lubrication increase the efficiency of machines? reduces friction
16. Why do efficient machines help to save energy? Less energy is needed to do work.
17. Why is it impossible to have an ideal machine? Friction is always present between surfaces in contact.

5:10 Compound Machines

What is a compound machine?

A compound machine contains two or more simple machines. The simple machines are connected so that the work each does is combined to do the job performed by the compound machine. One example of a compound machine is a mechanical pencil sharpener. The handle of the pencil sharpener is a wheel and axle. It rotates when someone applies a force to turn it. Connected to the axle are blades that shave the end of the pencil. Each of these blades is a wedge. The force applied to the handle is carried through the wheel and axle to the blades that do the work.

The complexity of a compound machine depends on the number of different kinds of simple machines.

One kind of mechanical can opener has three simple machines. The simple machines are the wedge, the lever, and the wheel and axle. The blade of the opener is a wedge. A lever on the handle is used to force the blade into the top of a can. The force needed to open the can is applied to the wheel and axle part of the opener. As the wheel and axle rotates on the edge of the can, the blade cuts around the top of the can. Working together the three simple machines make a compound machine that opens the can easily and efficiently.

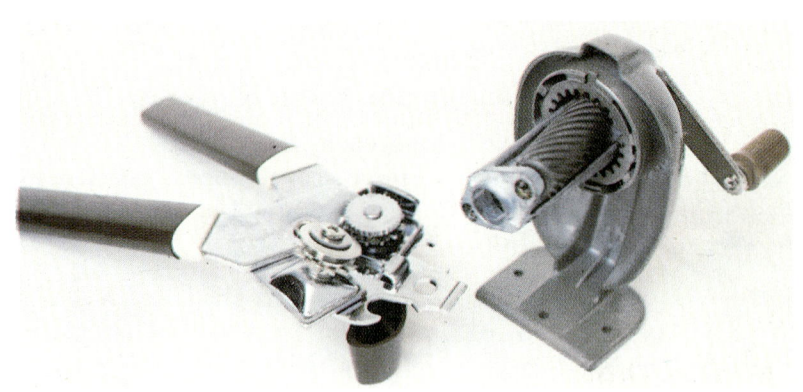

Obtain a pencil sharpener and a mechanical can opener. Point out the simple machines in each compound machine.

Kevin Fitzsimons

FIGURE 5–17. A pencil sharpener and a can opener contain simple machines.

Another example of a compound machine is a typewriter. Paper is fed into a typewriter by a long, round, hard rubber wheel called a platen. Inside the platen is an axle. Axle and platen are rotated by turning knobs located on each end of the axle. The

98   **Machines**

inked ribbon used to print on the paper is connected to two spools, one on each side of the machine. These spools feed the ribbon across the surface of the paper at the point where the printing takes place. Each spool is a wheel and each wheel rests on an axle. As the axles turn the wheels, the ribbon is fed out from one spool and wound up on the other. Some typewriters have anywhere from 45 to 50 levers in the center of the typewriter in front of the ribbon. These levers are metal bars that strike the ribbon to print letters, numerals, and other symbols. Each lever is connected to a typewriter key. At the top end of the bar is the type that prints the letter or numeral. At the opposite end or bottom, the bar connects to the typewriter. The point of connection is the fulcrum. Here the bar rotates when its key is depressed, causing it to strike the ribbon. Each bar is attached to a spring that snaps it back in place after the key is released.

A compound machine usually will not work if one of its simple machines is broken. Point out how the operation of a typewriter depends on the working of each of its simple machines.

FIGURE 5–18. A mechanical typewriter contains a number of levers and wheels and axles.

Eric Hoffhines

making sure

18. What simple machines in a mechanical pencil sharpener are in a mechanical can opener? wheel and axle, wedge
19. Name two simple machines that make a typewriter work. wheel and axle, lever
20. Why might a compound machine not work if one of its simple machines is broken? When one simple machine does not function, the machines connected to it cannot function.

5:10 **Compound Machines**

See Teacher's Guide.

PERSPECTIVES
skills

Organizing Notes in Outline Form

Outlining the information you read will help you to organize and remember it. Outlining is a way of taking notes. Taking notes leads you to think about what you have read. The outline form assists you in arranging the information in a sensible order. The outline of a chapter shows very quickly the relationship of the main ideas of the chapter and the smaller but important details.

A framework for an outline for part of Chapter 5 is shown below. Some of it has been done for you. For practice, complete the remainder of the outline. Note that many of the principal headings of the outline follow the section titles of the chapter. The section titles in heavy black print are part of the author's outline for the chapter.

I Machines (Note this is Section 5:1.)
   A. Definition: a device that can change the speed, direction, or amount of force.
   B. Types
      1. simple
      2. compound

II Simple Machines
   A. Levers (Note this is Section 5:2.)
      1. Definition: a bar that is free to move about a fulcrum
      2. Parts
         a. Fulcrum: the point at which a lever rotates
         b. Effort arm: the distance from the end of the bar to the fulcrum
         c. Resistance arm: the distance from the opposite end of the bar to the fulcrum
      3. Working with levers
         a. Effort force: the force applied to a lever
         b. Resistance force: the force that is overcome to do work
      4. Mechanical advantage of a lever (Section 5:3)
         a. Definition: the relationship of the force and the length of arm of the effort and the resistance
         b. Actual mechanical advantage (A.M.A.)
            (1) Definition: the number of times that a machine multiplies the effort force
            (2) Equation:
$$\text{A.M.A.} = \frac{\text{resistance force}}{\text{effort force}}$$
         c. Ideal mechanical advantage
            (1) Definition:
            (2) Equations:
   B. Pulley (Section 5:4)
      1. Definition:
      2. Types
      3. Mechanical advantage
   C. Wheel and axle (Section 5:5)
   D. Inclined plane (Section 5:6)

The form of your outline may vary depending on the type of information found in the chapter. One variation has been introduced to you in the partial outline given above—using section titles as the main headings of your outline. This variation makes it easier to refer to specific sections if you need to do so. Another variation might be to draw each simple machine and label the parts. Remember, taking notes in outline form should help you to (1) put the important information in your own words and (2) enable you to easily obtain the information when reviewing it.

main ideas

Chapter Review

1. Machines can change the amount, speed, or direction of a force.   5:1
2. The class of a lever is based on the relative positions of the effort force, resistance force, and fulcrum.   5:2
3. Ideal mechanical advantage of a machine is greater than its actual mechanical advantage.   5:3
4. A wheel and axle works on the same principle as a lever and pulley.   5:5
5. Less effort is needed to move an object up an inclined plane because the object is moved over a greater distance.   5:6
6. Turning a screw requires a small effort force to overcome a large resistance force.   5:8
7. The work output of a machine is less than the work input.   5:9
8. The job performed by a compound machine depends on the combined work of simple machines.   5:10

vocabulary

Define each of the following words or terms.

actual mechanical advantage (A.M.A.)
efficiency
effort force
fulcrum
ideal mechanical advantage (I.M.A.)
inclined plane
lever
machine
pulley
resistance force
screw
wedge
wheel and axle

study questions

DO NOT WRITE IN THIS BOOK.
A. True or False

Determine whether each of the following sentences is true or false. If the sentence is false, rewrite it to make it true.

F   1. Machines have ~~no~~ practical value.
T   2. A machine can change the direction of a force.
T   3. Every lever has two arms.
T   4. Levers are classified as first, second, and third class.
T   5. A wheel and axle and a wedge are forms of a lever.
T   6. The I.M.A. of a machine is usually greater than its A.M.A.

F 7. A short, thick wedge has a ~~greater~~ mechanical advantage smaller than a long, narrow wedge.

F 8. The work output of a machine is usually ~~more~~ than the less work input.

T 9. No machine is 100% efficient.

F 10. Mechanical advantage ~~is expressed in newtons.~~ has no units

B. Multiple Choice

Choose the word or phrase that completes correctly each of the following sentences.

1. Every lever has (*one, two, two or more*) arms.
2. A gear is an example of a(n) (*wheel and axle, wedge, inclined plane*).
3. A lever with the fulcrum between the resistance force and effort force is a (*first class, second class, third class*) lever.
4. If the I.M.A. of a machine is 1, the effort force is (*multiplied, decreased, unchanged*).
5. To find the I.M.A. of a machine (*size, movement, friction*) is not considered.
6. The total number of support strands is equal to the I.M.A. in a (*lever, wheel and axle, pulley*).
7. A circular inclined plane is called a (*wedge, screw, pulley*).
8. Work output and work input are needed to find the (*power, speed, efficiency*) of a machine.
9. Efficiency is measured in (*J, J/s, percent*).
10. The effort arm in a lever with an I.M.A. of 4 is (*2 times, 3 times, 4 times*) longer than the resistance arm.

C. Completion

Complete each of the following sentences with a word or phrase that will make the sentence correct.

1. The pivot point of a lever is called the _fulcrum_.
2. The effort arm of a lever is 4 m long, the resistance arm is 2 m long. The I.M.A. is __2__.
3. A block and tackle has four support strands. The I.M.A. is __4__.
4. When the resistance force for the pulleys in Question 3 moves through 5 m, the effort force moves through __20__ m.

5. A shovel is an example of a(n) _lever_.
6. A wedge is a type of _inclined_ plane.
7. Work output of a machine is _never_ greater than the input.
8. The efficiency of a machine is always _less than_ 100%.
9. When the effort force is equal to the resistance force in a lever, the I.M.A. is _1_.
10. In a can opener, a(n) _lever_ forces the blade into the top of a can.

D. How and Why See Teacher's Guide at the front of this book.
1. In what three ways are a lever, pulley, and wheel and axle alike?
2. Make a list of the six simple machines. Give an example of each simple machine and state its practical value.
3. Draw a diagram of a set of pulleys with an I.M.A. of 3. The effort force on the rope moves 2 m. How far will the resistance force move?
4. The work input of a machine is 200 J. The output is 50 J. Find the efficiency of the machine.
5. Make a list of five compound machines. Name one simple machine present in each of the compound machines.

challenges

1. Collect a variety of small hand tools. Examine each one to find how many simple machines it contains.
2. Find out how new cars are being designed to improve their efficiency. Learn how the weight, power, shape, engine, and other features of a car affect gasoline mileage.

interesting reading

Kerrod, Robin, *The Way It Works: Man and His Machines*. New York: Octopus, 1980.

Weitzman, David, *Windmills, Bridges, and Old Machines*. New York: Scribner, 1982.

Weiss, Harvey, *How to Be An Inventor*. New York: Crowell, 1980.

SIDE ROADS

The World of Robots

Dan McCoy/Black Star

Dan McCoy/Rainbow

Robot—the word is becoming more common all the time. What is a robot? What can it do? A robot is similar to an automated machine. It can do a specific task over and over. It never gets tired and does not need a lunch break. The difference between a robot and an automated machine is that an automated machine can only do one task. A robot can be taught to do many different tasks.

The control center or brain of a robot is a computer. If you want to teach a robot how to pick up an object, you program its computer brain. If you want the robot to do something else, you change the program.

In the next 5 years, more than 200 000 robots are expected to be working in industry. They will perform dangerous or often-repeated tasks. The automobile industry already uses robots to weld, paint, assemble parts, and move engines and doors from one place to another.

When you think of a robot, you may picture a shiny metal body on wheels. It may have arms with pincers and flashing lights. You may also picture a robot as a machine that talks and looks human. Neither of these ideas describes the modern robot. Most robots consist of an arm with one or two joints, and a two- or three-fingered hand. The arm is attached to a big box or cylinder that stands on the floor. The box contains the brain and mechanical parts that move the arm and hand. Such robots are limited to simple tasks because they cannot see or hear, and have no sense of touch.

Another type of robot looks like a motorized cart. This robot can deliver mail and papers between offices in a building. Some of these robots can bring you your morning orange juice or milk.

Today's robots can accomplish many tasks. Many scientists are working to make robots even more useful. For example, robots would do more jobs if they could see. Scientists have developed a computer that uses a television camera and lasers to study an area that requires a weld. A computer analyzes the situation, controlling the movements of the robot and the size of the weld. The ability to see also helps robots move. One type of robot can detect the body heat of a person and stop to avoid a collision.

Another improvement in robots is that they obey a voice command. Voice controlled robots are especially useful for the physically handicapped. A person who cannot use an arm could talk to a robot for help in picking up a glass of milk or a newspaper. The robot can even be instructed to turn the page of a book. Voice controlled robots are already used to make "smart" wheelchairs. A physically handicapped person tells the chair in which direction to move, and the chair moves.

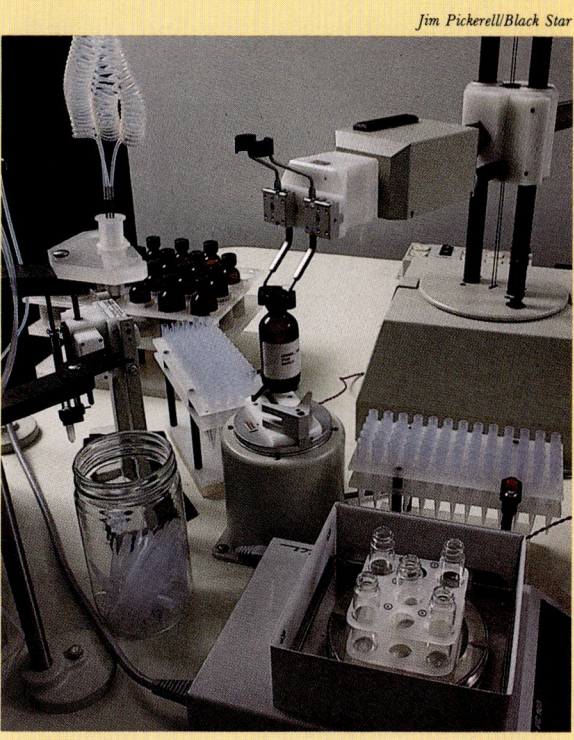

Doug Martin   *Jim Pickerell/Black Star*

For a robot hand to pick up a delicate piece of glass or turn a page, the sense of touch must be added. Scientists are working on robot hands that have plastic fingertips to hold objects. Within the fingers are pressure plates that tell the computer how tight an object is held. A computerized robot is used in Australia to shear sheep. The robot's computer has been programmed with the shape of a sheep's body. Sensors detect mild electric currents given off by a sheep's skin. The robot can shear more wool, at a faster rate, than a human.

As robots improve, they continue to do more dangerous and boring jobs. For example, robots can cut and slice foods, pour dangerous chemicals, and work in nuclear reactors. In the future, robots may work in fast food restaurants or run cash registers at supermarkets. Someday, you may have a robot butler that will take your coat when you get home. The "butler" may even look like a person.

A space shuttle has just been launched. Watching the launch of a spacecraft can be exciting. There is a feeling of great power and motion. What parts of the picture indicate motion? You experience motion in many ways. What is motion? What kinds of motion do you observe every day?

NASA

Motion

chapter 6

Introducing the chapter: Tie a heavy washer to a string one meter long to make a pendulum. Demonstrate the motion of the pendulum to the class. Ask students to identify when the pendulum is gaining speed and losing speed. Discuss the inertia of the pendulum when it is at rest and not swinging, and explain that a force is needed to put it in motion. Show that the arc of the pendulum continues to decrease over time as the pendulum loses momentum.

6:1 Inertia

When something is moving, it is in motion. The movies you see in a theater show motion because the film moves steadily through a film projector. You can hear the music on a tape when it is moving through a tape player. Without motion the hands of a clock would not indicate the time of day. For every motion there is a force that causes it.

A force is needed to start something moving or to change its direction. A force is also needed to stop motion. The tendency of matter to stay at rest or in motion, unless acted on by a force, is called **inertia** (ihn UR shuh).

A person riding in a car has inertia. Think of a car moving at a speed of 50 km/h. How fast is the person inside going? The person is moving with the car and is not left behind; therefore, the person must also be moving at 50 km/h. If the brakes are applied suddenly, what happens to the person in the car? The person continues to move forward even though the car is stopping. If the seat belt is unfastened, the dashboard or windshield may stop this forward motion.

If you are standing in a bus, you may be thrown off balance when the bus starts to move. Your body has inertia. It tends to remain in place as the bus

GOAL: You will gain an understanding of how forces produce, change, and stop the motion of objects.

What causes motion?

Define the term inertia.

Have students list examples of inertia based on the definition presented here. Let them cite personal experiences as examples. Stress that inertia is a property of all matter.

107

FIGURE 6-1. Passengers standing in a bus fall backward when the bus starts (a) and fall forward when the bus stops (b).

a Starting bus

b Stopping bus

begins to move. If the bus goes forward too fast, you may fall backward.

All matter has inertia. Inertia is a property of matter. The amount of inertia an object has depends upon its mass. The greater the mass of an object, the greater its inertia. A sofa of large mass has more inertia than a kitchen chair. It takes more force to move a sofa than to move a kitchen chair. It takes a larger force to start and stop a bus than to start and stop a small sports car.

activity See Teacher's Guide.

DEMONSTRATING INERTIA   Activity 6-1

Set a drinking glass on a desk or table. Lay a flat playing card on top of the glass. Place a coin on the center of the card. Pull the card away quickly with your fingers so it flies out from under the coin. Can you pull the card out fast enough so that the coin falls into the glass? How did this activity illustrate inertia? Because of inertia, the coin does not move with the card. Thus it falls into the glass.

FIGURE 6-2.

activity See Teacher's Guide.

INERTIA OF BOOKS   Activity 6-2

Tie a long, thin string around five books. Place the books on the floor. Raise the books by pulling upward very slowly on the string. Lower the books back to the floor. Pull up on the string with a sharp, jerking movement. Compare the effect of pulling slowly and suddenly. What did inertia have to do with your observation?

Slowly—force overcomes inertia; suddenly—inertia of the books causes the string to break.

making sure See Teacher's Guide.

1. How can seat belts help protect a passenger in a car?
2. Which has more inertia, a golf ball or a Ping-Pong ball? Explain.
3. A person is standing next to the first seat in a train moving at 150 km/h. The person jumps high in the air. Where will the person land? Why? Figure 6–3 will help you discover the answer.

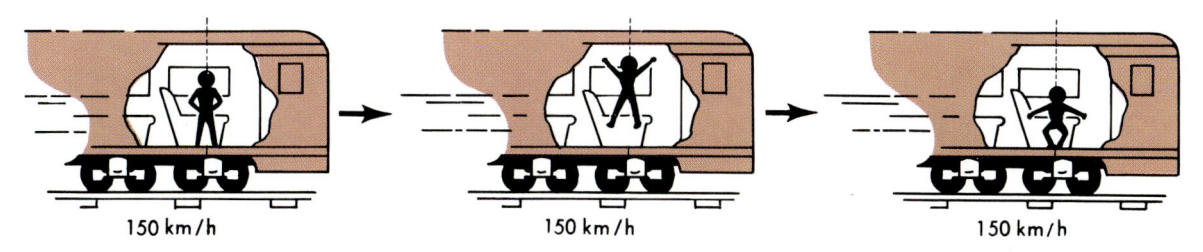

FIGURE 6–3.

6:2 Speed

One property of motion is speed. **Speed** refers to how fast an object is moving. It is defined as the distance an object travels per unit of time. For example, a car may travel at 88 km/h on a highway. At this speed the car will go 88 km in 1 h.

If a car moves at a steady 50 km/h, its speed is constant. It goes 50 km each hour it travels. In two hours the car will move a distance of 100 km. How far will it go in three hours? 150 km

FIGURE 6–4. The speed of a baseball can be determined by measuring the time it takes for the ball to travel the distance between the pitcher and the batter.

Joseph DiChello

Use the riding of a bicycle to illustrate speed and average speed.

How is average speed calculated?

Speed can be measured with a speedometer. Average speed is calculated.

A car is more likely to change speeds than to move at a constant speed. Thus, the speed of a car during a trip is really an average speed. Average speed is the total distance traveled divided by the time it takes to go that distance. The following equation is used to find average speed.

$$\text{speed} = \frac{\text{distance}}{\text{time}} \qquad S = \frac{d}{t}$$

Example

Find the average speed of a motorcycle that travels 16 km in 8 min. Give the speed in km/h.

Solution

Step 1: Write the equation for speed.

$$S = \frac{d}{t}$$

Step 2: Substitute the values for distance and time given in the problem.

$$S = \frac{16 \text{ km}}{8 \text{ min}}$$

Step 3: Divide to find the answer.

$$\frac{16 \text{ km}}{8 \text{ min}} = 2 \text{ km/min} \qquad S = 2 \text{ km/min}$$

Step 4: Change the answer to km/h by multiplying by 60 min/h.

$$\frac{2 \text{ km}}{\text{min}} \times \frac{60 \text{ min}}{\text{h}} = 120 \text{ km/h} \qquad S = 120 \text{ km/h}$$

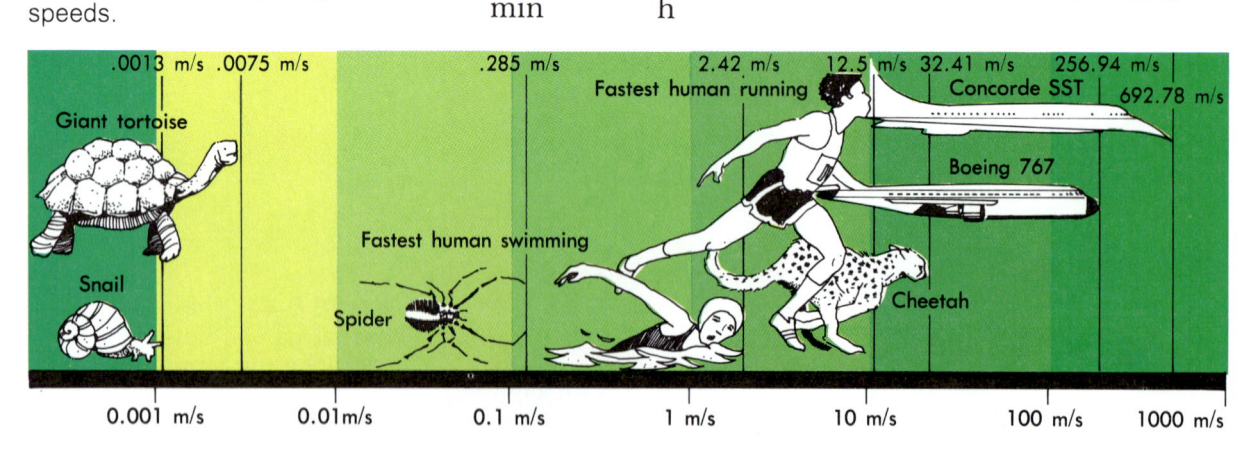

FIGURE 6–5. Comparing speeds.

making sure

4. Find the average speed of an airplane that flies 4500 km coast to coast in 5 h 15 min. 860 km/h
5. A motorcycle goes around a 1 km track 10 times in 2 min. Find its average speed. 5 km/min

6:3 Change in Speed

Every car has a "gas" pedal. This pedal is also called an accelerator. When a driver steps on the "gas" pedal, the speed of the car increases. The car accelerates. **Acceleration** (ak sel uh RAY shun) is the rate at which an object changes motion. An object accelerates whenever it changes speed or changes direction. A force is necessary to cause these changes.

To find the average acceleration of an object, divide its change in speed by the time it takes for the speed to change. The equation for finding average acceleration is

$$\text{acceleration} = \frac{\text{final speed} - \text{initial speed}}{\text{time}}$$

$$a = \frac{S_2 - S_1}{t}$$

The term acceleration actually describes a change in velocity. Velocity is a vector quantity having both magnitude and direction.

What happens when an object accelerates?

Write the equation for acceleration.

Like average speed, average acceleration is determined by the total time of motion under consideration.

Example

A car goes from rest to 79 km/h in 10 s. What is the average acceleration of the car?

Solution

Step 1: Write the equation for average accleration.

$$a = \frac{S_2 - S_1}{t}$$

Step 2: Substitute the values for final speed, initial speed, and time given in the problem.

$$a = \frac{70 \text{ km/h} - 0}{10 \text{ s}}$$

Step 3: Subtract, then divide to find the answer.

$$\frac{70 \text{ km/h} - 0}{10 \text{ s}} = 7 \text{ km/h/s} \qquad a = 7 \text{ km/h/s}$$

You may wish to introduce and explain the equation F = m × a, force = mass × acceleration, to show the relationship between these quantities.

The rate of acceleration depends on both the mass of an object and the force acting on it. An empty truck can accelerate faster than a loaded truck because it has less mass. A race car can accelerate faster than an average small car because it has a more powerful engine.

Which accelerates faster, a truck or a sports car? A truck may have a powerful engine, but it may also have a heavy load to move. The less powerful engine in a small sports car can accelerate the car faster because there is less mass to move.

Daytona International Speedway

FIGURE 6–6. A small race car can accelerate rapidly because of its low mass and powerful engine.

When an object that is moving comes to a stop, it decelerates. Deceleration is the rate at which the speed of an object decreases. A driver decelerates a moving car by applying the brakes. The brakes produce a force on the wheels that slows them down. As a result the speed of the car decreases.

List examples of deceleration.

FIGURE 6–7. A drag chute is used to decelerate a race car.

National Hot Rod Association

112 Motion

Deceleration (dee sel uh RAY shun) means negative acceleration. You can use the acceleration equation to calculate deceleration. The deceleration you calculate is always a negative number. Why? The initial speed is always greater than the final speed.

How is deceleration calculated?

Example
A sled traveling at 6 m/s slows to a stop in 3 s. Find the deceleration.

Solution

Step 1: Write the equation for average acceleration.

$$a = \frac{S_2 - S_1}{t}$$

Step 2: Substitute the values for final speed, initial speed, and time given in the problem.

$$a = \frac{0 - 6 \text{ m/s}}{3 \text{ s}}$$

Step 3: Subtract, then divide to find the answer.

$$\frac{0 - 6 \text{ m/s}}{3 \text{ s}} = -2 \text{ m/s}^2 \qquad a = -2 \text{ m/s}^2$$

To stop or slow a moving object, force must be applied. The rate of deceleration depends on mass and force. The greater the force applied in stopping something the faster it decelerates. The greater the mass of an object, the more slowly it stops. In the example problem above, friction and gravity were the two forces that caused the sled to stop.

making sure

6. In which of the following does acceleration occur? Explain your answers. See Teacher's Guide.
   (a) Air rushes out of an inflated balloon. yes
   (b) A clock's second hand moves in a circle. yes
   (c) An ocean current flows at 3 km/h. no
7. What is the acceleration of a skateboard whose speed changes from 5 km/h to 10 km/h in 5 s? 1 km/h/s
8. A parachute on a drag racer is opened. It slows the racer's speed from 260 km/h to 130 km/h in 10 s. Find the deceleration. -13 km/h/s

6:3 Change in speed

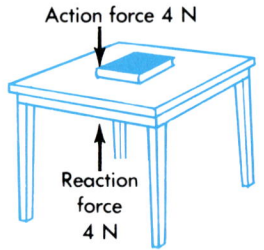

FIGURE 6–8. Every action force has a reaction force.

Describe one example of action and reaction forces.

FIGURE 6–9. In a Hero's engine, escaping steam causes the engine to spin.

6:4 Action and Reaction Forces

What happens to the person and chair when the person in Figure 6–8 pushes against the desk? The push on the desk is called an **action force.** As the desk is being pushed in one direction, the person and chair move in the opposite direction. This movement is caused by a force on the person called a **reaction force.** A reaction force is always equal in size and opposite in direction to an action force. In this case, the action force pushes against the desk. The reaction force makes the person and chair move away from the desk. Action and reaction always occur together. Every force has an equal and opposite force.

When a 4-N book rests on a table it exerts a downward action force of 4 N. The table pushes up on the book with a reaction force of 4 N (Figure 6–8). Action and reaction forces are always equal and in opposite directions.

In a Hero's engine (Figure 6–9), water is heated to steam. Then the steam moves out of the engine. As the steam escapes, it exerts a reaction force on the engine. This reaction force causes the engine to spin. It spins in a direction opposite to the direction in which the steam escapes.

In principle the operation of a rocket engine is somewhat like a Hero's engine. Hot gases escape from the rocket engine. These gases exert a reaction force on the engine. The reaction force is the push that makes the rocket move.

activity
See Teacher's Guide.

Activity 6-3 USING SPRING SCALES TO MEASURE ACTION AND REACTION FORCES

Tie a spring scale to the leg of a desk or table. Connect the hook of the spring scale to the hook of a second spring scale. Exert a force on the second spring scale. Keeping the force constant, read both scales. Repeat with different amounts of force. What did you observe? In each case, the force on both scales was the same.

activity

See Teacher's Guide.

AN ACTION AND A REACTION Activity 6-4

Inflate a toy balloon with air. Hold it tightly at the neck. Then release it. What happened to the balloon? Explain your observations.

The balloon flew around the room as it deflated. The force of the air leaving the balloon pushed it in the opposite direction.

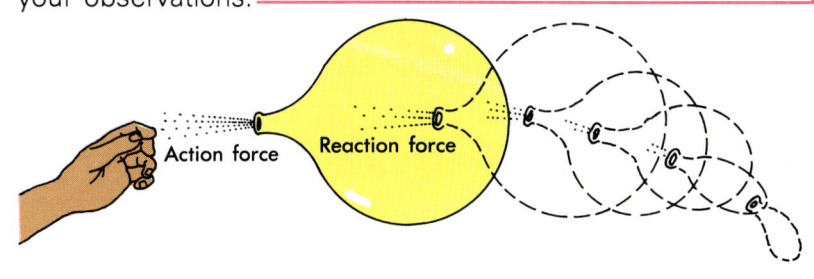

FIGURE 6–10.

making sure

9. Explain the following:
   (a) A person tries to jump from a boat to the dock. The boat moves away from the dock. *The reaction force on the boat is opposite to the person's direction.*
   (b) A water sprinkler rotates when the water is turned on. *The reaction force on the sprinkler makes it turn in the opposite direction from the water.*

6:5 Falling Objects

Falling is caused by the force of gravity. Near the earth a falling object accelerates at a rate of 9.8 m/s². That is, its speed increases by 9.8 m/s each second. After 2 s, it is moving at 19.6 m/s. After 3 s, it is moving at 29.4 m/s. The value 9.8 m/s² is called the **acceleration of gravity.** All falling objects near the

What is the acceleration of a falling object?

Students may not understand that the amount of force acting on an object is proportional to the mass of the object (a = F/m).

Table 6–1.
Speed of a Falling Object

Time of fall (s)	Speed (m/s)
1	9.8
2	19.6
3	29.4
4	39.2
5	49.0
6	58.8

As mass increases, the force of gravity increases and acceleration remains constant.

6:5 Falling Objects   115

earth accelerate at the same rate in a vacuum. Because there is no air or other matter in a vacuum, there is nothing to slow the fall of an object.

An object falling from a great height eventually stops accelerating. Its speed becomes constant. This point is called terminal speed. **Terminal speed is reached when the air resistance against the object equals the pull of gravity.** At this point, the force on the object is zero.

Define terminal speed.

FIGURE 6–11. Skydivers in free fall reach terminal speed when air resistance equals the force of gravity.

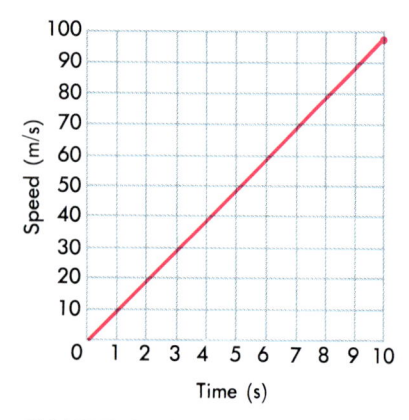

FIGURE 6–12.

activity
See Teacher's Guide.

GRAPHING TIME VS SPEED   Activity 6-5

On a sheet of graph paper make a graph using the data given in Table 6–1. Label the time on the horizontal axis from 0–10 s. Label the speed on the vertical axis from 0–100 m/s. (If graph paper is not available, use notebook paper. Allow 2 cm of space between each second. Use 1 cm of space between each 5 m/s of speed.) Plot the speed for each second of time as given in the table. Draw a line to connect the points you have plotted. How could you use your graph to find the speed at 8 s? What is the speed at 8 s and 10 s?   78.4 m/s   98 m/s

Continue the plotted line to 8 s and read the speed on the vertical (y) axis.

activity

See Teacher's Guide.

MEASURING THE SPEED OF FALLING OBJECTS
Activity 6-6

Motion along an inclined plane can be studied to learn something about falling objects. **(1)** Nail two boards, measuring 15 cm wide and 4 m long, together to form a trough. **(2)** Paste a 5-cm strip of aluminum foil on one side of the trough. **(3)** Paste 1-cm square pieces of foil along the opposite side as shown in Figure 6–13. The pieces of foil must be 2.5 cm apart. **(4)** Connect each foil piece to a wire and then connect one end of the wire to a battery. **(5)** Connect the battery to a bell and then connect the bell to the aluminum strip. "Liquid" or "cold" solder can be used to attach the wire to the foil. Tape may also be used. **(6)** Stand the trough up and tilt it at a slight angle. Run a metal ball down the trough. **(7)** Use a stopwatch to time the speed of the ball. Why did the bell ring more often as the ball got close to the bottom of the ramp? At what point was the ball's speed greatest? The ball accelerates as it descends the ramp. The ball's speed was the greatest at the bottom of the ramp.

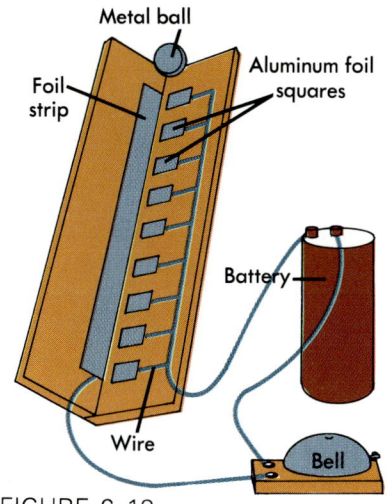

FIGURE 6–13.

Check with your school shop teacher for assistance in making this apparatus.

making sure

10. How does air friction affect the speed of a falling object? decreases the speed
11. Why might a sheet of paper fall more slowly than the same sheet crumpled into a ball? less air resistance when crumpled
12. The force of gravity on the moon is less than Earth's gravity. How is the speed of falling objects different on the moon? Speed will be less because acceleration is less.

6:6 Circular Motion

The motion of an object along a curved path is called circular motion. For example, a person riding on a merry-go-round rides in a circle. In order to keep an object moving in a circular path, a force must act toward the center of the circle. This force is called centripetal (sen TRIHP ut ul) force. A **centripetal force** keeps an object moving in a circular path.

Students are likely to have heard the term "centrifugal force" and may know about centrifuges.

How does a centripetal force keep an object moving in a circle?

6:6 Circular Motion 117

Point out that centrifugal force is a fictitious force and does not exist.

FIGURE 6–14. Centripetal forces cause people on this amusement park ride to move in a curved path.

List three examples of circular motion and centripetal force.

What happens when the car in which you are riding makes a fast turn around a corner? You may feel your body sliding toward the outside of the curve. The inertia of your body tends to keep you moving in a straight line while the car is turning. There must be a centripetal force acting toward the center of the circle to keep you moving around the curve. The door of the car prevents you from traveling in a straight line. The car door provides the centripetal force.

A bicycle rider leans inward on a curve to increase the centripetal force and make the bike move in a circle. Because of inertia, the bike tends to go in a straight line rather than in a curve. On a sharp curve, the inertia may cause the bike to tip over.

Most roads are banked on the curves. A banked road is higher on the outside edge than on the inside edge. Banked curves increase the centripetal force which pushes inward on a car as it moves around the curve.

118　Motion

Some washing machines use circular motion. After the laundry is washed, the wet clothes are spun to remove most of the water. Because of inertia, the water goes out through holes in the tub as the clothes are spun around.

Review inertia and explain how it applies here. Point out that the movement of the water out of the tub is not due to a "centrifugal force."

FIGURE 6–15. The banked curved of this bicycle track increases the centripetal force on the bicycles as they move around the curve.

Gerry Cranham/Photo Researchers, Inc.

See Teacher's Guide.

activity

CIRCULAR MOTION   Activity 6-7

CAUTION: This activity is to be done under the direct supervision of the teacher. Attach a ball to a string and swing it in a circle. Notice the motion your hand makes. Did you keep pulling on the string to keep it taut? Did you *yes* at the same time move the ball forward around the circle? *yes* If someone cut the string while the ball was moving, what would happen? *The ball would fly off in a straight line.*

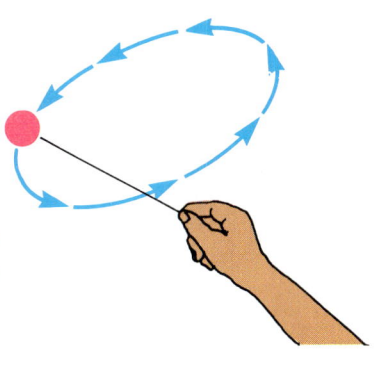

FIGURE 6–16.

See Teacher's Guide.

activity

AN EFFECT OF INERTIA   Activity 6-8

Obtain a thin metal pie pan. Cut out a 1/5 slice of the pan with sheet metal shears. Place a large marble along the inner edge of the pan. Move the marble swiftly in a circle toward the cut-out section. Which direction did the marble travel when it left the pan? Explain your observation. *Due to inertia, the ball continues in a straight path when the centripetal force (wall of pan) is removed.*

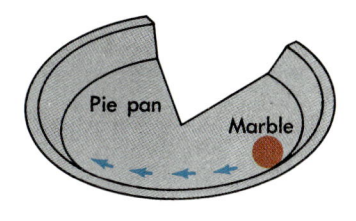

FIGURE 6–17.

6:6 Circular Motion

making sure

13. How does a banked road improve highway safety? increases centripetal force
14. Why might a car slide off an icy road when the car rounds a curve? lack of friction force on slippery surface
15. What keeps the moon traveling in a curved path around the earth? gravity of Earth
16. How does inertia affect the path of a moving object? Motion is in a straight line unless acted upon by a force.

6:7 Momentum

Every object in motion has momentum. **Momentum is the product of the mass of an object and its speed.** The more mass and speed, the greater the momentum. Which has more momentum, a truck moving at 60 km/h or a motorcycle moving at the same speed? The larger mass of the truck gives it more momentum. Does a fast-pitched or a slow-pitched softball have more momentum? The greater speed of the fast-pitched ball gives it more momentum. Momentum increases as either the mass or speed increases.

What is momentum?

Enrichment: Ask students to describe other examples of momentum.

FIGURE 6–18. Both the car and truck are traveling at the same speed. The truck has a greater momentum because it has more mass.

Janet Adams

The momentum of one object can change. But whenever one object gains momentum, another object loses an equal amount of momentum.

The **law of conservation of momentum** states that momentum cannot be created or destroyed. When one object gains momentum, another object loses it. For instance, if two roller skaters push against each other they will roll away in opposite directions. The skater with the larger mass moves slower than the skater with the smaller mass. The momentum for the two skaters is equal and opposite. When two billiard balls collide, there is no gain or loss of momentum. The speed of each ball may change. But the total momentum for the two balls does not change.

FIGURE 6–19. Momentum is transferred as the billiard balls collide.

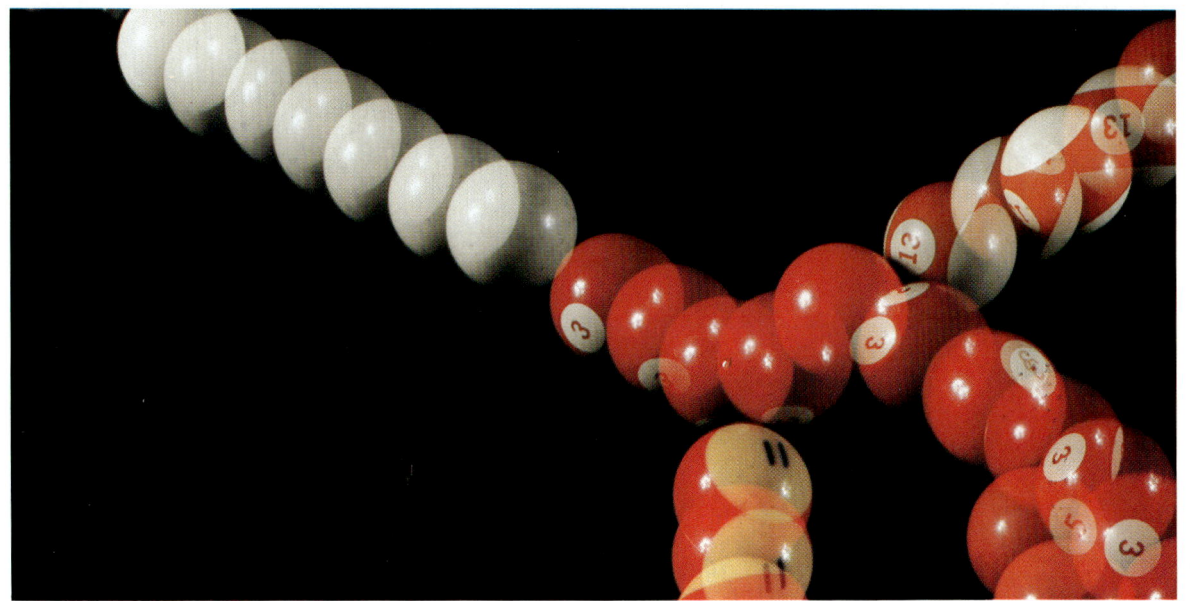

Hickson-Bender Photography

making sure

17. A student throws a Ping-Pong ball and a tennis ball as hard as possible. Which has more momentum? Why? Tennis ball has more mass and velocity.

18. Why will a bullet shot from a gun have more momentum than an arrow shot from a bow? The bullet has a very high velocity.

19. Why is it harder to stop a loaded truck than an unloaded truck? A loaded truck has more mass and more momentum.

20. It takes more than a kilometer for a heavily-loaded freight train to stop. Why? Its great mass gives it a large momentum.

See Teacher's Guide.
PERSPECTIVES
skills

Understanding Science Words

As you read *Principles of Science, Book One,* you will be introduced to many new science words. These words are from the different branches of science such as chemistry, biology, geology, and ecology.

There are several ways of finding the meanings of the new words. First a new word may be defined in your textbook when it is introduced. You can recognize the word easily because it will be shown in boldface print. Second, you can find the meaning in a dictionary or in the glossary. Third, you can think about what the word might mean by the way it is used in the sentence. A fourth way is to determine the word parts. Many science words are combinations of two or more word parts that form the science word we see. For example, the word *photometer* is a combination of two word parts. The word part "photo" is a Greek word part that means "light." The word "meter" is a Greek word part that means "measure." Thus, a photometer is an instrument used to measure light. The table below lists some word parts that can be combined to form more complex words.

Source	Word Part	Meaning
Latin	ager	field
Latin	anti	against, opposite
Greek	astro	star
Greek	auto	self
Latin	bi	two
Greek	bio	life
Latin	carn	flesh
Greek	chrom	color
Greek	chron	time
Greek	derm	skin
Greek	eco	environment
Greek	geo	earth
Greek	graph, gram	write
Greek	hemi	half
Greek	hydro	water
Greek	ism	manner or state
Greek	itis	inflammation of
Greek	logy	science of, study of
Greek	meter	measure
Greek	meteor	sky phenomenon
Greek	micro	small
Greek	mono	one, single
Greek	morph	form, shape
Greek	nomy	sum of knowledge
Greek	photo	light
Greek	pod	foot
Greek	poly	many, excessive
Greek	protos	early, first
Greek	pseudo	false
Greek	psych	mind
Latin	retro	backward
Greek	scop	see
Greek	techno	art, craft
Greek	tele	far
Greek	thermo	heat
Latin	ultra	beyond
Latin	vorous	feeding on, eating
Greek	zoion	animal

Use the table to determine the meanings of each of the following:

agronomy   chronometer   morphology
astronomy   dermititis   photography
biology   geothermal   technology
carnivorous   meteorology   telescope

Scientists combine root words when they need a new word. Therefore, knowing the meaning of the roots will make it easier for you to read and understand science. Scientists use mostly Greek and Latin roots when they need to label or explain a discovery.

main ideas
Chapter Review

1. The greater the mass of an object the greater its inertia. — 6:1
2. Average speed is found by dividing distance by time. — 6:2
3. It takes a force to change the motion of an object. — 6:3
4. The rate of deceleration depends on mass and force. — 6:3
5. A reaction force is always equal in size and opposite in direction to an action force. — 6:4
6. Speed of a falling object increases until it reaches terminal speed. — 6:5
7. Centripetal force must act toward the center of a curved path. — 6:6
8. The momentum of an object changes as either its mass or speed changes. — 6:7
9. Whenever one object loses momentum, another object gains an equal amount of momentum. — 6:7

vocabulary

Define each of the following words or terms.

acceleration	law of conservation	momentum
acceleration of gravity	of momentum	reaction force
action force	deceleration	speed
centripetal force	inertia	terminal speed
circular motion		

study questions

DO NOT WRITE IN THIS BOOK.

A. True or False

Determine whether each of the following sentences is true or false. If the sentence is false, rewrite it to make it true.

T 1. It takes a force to stop something that is moving.
T 2. Your body has inertia.
T 3. A basketball has more inertia than a tennis ball.
T 4. Centripetal force acts on a marble moving in a curved path.
T 5. The acceleration of gravity is 9.8 m/s².
F 6. Banked roads ~~decrease~~ the centripetal force acting on cars. increase
T 7. Every object has inertia.

Motion   123

F 8. On a trip, a ~~car always moves at its average speed.~~ car's speed varies
T 9. When a driver steps on the brakes the car decelerates.
T 10. When you start to run, your body accelerates.

B. Multiple Choice

Choose the word or phrase that completes correctly each of the following sentences.

1. The acceleration of a car moving at a constant speed of 30 km/h is *(30 km, 30 km/h, <u>0</u>)*.
2. A centripetal force acts on a person *(at rest, running in a straight line, <u>on a merry-go-round</u>)*.
3. To find the average speed of a bicycle, divide distance traveled by *(acceleration, speed, <u>time</u>)*.
4. The speed of a falling object *(<u>increases</u>, decreases, does not change)* until it reaches terminal speed.
5. *(5 km, <u>5 km/h</u>, 5 km/h/s)* is a speed.
6. *(<u>Force</u>, Inertia, Momentum)* is needed to stop a freight train.
7. As a moving van is loaded, its inertia *(<u>increases</u>, decreases, remains the same)*.
8. The speed of a ball *(<u>increases</u>, decreases, remains the same)* as it rolls down a 1-m ramp.
9. The acceleration of a ball *(increases, decreases, <u>remains the same</u>)* as it rolls down a 1-m ramp.
10. A falling object reaches terminal speed when it *(increases in speed, stops falling, <u>no longer increases in speed</u>)*.

C. Completion

Complete each of the following sentences with a word or phrase that will make the sentence correct.

1. An object remains at rest unless a(n) __force__ acts on it.
2. The amount of inertia an object at rest has depends on its __mass__.
3. A banked road is __higher__ on the outside than the inside.
4. If a car accelerates from rest to 50 km/s in 5 s, its acceleration is __10 km/s²__.
5. A __centripetal__ force exists when an object moves in a circle.
6. If you swing a ball attached to a string, the string exerts a(n) _____ force on the ball. centripetal

7. A car that travels 240 km in 4 h has an average speed of _60 km/h_.
8. A moving object that is slowing down is _decelerating_.
9. A parachute decelerates a drag racer by exerting a(n) _force_ on the racer.
10. Momentum decreases when either the mass or the _speed_ decreases.

D. How and Why   See Teacher's Guide at the front of this book.
1. An injury in an auto accident may be caused by the "second collision" of a person riding in the car. How?
2. Explain how inertia enables a clothes washer to spin water out of clothes.
3. How does the speed and acceleration of a parachute jumper change during a jump?
4. Give three examples of acceleration. What force produces each acceleration?
5. Why do all falling objects in a vacuum accelerate at the same rate?

challenges

1. Make a poster about motion using pictures from old magazines that show something moving. Beside each picture list the energy source that is used to produce motion. For example, gasoline is the energy source for a car. Organize the pictures into groups based on their energy source.
2. Obtain information from a library, auto club, or insurance company on automobile safety. Prepare a bulletin board poster that illustrates the connection between motion, forces, and automobile accidents.

interesting reading

Aylesworth, Thomas G., *Cars, Boats, Trains, and Planes of Today and Tomorrow.* New York, NY: Walker and Co., 1975.

Drake, S., "Galileo and The Rolling Ball; except from Galileo at Work." *Science Digest,* October 1978, pp. 74–76.

The shape of an airplane enables it to fly in air. Some airplanes are designed to use water surfaces to take off and land. Water and air are two examples of fluids. What are the special properties of fluids? How does an airplane use these properties to fly? How are forces in liquids used to do work? How does pressure in a liquid differ from pressure in a gas?

Johnny Johnson

Fluids and Pressure

chapter 7

Introducing the chapter: Teach your students how the use of performance objectives can aid their learning. Write the performance objectives for Section 7:1, page 66T, on the chalkboard or display them with an overhead projector. Have students read Section 7:1 and do the activity on page 129. Review the objectives with the class and relate them to the material covered. Follow up with a short quiz. You may wish to allow students to grade each other's papers under your direction and supervision.

7:1 Fluids

Liquids and gases are fluids. Both of these forms of matter are shapeless and can flow from place to place. Water and petroleum are the only liquids which occur in large quantities on the earth. The most plentiful gases in the air are nitrogen and oxygen.

A **fluid** is a substance that can move and change shape without separating. The properties of a fluid are due to the spaces and forces between particles in liquids and gases. In a solid, the particles are in fixed, rigid positions. In liquids and gases, the particles are farther apart and can move about each other. Fluids flow downhill, pulled by the force of gravity, and can be pumped through pipes.

When a liquid boils and turns into a gas, it may expand to many times its original volume. For example, boiling water may expand more than 1500 times as it forms water vapor. This expansion creates a force that can drive a steam engine. A gas can expand to a large volume or it can be compressed into a small space. When a gas expands the spaces between its particles increase. Particles in a liquid are packed closer together than the particles in a gas. Therefore, a liquid cannot be compressed like a gas.

GOAL: You will gain an understanding of fluids and how they are related to forces and motion.

What are fluids?

Explain how breathing and the flow of blood in the human body depend on the fluid properties of air and blood.

127

How can you observe surface tension?

Enrichment: You can demonstrate surface tension by preparing some soapy water and blowing bubbles with a bubble pipe. Surface tension holds the molecules of water together, forming the film.

Particles in a liquid are held together by forces. These forces among the particles on the liquid's surface create a **surface tension.** You can observe surface tension. Rub some grease on a needle and set it gently on the surface of water in a glass or dish. The needle floats because the forces between the water molecules keep them from spreading and allowing the needle to sink. The force or tension between the molecules is greater than the force of the needle pushing down. Water forms spherical drops when sprinkled on a waxed surface. Forces between the water molecules on the surface of a drop cause the molecules to pull together. The formation of spherical drops is a result of surface tension.

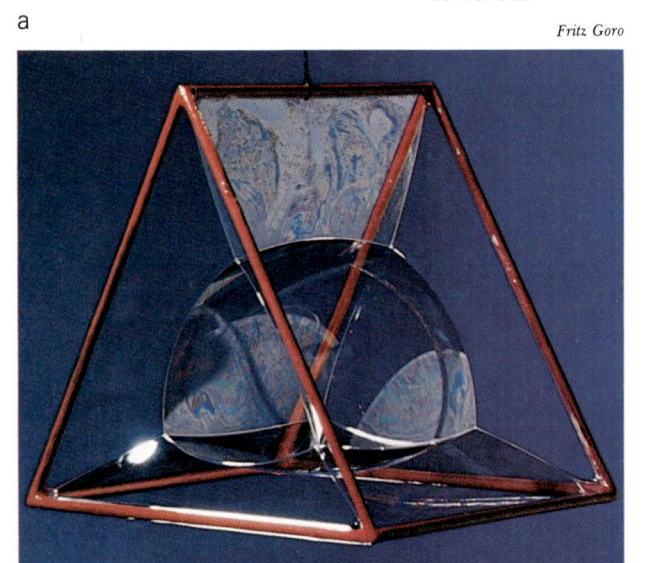

a *Fritz Goro* b *Ethyl Corporation*

FIGURE 7–1. A soap film forms because of surface tension (a). Surface tension allows a water strider to stand on the surface of water (b).

Air cooled engines depend on a flow of air over the engine for proper cooling.

Some fluids are used in machines for cooling. The cooling occurs when fluids carry heat away from an area that is hot. One way to cool an object is to blow air over it. Most automobile engines are cooled by pumping water through the engine. Water is also used to cool the fuel in a nuclear reactor so it does not melt. Oil and liquid sodium are two fluids used in some kinds of cooling devices.

128 *Fluids and Pressure*

activity

See Teacher's Guide.

WATER SEEKS ITS OWN LEVEL Activity 7-1

Obtain a piece of rubber tubing about 2 m long and two funnels. Insert a funnel into each end of the tubing. Hold each funnel at the same level. Pour water into one funnel until it fills the other funnel halfway. Take turns slowly raising and lowering each funnel. What happened to the water levels in the funnels when they were moved? The water level goes up or down in the funnels, but the height of the two levels stays the same.

FIGURE 7–2.

making sure

1. Name three fluids that are transported through pipes. Answers will vary. Some examples are water, oil, steam.
2. How does a hydroelectric dam use the fluid properties of water? Water flows through shafts and turns turbines.
3. Why might drops of water appear on the surface of a newly waxed car? Through surface tension, the water molecules pull into a sphere.

7:2 Pressure in Liquids

Pressure is the amount of force per unit area. Force is measured in newtons and area is expressed in square meters (m^2). The unit for pressure is the pascal. One pascal (Pa) is equivalent to 1 N/m^2. A kilopascal (kPa) is equal to 1000 pascals. Pressure in an automobile tire is about 196 kPa.

Point out to students that the tire pressure gauges with which they are familiar measure in lb/in^2.

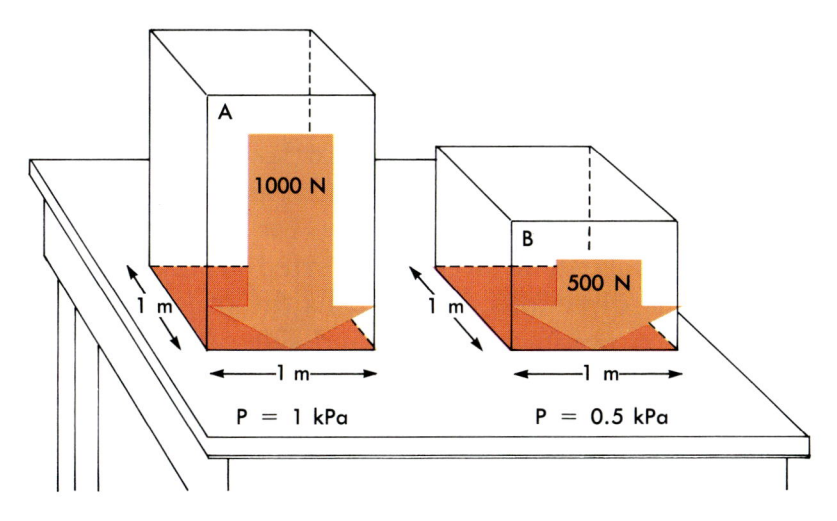

Remind students that force and pressure are related, one cannot exist without the other.

FIGURE 7–3. Block B weighs one-half as much as Block A and therefore, exerts one half the pressure of Block A.

7:2 *Pressure in Liquids* 129

How is pressure calculated?

To find pressure you must determine the force that is exerted on a surface. You must also determine the total surface area on which the force is exerted. The following equation is then used to find pressure.

$$\text{pressure} = \frac{\text{force}}{\text{area}}$$

$$P = \frac{F}{A}$$

Example

A liquid in a container exerts a force of 500 N on the bottom of the container that has an area of 2 m². Determine the pressure on the bottom of the container.

Solution

Step 1: Write the equation for pressure.

$$P = \frac{F}{A}$$

Step 2: Substitute the values for area and force given in the problem.

$$P = \frac{500 \text{ N}}{2 \text{ m}^2}$$

Step 3: Divide to find the answer.

$$\frac{500 \text{ N}}{2 \text{ m}^2} = 250 \text{ N/m}^2 \qquad P = 250 \text{ Pa}$$

Pressure in a liquid is directly related to the depth and density of a liquid. Any one who has dived to the bottom of a swimming pool or lake knows water pressure increases with depth. Submarines have been crushed when they descended too far below the ocean's surface. The pressure exerted at any depth is equal in all directions. This means the pressure is the same at any point, up, down, and sideways.

When pressure is applied at any point on a confined liquid, it is transmitted equally through the liquid. For example, suppose a jug is filled with water and corked. Then someone pushes down on

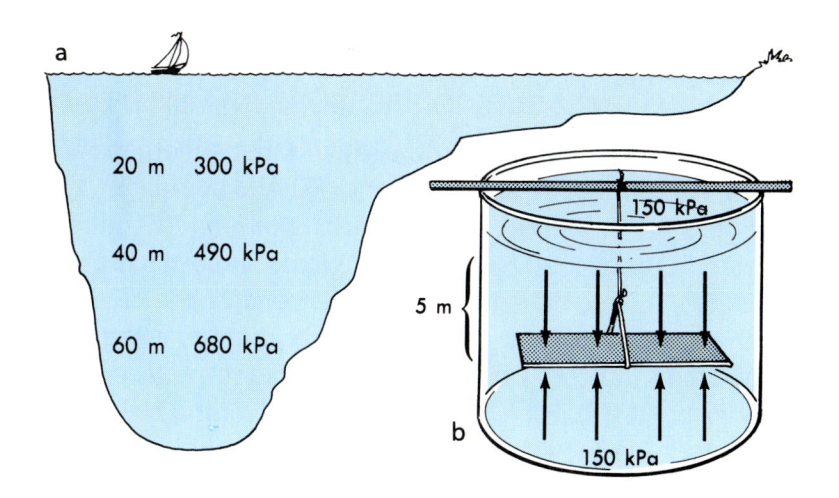

FIGURE 7–4. Water pressure within a lake increases with depth (a). Pressure exerted on a flat object placed in water will be the same on both the upper and lower surface (b).

the cork and exerts a pressure of 100 kPa. The pressure will be increased this amount at every point in the liquid. If the pressure of the cork becomes great enough the jug will break. Why?

Pressure exerted on the cork is transmitted through the liquid.

How does a hydraulic lift work?

Hydraulic lifts and other hydraulic equipment are based on the use of the pressure transmitted through liquids. Suppose two pistons are connected to a closed container of liquid. One piston has a small area and the other has a large area. A small force is exerted on the small piston. The force on the small piston causes pressure which is transmitted equally through the liquid. The pressure on the large piston creates a large total force because the surface area of the piston is so much greater. What if the area of the small piston is 1 cm² and the area of the large piston is 100 cm²? The area of the large piston is 100 times greater than that of the small piston. Therefore, the increase in force from the small piston to the large piston is 100 times.

FIGURE 7–5. In hydraulic equipment, pressure transmitted through a liquid is used to do work.

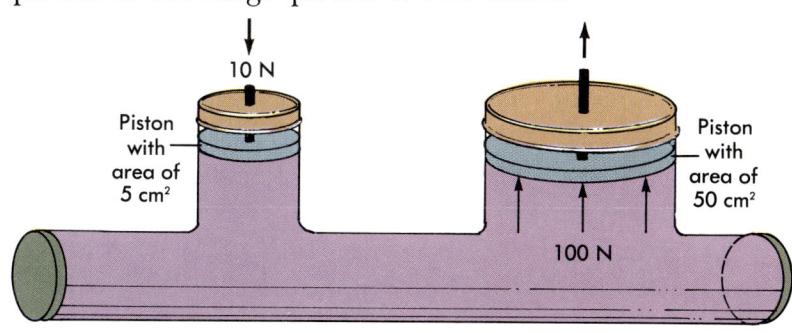

FIGURE 7–6. A force of 10 N exerted on the small piston will exert a force of 100 N on the large piston.

7:2 *Pressure in Liquids* 131

activity
Activity 7-2
See Teacher's Guide.

MEASURING PRESSURE IN A LIQUID

(1) Obtain a U-tube, thistle tube, thin rubber sheet, and a small piece of rubber tubing 1 m long. (2) Add some ink or colored water to the U-tube so it fills the curved bottom part. (3) Support the U-tube with a clamp. (4) Attach one end of the rubber tubing to the smaller end of the U-tube. Attach the other end to the thistle tube. (5) Cover the thistle tube with the rubber sheet and hold it in place with a rubber band. (6) Fill a tall cylinder with water. (7) Insert the thistle tube to different depths in the water. How did the level of liquid in the U-tube show pressure change?

The level rises as the depth of the thistle tube increases, showing an increase in pressure.

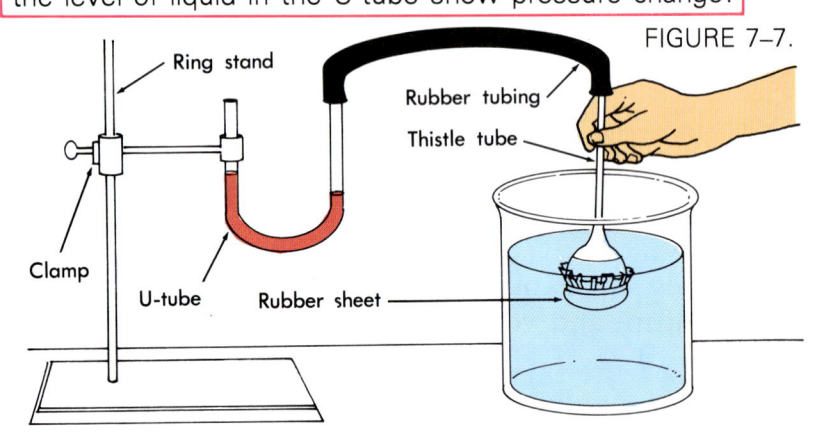

FIGURE 7-7.

making sure

4. Water has a density of 1 g/cm³ and mercury has a density of 13.6 g/cm³. In which liquid would the pressure be greater at a depth of 1 m. Why?

5. Why would a hydraulic lift not work if it was filled with air?

6. What is the pressure exerted by a force of 3200 N on an area of 3.2 m²? What is the pressure on each square centimeter? 1000 Pa, 0.1 N/cm²

7. A liquid in a container exerts a force of 270 N on the bottom of the container which has an area of 750 cm². Determine the pressure exerted on the bottom of the container. 3600 Pa

8. Why is a concrete dam along a deep river thicker at the bottom than at the top?

4. Mercury, the greater density means a greater weight at 1 m than for water.

5. Air would compress when a force was exerted on the piston.

8. Deep water has greater pressure than shallow water. Thus, the bottom of the dam must withstand greater force.

132 Fluids and Pressure

7:3 Gas Laws

If you could see molecules in a gas, what would you observe? You would see gas molecules in constant motion. The molecules move about rapidly in all directions because they have kinetic energy. When a gas is in a closed container, the gas molecules strike the walls of the container. The gas molecules exert pressure on the container walls. If you opened the container, the molecules would escape into the air.

What happens when you heat a closed container of gas? As the temperature of the gas increases, the molecules of the gas gain kinetic energy. The increase in kinetic energy causes the molecules to strike the walls of the container with more force. The pressure of the gas increases. If the container of the gas expands like a balloon, the increased pressure changes the volume. A hot air balloon, for example, is blown up by heating the air within it.

Charles' law explains the relationship between temperature and volume of a gas. According to Charles' law, the volume of a gas increases when it is heated. What happens to the volume if the gas is allowed to cool? The volume decreases.

Suggest as a model of a gas a large box with Ping-Pong balls flying about in random, straightline paths, striking and bouncing off the walls of the box. A pressure cooker illustrates Charles' law.

State Charles' law.

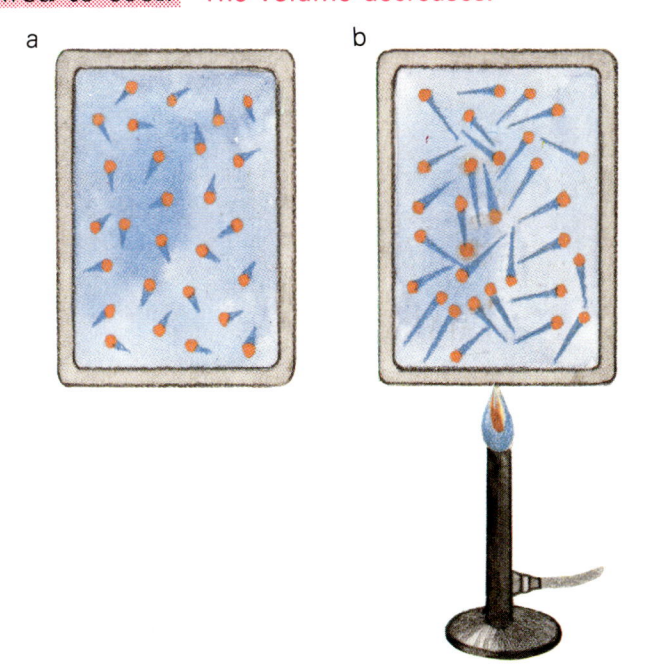

FIGURE 7–8. Molecules of a gas have kinetic energy and exert a pressure on the walls of a closed container (a). When the gas is heated, the kinetic energy increases and the pressure on the container walls increases (b).

Changes in pressure and size of an inflated bicycle tire illustrate Boyle's law.

State Boyle's law and describe one example.

If you push down on the handle of a tire pump, you are decreasing the volume of air inside the pump. The air in the pump cylinder is squeezed or compressed. What happens to the particles of air inside the pump? Since the particles have less space in which to move, they strike the cylinder walls more frequently. The pressure inside the pump increases if there is no change in the temperature of the air. The relationship between pressure and volume for a gas is explained in **Boyle's law.** The law states that the pressure of a gas increases as its volume decreases.

making sure   See Teacher's Guide.

9. Explain each observation.
   - (a) An automobile tire has a blowout on a very hot day.
   - (b) Pressure increases when you heat water in a pressure cooker.
   - (c) A tea kettle whistles when the water in it boils.
   - (d) An inflated balloon decreases in size when held in ice water.
   - (e) The gas discharged from a fire extinguisher feels cold.
   - (f) Heavy steel tanks are used to store compressed gas.

7:4 Buoyancy

If you drop a piece of metal such as a coin into water, it sinks. Yet you know that large ships are made of steel. Why do some objects float? An object floats when the force pushing up on it is equal to the object's weight. The upward force exerted by a fluid on objects on or within the fluid is called **buoyancy.** Due to buoyancy, solids such as cork float on water. Helium-filled balloons float in air.

An object's ability to float is related to the volume of fluid it displaces. When an object is placed in a liquid, for example, it pushes the liquid aside. Have you noticed how the level of water in a bathtub rises

FIGURE 7-9. A block of wood floats on water because the upward buoyant force of the water equals the weight of the block.

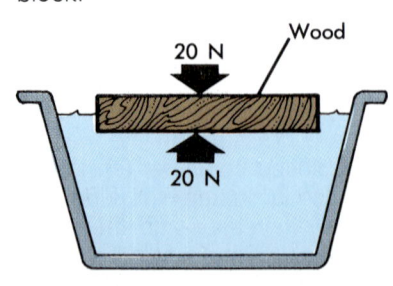

134 Fluids and Pressure

when you sit down in it? Your body displaces the water. When an object floats in water, only part of it displaces the water. The other part of the object remains above the water. Before an object can float, a definite amount of water must be displaced.

When the weight of the displaced liquid is equal to the weight of the object, the object floats. This discovery was made in 250 B.C. by Archimedes, an early Greek philosopher. When a block of wood is placed in water it begins to sink. It sinks until it displaces a volume of water that has a weight equal to the weight of the wood. The water beneath the wood exerts a force to hold it up. This upward force is the buoyant force. **Archimedes' principle** states that the buoyant force is equal to the weight of the fluid that is displaced by an object.

In Figure 7–10, each object sinks through the different liquids until it displaces a volume of liquid whose weight is equal to its own. At this point the object floats. The buoyant force is able to support the object.

Define the term buoyancy.

Why does a ship float?

State Archimedes' principle and describe one example.

Air	0.001 g/cm³
Wood (oak)	0.710 g/cm³
Corn oil	0.925 g/cm³
Water	1.000 g/cm³
Plastic	1.170 g/cm³
Glycerin	1.260 g/cm³
Rubber	1.340 g/cm³
Corn syrup	1.380 g/cm³
Metal alloy	7.810 g/cm³

Use Figure 7–10 to clarify the explanation of floating. A steel ship floats because it displaces a volume of water equal to its own mass. The buoyancy of the water beneath it supports the ship.

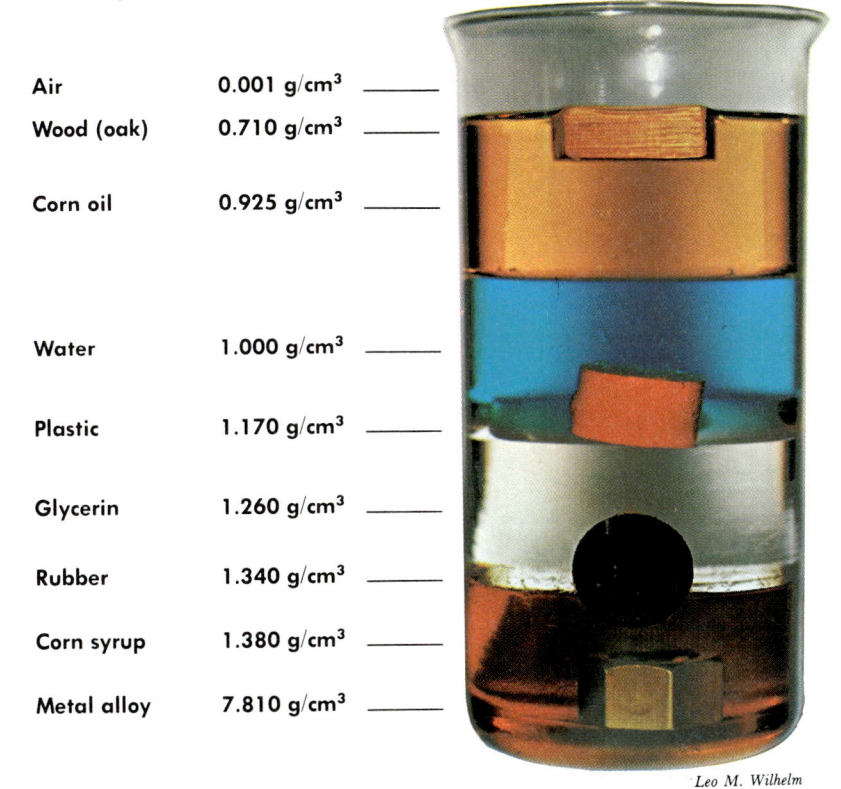

FIGURE 7–10. Floating objects and liquids of different densities illustrate Archimedes' principle—matter sinks until it displaces its own weight.

7:4 Buoyancy 135

The buoyancy of a fluid depends on its density. A more dense fluid has a greater buoyancy than a less dense fluid. For example, a block of aluminum sinks in water. The same block floats in mercury, which has more than 13 times the density of water. Because ocean water is more dense than fresh water, it supports floating objects better. The density of fresh water is 1 g/cm³, while the density of ocean water is 1.03 g/cm³.

A block of iron placed in water will never displace a volume of water having the same weight as the iron. The block is too dense. When an iron block is placed in water, it will sink. What would happen if the block is stamped into the shape of a pan? If the pan is placed in water, it will displace more water than the block. The pan will sink until it displaces a volume of water having the same weight as the iron. The iron will then float.

Why does a balloon containing helium float in air? Helium is a very light gas whose density is much less than air. A helium-filled balloon weighs less than the volume of air it displaces. As a result, the force of buoyancy pushes the balloon upward. A balloon filled with hot air also rises. Why? The air inside the balloon expands in volume when heated. Therefore, the density of the hot air inside the balloon is less than the density of the outside air. Since the volume of hot air weighs less than an equal volume of cooler air, the balloon rises.

Stress the similarity between floating in air and floating in water.

Why does a hot air balloon rise?

FIGURE 7–11. Balloon A rises because the warm air inside the balloon is less dense than the outside air. Balloon B sinks because the densities of the cool air and the balloon are greater than the outside air.

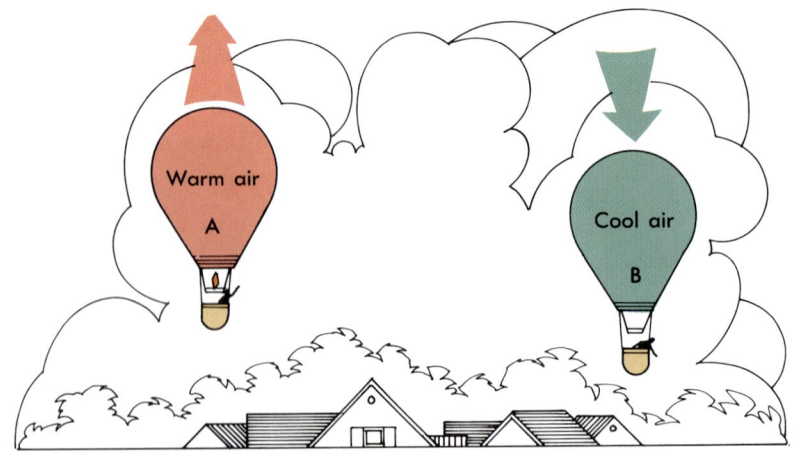

136 *Fluids and Pressure*

activity

See Teacher's Guide.

BUOYANCY AND DENSITY   Activity 7-3

Pour 15 mL of water into a 25-mL graduated cylinder. Tie one end of a piece of string tightly around a small stone. Hold the opposite end of the string and lower the stone into the water until it is completely submerged (Figure 7–12). What is the volume of the stone? Find the mass of the stone on a balance and then calculate its density. Compare it to the density of water (1 g/mL). Did the stone sink because it was more dense or less dense than water? more dense

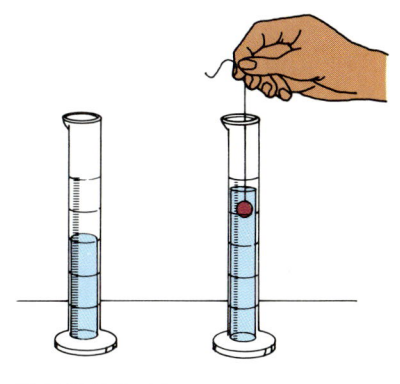

FIGURE 7–12.

activity

See Teacher's Guide.

FLOATING OBJECTS   Activity 7-4

(1) Place a displacement can on a table or desk top. Set a small beaker under the arm of the can. (2) Fill the can with water until water begins to run into the beaker. Empty water from the beaker and set it back under the arm of the filled can. (3) Fold a piece of heavy aluminum foil into the shape of a saucer. Place several small coins or washers in the saucer. (4) Find the mass of the saucer and contents on a beam balance. (5) Then put the saucer in the displacement can (Figure 7–13). What happened to the saucer? floats Catch the displaced water in the small beaker. Measure the volume of the displaced water in a graduated cylinder. (6) Then find the mass of the water (1 mL = 1 g). (7) Now fold the aluminum saucer tightly around the coins. Put it into the refilled displacement can. What happened? sank (8) Find the volume and mass of the displaced water. Compare the mass of the water to the mass of the saucer. See Teacher's Guide.

Displacement can

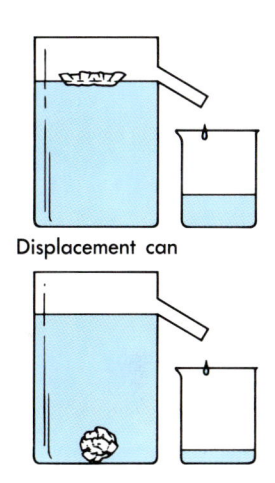

FIGURE 7–13.

making sure

10. Would a block of wood float higher or lower in fresh water than in ocean water? lower
11. The density of alcohol is about 0.8 g/mL. Would a block of wood float higher or lower in alcohol than in water? lower
12. Explain why a steel washer will sink in water but float in mercury. See Teacher's Guide.

7:5 Bernoulli's Principle

You may wish to provide strips of paper for this activity.

How can you demonstrate Bernoulli's principle?

Here is a simple activity you can do. Fold a piece of notebook paper in half lengthwise. Tear it along the crease. Take one piece of paper and insert about 4 cm of it between the pages at the top of this book. Hold the book upright in front of your mouth with the paper hanging down and away from you. Blow hard across the top of the book over the paper. What do you observe? What happens when you stop blowing? Paper rises into the stream of air. Paper falls back down.

The activity described above illustrates **Bernoulli's principle.** This principle states that the pressure in a moving stream of fluid is less than in the fluid around it. The paper rises when you blow because air pressure in the stream of breath is less than in the air around it. The harder you blow the lower the pressure. The air pressure under the paper is greater than the pressure above the paper. The paper is pushed up into the moving air above it. Whenever there is a flowing stream within a fluid, the fluid around it is forced into the stream.

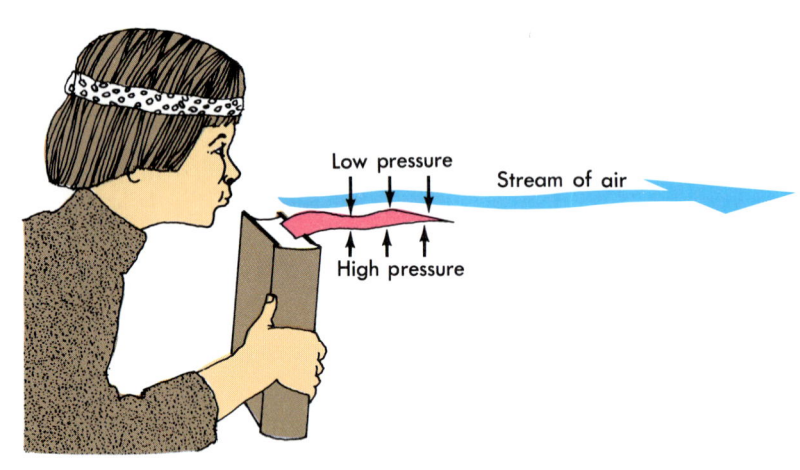

FIGURE 7-14. Blowing a stream of air across a piece of paper demonstrates Bernoulli's principle.

Draw a diagram of an atomizer on a chalkboard and explain how it works.

How does an atomizer work?

Bernoulli's principle explains how an atomizer works. When someone squeezes the rubber bulb of the atomizer, a spray of liquid comes out of it. Squeezing the bulb forces air through the tube attached to the bulb. This causes liquid in the container underneath to be pushed up into the stream of air and sprayed out of the atomizer.

activity

See Teacher's Guide.

DEMONSTRATING BERNOULLI'S PRINCIPLE Activity 7-5

Hold a funnel with the spout down and place a Ping-Pong ball in the funnel. Hold the funnel over your head. Keeping the funnel vertical, lower it to your mouth and blow hard through the stem. Try to blow the ball out. Did you succeed? Where was the air pressure greater when you blew hard, under the ball or on top? How does this fact explain your observation.

on top

Pressure on top of the ball is greater than the pressure of the moving air under the ball. Thus the ball cannot be blown out.

7:6 Lift

An airplane can fly even though it is heavier than air. An upward force on the wings, called lift, keeps it up. Lift is created by the shape of the wing. A wing is round in front and thickest in the middle. It tapers to a narrow edge at the back. When an airplane moves forward, air flows over the wing and lifts the airplane (Figure 7–15).

The shape of the wing causes air to flow faster over the top than the bottom of the wing. This difference in the speed of the air flow causes air pressure to be greater on the bottom wing surface than on the upper surface. The unequal pressure pushes upward causing the lift that makes the airplane fly.

What causes an airplane wing to produce lift?

How does lift make an airplane fly?
Relate this principle to the flight of a paper airplane.

FIGURE 7–15. A wing splits the air into two streams. The stream above the wing moves faster than the stream below it (a). Lift results because greater air pressure below the wing exerts an upward force (b).

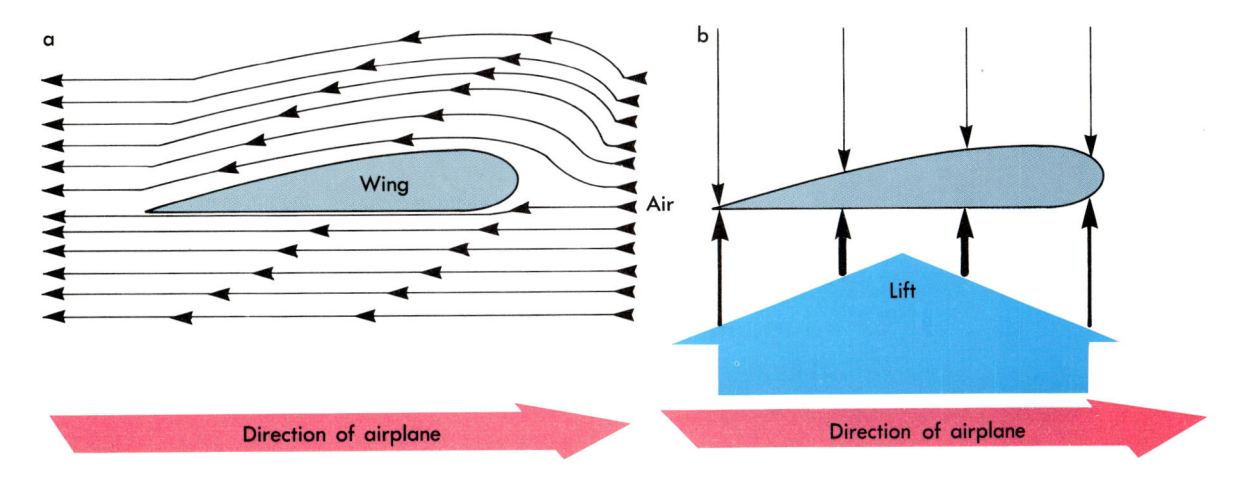

7:7 Flight

In order for an airplane to take off, it must overcome the force of gravity. In other words, the lift must be greater than the weight of the airplane. When the airplane is in level flight, the lift force on the wings is equal to the weight of the airplane.

FIGURE 7–16. Wings differ in size and shape. The wings of the glider are designed for soaring (a). Wings of jet aircraft are designed for high-speed flight (b).

The forward motion of the airplane caused by the pull of the propeller creates an air flow across the wing resulting in lift.

How is the flight path of an airplane controlled?

The operation of an airplane's rudder is similar to the rudder of a boat.

Recall that for lift to occur, air must be moving across the wings. Therefore, the airplane must be in motion. During takeoff, the airplane accelerates until there is enough lift to force it upward. A propeller is used in some airplanes to drive them forward. A gasoline engine may be used to turn the propeller. A propeller is like a screw. As it turns it exerts a force against the air and pulls the airplane forward. Many airplanes use jet engines. Expanding gases inside the engine push the airplane forward.

Once the airplane is in the air, the pilot must be able to control its path. The pilot directs the path using ailerons (AY luh rahnz), elevators, and rudder. Moving these separate parts cause the amount of lift on the wings and tail to change. Ailerons are movable sections on each wing that enable the airplane to tilt from side to side. Elevators on the tail cause the nose of the airplane to go up or down. Turning right and left is controlled by the rudder on the tail. When taking off or landing, the pilot uses

140 *Fluids and Pressure*

FIGURE 7–17. A pilot controls an airplane by using the movable parts on the wings and tail.

the elevators and ailerons. When turning in flight, the rudders and ailerons are used.

If you try running while holding a large piece of cardboard, you can feel air pushing against the cardboard. The faster you run the harder the air pushes against the cardboard. The pushing you feel is called drag. **Drag** is the name given to the force of air against any object that moves through it. A moving airplane experiences drag. The amount of drag increases as the speed of the airplane increases. A larger drag slows down an aircraft or makes it necessary to use more fuel per kilometer traveled.

Airplanes are streamlined to reduce drag. The bodies and parts are shaped to allow a smooth flow of air to pass. Streamlined aircraft are pointed in the front and have a thin tail. How many things can you name that have streamlined bodies that reduce drag? fish, birds, ships, submarines, automobiles, etc.

Relate the shape of a car to drag force and fuel efficiency.

FIGURE 7–18. Wind-tunnel tests are done to determine the effects of drag on an aircraft.

making sure

13. Why would ailerons, elevators, and rudders not work on a spacecraft? There is no air in space to create a force on these parts.
14. How is a jet engine similar to a rocket engine? Both use escaping gas to produce force.
15. What is the best shape for a submarine? Why? Streamlined shape reduces drag as it moves through water.
16. Why are some airplanes tied down during strong winds? So they are not lifted or blown over by the force of the winds.
17. Why are high-speed aircraft more streamlined than slower aircraft? At high speeds there is more drag.

See Teacher's Guide.

PERSPECTIVES
people

First Flight

Bob entered the plane. It was about one hour before take-off. This flight was his first as a flight attendant. The other flight attendants were arriving. They checked all passenger areas and areas for passenger services. Everything and everyone had to be ready and sparkling clean for the passengers' arrival. Bob remembered that every passenger should be treated as he would treat a guest in his home.

As he worked, Bob thought about how his training began. First, he talked with his college counselor about applying for the position with an airline. He asked about the skills and personal qualities needed to be a flight attendant. The most important thing was liking people and relating to them. Enjoying travel was important, too. Organizing his thoughts and communicating them clearly was necessary. Excellent physical health and a well-groomed appearance were also important.

Before he entered the training school, Bob was interviewed by airline personnel. He had to meet certain height and weight requirements. He also needed a high school diploma and two years of college or business training. Bob met all of the requirements. He prepared carefully for the interview, thinking through possible answers to questions he would be asked.

After talking with the interviewer, Bob talked with several flight attendants. He asked what they found most enjoyable about their work. Traveling and meeting new people was their answer. He asked what seemed most difficult about their work. They felt that working irregular hours such as 2:00 A.M. flights was difficult. Feeling lonely was also

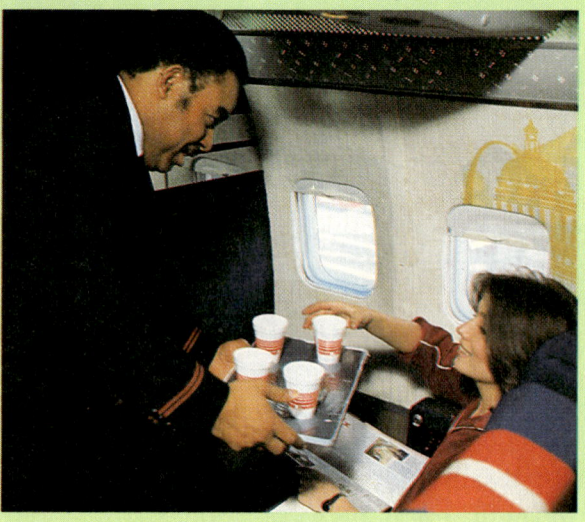

Doug Martin, Courtesy of TWA

difficult. The loneliness was a result of never staying in one place very long.

Bob had gone to the school sponsored by his airline six weeks ago. The airline paid for his training, room, and food while he was there. Classes began at 7:00 A.M. and ended at 5:00 P.M. each day. The training was intensive. At times he wondered how he would remember everything. The classes involved many areas—psychology, safety and first aid procedures, how to prepare and serve food, and parts of an airplane. Bob had to maintain good grades.

Suddenly, Bob realized that the First Flight Attendant was demonstrating safety procedures for the passengers. After checking the passengers, the flight attendants sat down and fastened themselves in. Bob sat back, he was really looking forward to his career as a flight attendant.

Chapter Review

main ideas

1. Liquids and gases are fluids.	7:1
2. The pressure at any point in a fluid is the same in all directions.	7:2
3. The pressure of a gas in a container increases when its temperature increases.	7:3
4. As the pressure of a gas increases its volume decreases.	7:3
5. Buoyancy causes certain objects to float in liquids and gases.	7:4
6. The pressure within a moving fluid is related to the speed of the fluid.	7:5
7. Air pressure produces the lift force on an airplane.	7:6
8. A change in pressure on the control surfaces of an airplane directs its path in flight.	7:7

vocabulary

Define each of the following words or terms.

Archimedes' principle
Bernoulli's principle
Boyle's law
buoyancy
Charles' law
drag
fluid
lift
pressure
surface tension

study questions

DO NOT WRITE IN THIS BOOK.

A. True or False

Determine whether each of the following sentences is true or false. If the sentence is false rewrite it to make it true.

T 1. Water and petroleum are two liquids that occur in large quantities on earth.

F 2. Nitrogen and ~~helium~~ are plentiful on the earth. oxygen

T 3. Molecules in a liquid are closer together than in a gas.

F 4. When water boils, its volume ~~decreases~~. increases

F 5. Surface tension is a property of ~~gases~~. liquids

F 6. Fluids are ~~seldom~~ used to cool things. often

T 7. Water expands when it changes into a gas.

Fluids and Pressure

T 8. One newton per square meter = 1 pascal.
T 9. Pressure in a liquid increases as the depth increases.
T 10. When you blow up a balloon, the pressure inside increases.

B. Multiple Choice

Choose the word or phrase that completes correctly each of the following sentences.

1. The (*elevator*, *rudder*, *propeller*) of an airplane is used to increase or decrease altitude.
2. When you squeeze the bulb of an atomizer, you are using (*Charles' law*, *Boyle's law*, *Bernoulli's principle*).
3. An increase in the drag force on an airplane causes an (*increase*, *decrease*, *no change*) in the amount of fuel needed to fly it.
4. When a gas is heated in a closed container, its pressure (*increases*, *decreases*, *stays the same*).
5. Surface tension is caused by the (*size*, *number*, *forces*) of molecules.
6. A pressure of 1 Pa is (*1 N*, *1 N/m*, *1 N/m²*, *1 N/m³*).
7. The force of a hydraulic lift is produced by an increase in (*volume*, *pressure*, *temperature*) in the cylinder.
8. The density of air in a hot air balloon is (*less*, *the same as*, *greater*) than in the air around it.
9. (*Buoyancy*, *Lift*, *Drag*) is the force that makes an object float.
10. A solid piece of plastic will float if its density is (*greater*, *less than*,) the density of water.

C. Completion

Complete each of the following sentences with a word or phrase that will make the sentence correct.

1. If a gas in a sealed container is cooled, its pressure __decreases__.
2. Pressure __increases__ when you heat water in a pressure cooker.
3. A floating block of wood displaces water equal to its own __weight__.
4. Ocean water is __more__ dense than fresh water.
5. Objects float __higher__ in salt water than in fresh water.
6. The __weight__ of the water displaced by a boat is equal to the weight of the boat.

7. Pressure in a moving stream of air is __less__ than in the air around it.
8. __Lift__ is the upward force on an airplane in flight.
9. Lift on an airplane is caused by __greater__ pressure on the lower surface of the wing.
10. The __rudder__ on an airplane's tail is used to make it turn right and left.

D. How and Why See Teacher's Guide at the front of this book.
1. How can a gas be compressed?
2. How does a hydraulic lift work?
3. Why does a ship float?
4. How is the altitude and direction of an airplane controlled?
5. Why does an airplane require more speed to take off when the air temperature is high?

challenges

1. Have a paper airplane contest. Devise and agree on a set of rules for making airplanes and flying them. Conduct tests to find out which paper airplane flies farthest.
2. Obtain an atomizer and study its workings. Make a poster size drawing of an atomizer and label its parts. Add information that shows how it operates.
3. Obtain a kit for building a model airplane. Learn how its parts fit together and make it fly.
4. Obtain information on the design of submarines. Find out how the weight of a submarine can be changed so that it can descend or rise in the ocean.
5. Research and report on the operation of a helicopter.

interesting reading

Azarin, Beverly, "Return of The Clipper Ship." *Science 81,* March 1981, pp. 80–87.
Cook, Brian, *Gas*. New York, NY: Watts, 1981.
Earle, Sylvia, "The Descent of Man." *Science 81,* September 1981. pp. 44–53.
Joss, John, "The Wright Stuff." *Science 81,* April 1981. pp. 54–61.

The Earth is constantly changing. Forces at and below the earth's surface cause continents to move and mountains and valleys to form. Oceans and rivers erode rock and soil. Weather changes from day to day. How have climates changed over many years? What changes produced the Grand Canyon? How does the moon change Earth's oceans?

Vince Streano

Earth Science
unit 3

Earth as viewed from the space shuttle orbiter looks far different from Earth as you see it. Both viewpoints can be important in observing and learning about the Earth. What are Earth's different movements? What causes the seasons? How can you locate a specific place on the Earth's surface?

The Earth

chapter 8

Introducing the chapter: Obtain a globe and demonstrate its main features. Ask students why the globe is tilted on its axis and in what direction Earth rotates. Have students identify the north and south poles. Have them locate the north magnetic pole. Use a compass to locate the direction of magnetic north from the classroom and mark the direction on the floor with an arrow made of masking tape.

8:1 Earth's Shape and Size

Most people have seen pictures of the earth taken by astronauts who traveled to the moon. These pictures show the earth as a round ball with large continents and oceans. Although the earth as viewed from space appears to be a sphere, it is not perfectly round. The earth's circumference through the North and South Poles is about 43 km less than around the equator. Also, the Southern Hemisphere is slightly larger than the Northern Hemisphere. In other words, the earth is slightly fatter in the middle and somewhat pear-shaped.

About 200 B.C. a Greek astronomer, Eratosthenes, made the first measurements of Earth's size. He compared the angles at which sunlight struck the earth at noon in two cities in Egypt. The cities were 925 km apart. In one city, sunlight struck the earth at an angle. The sun's rays produced a shadow from a stick stuck in the ground. In the second city, the sun shone directly into a deep well and did not cast a shadow. There was no shadow because the sunlight was perpendicular to the earth's surface. Eratosthenes reasoned that the angle of sunlight in the first city was caused by the earth's curved surface. If the earth were flat, there would be no shadow in either city.

GOAL: You will learn some of the features of the Earth's surface and how maps are used to show them.

Describe the shape of the earth.

Enrichment: Ask students when and where they are likely to see their shadows.

Early Egyptians understood geometry and surveying methods.

149

FIGURE 8-1. Eratosthenes determined the size of the earth by measuring the angles that sunlight strikes the earth at two cities in Egypt.

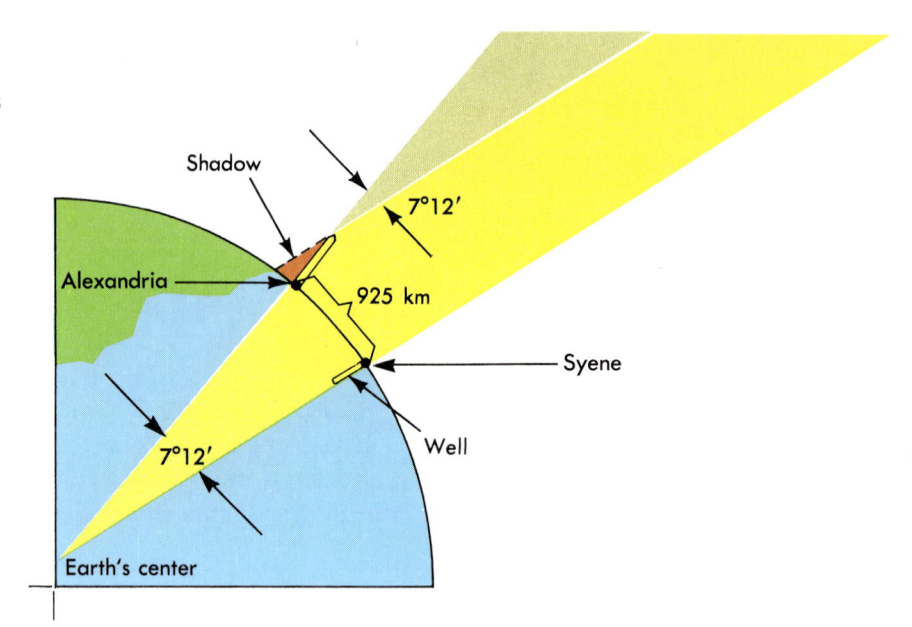

What is Earth's circumference?

Eratosthenes measured the angle of the sunlight. He used this value to determine the earth's circumference. His answer was close to the correct value of 40 000 km.

What is the mass of the earth? There is no way to put the earth on a balance to measure its mass. The answer must come through indirect measurements using gravity.

Review the concepts of mass and weight.

All matter has gravity. The gravity of an object is directly related to its mass. The more mass an object has, the greater its gravity.

One of the first attempts to measure the earth's mass using gravity occurred during the nineteenth century. Two objects of the same mass were placed on a balance. The objects were in balance because the force of Earth's gravity was the same on both objects. A lead weight was placed under one of the objects. The force of gravity of the lead attracted the object and caused the two objects to become unbalanced (Figure 8-2a). The force of gravity of the lead was determined by measuring the mass needed to balance the objects again (Figure 8-2b). By comparing the gravity of the earth with the gravity of the lead, the mass of the earth was calculated. The earth's mass was found to be about 6×10^{24} kg. Today, measurements are made using satellites.

How was the mass of the earth measured using gravity?

150 The Earth

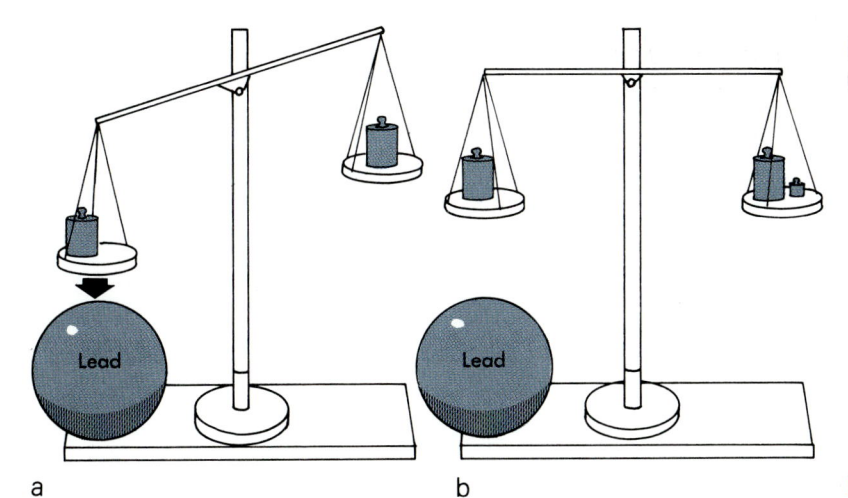

FIGURE 8–2. The experiment used to find the earth's mass.

8:2 Earth's Motions

How do we know the earth moves? One way to find out is through the use of a Foucault pendulum. This kind of pendulum consists of a heavy weight suspended from a wire. The wire, which may be as long as 60 m, is attached so it is free to rotate and swing back and forth. The path of the swinging pendulum is marked as a straight line on the floor as in Figure 8–3. One hour later, the path is marked again. It is observed that the second path is different from the first. Even though the pendulum is swinging in the same direction, its path on the floor has changed. The path changed during the hour because the earth moved slowly beneath the pendulum. This movement demonstrates the earth's rotation.

Rotation is the turning of the earth about its axis. The axis is an imaginary line that runs through the center from pole to pole. The earth rotates slowly from west to east. It takes one day or about 24 h for one complete rotation. You can judge the passing of time by observing the position of the sun in the sky during the day. The sun appears to rise in the east in the morning and to set in the west in the evening. The earth's rotation causes the sun to rise and set at different times in different parts of the world. When the sun is setting in Chicago, it is not setting in San Francisco. The sun has already set in New York and will set about two hours later in San Francisco.

How does a Foucault pendulum indicate Earth's rotation?

Define the terms rotation and revolution.

FIGURE 8–3. The path of the swinging pendulum along the floor changes as the earth rotates (a). Night and day are caused by Earth's rotation (b).

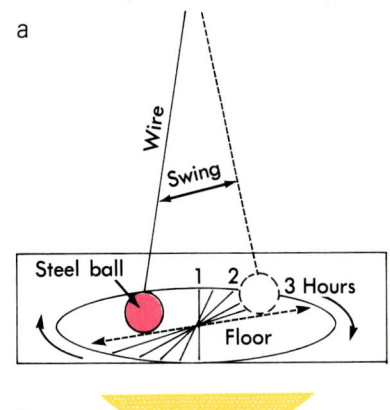

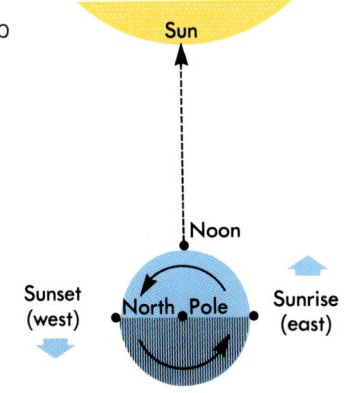

8:2 Earth's Motions 151

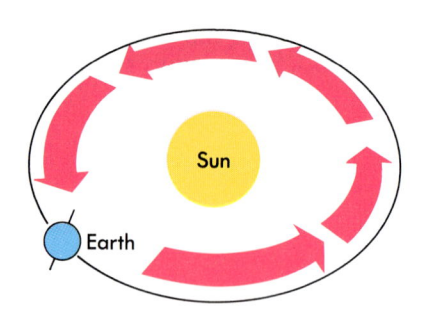

FIGURE 8–4. The earth revolves in a curved path around the sun.

*Measurements represent equatorial dimensions.

The earth also moves in a curved path or orbit around the sun. This movement is called **revolution**. It takes 365.25 days or one year to complete one revolution. As the earth revolves, the distance between the earth and the sun changes. On January 4, the earth is closest to the sun. On July 5, it is the farthest from the sun. The average distance between the earth and the sun is about 150 000 000 km.

Table 8–1.
Facts About the Earth

Mass	6×10^{24} kg
Circumference *	40 000 km
Diameter *	12 756 km
Time of rotation	23 h 56 min
Time of revolution	365.25 days
Average density	5.52 g/cm^3
Average distance from the sun	1.49×10^8 km

activity

CHANGING SHADOWS Activity 8-1

Obtain a stick about 1 m long. In the morning of a bright, sunny day, push the stick into the ground so it stands perfectly straight. Place three pebbles or a small stick on the shadow to mark its location. Return every hour to observe any changes in the position, direction, and length of the stick's shadow. For each observation, mark the location of the shadow. Draw a diagram that shows the stick and the positions of the shadows you observed. See Teacher's Guide.

8:3 Seasons

Many regions have major changes in climate from season to season. Winter is cold. Summer is warm. Spring and fall are mild.

Seasonal changes depend on two factors. These factors are the number of daylight hours and the angles at which the sun's rays strike the earth. Both factors change throughout the year because the earth's axis is titled at a 23½° angle.

As the hours of daylight increase from winter to summer, the total solar radiation per day increases. During summer days in the Northern Hemisphere, the sun shines for a long period of time. The sunlight also strikes the earth more vertically. Thus more radiant energy reaches the earth. In the winter, the Northern Hemisphere is tilted away from the sun. The sunlight strikes the surface less vertically and the energy is spread over a larger area (Figure 8–5).

On June 21, the noon sun is directly above the Tropic of Cancer, an imaginary line 23 1/2° north of the equator. June 21 is the beginning of summer in the Northern Hemisphere. There are more daylight hours in the Northern Hemisphere on June 21 than on any other day.

September 22 is the first day of fall. At noon on this day, the sun is directly above the equator. The earth's axis is tilted neither toward nor away from the sun. There are exactly 12 hours of daylight and 12 hours of darkness at every place on the earth except at the North and South Poles.

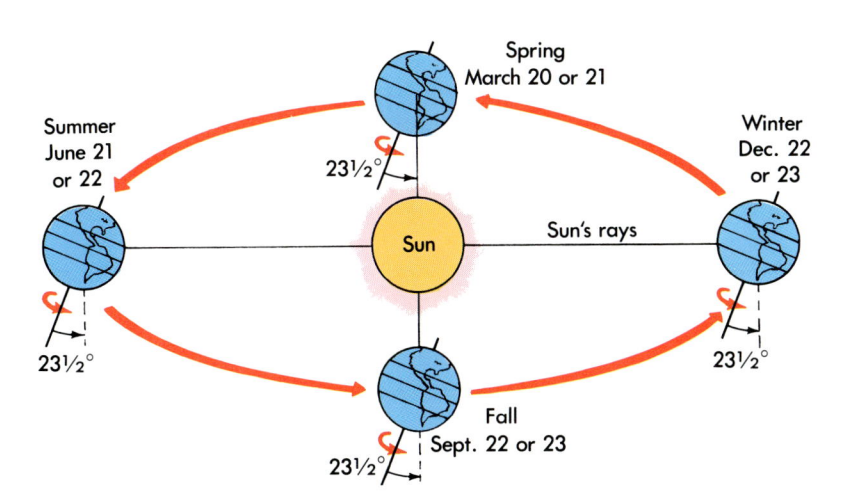

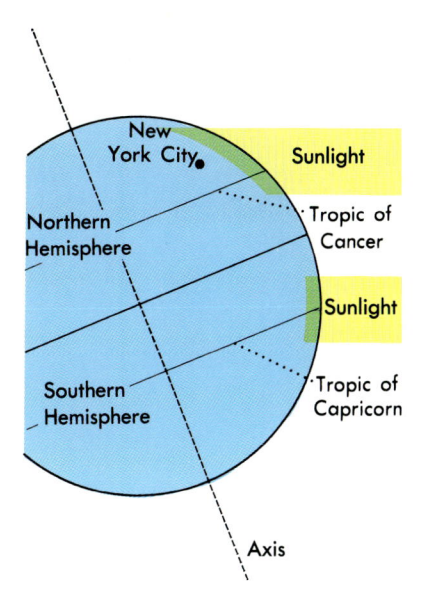

FIGURE 8–5. During winter, solar radiation strikes the Southern Hemisphere more vertically than the Northern Hemisphere.

Why is the Northern Hemisphere colder in January than the Southern Hemisphere?

Relate the change in seasons to solar energy and temperature.

FIGURE 8–6. Four distinct seasons are caused by the 23½° tilt of the earth on its axis.

8:3 Seasons 153

What is the Tropic of Capricorn?

Winter in the Northern Hemisphere begins on December 21. On this date, the noon sun is directly above the Tropic of Capricorn. This is an imaginary line 23 1/2° south of the equator. In the Northern Hemisphere, there are fewer hours of daylight on December 22 than any other day.

Bill O'Connor/Peter Arnold, Inc.

FIGURE 8-7. During spring and summer the sun may never set for people living near the North Pole.

March 21 is the first day of spring. On this date, the earth is directly opposite its position on September 22. The earth's axis is tilted neither toward nor away from the sun. Again there are 12 hours of daylight and 12 hours of darkness at every place on the earth except at North and South Poles.

activity

DEMONSTRATING EARTH MOTIONS Activity 8-2

(1) Obtain a sphere such as a classroom globe or a large rubber ball and a flashlight. (2) In a darkened room, hold the lighted flashlight about 2 m from the sphere. (3) Observe that one side of the globe is bright and one side is dark. How is the difference related to night and day on Earth? (4) Have someone turn the sphere slowly on its axis. Did the portion of the sphere that was in light change when it was rotated? (5) Tilt the top of the sphere away from the light so its axis is at a 23 1/2° angle. (6) Have someone move the sphere in a path circling around the light while keeping the flashlight pointed directly at it. How did this motion show the change in seasons on Earth?

Annotations:
- It is a model showing how daylight and darkness are produced on Earth.
- no
- Portions of the globe that are in light are changed just as portions of the earth vary in the amount of sunlight they receive during the year.

154 The Earth

8:4 Latitude and Longitude

Different places on the earth's surface are located using imaginary lines. On a map or globe of the earth, these lines cross each other. Lines of **latitude** (LAT uh tewd) run east and west parallel to the equator. Lines of **longitude** (LAHN juh tewd) run north and south through the poles. The location of a particular place in the earth is identified by the point at which lines of latitude and longitude cross.

Lines of latitude are measured in degrees north and south of the equator. The equator is 0° latitude and lies halfway between the poles. The North Pole is 90° north latitude and the South Pole is 90° south latitude. The city of Paris, for example, lies at about 49° north latitude.

Each degree of latitude is further divided into 60 minutes and each minute is further divided into 60 seconds. The distance between two lines of latitude 1° apart is about 111 km. A difference in latitude of 1 minute equals about 1.9 km.

Lines of longitude are often called **meridians** (muh RIHD ee unz). Longitude is measured in degrees east and west from the prime meridian. The **prime meridian** passes through Greenwich, England and is 0° longitude. A place halfway around the world on the other side of the earth from the prime meridian lies at 180° longitude. In the United States, lines of longitude are west of the prime meridian.

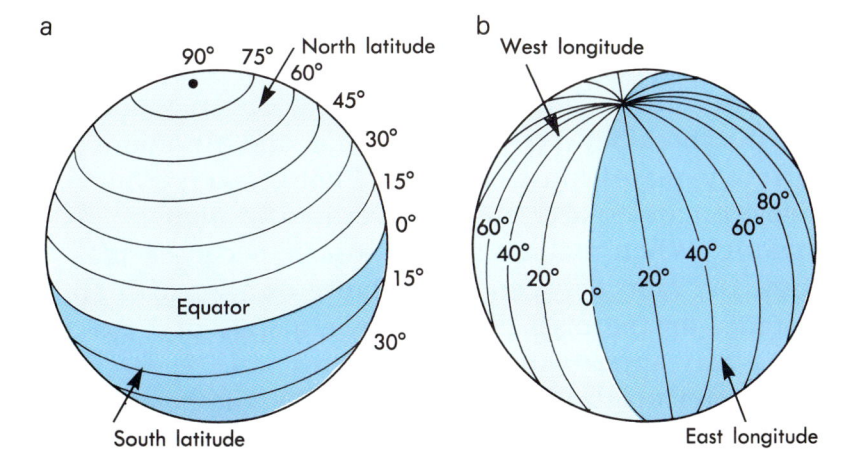

What are latitude lines?

What are longitude lines?

FIGURE 8–8. Lines of latitude are parallel to the equator (a). Lines of longitude pass through the poles (b).

FIGURE 8-9.

activity

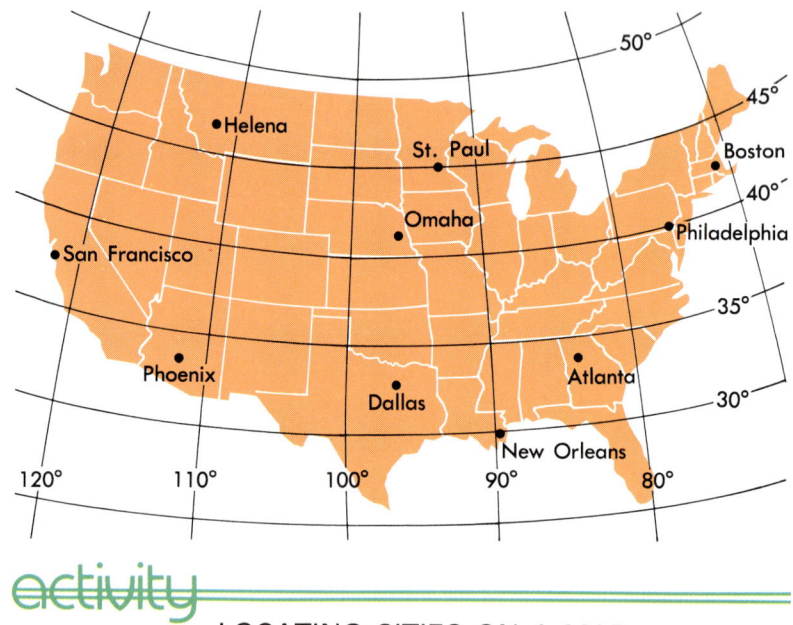

LOCATING CITIES ON A MAP Activity 8-3

Identify the lines of latitude and longitude in Figure 8–9. What is the latitude and longitude for each city?

See Teacher's Guide.

making sure

1. What is the degree of latitude for a place exactly halfway between the equator and the North Pole? 45°

8:5 Mapping the Earth's Surface

Maps are small scale pictures of all or part of the earth's surface. A map is drawn to a scale that represents each part of the earth at a smaller size. Maps are made from information gathered by scientists and technicians. Careful measurements of the distances between certain points and elevations of different places must be made. Photos taken from airplanes and space satellites are used in map making. When making a map, a scale is chosen to show the ratio between the distance on the earth and distance on the map. For example, a scale of 1 cm = 1 km may appear on a map. On this scale, each centimeter on the map is equal to one kilometer on the earth's surface. How many kilometers equal 4.7 cm on the map? 4.7

FIGURE 8-10. Satellite images are often used to make maps.

EROS Data Center

156 The Earth

Features on the land such as roads and rivers can be shown on a map through the use of colors and symbols. The color green, for example, may be used to show a forest. Small symbols on the map sometimes show the location of hospitals and schools. An airport is frequently indicated by a small airplane symbol.

One type of map, called a **topographic** (tahp uh GRAF ihk) **map** also shows the landscape. Topographic maps include heights of mountains and depths of valleys. The height above or below sea level is called elevation. Contour lines are used to show elevation. A **contour line** is a line drawn on a map joining all points of the earth's surface having the same elevation. A contour line is indicated in numbers and units. Study Figure 8–12. Contour lines representing a mountain are drawn through points of the mountain that have the same elevation. Contour lines spaced close together represent a steep slope. In the same way, contour lines widely spaced represent a gradual slope.

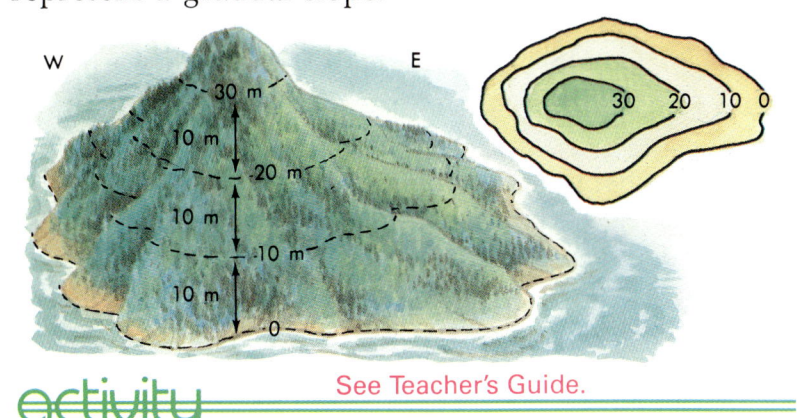

FIGURE 8–11. Symbols on a map are used to show roads and other features.

How do contour lines show elevation on a topographic map?

Review the concept of ratios. The principle of scaling in map making or model making depends on an appropriate ratio.

FIGURE 8–12. On topographic maps, contour lines connect points of equal elevation.

activity

See Teacher's Guide.

CONSTRUCTING A MAP SCALE Activity 8-4

Make a scale that could be used in constructing a map of your state. The scale should be of the proper ratio so the map will fit on a piece of paper the size of this page. How many kilometers would equal 5 cm on the map? Make a scale to use in constructing a map of your neighborhood. Calculate the number of meters that equal 5 cm on the map. Why did you choose two different scales? Which map would have the most detail? neighborhood

Remind students that they are not to draw maps. Point out the relative difference in the size of the state and neighborhood.

The state and neighborhood are very different in area.

8:5 *Mapping the Earth's Surface*

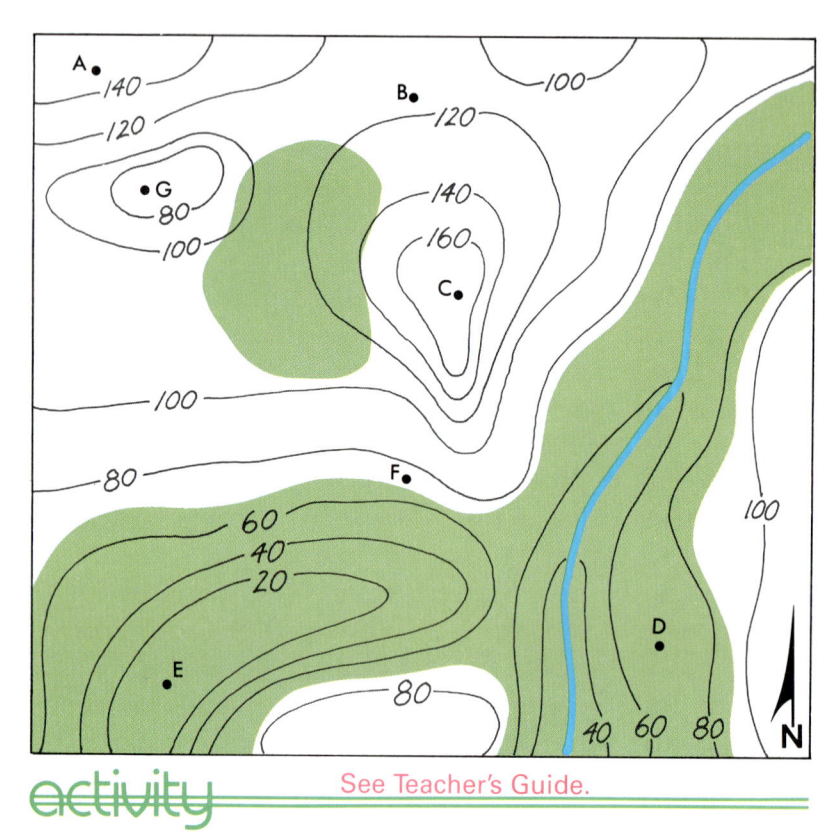

FIGURE 8-13. Topographic map. Elevations are in meters.

(1) C
(2) E
(3) South, the elevation increases to the north.
(4) 70 m
(5) 80 m
(6) Steeper between C and F because the contour lines are closer.
(7) from point E eastward through the valley
(8) Point G is lower than the land around it.
(9) no difference in elevation.

activity

See Teacher's Guide.

USING A TOPOGRAPHIC MAP Activity 8-5

Use Figure 8–13 to answer the following questions.
(1) What point is the highest elevation shown on the map? **(2)** What point is the lowest elevation? **(3)** In which direction does the stream flow? How can you tell? **(4)** What is the approximate elevation of Point *D*? **(5)** What is the difference in elevation between Points A and *G*? **(6)** Is the slope steeper between Points *C* and *F* or between Points *C* and *B*? Explain. **(7)** Where on the map might a stream once have been located? **(8)** Point *G* sometimes is flooded during a heavy rainfall. Explain why this is so. **(9)** What is the difference in elevation between Points *F* and *D*?

making sure

2. How can you tell the difference between a valley and a hill on a topographic map? See Teacher's Guide.

158 The Earth

8:6 Time Zones

People flying across the United States from coast to coast must reset their watches. Otherwise they will have the wrong time when they land. The United States is divided into four standard time zones: Pacific, mountain, central, and eastern. **Time zones are geographical regions within which the time is the same.** If you travel east you lose one hour as you cross into another time zone. Going west you gain one hour. For example, when it is 5:00 P.M. in New York City, it is 4:00 P.M. in Chicago, 3:00 P.M. in Denver, and 2:00 P.M. in San Francisco.

Relate time zones to viewing sports events on television at a local time which is different from the time at the location of the event.

What are time zones?

FIGURE 8–14. Time zones of North America.

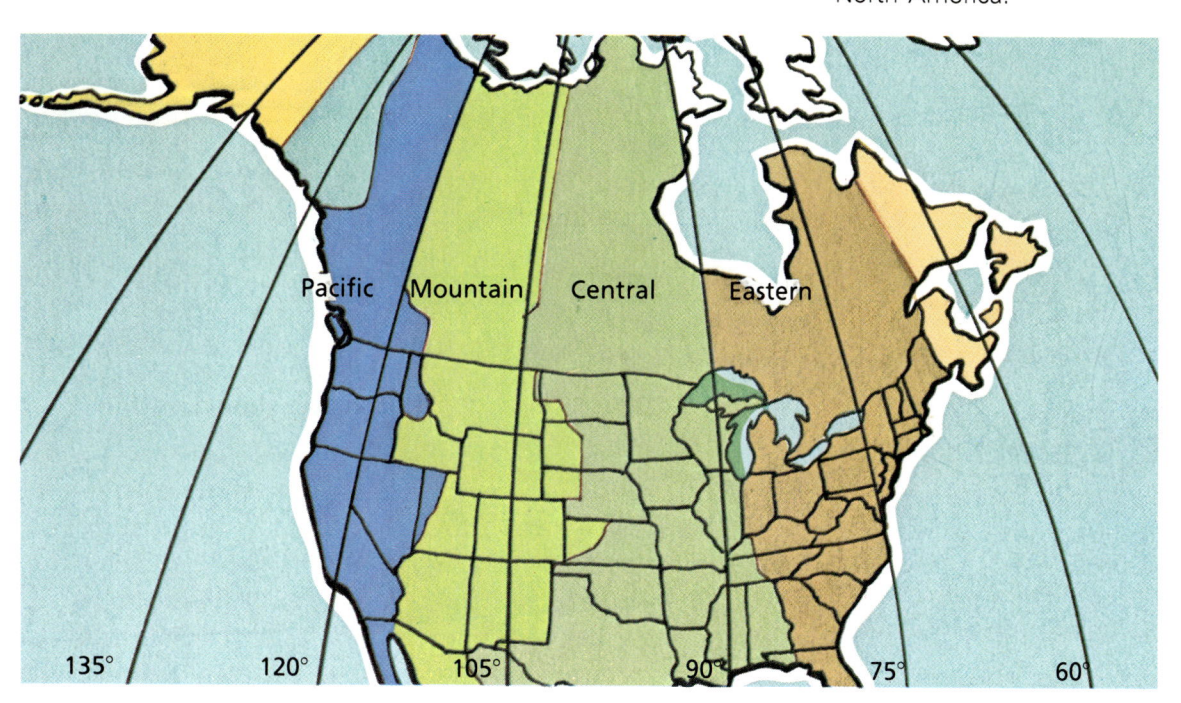

Pacific Mountain Central Eastern

135° 120° 105° 90° 75° 60°

Time zones in the United States are part of the series of twenty-four International Time Zones. These twenty-four zones cover the total surface of the earth. Each zone is 15° of longitude wide. The width of a time zone is based on the speed of the earth's rotation which is 15° of longitude per hour.

Boundaries for the time zones run from the North Pole to the South Pole. These boundaries are adjusted to fit the boundaries between states and countries. If not, different parts of some states and cities would have different times. Every place within each time zone has the same time. For instance, when it is 3:00 P.M. in Washington, D.C. it is also 3:00 P.M. in Lima, Peru. Both Washington and Lima are in the same time zone.

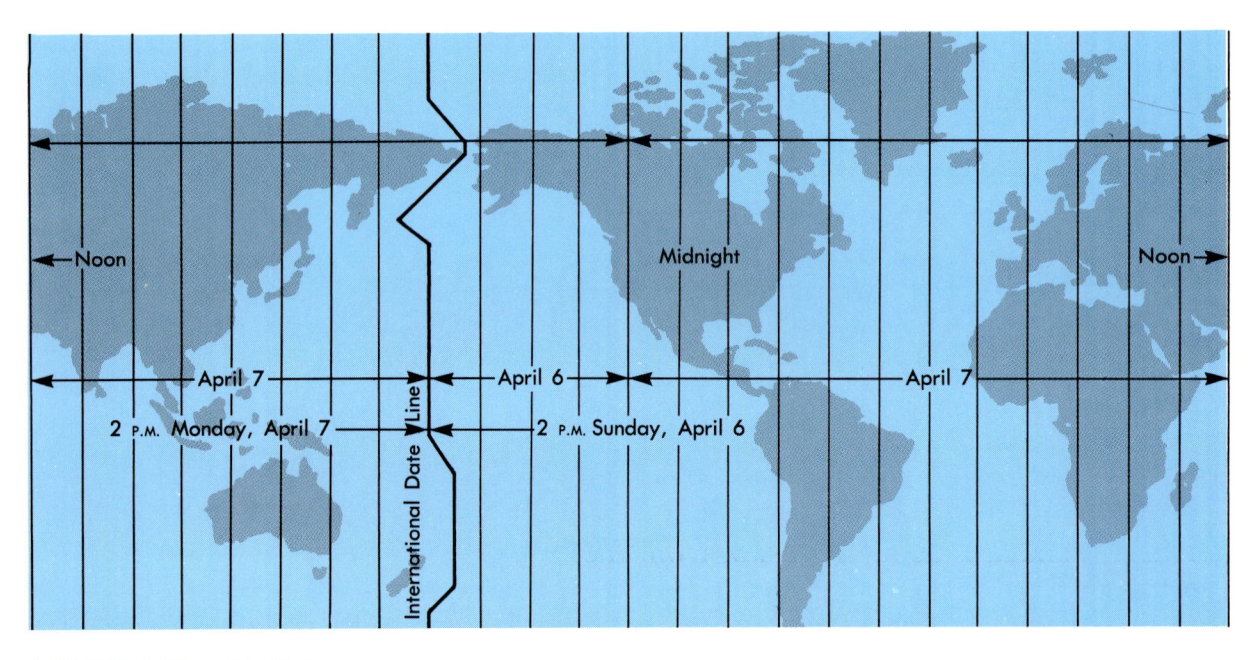

FIGURE 8–15. World time zones are determined by longitude.

A clock setting mnemonic is spring forward—fall back.

Adjustments sometimes are made within a time zone. For example, daylight savings time is used to take advantage of extra daylight hours during the summer. When switching to daylight savings time, clocks are set one hour ahead. If the sun sets at 8:00 P.M. standard time, it would set at 9:00 P.M. daylight savings time. One hour of daylight is added to the evening.

Why is it necessary to have time zones? If there were no time zones, every city and town might have a different time. It would be noon in New York, five minutes before noon in Philadelphia and ten minutes before noon in Baltimore. This confusion was actually the case about 100 years ago before the time zones were established.

The **International Date Line** is on the opposite side of the earth from the prime meridian. This imaginary line follows, for the most part, the 180° line of longitude. The International Date Line has been adjusted to fit around islands and other populated land areas. The days are different on opposite sides of the date line. It is always one day later west of the date line. For instance, when it is Sunday on the east side of the line, it is Monday on the west side.

Where is the International Date Line?

8:7 The Earth as a Magnet

Have you ever used a toy magnet to pick up tacks or other metal objects? A magnet is made of iron, cobalt, or nickel. Each of these metals has magnetic properties.

The earth is a giant magnet. One way to detect the earth's magnetism is with a compass. A compass consists of a small magnet. Its needle points north and south toward the earth's magnetic poles. The **magnetic north pole** near Bathurst Island in northern Canada is about 1670 km from the geographic north pole. The **magnetic south pole** is near the coast of Antarctica about 2670 km from the geographic south pole.

How are the magnetic poles of the earth different from the geographic poles?

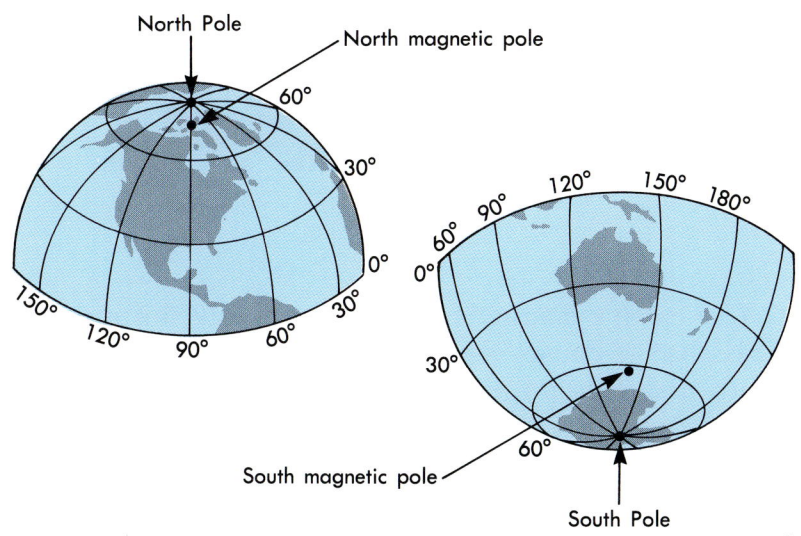

FIGURE 8–16. The north and south magnetic poles.

8:7 *The Earth as a Magnet* 161

Earth's magnetism is also detected with a magnetometer (mag nuh TAHM ut ur). A magnetometer measures the strength of Earth's magnetic field. Measurements made with this instrument show the earth's magnetic field varies slightly from place to place. This variation is caused by the rock structures below the earth's surface. An airplane trailing a magnetometer is often used for mapping the earth's magnetism since the airplane can cover a wide area in a short time.

From some places, a compass points directly to the geographic north pole. In most places, however, a compass points to the magnetic north pole forming an angle with the geographic north pole. This angle is called magnetic **declination** (dek luh NAY shun) or variation. Magnetic declination is different at different points on the earth. When using a compass, you must add or subtract this angle to get true directions. The magnetic declination is shown by a symbol on a topographic map.

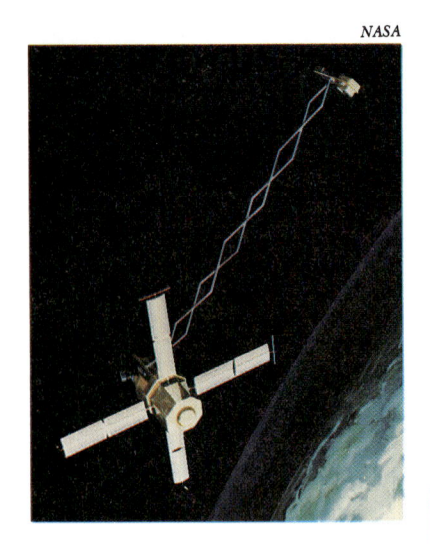

FIGURE 8-17. Mapping of the earth's magnetic field may be done through the use of satellites.

What is magnetic declination and how is it used to find directions?

Have students use a compass to determine directions from their school site.

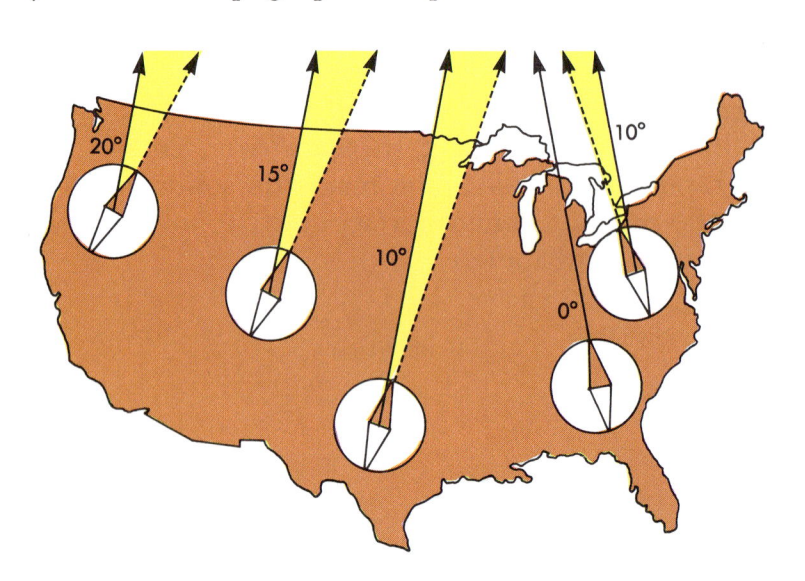

FIGURE 8-18. A compass points to the magnetic north pole which, for most locations, is a different direction than the geographic north pole.

The earth's magnetic field is constantly shifting. This change causes the positions of the north and south magnetic poles to change. Therefore, persons using magnetic compasses use up-to-date charts that show the angle of magnetic declination for an area.

What causes the earth's magnetism? As yet there are no satisfactory explanations. One theory is that the earth's magnetism is caused by the slow movements of molten iron and nickel in the earth's core.

Every magnet has a magnetic field around it that contains lines of magnetic force. These lines of force cause the magnet to attract iron and certain other metals. Also, the lines of force in the magnetic field cause the ends of two magnets to attract or repel each other. Lines of force in the earth's magnetic field pass outward from the poles into space. The earth's magnetism extends far beyond the atmosphere to somewhere between 64 000 km to 130 000 km above the surface.

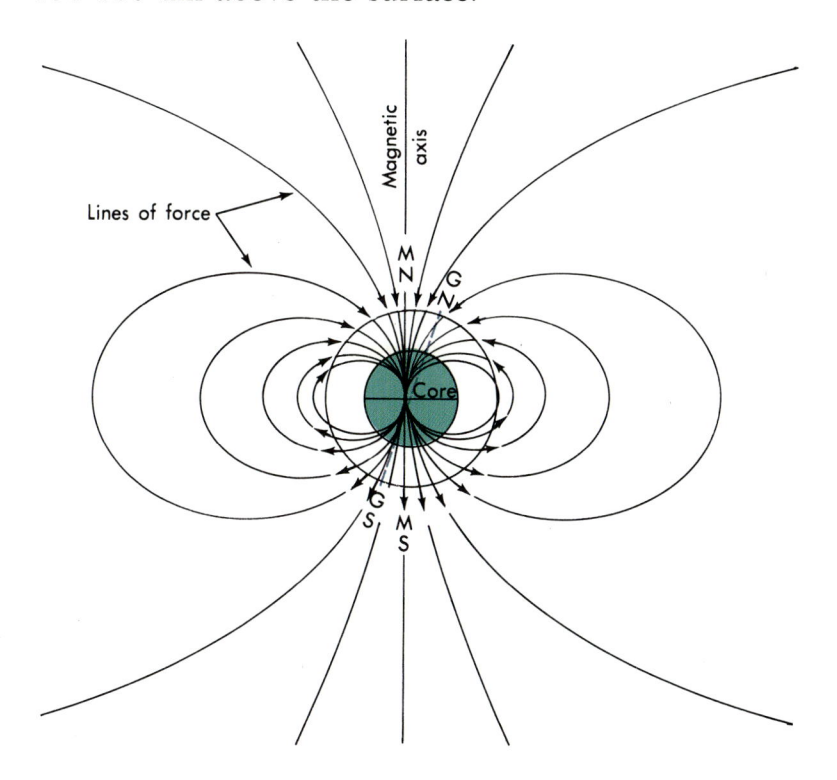

Enrichment: Cover a bar magnet with waxed paper. Sprinkle iron filings over the paper. Point out the magnetic field lines indicated by the pattern of filings.

FIGURE 8–19. Earth's magnetic field.

making sure

3. How would a compass be used in making a map?
4. Why is the angle of magnetic declination at your school different from the magnetic declination 200 km west of the school? Answers will vary.

3. A compass would be needed to show directions on the map.

8:7 The Earth as a Magnet

See Teacher's Guide.

PERSPECTIVES

skills

Reading Maps

Maps are scale drawings of the whole earth or part of it. They show many kinds of information, but no one map shows everything. Most maps are made to show specific information. For instance, you might draw a map showing a classmate how to get from school to your home. On such a map, you would likely show streets, traffic lights, and any important landmarks which would help to show were your home is. Of course, if you were asked by your teacher to draw a map of the United States, you would not include streets and landmarks. Rather, you would probably show the states, the principal cities, and main geographical features such as rivers and mountains. Let us look at some of the maps in Chapter 8 and see the kinds of information we can get from them.

Use Figure 8–9 to answer the following questions.
1. By noting many of the larger cities in the United States, locate approximately where you live.
2. Estimate the latitude and longitude of where you live.
3. Where is north on the map? Would Helena be north or south of Boston? Do you live north or south of San Francisco?
4. Using the scale 1 cm = 450 km, determine the distance between Dallas and St. Paul. What would be the distance between where you live and the nearest city shown on the map?
5. What ocean borders the western coast of the United States?
6. Does the United States occupy an entire continent or only part of one? What is the name of the continent?

Use Figure 8–13 to answer the following questions.
1. How can you tell the difference between a stream and a contour line on the map?
2. The vertical distance between contour lines is called the contour interval. What is the contour interval on the map?
3. Does any part of the land surface appear to be at sea level? How would sea level be indicated?

Use Figures 8–14 and 8–15 to answer the following questions.
1. How many time zones are found in the United States? If it is 9:00 A.M. in New York, what will be the time in San Francisco?
2. Knowing that 180° is one half of a circle, the International Date Line is opposite from what well known meridian?
3. How many time zones would you cross traveling from London, England to Phoenix, Arizona? What time would you land in Phoenix if your airplane took off at 1:00 A.M. and the flight lasted eight hours?

Use Figures 8–16 and 8–18 to answer the following questions.
1. What two hemispheres are shown in Figure 8–16? How are these maps like a globe?
2. What would you estimate the magnetic declination of your home to be?
3. Do the dashed arrows in Figure 8–18 show the direction of geographic north pole or magnetic north pole?

In chapter 8, there are maps to show four kinds of information. Maps can show many other types of information as well. Look through other chapters in this book and in your social studies book and see what other information can be shown on maps.

main ideas

Chapter Review

1. The earth is slightly fatter at the equator than at the poles. 8:1
2. Earth rotates on its axis and revolves in its orbit around the sun. 8:2
3. Many regions of the earth have climates that result from seasonal changes. 8:3
4. Lines of latitude and longitude are used to locate positions on the earth's surface. 8:4
5. Topographic maps show many landscape features. 8:5
6. The earth is divided into 24 time zones, each 15° of longitude in width. Time differs by one hour between each zone. 8:6
7. Earth is a natural magnet having a magnetic field that extends from the poles into space. 8:7

vocabulary

Define each of the following words or terms.

contour line
declination
International Date Line
latitude
longitude
magnetic north pole
magnetic south pole
meridian
prime meridian
revolution
rotation
time zone
topographic map

study questions

DO NOT WRITE IN THIS BOOK.

A. True or False

Determine whether each of the following sentences is true or false. If the sentence is false, rewrite it to make it true.

T 1. There are 24 different time zones.
T 2. It is possible to measure the mass of the earth by indirect methods.
T 3. A Foucault pendulum demonstrates the earth's rotation.
F 4. One ~~rotation~~ of the earth takes about a year. revolution
F 5. Sunlight strikes the earth most directly at the ~~North and South Poles.~~ equator
T 6. Night and day are caused by the rotation of the earth.
T 7. The seasons are caused by the tilt of the earth's axis and the earth's revolution.

The Earth 165

F 8. In the United States, the degrees of latitude are ~~east~~ of the prime meridian. west
F 9. The equator is 0° ~~longitude~~. latitude
T 10. The geographic north pole is 90° north latitude.

B. Multiple Choice

Choose the word or phrase that completes correctly each of the following sentences.

1. Lines of longitude are measured in *(kilometers, meters, <u>degrees</u>)*.
2. The prime meridian is *(<u>0°</u>, 90°, 180°)* longitude.
3. A scale of *(1 cm = 1 m, <u>1 cm = 500 m</u>, 1 cm = 100 km)* could be used to make a map of a large city.
4. On a topographic map, a steep mountain is indicated when the contour lines are spaced *(<u>close together</u>, evenly, far apart)*.
5. A contour line runs through all points having *(different, <u>the same</u>, the highest, the lowest)* elevation(s).
6. Information from *(<u>contour</u>, longitude, latitude)* lines indicates the shape of the landscape.
7. Pacific, mountain, central, eastern are the names of *(meridians, boundaries, <u>time zones</u>)*.
8. The hands of a clock are moved *(forward, <u>backward</u>, not at all)* when switching from daylight savings time to standard time.
9. Chicago is located *(north, south, east, <u>west</u>)* of the prime meridian.
10. The difference in time when crossing the International Date Line is *(1 hour, 12 hours, <u>1 day</u>)*.

C. Completion

Complete each of the following sentences with a word or phrase that will make the sentence correct.

1. The _____ is on the opposite side of the world from the prime meridian. International Date Line
2. A compass needle points towards the _____ north and south poles. magnetic
3. The strength of the earth's magnetic field can be measured with a _____. magnetometer

4. The angle of _____ is the angle between the magnetic north pole and the geographic north pole. **declination**
5. A _____ will not be cast by a stick if sunlight strikes from directly above it. **shadow**
6. The _____ of the earth is about 40 074 km. **circumference**
7. If the earth had a larger **mass** its gravity would be greater.
8. As time passes, the path of the Foucault pendulum _____. **changes**
9. The sun rises in the **east** and sets in the **west**.
10. Every day the sun sets in San Francisco **3** hours later than in New York City.

D. How and Why See Teacher's Guide at the front of this book.
1. How can a Foucault pendulum be used to show the earth moves?
2. What causes the four seasons of the year?
3. Why are time zones needed and why are they 15° in width?
4. How are a distance scale and a compass used in making a map?
5. When does a compass point to the geographic north pole?

challenges

1. Obtain a book about sundials. Make a sundial and use it to tell the time of day.
2. Do library research on different timepieces that have been used through history. For example, compare an hour glass to a pendulum or grandfather clock. Prepare a report that includes drawings of the timepieces.
3. Obtain a topographic map for your area from the U.S. Geological Survey, state geology department, or nearby college geology department. Display the map in your classroom and locate major features of the local landscape.

interesting reading

Challand, Helen, *Activities in the Earth Sciences.* Chicago: Children's Press Inc., 1982.

Jesperson, James and Jane Fritz-Randolph, *Time and Clocks for The Space Age.* New York, NY: Atheneum, 1979.

Rocks and rock formations are sources of geologic information. Many of the features in the photograph were formed by wind erosion. Wind erosion is one of several processes by which the landscape is changed. How are rocks formed? From what materials are rocks made? What geologic forces are at work within the earth?

David M. Dennis

Geology

chapter 9

Introducing the chapter: Have students bring or provide them with pencils or markers. Have students draw a cross section of the earth based on Figure 9-1, page 170. Have students color and label each layer. Have students correlate their drawings with the information in Section 9:1.

9:1 Structure of the Earth

Geology is the scientific study of the earth. Scientists in this field are called geologists. Geologists believe that the inside of the earth can be divided into three sections—core, mantle, and crust.

The **core** is the center section of the earth and consists of an inner and an outer region. The inner core begins at the center of the earth, 6400 km below Earth's surface. It has a radius of about 1300 km. Geologists believe that the inner core is mainly solid iron and nickel. This solid metal exists under great pressures and very high temperatures. The temperature may be as high as 4300°C.

The outer core surrounds the inner core and has a thickness of about 2250 km. Scientists think that the outer core is made of liquid iron and nickel. The outer core is slightly cooler than the inner core. In its upper part, it is about 3700°C.

GOAL: You will gain knowledge about the properties of rocks and minerals and the changes that occur in the earth's surface.

Name the three sections of the earth.

Iron and nickel are two metals that can be magnetized.

Describe the main features of each section of the earth.

169

The **mantle** surrounds the outer core and has a thickness of 2885 km. It is thicker and has more matter in it than any other part of the earth. The mantle is made of silicon, oxygen, aluminum, iron, and magnesium. Although most of the mantle consists of solid rock, high temperature and pressure cause certain parts of it to be somewhat fluid and to "flow" very slowly. Temperatures in the mantle vary from about 3700°C at its deepest part to about 870°C at the outer edge.

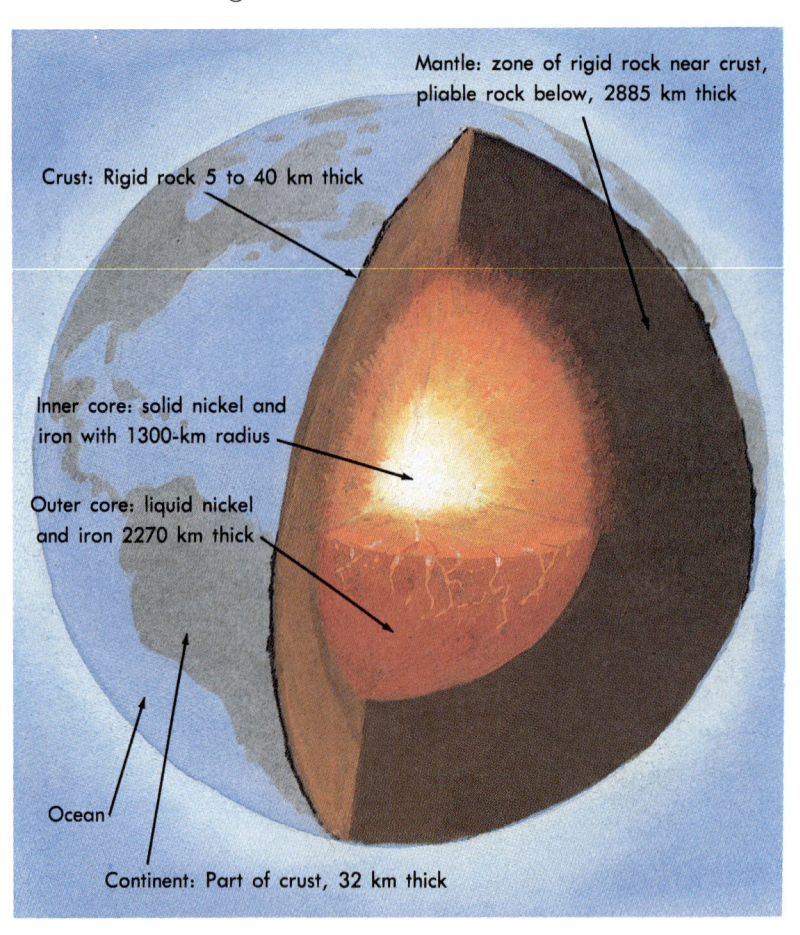

FIGURE 9–1. The earth consists of three major sections.

Above the mantle is the crust. The **crust** is like a thin rocky skin around the mantle. Its thickness varies from 5 to 40 km. The crust consists of all the continents and ocean floors of the earth's surface. Temperatures within the deepest parts of the crust may be hot enough to melt rock.

Near the surface of the earth, the crust is solid rock. Usually the rock is covered with a layer of soil. The layer of soil is often a meter or more deep.

making sure

1. Why would it be impossible to send a person to explore earth's mantle and core? It is too hot and the pressure is too great for human life.

9:2 Plate Tectonics

Plate tectonics (tek TAHN ihks) is a theory that explains major features of the earth such as earthquakes, volcanoes, and mountains. According to this theory, the earth's surface is divided into six major plates. There are also several minor plates. A **plate** is a large section of the earth's crust that is slowly moving. For instance, North and South America are part of the American plate that is drifting westward. Most geologists believe plate movement is caused by flowing currents of hot rock within the mantle.

Point out that the concept of plate tectonics is a theory based on scientific reasoning.

Explain the theory of plate tectonics.

FIGURE 9–2. According to theory, the earth's surface is divided into six major plates and several smaller ones.

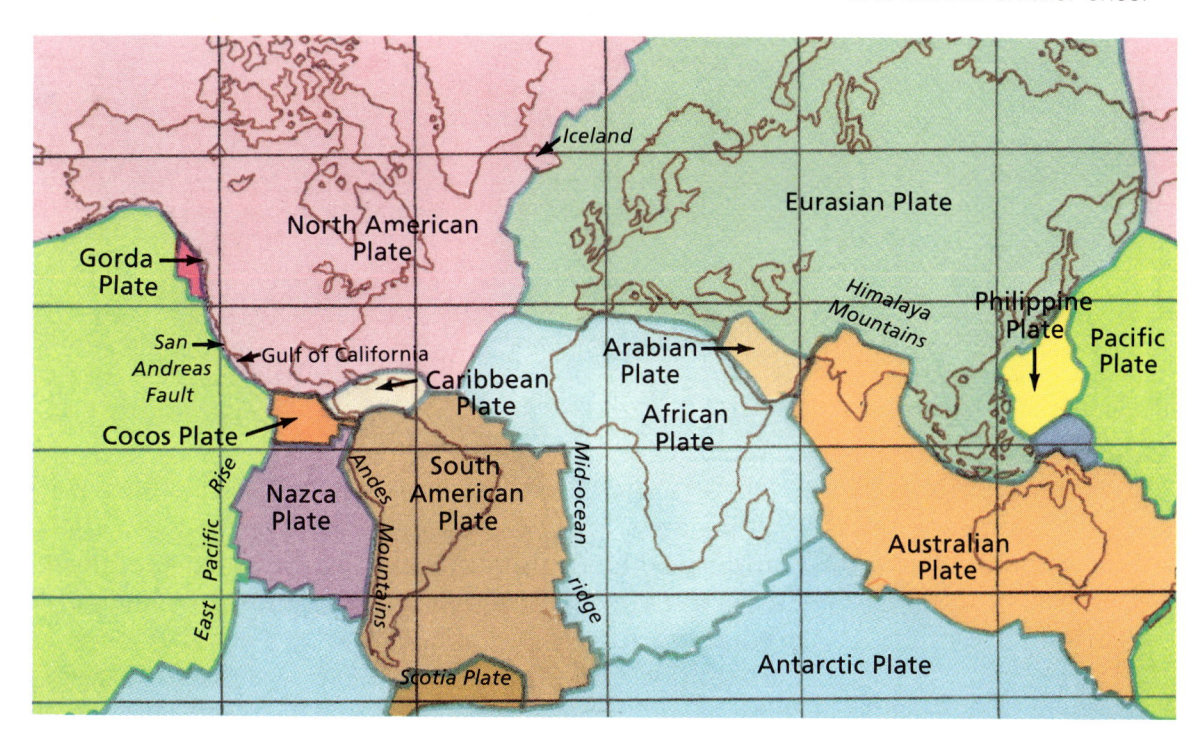

9:2 Plate Tectonics 171

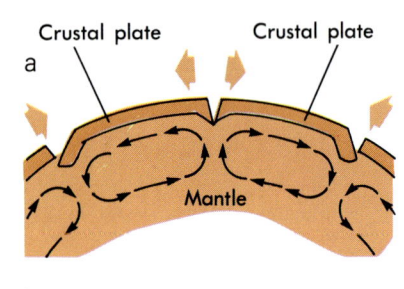

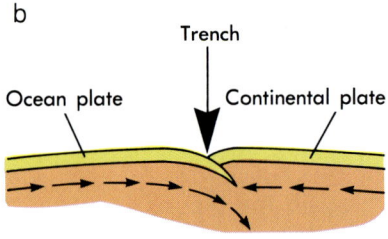

FIGURE 9–3. The movement of crustal plates is thought to be a result of convective currents of hot rock in the mantle (a). A trench forms when two plates collide, with one plate moving beneath the other (b).

What changes may occur when two plates meet?

FIGURE 9–4. The Pacific plate and the American plate slide past each other along the San Andreas fault in California.

Movement of the crustal plates creates stress and strain within the earth. When two adjoining plates are moving toward each other, one plate may pass under the edge of the other plate. The bottom plate bends downward to enter the mantle. This bending causes trenches to form. A trench is a deep, long valley in the ocean floor. The rubbing of two plates against each other produces tremendous heat and friction.

The Pacific and American plates meet along the western coast of North America. Instead of one plate passing under the other, the two plates slide past each other. The Pacific plate moves about 5 cm per year in a northwest direction. The American plate moves in a southeast direction. Movement of the two plates does not occur smoothly. The rocks tend to stick and then slip. When the rocks slip suddenly, shock waves are sent out through the surrounding area. The shock waves often go unnoticed by most people. But when they are strong enough to be felt, the event is called an earthquake.

Aerial photo by Collier/Condit

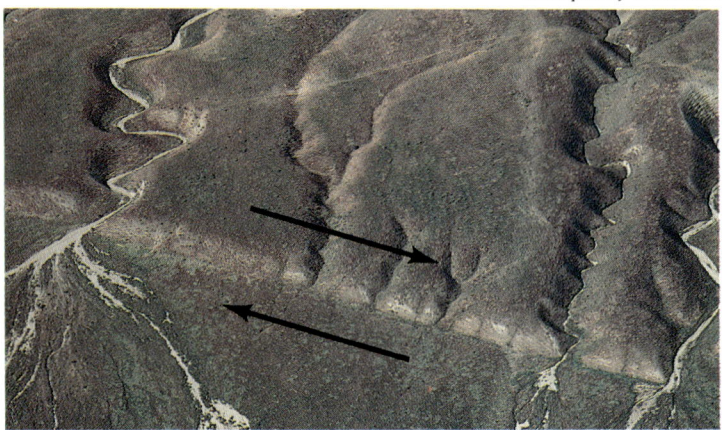

While some plates are moving together, others are moving apart. Molten material flowing up from the mantle pushes the plates apart at the surface. The Eurasian (Europe and Asia) and American plates provide an example of plates moving apart. Under the Atlantic Ocean, between these two plates, a giant ridge of mountains has formed. The mountains extend north to south along the center of the Atlantic Ocean floor. They are called the Mid-Atlantic Ridge. Iceland is a part of this ridge that has been pushed up through the ocean's surface.

According to geologists, all of the earth's continents were locked together about 200 million years ago (Figure 9–5). One by one the continents drifted away from each other. The Atlantic Ocean was formed when North America moved west away from Europe. India was a plate next to Africa that moved northward to the continent of Asia. As it moved away from Africa, the Indian Ocean was formed. The collision of India with Asia formed the Himalaya mountains. The Alps mountains were born in the collision of the plate that is now Italy with the southern edge of Europe.

FIGURE 9–5. Millions of years ago, the continents were joined as one supercontinent (a). They gradually split up and moved apart to their present positions (b).

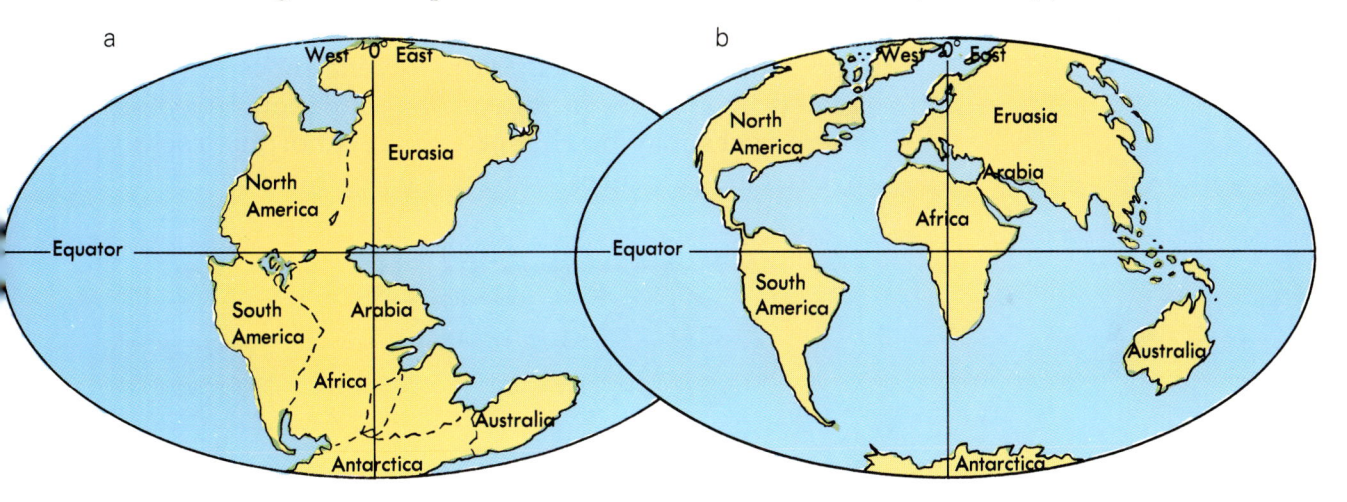

activity

COMPARING THE SHAPES OF CONTINENTS
Activity 9-1

Obtain two pieces of cardboard or heavy construction paper. On one, draw an outline of North and South America. On the other, draw an outline of Europe and Africa. Cut the paper along the outline. Then try to fit the two outlines together. How do the shapes compare? How does the plate tectonics theory fit your observations? The edges tend to fit together. North and South America drifted west away from Europe and Africa.

9:2 **Plate Tectonics**   173

9:3 Rocks and Minerals

How are rocks identified?

Scientists learn many things about the earth by studying rocks. There are many different kinds of rocks. Rocks can be identified by their physical properties. Color, shape, hardness, and texture are some physical properties. The region in which a rock is found is also helpful in identifying it. Hard rocks, for example, are often found in regions of tall mountains. Soft rocks are usually found in valleys.

Into what three groups are rocks divided?

Rocks may be classified into three groups—igneous (IHG nee us), sedimentary (sed uh MENT uh ree), and metamorphic (met uh MOR fihk). Igneous rocks are formed from hot liquid matter forced up from the mantle or lower crust. These rocks make up about 95 percent of the earth's crust. Sedimentary rocks result from the cementing of tiny rock particles. For example, beds of clay, sand, gravel, or limestone shells may cement together as time passes to form solid sedimentary rocks. Metamorphic rocks are created by changes in igneous or sedimentary rocks. These rocks may be changed to metamorphic rocks by extreme pressures and high temperatures.

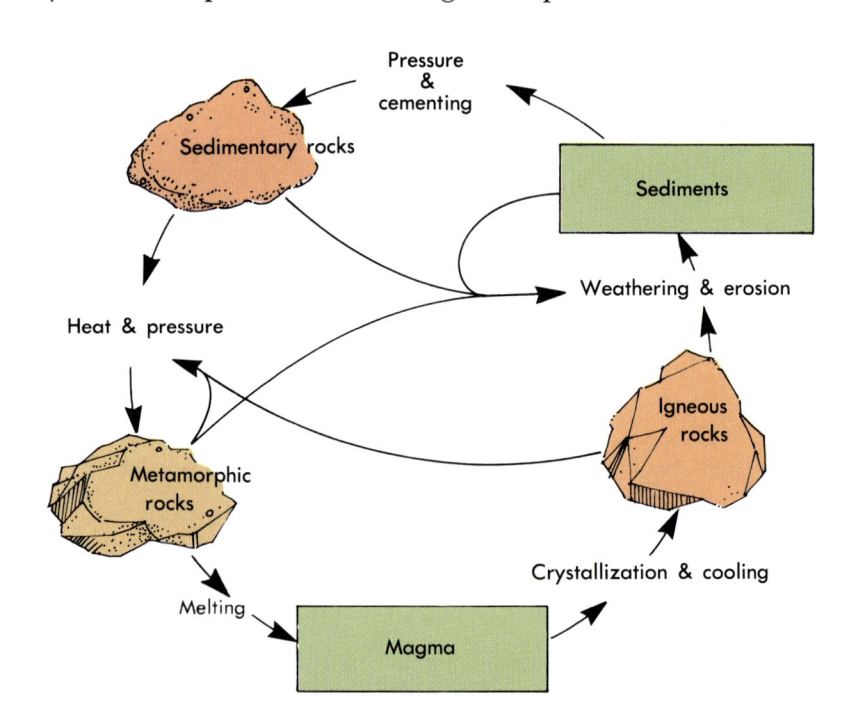

Have students make a chart in their notebooks listing the three kinds of rocks and their features.

FIGURE 9–6. The rock cycle describes the processes by which rocks change from one form to another.

174 Geology

All rocks are made of minerals. A mineral is a naturally occurring substance that has a definite chemical composition and a crystal form. The crystal form is produced by the way the atoms in the mineral are bonded together. The crystal form also determines some other characteristics such as how the mineral breaks. Some rocks contain only one mineral. For instance, limestone contains only calcite. Other rocks contain two or more minerals. Granite contains quartz, feldspar, and small amounts of other minerals.

More than 2000 minerals have been named. Most common minerals can be identified by properties such as luster, color, hardness and crystal form. The hardness of minerals can be found by scratching one against another. The harder mineral will always scratch the softer one.

FIGURE 9–7. Calcite.

FIGURE 9–8. Fluorite.

FIGURE 9–9. Apatite.

FIGURE 9–10. Quartz.

FIGURE 9–11. Topaz.

Table 9–1.
Mohs Scale of Mineral Hardness

1–Talc
2–Gypsum
3–Calcite
4–Fluorite
5–Apatite
6–Orthoclase
7–Quartz
8–Topaz
9–Corundum
10–Diamond

Table 9–2.
Approximate Scale of Mineral Hardness

1–Soft greasy flakes on fingers
2–Scratched by fingernail
3–Cut easily by knife, scratched slightly with penny
4–Scratched easily by knife
5–Not scratched easily by knife
6–Scratched by file
7–9–Scratches glass easily
10–Scratches all other materials

FIGURE 9–12. Talc.

FIGURE 9–13. Gypsum.

FIGURE 9–14. Corundum.

9:3 Rocks and Minerals

Every mineral is composed of elements. For example, quartz is made of silicon and oxygen.

Table 9–3.
Properties of Some Minerals

Mineral	Luster	Color	Hardness	Density (g/cm³)
Talc	Pearly to dull or greasy	Bright green to white, gray	1	2.8
Gypsum	Pearly, silky, dull, glassy	White, gray, brown	2	2.3
Calcite	Dull or pearly	White	3	2.7
Fluorite	Glassy	White, green, violet, blue, brown, yellow	4	3.1
Orthoclase	Glassy to pearly	White to gray, red, green	6	2.5
Quartz	Waxy to dull	Gray, brown, black, and so on	7	2.6
Topaz	Glassy	Yellow, white, blue, red, green	8	3.5
Corundum	Glassy to diamondlike	Gray, brown, red, yellow, blue, black, pink	9	4.0
Galena	Metallic	Dark lead gray	2.5	7.5

The main purpose of this activity is to call attention to the ways in which different rocks are similar and different.

activity

IDENTIFYING ROCKS   Activity 9-2

Collect a dozen different small rocks. Use an egg carton as a container for the rocks (Figure 9–15). Study all of the rocks carefully and classify them. How many are light in color? How many are dark? Are some smooth and others rough? Examine the texture, or grain, of each rock. Some are coarse-grained and some are fine-grained. How many of the rocks can you identify? Compare them with the pictures on Pages 180–183. If a collection of labeled rock samples is available, it will be very helpful in identifying your rock samples. See Teacher's Guide.

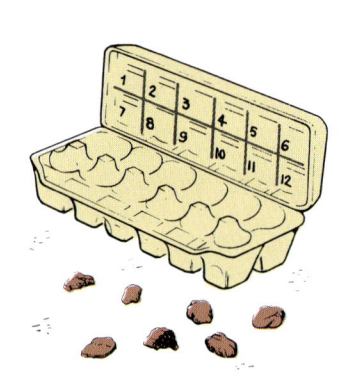

FIGURE 9–15.

176   Geology

activity

Activity 9-3

TESTING THE HARDNESS OF MINERALS

Obtain samples of talc, gypsum, calcite, fluorite, orthoclase, quartz, topaz, and corundum. Test the hardness of each mineral by trying to scratch it with your fingernail and with a knife blade. Try to scratch each of the minerals with a copper penny. Which of the minerals will scratch a glass plate? Compare your results with the scales of mineral hardness (Tables 9–1 and 9–2).

Hardness is the resistance to being scratched.

quartz, topaz, and corundum.

making sure

2. A 10-cm³ mineral sample has a mass of 23 g.
    (a) Calculate the density of the mineral. 2.3 g/cm³
    (b) Use Table 9–3 to identify this mineral. gypsum
3. An unknown rock sample scratches glass easily. It does not scratch other materials. What is the hardness of the sample? See Table 9–2. 7-9

9:4 Weathering and Erosion

Rocks and minerals are broken into smaller pieces by weathering. **Weathering** is the breaking down of rock by the action of water, ice, plants, animals, and chemical changes. Continued weathering of large pieces of rock produces the tiny particles that make up soil.

How does weathering produce soil?

FIGURE 9–16. Caverns and cave formations result from a type of weathering.

Gene Frazier

Weathered rock pieces are carried from one place to another by wind, moving water, and moving ice. This process by which rocks and soil are worn away is called **erosion.** Erosion in a desert may be caused by winds blowing sand that scratches exposed bedrock. Streams and rivers can carry rocks and soil for long distances. Many regions contain rocks that were carried there by glaciers a long time ago.

FIGURE 9–17. Shoreline features are the results of wave erosion (a). In an arid region such as Bryce Canyon National Park, Utah, stream erosion produces steep cliffs (b).

activity

THE EFFECT OF FREEZING ON THE WEATHERING OF ROCK   Activity 9-4

Fill an empty milk carton with water and seal the top. Place it in a freezer. After one day remove the carton. What change occurred in the water? *The water froze.* Did the shape of the container change? Why? *Yes, the water expanded when it froze.* What might have happened if the container were glass instead of cardboard? *The glass would crack.* Compare your results with the effect of freezing water in the cracks of rocks. *When the water freezes, it expands and makes the crack bigger.*

making sure

4. How does the speed of wind and water affect the rate of erosion?

   the faster the speed, the greater the rate of erosion.

5. Suppose you found a pebble that was very smooth. Do you think it was made smooth by wind, ice, or water? Explain.

   Wind and ice roughen the surface of a pebble. A pebble bouncing around in a stream of water is smoothed.

9:5 Igneous Rock

Granite, basalt, and olivine are examples of igneous rocks. Igneous rocks are formed by the cooling of molten rock material from within the earth. Molten rock material within the earth is called **magma.** Due to great pressure, extremely hot magma may be forced out of the lower crust and mantle. As it moves upward through the earth's crust, it cools and solidifies to form igneous rocks.

Igneous rocks may be divided into two types—extrusive (ihk STREW sihv) and intrusive (ihn TREW sihv). Extrusive igneous rocks are formed from magma that flows out over the earth's surface. The magma may flow from volcanoes or from deep cracks in the earth's crust. For instance, basalt is an extrusive igneous rock formed by the cooling of magma which flows from a volcano.

Intrusive igneous rocks are formed from magma that cools and crystallizes beneath the earth's surface. Crystallize means to form a crystal structure. Granite is an example of an intrusive igneous rock. The magma which forms intrusive igneous rock cools more slowly than magma released at the surface.

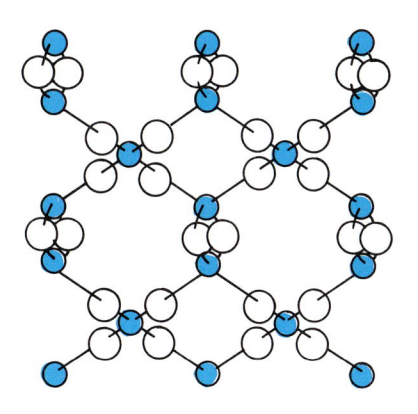

FIGURE 9–18. Quartz consists of silicon and oxygen atoms arranged in a crystal pattern. Quartz forms in granite when magma crystallizes beneath the earth's surface.

What is the difference between intrusive and extrusive igneous rocks?

FIGURE 9–19. Devil's Tower (a) consists of the hard interior rock of a volcano that was exposed when the outer layers were eroded (b).

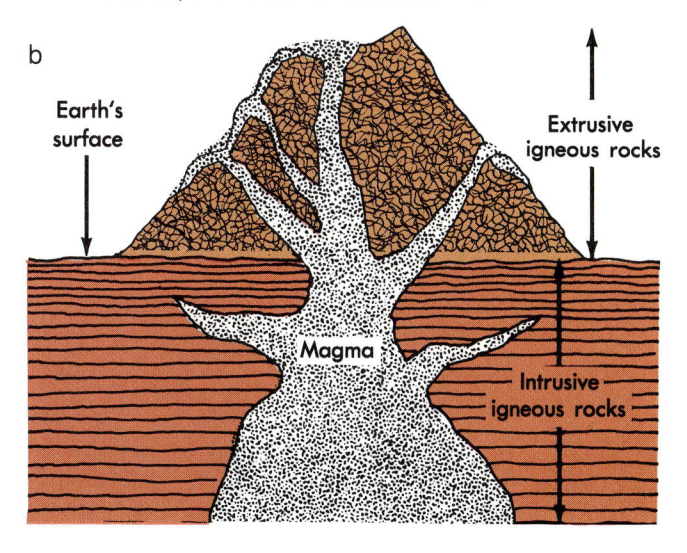

FIGURE 9–20. Large grains are formed when molten rock cools slowly underground (a). Fine grains form during cooling on the surface (b).

How are sedimentary rocks formed?

Enrichment: Display a piece of broken concrete and direct students to observe its structure. Explain how concrete is made from sand, gravel, cement, and water. A sidewalk is an artificial "rock" similar to a bed of sedimentary rock.

The speed with which an igneous rock changes from liquid to solid affects its texture. Texture refers to the size of the crystals, or grains, of rock. Most extrusive rocks are fine-grained because they cool quickly. Their texture may be so fine that the separate rock crystals can only be seen with a microscope. Intrusive rocks are coarse-grained and hard because they cool very slowly beneath the earth's crust. Their crystals are usually large, measuring from 0.4 to 1.3 cm across.

activity

TEXTURE OF IGNEOUS ROCKS Activity 9-5

Examine specimens of igneous rock. Compare the size of the crystals in the rocks. Divide the rocks into two groups—intrusive and extrusive. **Divide the rocks on the basis of crystal size.**

9:6 Sedimentary Rock

At one time, the entire crust of the earth was made of igneous rocks. Today, about one tenth of the total crust is made of sedimentary rocks. Sedimentary rocks are formed when particles of eroded rock are deposited together and then become cemented. These deposited rock particles are called sediments. The rock particles in sediments are produced by weathering and range in size from coarse gravel to fine powdered clay.

FIGURE 9–21. Power shovels are used to remove sediments that accumulate at the bottom of a river to allow for the safe passage of ships.

From the place where rock is weathered, particles are carried away by erosion. Wind, waves, rivers, and glaciers carry particles to places often far away from where they began. As particles are carried they are deposited to form sediments.

Sediment can accumulate in great deposits in oceans, rivers, streams, and lakes. How? When a river flows into an ocean or lake, it slows down in speed and drops the particles it is carrying. This process is called settling. The particles are deposited in layers on the ocean or lake floor. The largest and heaviest rock pieces are usually the first to settle. Then lighter and lighter pieces settle in layers above.

After thousands of years, sediment may become many layers thick. The particles of sediment in lower layers are under pressure from the sediment above. This pressure causes small particles to stick together. Larger rock fragments may be cemented together by dissolved minerals. When minerals like calcium carbonate or silica are deposited with fragments of sand or gravel, they cement the fragments together. Sedimentary rock is produced by both mineral cementing and by pressure from above.

As sedimentary rock forms, it makes strata, or distinct layers of rock. The strata are nearly

Many different kinds of fossils are found in sedimentary rocks.

FIGURE 9–22. Sedimentary rocks form in distinct layers or strata.

9:6 **Sedimentary Rock** 181

Name three sedimentary rocks.

horizontal when first hardened. But, many of the layers are eventually tilted or folded by forces within the earth's crust.

Sandstone, shale, limestone, and dolomite are common sedimentary rocks. A conglomerate is a sedimentary rock that contains rock fragments mixed with sand or clay. The fragments may range in size from pebbles to boulders. Sandstone, as its name suggests, is formed from particles of sand. Sandstone is composed mainly of quartz. It is highly resistant to wear and decay. Because of its many attractive colors, sandstone is often used as a building stone. Shale is rock formed from layers of clay or silt. Limestone is largely calcium carbonate. Dolomite is composed of both calcium carbonate and magnesium carbonate.

FIGURE 9–23. Conglomerate (a), shale (b), and sandstone (c) are common sedimentary rocks.

a

Russ Lappa

b

William Ferguson

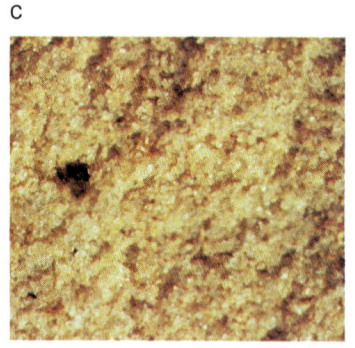

c

Carolina Biological Supply Co.

activity

See Teacher's Guide.

ROCKS THAT FIZZ   Activity 9-6

Place a drop of dilute hydrochloric acid on a few rocks. **CAUTION:** Use care when working with acid. Wear eye and clothing protection. If any one of the rocks contains a carbonate, the rock will fizz. The fizzing is produced by the release of bubbles of carbon dioxide gas produced in a chemical reaction between the acid and the rock. The word equation for the reaction is

Calcium carbonate + hydrochloric acid $\longrightarrow$ carbon dioxide + water + calcium chloride

activity

See Teacher's Guide.

FINDING THE COMPOSITION OF A SEDIMENT Activity 9-7

Spread 1 mL of coarse sand on millimeter-lined graph paper. Using the millimeter squares as a scale, measure the diameters of the sand particles. Classify the sand particles by sizes. View the minerals through a hand lens. How many can you identify? What fraction of the sample is quartz? Answers will vary. See Teacher's Guide.

making sure

6. How do sedimentary rocks form? Particles of buried sediment cement together due to great pressure.
7. How is sandstone different from shale? Sandstone is composed of quartz, while shale is formed from layers of clay or silt.
8. Why is sandstone sometimes used for building? highly resistant to wear, attractive colors

9:7 Metamorphic Rock

How are metamorphic rocks formed?

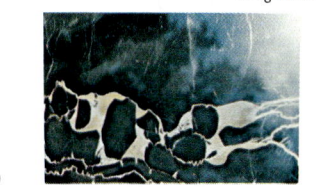

Metamorphic rocks are formed from sedimentary or igneous rocks. Metamorphic rocks are produced by extreme pressures or high temperatures below the earth's surface. Metamorphic rocks are generally the hardest and densest rocks.

A type of metamorphic rock is derived from each sedimentary and each igneous rock. Slate is a metamorphic rock derived from shale. Because slate splits lengthwise to form sheets, it is sometimes used for roofing and walkways. Quartzite is formed from sandstone. It is the hardest and most resistant common rock. Marble is a metamorphic rock formed from limestone and dolomite. It can be easily crushed into small pieces.

making sure

9. Explain why a drop of lemon juice fizzes when placed on marble. The acid in lemon juice reacts with the marble releasing carbon dioxide.
10. How can you distinguish sandstone from quartzite? Sandstone contains sand grains.

FIGURE 9–24. High pressure and temperature formed slate (a), marble (b), and quartzite (c).

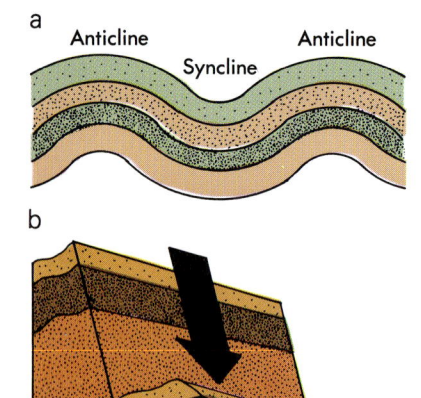

FIGURE 9–25. Forces within the earth's crust cause rock layers to fold, producing anticlines and synclines (a). A vertical fault forms when the crust breaks and part of the crust slips upward or downward (b).

What is a fault in the earth's crust?

FIGURE 9–26. These rock layers were once horizontal. Intense pressure from above pushed the layers downward.

9:8 Rock Formations

The earth's crust is always changing. In some places, the crust is being slowly pushed upward due to forces below the surface. In other places, hills and mountains are being worn down by weathering and erosion. Volcanoes and earthquakes are also changing the earth's surface. These changes within the earth produce mountains, valleys, caves, cliffs, and river deltas. They also produce special rock formations. A rock formation is a group of rocks arranged in a certain way.

Changes in the crust occur when pressure causes rock layers to bend or fold. Folded rocks may sometimes be seen along the sides of roads or streams. The folds in the rocks can be gentle or quite steep. When the rock layers are folded upward like an arch, the formation is called an **anticline** (ANT ih kline). A downward fold or trough is called a **syncline** (SIHN kline). The Appalachian Mountains are made of many anticlines and synclines. The folds were caused by forces that resulted when the Eurasian plate collided with the North American plate.

If pressure on rock layers becomes too great, they may break. A break in the crust along which movement occurs is called a **fault.** In some cases, the crust slips upward or downward along the break. The Grand Teton Mountains in Wyoming were formed by this type of faulting.

James Westwater

Sometimes, parts of the earth's crust slides past each other. This type of movement occurs along the San Andreas fault in California. The San Andreas fault is the zone where the Pacific and North American plates meet.

FIGURE 9–27. The Grand Teton Mountains in Wyoming were formed by faulting.

activity   See Teacher's Guide.

ANTICLINES AND SYNCLINES   Activity 9-8

(1) Obtain four large pieces of modeling clay in different colors. Press each piece into a rectangle about 2 cm thick. (2) Lay the pieces of clay on top of each other so they are like rock layers. Fold the combined layers into the shape of an anticline. (3) Cut off the tip of the rounded part of the fold so you can see the rock layers (4) Make a drawing of the layers as they appear from above. Which layer (color) of clay represents the oldest rock layer in an anticline? inner Youngest rock layer? Label these layers on your drawing. outer (5) Place four layers of clay together as you did in Steps 1 and 2. (6) Fold them into the shape of a syncline. Which layer (color) of clay represents the oldest rock layer in a syncline? outer Youngest rock layer? (7) Make a diagram of the inner layers when viewed from above. Label the oldest and youngest layers.

CAUTION: Remind students to be careful when using the knife to cut the clay.

making sure

11. How do eroded rock layers show the location of anticlines and synclines? Layers are curved rather than horizontal.

What causes an earthquake?

A heavy truck rolling down a street sends shock waves through the ground that may cause windows to rattle.

While going over this scale, call attention to the increased visible effects of the shock waves.

FIGURE 9–28. Destruction of a highway after an earthquake in Montana.

9:9 Earthquakes

An earthquake is caused by a sudden slippage of rock along a fault. Movement of the rocks can be vertical or horizontal. If you take hold of a stick and snap it quickly you can feel the force used in breaking it. Breaking the stick is similar to the slippage of rocks that causes an earthquake. The sudden release of energy sends out shock waves. A shock wave is a vibration that travels through the earth's interior.

Table 9–4.
Scale of Earthquake Intensity

1 Recorded by instruments.
2 Noticed only by few persons at rest.
3 Vibrations felt like passing truck; felt by people at rest, especially on upper floors.
4 Felt by people while walking; rocking of loose objects.
5 Felt by nearly everyone; most sleepers are awakened; bells ring.
6 Trees sway; suspended objects swing; damage by overturning and falling of loose objects.
7 Walls crack; plaster falls.
8 Difficult to drive cars; chimneys fall; poorly constructed buildings are damaged.
9 Some houses collapse where ground begins to crack; pipes break open.
10 Ground cracks badly; many buildings destroyed; railway lines bent; landslides on steep slopes.
11 Few buildings remain standing; bridges destroyed; great landslides and floods.
12 Total destruction; objects thrown into the air; ground rises and falls in waves.

A **seismograph** (SIZE muh graf) is used to record the shock waves sent out by earthquakes. Seismographs can record shocks that occur thousands of kilometers away. A seismograph contains a rotating recording drum on a base that is attached securely to the earth. A heavy weight supported by a

186 Geology

long thin wire contains a pen that inks a line on the recording drum. What happens to the weight when the earth around it shakes? The weight remains motionless. Its inertia causes it to stand still. The recording drum vibrates however, with each earth movement. This causes the needle to ink a line of waves on the recording drum.

A recording made by a seismograph needle is shown in Figure 9–30. The recording is called a seismogram. Up and down strokes of the pen are caused by vibrations produced by an earthquake. The more severe the vibrations, the higher the up and down strokes.

The seismograph described here is too simple to work well. Modern seismographs use magnetic and electronic devices to pick up and record earth vibrations.

The first indication of an earthquake on a seismogram is a series of primary waves (P-waves). As these waves start to fade, they are followed by a burst of secondary waves (S-waves). P-waves travel 1.7 times faster than S-waves. Following these two waves, surface waves are recorded. Surface waves travel parallel to the earth's surface and are the most destructive. They are the last waves to reach the seismograph.

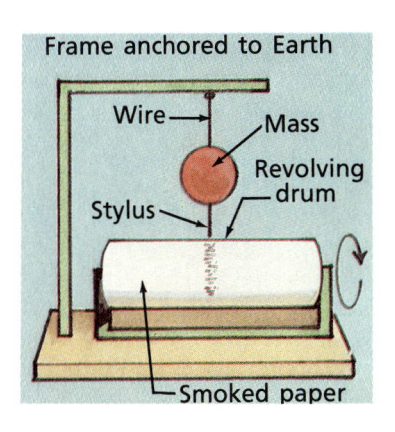

FIGURE 9–29. An earthquake is recorded when shock waves cause the drum of a seismograph to move.

Review inertia and relate it to the operation of a seismograph.

Name the three kinds of waves produced by an earthquake.

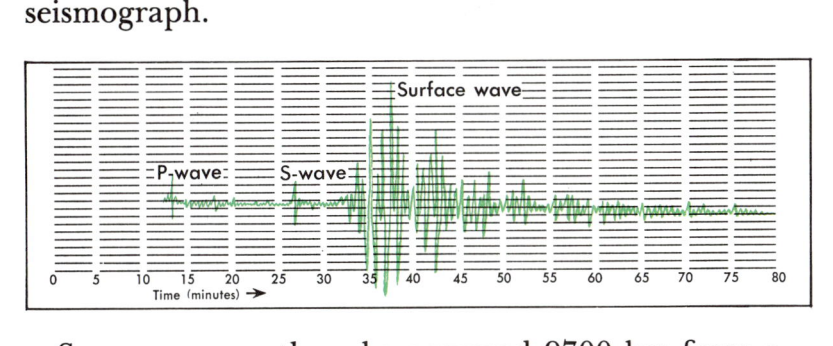

FIGURE 9–30. This seismogram is a record of both the time and intensity of an earthquake's shock waves.

Suppose an earthquake occurred 9700 km from a seismological station. It would take about 13 minutes for P-waves to reach the station. S-waves would arrive about 10 minutes after the P-waves. The time between the arrival of the P-waves and arrival of the S-waves depends on the distance between the recording station and the center of the earthquake. The greater the distance, the longer the time.

9:9 Earthquakes

How are seismographs used to locate the epicenter of an earthquake?

Finding the exact center of an earthquake requires reports from three or more recording stations. After the distance of each station from the earthquake center is found, a circle is drawn around each station. The radius of the circle gives the earthquake's distance. The point where the three circles meet gives the exact location of the earthquake center. This center is called the **focus**. The point directly above the focus on the earth's surface is called the earthquake's **epicenter**.

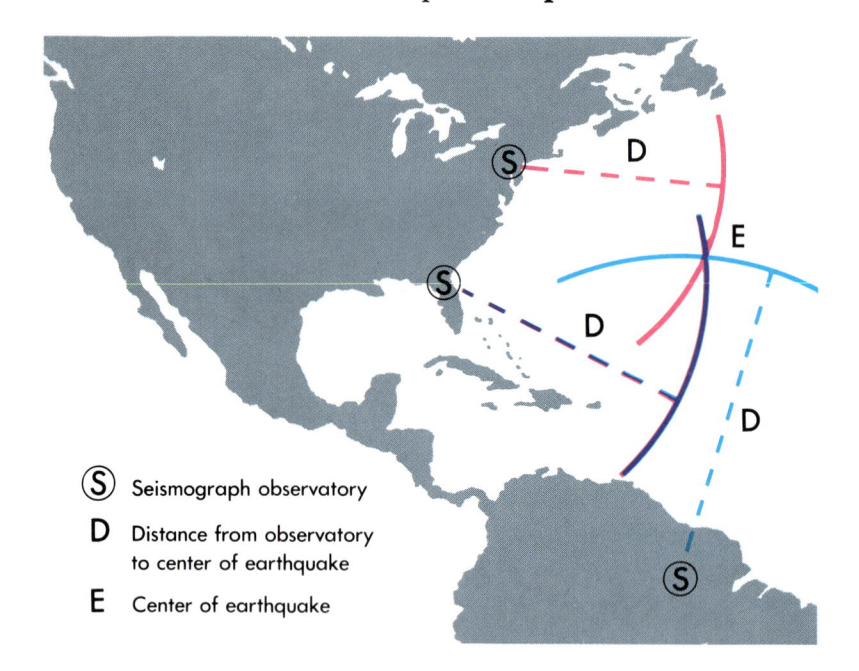

Most earthquakes are from faults within 75 km of the earth's surface. Most of the large earthquakes are related to the boundaries between plates.

Point out the difference between the focus and epicenter of an earthquake.

FIGURE 9–31. Exact location of an earthquake is determined by drawing circles around three or more earthquake stations. The radius of each circle is the distance of the earthquake from the station.

- **S** Seismograph observatory
- **D** Distance from observatory to center of earthquake
- **E** Center of earthquake

making sure

12. Refer to the seismogram in Figure 9–30 to answer the following questions.
    (a) At what time was the first earthquake shock recorded? *13 minutes after beginning of tape*
    (b) Name the three kinds of waves produced by an earthquake. *P, S, and surface*
    (c) Which waves are more severe? *surface waves*
    (d) What was the time interval between the arrival of the first P-wave and the first S-wave? *about 1 minute*
    (e) What was the time interval between the arrival of the first S-wave and the first surface wave? *about 14 minutes*

188 Geology

9:10 Volcanoes

Volcanoes form when magma is squeezed from the earth's interior to the surface. Volcanoes form on continents as well as on the ocean floor. Most volcanic activity occurs where plates collide or move apart. Volcanism refers to the process by which magma is produced and moved to the surface. Magma that reaches the surface is called **lava.** Volcanism also produces and expels hot gases and solid debris from openings in the crust.

Where do volcanoes form?

Volcanoes tend to be located along the earthquake belts.

What is lava?

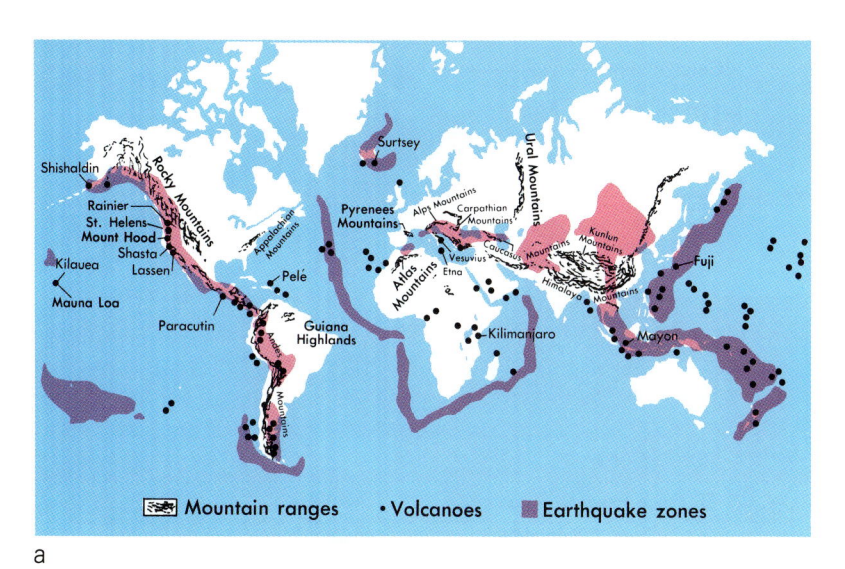

Not all volcanoes erupt violently. Sometimes lava pours from great cracks in the earth's surface. The flowing lava does not build a volcano. Instead, the lava floods a large area and forms a high plateau. Parts of North America were formed in this manner.

FIGURE 9–32. Locations of the world's mountain ranges, volcanoes, and earthquake zones are related to the movements of Earth's crustal plates (a). This volcano in Iceland erupted along the Mid-Atlantic Ridge where two crustal plates are moving apart (b).

FIGURE 9–33. The Columbia Plateau in northwestern United States formed when basalt lava erupted from large cracks in the earth's surface.

What are the three types of volcanoes?

When magma rises to the surface through a vent or pipe, a volcano forms. See Figure 9–34. Some volcanoes have magma that flows quietly onto the surface. Other volcanoes explode violently. During the explosion, rock fragments and lava are thrown high into the air. Also, a hill or cone builds around the vent. Some volcanoes form from quiet lava flows and explosions. Mt. St. Helens, built up over many years, is an example of this type of volcano.

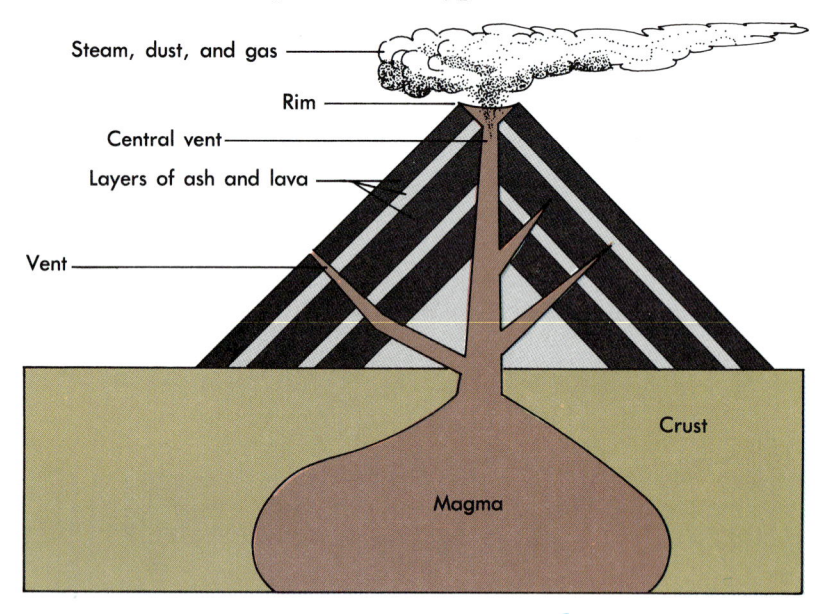

FIGURE 9–34. Some volcanoes may contain layers of ash and lava.

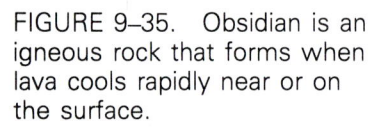

FIGURE 9–35. Obsidian is an igneous rock that forms when lava cools rapidly near or on the surface.

Eric Hoffhines/Courtesy of Carlton Davis

When first expelled, lava is red or white-hot. It quickly cools from a temperature of about 1000°C. Sometimes it cools to form a volcanic glass called obsidian. Obsidian cools so quickly that it has no crystals.

Lava foams when it contains gases under high pressure. The resulting material looks somewhat like shaving cream. As the foam lava hardens, it produces a rock called scoria. Scoria is filled with tiny bubbles. Pumice is a fine-grained, light-colored scoria rock.

making sure

13. Compare the locations of major volcanoes with the earth's earthquake belts. How do they compare? Volcanoes and earthquakes generally occur in the same areas.

main ideas

Chapter Review

1. The earth is made of a core, a mantle, and a crust. 9:1
2. Plate tectonics explains rock movements and changes in the earth's crust. 9:2
3. Rocks can be classified into three main groups—igneous, sedimentary, and metamorphic. 9:3
4. Rock is broken up and worn away by weathering and erosion. 9:4
5. Igneous rocks may be extrusive or intrusive. 9:5
6. Sedimentary rocks form in layers or strata. 9:6
7. Metamorphic rocks are formed by extreme heat and pressure. 9:7
8. Plate movements cause earthquakes and volcanoes. 9:9, 9:10

vocabulary

Define each of the following words or terms.

anticline	fault	magma	plate tectonics
core	focus	mantle	seismograph
crust	geology	mineral	syncline
epicenter	lava	plate	weathering
erosion			

study questions

DO NOT WRITE IN THIS BOOK.

A. True or False

Determine whether each of the following sentences is true or false. If the sentence is false, rewrite it to make it true.

F 1. The crust of the earth consists entirely of ~~liquid~~ rock. solid
F 2. The thickness of the ~~mantle~~ varies from 5 to 40 km. crust (or is about 2885 km)
F 3. ~~All~~ rocks have a coarse-grained texture. Some
T 4. A mineral is an inorganic substance which has a definite chemical composition and a crystal form.
T 5. The hardness of minerals can be determined by scratching them.

T 6. Rocks are made of minerals.
F 7. Igneous rocks are always formed beneath the earth's crust. *(Intrusive, inserted above "always")*
T 8. Coarse-grained intrusive rocks result from the slow cooling of magma.
T 9. Trenches are sometimes formed when two adjoining plates move toward each other.
T 10. Shock waves are produced by earthquakes.

B. Multiple Choice

Choose the word or phrase that completes correctly each of the following sentences.

1. Sedimentary rock is formed from (*loose rock particles*, liquid rock, soil).
2. Continued weathering of rock will eventually produce (marble, sandstone, *soil*, shale).
3. (Shale, Sandstone, *Limestone*, Quartz) will react with acid to produce carbon dioxide gas.
4. (Soil, *Slate*, Sandstone, Granite) is a metamorphic rock.
5. Marble is a metamorphic rock formed from (shale, sandstone, *limestone*, quartz).
6. Geologists believe the earth's (*inner core*, outer core, mantle) consists of solid nickel and iron.
7. As a mountain is weathered and eroded, its size (increases, *decreases*, remains the same).
8. Anticlines and synclines are produced by the (*folding*, faulting, splitting, melting) of bedrock.
9. According to plate tectonics, Iceland was formed by the (sinking, *rising*, faulting) of the Mid-Atlantic Ridge.
10. A (syncline, cone, *fault*, plate) is a break in the earth's crust along which movement can occur.

C. Completion

Complete each of the following sentences with a word or phrase that will make the sentence correct.

1. The ———— theory states that the earth's crust is divided into six major plates. *plate tectonics*
2. Liquid rock material within the earth is called ————. *magma*
3. The shock waves of an earthquake may be recorded by a(n) ————. *seismograph*

192 Geology

4. ____, ____, and ____ are three kinds of earthquake waves. Surface waves, S-waves, and P-waves
5. Most earthquakes are located on the ____ of the plates. boundaries
6. Quartz is a mineral found in rocks.
7. Magma is molten rock that flows from a volcano and turns into solid rock when it cools. Lava
8. Bedrock is usually covered with a layer of soil .
9. A fault is a break in the rocks, within the earth's crust.
10. ____ is the breaking down of the earth's crust by wind, water, ice, plants, animals, and chemical changes. Weathering

D. How and Why See Teacher's Guide at the front of this book.
1. How could you identify a rock or mineral?
2. How is sedimentary rock formed?
3. Why is the earth's crust constantly changing?
4. Why is it not possible to locate the center of an earthquake with a single seismograph?
5. How were the Alps and Himalaya mountains formed?

challenges

1. Have you ever thought about becoming a rock hound? Obtain a book about rock and mineral collecting. Make a collection of different rocks and minerals. Find out about chemical tests for identifying minerals.
2. Obtain information about research that scientists are doing to predict the occurrence of earthquakes. Prepare a report for your class.
3. Obtain a book from the library on crystal growing. Learn how to grow the crystals of several kinds of minerals. Prepare an exhibit of the crystals along with a description of the process you used to grow them.

interesting reading

Deeker, Robert and Barbara, *Volcanoes*. San Francisco, CA: Freeman, 1980.

Dietrich, R. V., *Stones: Their Collection, Identification, and Uses*. San Francisco, CA: Freeman, 1980.

Scientists are working carefully to remove a fossil from the digging site. Fossils and rock layers are clues to events that took place millions of years ago. How can the ages of rocks and fossils be determined? How can geological information be used to predict events in the future?

Earthwatch

Earth History

chapter 10

Introducing the chapter: Prepare a large wall chart or overhead projection based on Table 10–1. Display the table for the class and go over the concept of geologic history, distinguishing the eras, periods, life forms, and age estimates. Ask students what kinds of scientific information were needed to construct a table of this kind. Discuss the significance of fossils as records of Earth's history.

10:1 The Grand Canyon

How would you find the date of the first expedition to the South Pole? How would you find the date the steam engine was invented? These dates were recorded by people when the events took place. They may be found in a history book.

Geologists search for clues to geologic history "written" in the earth's rocks. Geologic history is an account of Earth's past recorded in rocks. The Grand Canyon in northern Arizona is a good place to study geologic history. The Grand Canyon is located on a plateau 2300 to 2800 m above sea level. A plateau is a level land area raised above the land around it. The canyon is 349 km long, 6 to 30 km wide, and more than 1.5 km deep. The Colorado River flows in a winding path through the canyon bottom. The Grand Canyon walls are made of sedimentary rocks—sandstone, limestone, and shale. Layer upon layer of rocks lie in an almost horizontal position above folded metamorphic rocks.

How did the Grand Canyon form? How old are the rocks that make up its walls? Are all the rock layers of the same age? What do the rocks reveal about events of the past? These are a few of the questions that interest geologists.

GOAL: You will learn how the earth and its life has changed during geologic time.

Explain the phrase "written in rocks" as an analogy to written in books. Photos or slides of the Grand Canyon can be used to motivate study of this section. Point out that this area is just one of many that are rich in geologic history. The same principles of geology and paleontology apply to the local area. Have students distinguish between recorded and prerecorded history.

FIGURE 10-1. Many sedimentary rock layers are visible in this view of the Grand Canyon.

The Grand Canyon is emphasized in this text because it vividly illustrates principles of geologic history. You may illustrate the same pinciples with local geological features, such as canyons or seashores.

How was the Grand Canyon formed?

The Grand Canyon was cut out of the Colorado Plateau by the Colorado River a long time ago. During that time, the plateau was slowly being uplifted. Year after year, the force of the speeding river waters eroded the rock to cut the canyon. The forces that formed the Grand Canyon continue today. As the plateau is uplifted, erosion cuts it deeper and deeper.

What factors determine the rate of rock weathering and erosion?

The rate at which rock is worn away is based on two factors—the speed of the running water and the type of rock. As the speed with which water runs across a rock increases, its wearing action also increases. Thus, the faster the water flows, the more

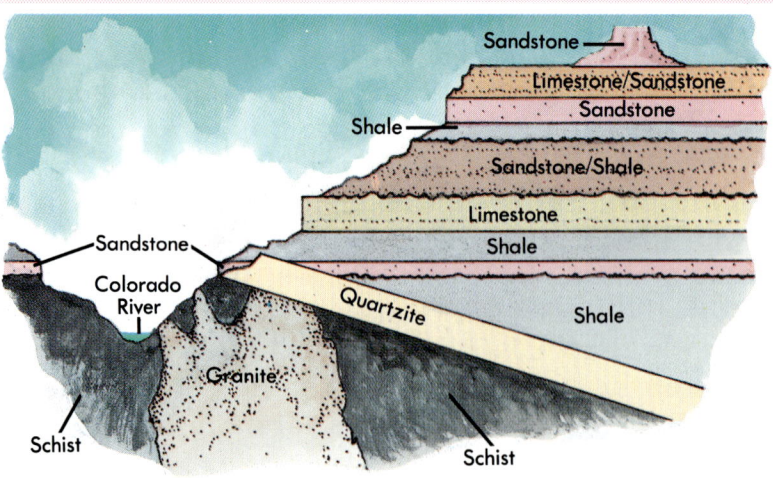

FIGURE 10-2. The sedimentary rock layers of the Grand Canyon lie above granite intrusions and tilted or folded metamorphic rocks.

196 Earth History

rapidly the rock is broken down. Sedimentary rock is worn away faster than metamorphic rock. The average rate of wear is about 1 m of rock for every 15 000 years.

Many fossils have been found in the rocks of the Grand Canyon. These fossils indicate that the land once lay under a shallow sea. It was at this time that the layers of sedimentary rock were formed. Later, over long periods of time, the land was uplifted from the sea. Desert conditions were present and land animals roamed the area. Fossils of animals, including reptiles are preserved in sand sediments formed by the blowing desert winds.

The rock history of the Grand Canyon is not complete. There are some time periods from which no rocks remain. The absence of rock layers is the result of erosion. A surface of eroded rock that separates younger rock layers from older layers is called an **unconformity**. Erosion took place when layers of rock were uplifted from beneath the sea and exposed to rain and wind. Sometimes a whole layer of rock was eroded away. See Figure 10–3.

According to the **principle of uniformity,** the processes that changed the earth in the past still exist today. The processes of weathering and erosion, which are evident today, have produced many of the earth's features.

Forces that change the earth's surface have always been the same. However, the magnitude of the forces and the amount of time in which they act varies in different places and in different times past.

Define unconformity.

Explain the principle of uniformity.

FIGURE 10–3. Sedimentary deposits beneath the ocean formed sedimentary rocks. Magma cut through the rock layers and formed a granite intrusion (a). When the rock layers were uplifted, erosion removed the upper layers. As ocean waters covered the rocks again, new sedimentary rock layers were deposited (b).

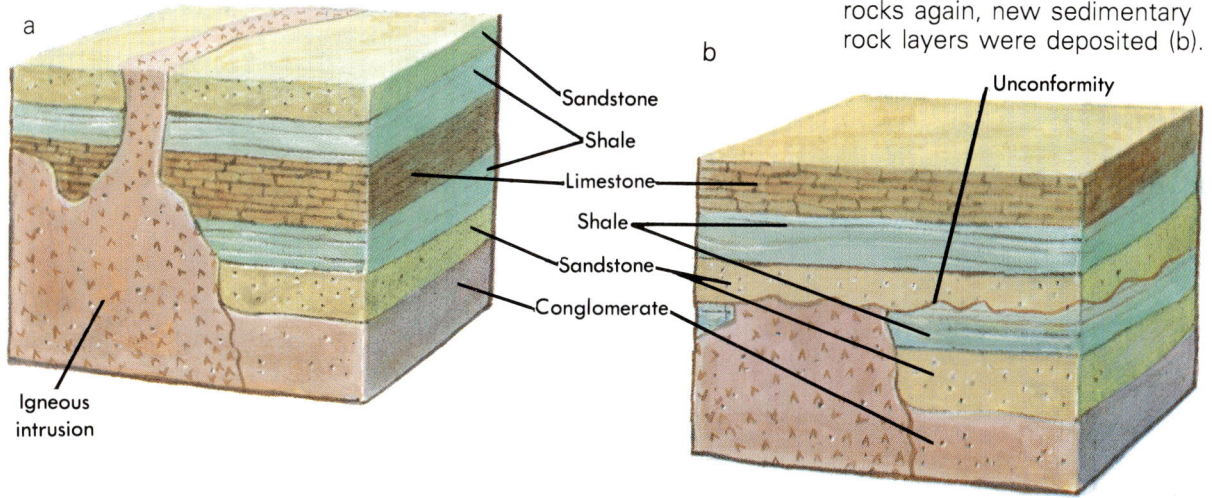

10:1   The Grand Canyon   197

Rivers presently appear to carry sediment to the sea much faster than during many past periods, and we know that climates and topography have not been constant.

These same forces are still at work changing the earth's surface. Most geologic forces do not appear to alter the landscape much during a person's life span. Yet they produce great changes over many centuries.

The forces that carve and mold the earth's surface have not always operated at their present rate. Sometimes their intensity has been greater, and sometimes less. For example, volcanoes once existed in many places where they are no longer found. The rate of weathering, erosion, and sedimentary rock formation has varied. Also, at times the climates of the world were much colder than they are today. During these periods, glaciers caused widespread erosion.

a
Dan McCoy/Rainbow

b
Sidney White Collection

FIGURE 10–4. Ancient sand dunes buried under other layers of sediment eventually become sandstone. Studying these sandstone formations provides evidence of climates of the past (a). Ocean waves gradually erode a rocky coastline (b).

making sure

1. By using a rate of erosion of 1 m/15 000 yr and a depth of 2000 m, find the approximate age of the Grand Canyon. **30 million years**

10:2 Age of Rocks

How is the age of a rock determined?

Layers of rocks are often arranged in order of age. In a series of level rock beds, the lowest layer is the oldest. Each higher layer is younger. The youngest rocks in the Grand Canyon are those near the surface. Deeper into the canyon, you see older and older layers of rock.

198 Earth History

It is not possible to tell the age of a rock bed from its position alone. One method for finding the age of rocks is **radioactive dating.** Radioactive dating is based on the rate of decay of certain radioactive elements. A radioactive element is one that gives off nuclear radiation. In releasing the radiation the element decays, which means it breaks down into lighter elements. Each radioactive element has a definite, fixed rate at which it decays. For instance, the rate at which radioactive uranium atoms break down into atoms of other elements is constant. Hence, uranium atoms act as a geologic clock that keeps track of the passing years.

As time passes, the amount of uranium in a given mass of rock decreases. The products of the radioactive decay, helium and lead, increase. One gram of uranium produces 1/7 600 000 000 g of lead each year. Such small amounts of lead can be detected with modern equipment.

Nuclear reactions are not influenced by factors such as pressure and temperature.

What is a radioactive "clock?"

Improved data on decay rates have changed early estimates of geologic time.

Half-life is the time it takes ½ of a radioactive element to decay and form other elements.

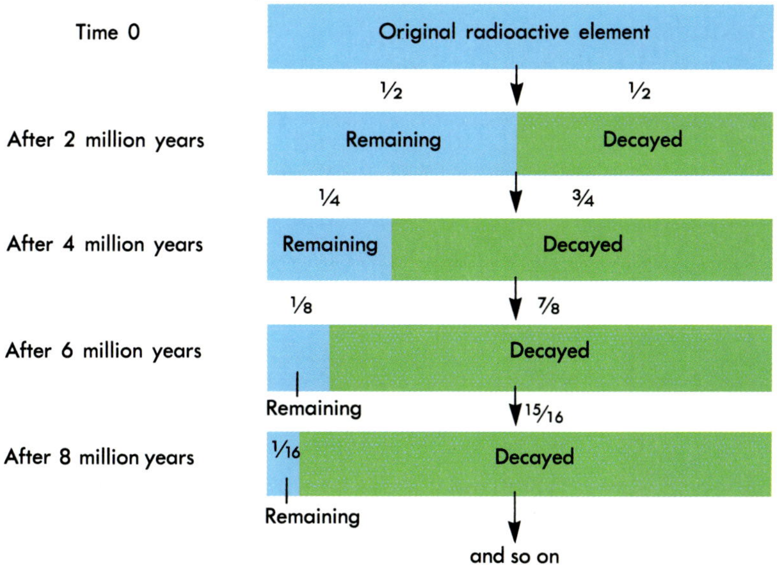

FIGURE 10–5. The fixed rate at which a radioactive element decays is called half-life. An element with a half-life of 2 million years means that one-half of the radioactive element decays after each 2-million-year period.

To discover the "birthdate" of a uranium-bearing igneous rock, the amount of lead present is compared to the amount of uranium present. Using this ratio, the rock's age may be computed. Rocks have been dated as old as 4.45 billion years.

Uranium 238 decays to lead 206 through a long sequence of steps that involve the emission of both alpha and beta particles.

10:2 Age of Rocks 199

If there is a river, stream, falls (anyplace where sedimentary rock may be exposed) in the area, you might take students there to look for fossils.

Define fossil and paleontology.

Why are animal fossils more plentiful than plant fossils?

An insect and any other object imbedded in a block of clear plastic resembles an insect fossil trapped and preserved in amber. Students may have samples of petrified wood at home that they may bring to school for display. Density of petrified wood is about 2.69 g/cm³.

FIGURE 10–6. Fossils of animals are preserved in areas where tar pits (a) and water holes (b) existed.

10:3 Fossils

Fossils are preserved remains or traces of plants and animals which lived long ago. A **fossil** is any evidence of past life. The study of prehistoric living things through fossils is called **paleontology** (pay lee ahn TAHL uh jee).

Most fossils consist of the hard parts of animals such as bones, teeth, and shells, which are preserved in rocks. When an organism dies, the soft fleshy parts usually decay. Plant fossils are scarce because plants contain few hard parts. Rarely does a whole animal or plant become a fossil. Exceptions to this are insects trapped in hardened plant sap, woolly mammoths buried and frozen in ice, and saber-toothed tigers trapped in tar pits.

Sometimes plant or animal tissue decays and is replaced with minerals. As a result the whole organism is cast in rock. Petrified wood has been formed in this way. As time passed, organic matter in the wood was replaced by silica and other minerals.

Footprints and other imprints left by organisms are another type of fossil. Tracks and footprints may be left in soft mud and preserved as the mud hardens. A fossil is also formed when an organism decays and leaves behind an empty space. If this empty space becomes filled with minerals or sediments, the fossil is called a cast.

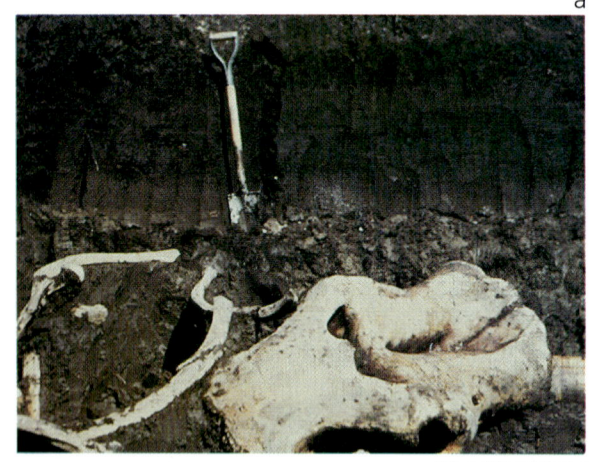

Sidney White Collection

Field Museum of Natural History, Chicago

FIGURE 10–7. Petrified wood (a), an imprint of an insect (b), and a dinosaur footprint (c) are fossils.

Only a small fraction of all past living things have been preserved as fossils. Most are not preserved because they are attacked by bacteria and eaten by scavengers. But living things that die in places where sediments are collecting may be preserved. Very few fossils, if any, form where erosion takes place. The remains of living things are most often preserved in ocean sediments or in lake deposits, where quick burial occurs. You are most likely to find fossils of clams that lived along muddy coastlines than fossils of flying insects. The leaves of a tree on a lakeshore are more likely to be preserved than leaves of a tree growing on a mountainside. Most of the known fossils have been found in sedimentary rocks.

Fossils are often found in unexpected places. For instance, fossil shark teeth have been found in the rocks of mountains located far from the ocean. These fossils were formed in rock at the bottom of a sea. But, through movements of the earth plates over millions of years, the rock was uplifted to form mountains.

Why is there little chance of an animal becoming a fossil when it dies?

An organism trapped in a tar pit may be preserved by the tar. Tar is a petroleum residue which is slightly acidic. Some foods are preserved in vinegar, a weak acid.

Petrification may be demonstrated by soaking a cellulose sponge in glue. When the glue hardens, the sponge is hard and dense, and more resistant to chemical and physical change.

See Teacher's Guide.

PETRIFIED WOOD AND OAK   Activity 10-1

Place a piece of oak and a piece of petrified wood, both about the same size, on opposite ends of a balance. Which has the most mass? Why? Calculate the density of petrified wood. Use the procedure described in the Activity on Page 28.

petrified wood contains minerals

10:3   Fossils   201

FIGURE 10-8.

activity
CASTING A LEAF   Activity 10-2

imprint made by part of an organism

(1) Mix some plaster of paris with water in an empty aluminum pie pan or saucer. Add just enough water to make a thick paste. The mixture should be about 3 cm deep. The top should be flat and smooth. (2) Lay a leaf on the surface of the mixture. Press down over the entire leaf to leave an imprint. (3) Remove the leaf carefully and set the plaster in a safe place until it hardens. How is the leaf imprint similar to a fossil? Is the hardened plaster of paris most like igneous, metamorphic, or sedimentary rock?

See Teacher's Guide.   sedimentary rock

activity   Activity 10-3
PREDICTING FOSSIL FORMATIONS

Obtain a map of the United States. Estimate in square kilometers the area of the seafloor where fossils may now be forming off the coasts of the United States. Use 48 km for the width of shallow seas along the Atlantic and Gulf Coasts. Use 16 km for those along the Pacific Coast. List five animals that might be preserved in each region.

See Teacher's Guide.

10:4 Fossil Records

Fossils reveal much about the past. Very old fossils are less complex than fossils of plants and animals that lived more recently. Less complex means they do not have as many different kinds of body parts. For example, shellfish are less complex than dinosaurs. A fern is less complex than a tree.

The fossil record also shows that many species of the past have become extinct. Extinct means that they no longer exist on the earth. Major changes in climate can cause a species to become extinct. Many findings indicate that climates and landforms have changed. For instance, dinosaur tracks found in hardened mud in New Jersey hint that this area was

One way a species may become extinct is when its environment changes and the species cannot adapt to the new environment. Many species are becoming extinct each year due to changes taking place in different parts of the world.

Why have many species become extinct?

FIGURE 10–9. Crinoids (sea lilies) were abundant in shallow waters about 500 million years ago. They still live today, but only in water greater than 200 m in depth.

swampy at one time. Woolly mammoths, suited for life in a bitterly cold climate, have left their fossils in New York State. Fossils of tropical plants have been discovered close to the Artic Circle. Fossils of corals, tropical marine animals, occur in almost all regions of the world.

Fossils are used to determine the age of sedimentary rocks. Some fossil species are only present in rocks of a certain age. Thus, the presence of a single species can be used to identify rocks of the same age throughout the world. The trilobite (TRI luh bite), an ancient marine animal, is one of these fossils. Trilobites were most plentiful 500 to 600 million years ago. Thus, any rock layers rich in fossils of certain trilobite species are likely to be between 500 and 600 million years old.

A fossil used to identify specific rock layers is called a guide fossil or **index fossil**. To be an index fossil, an organism must have lived during a brief period of geologic time. It must have traits that make it distinct from other species so it can be easily identified. Also, it must be found in rocks in different parts of the earth.

Index fossils help scientists discover how rocks of different regions are related. For instance, there are trilobite-bearing rocks in the Grand Canyon. These rocks are like trilobite-bearing rocks found in New York State. Although they are more than 3200 km apart, the rocks are of the same age.

Index fossils are distinctive and usually have no indication of ancestry. These widespread and easily recognized fossils are found in well cores as well as in widely separated outcrops and in complex structures. The best index fossils are swimming or floating organisms that evolved rapidly.

Why is a trilobite a guide fossil?

FIGURE 10–10. Trilobite fossils are index fossils because the trilobite lived for only a certain period of time in geologic history.

10:4 Fossil Records

Introduce this section with a brief overview of Table 10–1. Have students develop a similar table in their notebooks based on the information in sections 10:5–10:9.

What are the four major geologic eras?

10:5 Geologic Time

The history of the earth has been classified into four major divisions called **eras.** These are known as Precambrian (pree KAM bree un) Era, Paleozoic (pay lee uh ZOH ihk) Era, Mesozoic (mez uh ZOH ihk) Era, and Cenozoic (sen uh ZOH ihk) Era (Table 10–1). Each of the eras is further divided into time sections called **periods.**

Table 10–1.
Geologic History and Characteristic Life

Era	Period	Major Life Forms	Age Estimate—Absolute (years before present)
Cenozoic	Quaternary	Modern mammals	
	Tertiary	Mammals dominant Birds	65 million years (to present)
Mesozoic	Cretaceous	Massive extinction of reptiles	135–65 million years
	Jurassic	Reptiles dominant Conifers and cycads	193–135 million years
	Triassic	First mammals	225–193 million years
Paleozoic	Permian	Great extinction of marine invertebrates	280–225 million years
	Carboniferous	First reptiles Lycopod trees Amphibians dominant	345–280 million years
	Devonian	First amphibians Age of fish	395–345 million years
	Silurian	First land plants Age of corals	435–395 million years
	Ordovician	First vertebrates	500–435 million years
	Cambrian	Invertebrates dominant (trilobites, brachiopods)	570–500 million years
Precambrian		Primitive plants Sponge spicules Bacterialike microscopic forms and algae	4.5 billion or more years to 570 million years

204 Earth History

How do geologists divide time into eras and periods? The rocks of each era have been found to contain certain kinds of plant and animal fossils. These fossils are very different from those found in other eras. Every era has ended with major changes in landscape and climate throughout most of the world. Major changes in plants and animals also took place. Some species completely disappeared, while new species developed. Thus, the fossils in the rocks of each time period differ. Rocks from each time period contain fossils of organisms that were dominant at that time in the history of the earth.

How do the rocks of the various geologic periods differ?

making sure

2. The clock in Figure 10–11 represents 4 500 000 000 years of geologic time. Calculate the number of years that represent 1 min on the clock. Approximately 6 250 000 years
3. How are eras and periods different? Eras are longer times that are divided into periods.

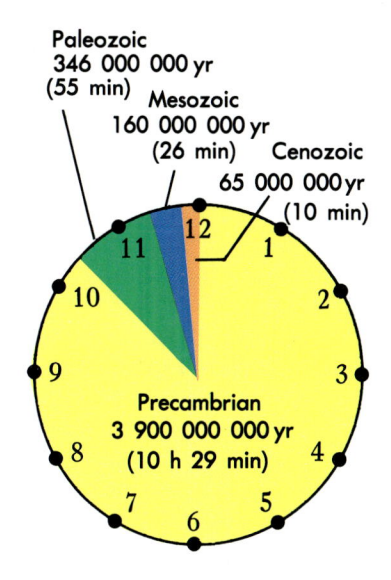

FIGURE 10–11.

10:6 Precambrian Era

Precambrian time began when the earth was first formed over 4.5 billion years ago. It lasted for 4 billion years. This era covers about 85 percent of all geologic time. Rocks from Precambrian time are found on every continent.

Precambrian rocks contain very few fossils. No hard parts of animals or plants have been preserved in these rocks. This lack of fossils is used to tell the difference between Precambrian rocks and more recent rocks. Because details about the Precambrian Era are few, this era is not divided into periods.

Since Precambrian rocks contain very few fossils, does this mean that life was absent on earth during these early years? Probably not. Remains of algae, bacteria, sponges, and the casts of worm burrows have been found. In addition, the presence of marble and graphite serves as evidence for the existence of some living things. Marble is formed from limestone, which may be formed by the deposits of certain living things. Graphite is derived

When did Precambrian time begin?

Explain the meaning of Pre-(prior) in Precambrian.

Schists and gneisses below the sedimentary rocks at the bottom of the Grand Canyon are examples of Precambrian rocks.

What observations indicate that life was present on Earth during the Precambrian Era?

Stress that continental changes such as submergence and mountain building are mostly a result of tectonic activity.

from either plant or animal matter. Graphite is a black, crystal form of carbon made by metamorphic forces acting on buried organic matter. Extreme pressures and high temperatures compress the organic material forcing the carbon atoms to rearrange into graphite crystals.

Large mineral deposits are found in some Precambrian rocks. Huge iron ore deposits are found in the Great Lakes region of the United States. Gold, silver, copper, and nickel ores in eastern Canada were formed during Precambrian time. The uranium ores in northwestern Canada are Precambrian deposits. Some granite and marble used today in buildings were also formed during Precambrian time.

Name several important mineral deposits discovered in Precambrian rocks.

FIGURE 10-12. The great iron ore deposits found in Minnesota were formed during Precambrian time.

Paul W. Nesbit

Go over the guide questions for sections 10:7–10:9. Have students read each section and write the answers to the guide questions in their notebooks. Then review the features of the eras and periods.

Why do sedimentary rocks of the Paleozoic age contain many fossils?

10:7 Paleozoic Era

The Paleozoic Era began about 570 million years ago and lasted for 345 million years. At times, shallow seas invaded the land, then receded. Although climates tended to be mild, they varied at times from humid tropical in some regions to desert in others. There were large numbers of plants and animals. Therefore, the sedimentary rocks of the Paleozoic age contain many fossils.

During the closing stages of the Paleozoic Era, very high mountain ranges were formed, including the Appalachians and the Alps. Glaciers also formed in Australia, Africa, India, and South America.

206   Earth History

The Paleozoic Era has been divided into six time periods. Each time period is marked by the appearance of new plant and animal forms.

The *Cambrian* (KAM bree un) *Period* began about 570 million years ago. During the Cambrian Period, large parts of the continents were covered by shallow seas. Great numbers of marine animals without backbones lived in these waters. Fossil traces of trilobites, jellyfish, sponges, brachiopods, snails, and cephalopods have been discovered. All of the plant fossils found in the Cambrian rocks are seaweeds.

Trilobites were the most plentiful forms of sea life. Some were only a fraction of a centimeter long and others were as long as 68 cm. Trilobites had segmented bodies with a tough outer skeleton. They were somewhat like today's lobsters and crayfish.

The *Ordovician* (ord uh VIHSH un) *Period* followed the Cambrian period. In the first part of this period, much of the land was still covered by shallow seas. There was a wider variety of marine life than existed in the Cambrian Period. The first fishlike vertebrates began to appear. Great limestone, oil, and natural gas deposits were formed.

Repeated submergence and emergence of parts of the continents have made it possible for geologists to divide the Paleozoic era into six periods.

Name a major characteristic of the Cambrian Period.

Point out to students that the scenes depicted in this chapter and in museums are the conceptions of artists based on scientific evidence and theory.

During which period did fishlike vertebrates begin to form?

FIGURE 10–13. A great number of marine animals flourished in the large shallow oceans of the Cambrian Period.

Field Museum of Natural History

During which period did the first land plants and the first air-breathing land animals appear?

The *Silurian* (suh LOOR ee un) *Period* had much volcanic activity. In some areas, lava and ash were deposited to depths of 1200 to 3000 m. Seas disappeared from parts of the continents. Fossils of the first land plants were found in the Silurian rocks of England and Austria. The first air-breathing land animals, the scorpions, also left fossils in Silurian rocks.

FIGURE 10-14. During the Silurian Period, *Eurypterids* were common (a). Many species of trilobites inhabited the oceans of the early Paleozoic (b).

Which period is known as the "Age of Fishes?"

The *Devonian* (dih VOH nee un) *Period* had a wide variety of fish, including the first backboned fish. This period is known as the "Age of Fishes." Land plants growing as tall as 12 m were plentiful. The first amphibians appeared. At least 700 kinds of brachiopods have been found in the rocks of this period. Trilobites, once so plentiful, were slowly becoming extinct. The first fernlike seed-bearing trees appeared.

Some time scales divide the Carboniferous Period into the Mississipian (Lower Carboniferous) and Pennsylvanian (Upper Carboniferous).

The *Carboniferous* (kar buh NIHF rus) *Period* followed the Devonian Period. Major coal fields date from the Carboniferous Period. The term Carboniferous refers to these coal deposits.

What geologic period was most favorable to plant growth?

The coal beds in Carboniferous rocks tell much about the earth during this time. Coal was formed from the carbon in plant matter. Thus, conditions during Carboniferous time must have favored plant growth. The land was low in many parts of the world. The climate was moist and warm. These conditions were ideal for the formation of swamps filled with giant plants. Several thousand species of

FIGURE 10–15. Some dragonflies of the Carboniferous Period had wingspans of 60 cm.

Enrichment: Have students construct a time line based on the years listed in Table 10-1. Use a scale of centimeters: years that fits the size of the time line to be drawn. You may wish to construct a time line on the chalkboard or on a bulletin board.

plants that lived at this time have left their fossil traces. Most of the species were ferns and cone-bearing plants.

Coal is formed in the following way. Large deposits of plant matter build up in an area. Much of the vegetation is preserved in the swampy environment. The area slowly sinks down and sea water flows in to cover it. Layers of sand and gravel are deposited at the bottom of the sea. Through millions of years, the sediment layers build upward. Intense heat and pressure change the sediments to rock and the vegetation to coal. Petroleum deposits were also formed during the Carboniferous Period. Petroleum, like coal, is a fossil fuel.

Coal is our most abundant fossil fuel. Relate coal to modern day energy needs, pointing out that the chemical energy in coal came from solar energy that was trapped by plants millions of years ago.

The *Permian* (PUR mee un) *Period* followed the Carboniferous Period. During this period climates became drier and colder. The land began to rise from below the sea. Because of the dry climate, the plants present during the coal ages could not survive. Flowering plants replaced the ferns that had been the main plant life for so long. All trilobites and nearly all the amphibians died out during this period.

What major climatic changes occurred during the Permian Period?

making sure

4. How might fossils be a helpful clue in the search for petroleum? Presence of fossils of plants from the Carboniferous period may indicate the presence of petroleum.

10:7 Paleozoic Era   209

Which era is known as the "Age of Reptiles?"

10:8 Mesozoic Era

The Mesozoic Era began about 225 million years ago and lasted about 160 million years. Mesozoic means middle-life. Reptiles and other dinosaurs became the main forms of animal life during this era. Thus it is known as the "Age of Reptiles." Flowering plants became the major type of plant life. Ammonites, animals that could float on the sea, were widespread. The Mesozoic era is divided into three time periods.

During the *Triassic* (tri as ihk) *Period*, the first period of the Mesozoic Era, much of the land was above the sea. Climates were dry. Volcanic activity occurred in many places. Red sandstone and shale, like that of the Painted Desert in Arizona, were formed. Primitive mammals and many types of dinosaurs developed. Some dinosaurs lived on the open plains. Others lived in forests, while many lived in swamps. Some dinosaur species were flesh-eaters while others ate only plants. Dinosaurs varied in size. Most were small and fast but some were huge and slow. The largest was over 24 m in length and had a mass of over 41 metric tons.

Tropical forests were composed largely of ferns and palmlike plants. In drier, cooler regions, forests were composed of huge conifers.

FIGURE 10-16. Ammonites, the most common marine organism of the Mesozoic, had shells consisting of many partitions (a). The *Stegosaurus* of Jurassic age was about 6 m long and had two rows of large, bony plates along its backbone (b).

210   Earth History

The *Jurassic* (joo RAS ihk) *Period* is the middle period of the Mesozoic Era. Landforms in the Jurassic Period were like those of the Triassic. The major life form on land was the dinosaur. Brontosaurus and Stegosaurus were both plant-eaters.

Archaeopteryx, which was about the size of a pigeon, also lived during this time. This creature had the skin and teeth of a reptile, but the feathers and wings of a bird. Evergreens, tree ferns, and palms were the main plants in the forests.

Widespread rock folding and faulting brought the Jurassic Period to an end. The Sierra Nevada mountains and the Coast Ranges of California were pushed up at this time. Gold ore was formed in the rocks of these mountains. Millions of years later, this gold was mined by the "forty-niners" during the Gold Rush of 1849.

The *Cretaceous* (krih TAY shus) *Period* is the last period of the Mesozoic Era. During this time, the sea in North America extended from the Gulf of Mexico to the Arctic Ocean. Sediments deposited in the seas were rich in limestone. The famous White Cliffs of Dover, England, were formed at this time. Intense volcanic activity was widespread. The snail and clam fossils of the Cretaceous Period are like the modern animals. Other rocks of this period contain fossils of trees such as fig, poplar, willow, magnolia, and maple. Grasses, grains, and fruitbearing plants were also present.

Large numbers of *Tyrannosaurus* roamed the area. *Tyrannosaurus* was the largest of all flesh-eating dinosaurs. This giant beast had a skull over 1 m long and a body over 6 m high. But, like all dinosaurs, its body may have been too large for its brain. The brain of *Tyrannosaurus* was about the size of a chicken egg. Dinosaurs became extinct during the last 10 to 15 million years of the Cretaceous Period.

making sure

5. How was the *Archaeopteryx* similar to both birds and reptiles? *It had the skin and teeth of a reptile but the features and wings of a bird.*

FIGURE 10-17. *Archaeopteryx*, the earliest known bird resembled early reptiles.

What three periods make up the Mesozoic Era?

FIGURE 10-18. *Tyrannosaurus*, the largest land-living, flesh-eating dinosaur, lived at the end of the Cretaceous Period.

10:9 Cenozoic Era

How was the major animal life of the Cenozoic Era different from that of the Mesozoic Era?

What two periods make up the Cenozoic Era?

The Cenozoic Era is the most recent in time, it began about 65 million years ago and extends into the present time. During the Cenozoic Era, mammals and birds became plentiful. The dinosaurs and ammonites had disappeared.

The *Tertiary* (TUR shee er ee) *Period* found the continents rising. Inland shallow seas did not exist. Fossils show that mammals were the main class of animal life.

Flowering plants, rodents, and the first hoofed animals have left fossil remains in Tertiary rocks. *Eohippus*, an early ancestor of the modern horse, appeared at this time. *Eohippus* was about the size of a small dog. Members of the camel family roamed North America. Mastodons and mammoths, which were similar to the modern elephant, were also present at the end of the Tertiary Period.

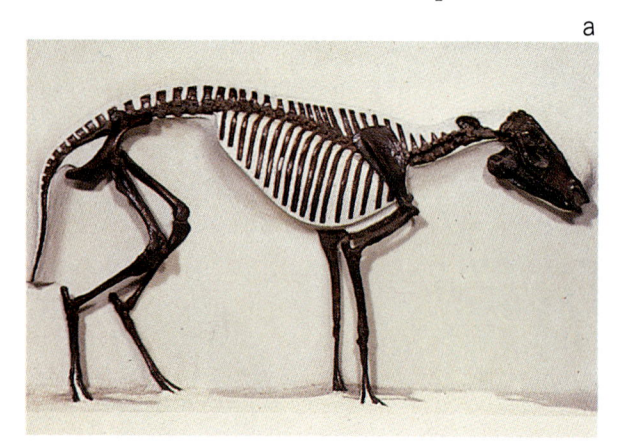

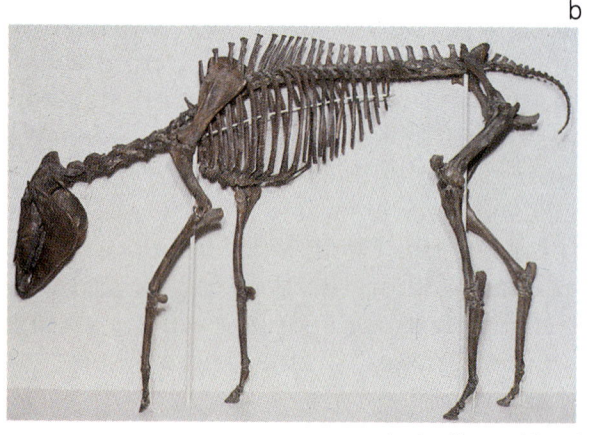

FIGURE 10–19. *Eohippus*, the four-toed horse, was the size of a collie dog (a). *Pliohippus* appeared later was larger and had a single toe equipped with a solid hoof (b).

The rock movements that began during the Tertiary Period are still occurring. They are responsible for the Coast Range mountains and frequent earthquakes in California. Why are these rock movements occurring? related to plate tectonics

The *Quaternary* (KWAHT ur ner ee) *Period* includes the present time. It is only a tiny fraction of the earth's history. Suppose all of the geologic history of the earth were represented by 1000 days. Quaternary time would cover only about six hours out of the 1000 days.

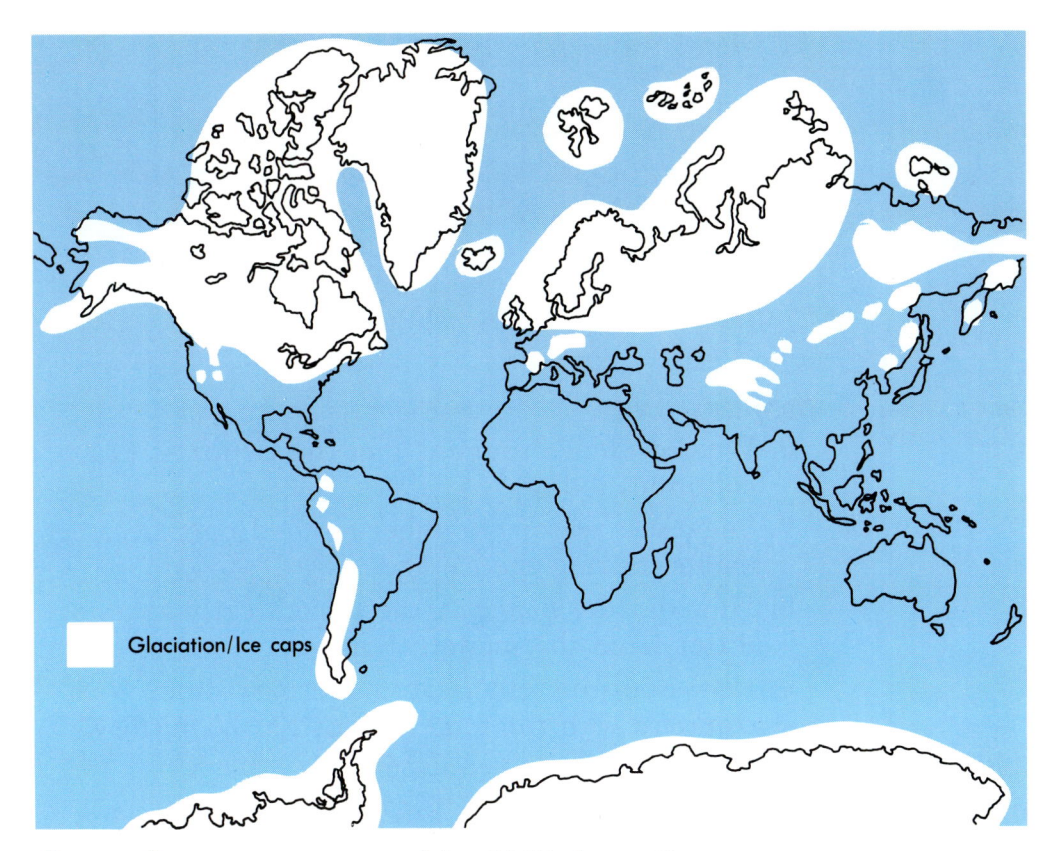

In the early part of the Quaternary Period, mastodons and woolly mammoths roamed in herds. Flesh-eating mammals such as the saber-toothed tiger were alive. Each of these species became extinct during the ice ages.

Ice ages have occurred several times during the Quaternary Period. Huge glaciers flowed south from the Arctic. The glaciers covered the northern portions of Europe, Asia, and North America. Some sections were over 1.5 km thick. The glaciers scratched and grooved the rock surfaces. They changed valleys and river courses. In many places, large deposits of sand and gravel were left when the glaciers receded. The last glacial ice age was about 10 000 years ago.

FIGURE 10–20. This map shows the maximum extent of glaciation during the Cenozoic.

What effect did glaciers have on the land?

FIGURE 10–21. Receding glaciers left grooves and scratches in rocks.

Sidney White Collection

10:9 Cenozoic Era   213

16,000 years ago 14,000 years ago 9000 years ago

FIGURE 10–22. The Great Lakes in North America were formed during the Quaternary Period as a result of glaciation.

making sure   See Teacher's Guide.

6. What evidence suggests that the Cenozoic Era continues today?
7. Suggest a reason the saber-toothed tiger became extinct during the ice ages.
8. How did the *Eohippus*, the horse's early ancestor, differ from the modern horse?
9. How does the length of Quaternary time compare with the total time of Earth's history?

Describe five geologic changes that took place in the past.

Enrichment: Have students write their predictions of the earth's future.

Enrichment: Conduct a discussion using the following questions. Could modern people survive a new ice age? How? Perhaps people could build an underground city in advance of the ice age and use geothermal energy for light and heat. Provisions would have to be made for food, oxygen, and removal of wastes.

10:10   Earth's Future

What will the earth be like in the future? Scientists predict future events based upon the earth's history over 4.5 billion years. During these years of geologic time huge mountains were pushed up and worn down many times. Huge rock beds were bent, folded, and crushed. Seacoasts rose and fell as shallow seas invaded the continents. Volcanic explosions and earthquakes occurred in all parts of the world. Glaciers moved down from the north and froze areas that had been warm. The glaciers extended south far enough to cover parts of what are now the states of Missouri, Illinois, Indiana, Ohio, and Pennsylvania. Times, when much of the earth was covered with ice, are called the ice ages.

Today's world differs from past geologic ages. In the future the earth's surface may be very different than it is today. Most scientists believe the ice ages of the past will return again, but they disagree on when. Perhaps the next ice age will be here in a few thousand years or in a few hundred years. There is no certain answer.

If history repeats on schedule a new ice age should now be approaching. Yet, the impact of people on the earth's climates must be considered. Great amounts of fossil fuels are being burned, which warm our planet. Carbon dioxide produced by burning goes into the atmosphere and acts as a blanket that prevents heat loss. Another view is that the earth may actually be warming. If so, and the warming continues, the ice caps will start melting and raise sea levels around the world. Many seacoast cities and towns will be flooded.

Tides on the earth produced by the sun and moon are causing the earth to spin more slowly. It is predicted that in five million years from now, a day will be 36 hours long. Daytime hours may be much hotter and nighttime hours may be much cooler.

Will the earth last forever? Certainly not in its present form. The sun, around which the earth moves, is a star just like most other stars in the universe. Observations of stars similar to our sun reveal that stars burn out when they get old. Before the end, our sun will become many times larger than it is now. Temperatures on the earth will be hot enough to boil away the oceans as steam. Eventually, the sun will cool and shrink down to a size smaller than the earth. Do not worry! This event is predicted to take place in about 5 billion years.

How may the earth be different in the future?

making sure

10. Give three examples that show the earth is still changing.
11. How can the burning of fossil fuels cause changes on the earth?
12. List five cities that would be affected if the ice caps melted.
13. How does the sun and moon affect the earth's time of rotation?

10. earthquakes, volcanoes, erosion
11. Carbon dioxide acts as a blanket to hold heat in and may cause the earth to warm up. Pollution affects plant and animal life.
12. Answers will vary, any city on the coast.
13. produce the tides that slows radiation.

10:10 Earth's Future

See Teacher's Guide.

PERSPECTIVES

people

Mike Hansen: Earth Scientist

Mike Hansen is a state geologist for the Ohio Geological Survey. His work involves a wide variety of activities. In the course of a day, he may spend time working in a laboratory, examining a rock outcrop near a river, and talking to newspaper reporters.

One of the major responsibilities of a state geological survey is to determine the location and extent of the state's mineral resources. This task is accomplished by several steps. The first step includes a thorough study of exposed rocks throughout the state. All surface rocks are identified, their thickness is determined, and their locations recorded. From all this data, maps are drawn showing the distribution and structure of rocks throughout the state.

Geologic maps are valuable to many people. A mining company needs to know the best areas in the state for mineral deposits. If a company needs limestone for their manufacturing process, for example, they would benefit by building close to limestone quarries that are not often flooded by underground water.

Part of Mike Hansen's job takes place in the laboratory. If unidentified rocks and minerals are found, it is necessary to determine what they are. The types of rocks may indicate how an area of the state may have looked millions of years ago. Such information is useful in looking for valuable earth minerals.

Most of Mike Hansen's time is spent with public service activities. People frequently contact the geological survey for information. A collector may wish to know the best areas of the state to look for a specific type of fossil. Landowners might inquire about the possibility of valuable mineral deposits on their land. If an earthquake occurs in the state, the news media will contact the geological survey.

A geologist's job also involves research. Mike Hansen is currently studying the different types of rocks associated with coal deposits. By examining the characteristics of these rocks, it is possible for him to determine the conditions that were associated with the formation of the coal beds. He will then be able to look for similar structures elsewhere, and possibly locate more coal.

Mike Hansen is enthusiastic about learning. He knows that his work is valuable. Our electricity may be generated from burning coal that a state geologist helped discover. Geologists also play an important role in the discovery of petroleum and natural gas. Before your school was built, a geologist was probably consulted to determine if the underlying bedrock would be able to support a building.

Doug Martin

main ideas
Chapter Review

1. Geologic forces that shape the earth have not always operated at the same rate.   10:1
2. The age of some rocks may be found by radioactive dating.   10:2
3. Most fossils have been found in sedimentary rock.   10:3
4. Fossils reveal that many living creatures of the past are no longer living today.   10:4
5. The geologic history of the earth has been classified into four major divisions called eras. Eras are subdivided into periods.   10:5
6. The Precambrian covered about 85 percent of geologic time.   10:6
7. The Paleozoic Era was a time of variations in climate; there were large numbers of plants and animals; glaciers and mountains were formed.   10:7
8. The Mesozoic Era is referred to as the "Age of Reptiles"; flowering plants became the major form of plant life.   10:8
9. The Cenozoic Era began 65 million years ago and extends into the present time.   10:9
10. Based on its geologic history the earth will continue to change in the future.   10:10

vocabulary

Define each of the following words or terms.

era	paleontology	radioactive dating
fossil	period	unconformity
index fossil	principle of uniformity	

study questions

DO NOT WRITE IN THIS BOOK.

A. True or False
Determine whether each of the following sentences is true or false. If the sentence is false, rewrite it to make it true.

T   1. The Grand Canyon is formed mostly from sedimentary rocks.
T   2. The lowest bed in an undisturbed series of rock layers is the oldest.
T   3. The speed of running water affects the rate at which it wears away rock.

T   **4.** The Grand Canyon is being uplifted.

F   **5.** Geologic forces of the past ~~were different from~~ those of today. *are the same as*   *decay of radioactive elements*

F   **6.** Radioactive dating is based on the use of ~~radio waves.~~

F   **7.** Mammals appeared on the earth ~~less~~ than 1 million years ago. *more*

T   **8.** Fossils of sea-living animals have been discovered in rocks hundreds of kilometers from the sea.

T   **9.** Petrified wood is a fossil.

F   **10.** Trilobites are ~~plentiful and widespread~~ today. *extinct*

B. Multiple Choice

Choose the word or phrase that completes correctly each of the following sentences.

1. (Petrified, <u>Index</u>, Old, Radioactive) fossils are used to identify the geologic period of a rock layer.
2. Rocks of different time periods contain *(the same, <u>different</u>, closely related)* fossils.
3. A fossil is most likely to be formed in a *(desert, field, glacier, <u>shallow sea</u>)*.
4. The geologic history of the earth is classified into *(2, 3, <u>4</u>, 5)* major divisions called eras.
5. The era in which we now live is *(Precambrian, Paleozoic, <u>Cenozoic</u>)*.
6. The geologic time scale covers about *(1, 2.5, 3.4, <u>4.5</u>)* billion years.
7. Precambrian rocks contain *(many, <u>few</u>, reptile, mammal)* fossils.
8. Paleozoic rocks contain *(<u>many</u>, few, reptile, mammal)* fossils.
9. Backboned animal fossils first appeared in the *(Cambrian, Ordovician, Silurian, <u>Devonian</u>)* Period.
10. The *(Cambrian, Ordovician, Silurian, <u>Devonian</u>)* Period is known as the "Age of Fishes."

C. Completion

Complete each of the following sentences with a word or phrase that will make the sentence correct.

1. Huge coal deposits were formed during the _____ period. *Carboniferous*
2. Coal is formed from <u>plants</u>.

3. The _____ Era was the age of reptiles. Mesozoic
4. Fossils of *Archaeopteryx* first appear in the _____ Period. Jurassic
5. During the _____ Period, the Sierra Nevada Mountains of California were uplifted. Jurassic
6. The last ice age was about 10 000 years ago.
7. During the ice ages, _____ scratched rocks and changed the courses of rivers. glaciers
8. We are living in the _____ Era. Cenozoic
9. If the total geologic record covered 1000 days, the Quaternary Period would cover about 6 h .
10. During the Tertiary Period, _____ were the main class of animal life. mammals

D. How and Why See Teacher's Guide in the front of this book.
1. Why is the Grand Canyon a good place to study geologic history?
2. How is the age of a rock formation determined?
3. How do fossils aid in discovering events of the past?
4. How is the geologic time scale divided into eras and periods?
5. Why is little known about the events of Precambrian time?

challenges

1. Make a scale model of the Grand Canyon.
2. Make a fossil collection by searching and collecting them in your home area. Identify your fossils and use them to learn about the geologic history of your area.
3. Do library research on the geologic history of your area. Prepare a report for your class.

interesting reading

Fodor, R. V., *Frozen Earth: Explaining the Ice Ages.* Hillsdale, NJ: Enslow, 1981.

Lambert, Mark, *Fossils.* NY: Arco, 1979.

Motz, Lloyd, "Earth: Final Chapters." *Science Digest.* August 1981, pp. 76-85.

Clouds are a result of changes in the atmosphere. Special gliders are sometimes used to collect information about how some clouds become thunderstorms. Why does warm air rise? What are the jet streams? How is fog different from clouds? What is used to measure water vapor in the air?

National Center for Atmospheric Research

The Atmosphere   chapter 11

Introducing the chapter: Begin a discussion of the importance of air by asking students what they would need to inflate a balloon and fly a glider plane. Ask students to explain how a parachute is used to slow the rate of descent of a falling object. Proceed to a discussion of other activities dependent upon air.

11:1 Air

You live at the bottom of an ocean of air that surrounds the earth. Breathing, cooking, flying, and sailing, are just a few of the activities that depend upon air. Without air it is likely there would be no life at all on Earth.

Air is a mixture of gases, fine dust particles, and water vapor. It occupies space and has mass. About 78 percent of air is nitrogen. About 21 percent is oxygen. The other 1 percent is composed of argon, carbon dioxide, and tiny amounts of other gases (Table 11-1).

Nitrogen does not combine readily with other substances. Nitrogen dilutes the oxygen in air. This fact is important because oxygen enters readily into chemical reactions such as rusting and burning. If oxygen were not diluted by nitrogen, these chemical reactions would occur at a very rapid rate. For example, a forest fire burning in pure oxygen would be difficult to put out.

GOAL: You will learn the major features of the atmosphere and how winds, clouds, and precipitation are formed.

What is the composition of air?

Remind students of the no smoking signs in hospital rooms that have oxygen tents.

221

Note that Table 11–1 shows percent by volume, not weight.

Other gases of the lower atmosphere include nitrous oxide, radon, sulfur dioxide, nitrogen dioxide, ammonia, and carbon monoxide.

What is the atmosphere?

Emphasize the fact that there is not a precise outer boundary of the atmosphere.

The height of the troposphere slopes from about 6 km over the poles to about 20 km over the equator.

FIGURE 11–1. Data about the atmosphere such as temperature and wind may be obtained through the use of balloons.

Table 11–1.
Composition of Air

Gas	Percent by volume
Nitrogen	78.08
Oxygen	20.95
Argon	0.93
Carbon dioxide	0.03
Water vapor	0 to 5.0*
Neon	trace
Helium	trace
Methane	trace
Krypton	trace
Xenon	trace
Hydrogen	trace
Ozone	trace

*As water vapor content increases other gases decrease proportionately.

11:2 Parts of the Atmosphere

The air that surrounds the earth is called the **atmosphere** and extends about 900 km above the earth's surface. The atmosphere can be divided into five regions or layers. The layers vary in depth and blend into each other.

You live in the **troposphere** (TROHP uh sfihr). This is the first layer and extends from the ground to an average height of 11 km above the earth. The exact

National Science Foundation

222   Atmosphere

height depends on the season of the year and the latitude. Temperatures in the troposphere drop about 6.5°C for each kilometer of increase in altitude. Almost all weather activity occurs in this layer.

The **stratosphere** (STRAT uh sfihr) lies above the troposphere. It extends from about 11 km to 50 km above the earth. In the lower part of the stratosphere, the temperature does not change with height. But, in the upper part, the temperature increases with height. At 50 km, the temperature is about 7°C. Air in the stratosphere is dry and thin, and there are very few clouds.

A form of oxygen called ozone is produced in the stratosphere by a reaction involving the sun's energy. Ozone absorbs harmful ultraviolet rays. The ultraviolet radiation changes to heat and warms the upper stratosphere.

The **mesophere** (MEZ uh sfihr) is above the stratosphere. It extends from about 50 km to 80 km. Here the temperature decreases with height. At the top of the mesosphere, the temperature falls to about −100°C. This region is the coldest part of the atmosphere.

The **thermosphere** (THUR muh sfihr) extends from 80 km to about 600 km about the earth. In the thermosphere, temperature increases rapidly with height and may reach 700°C.

The **exosphere** (EK so sfihr) is the layer of the atmosphere farthest from the earth. It begins at an altitude of about 600 km and extends outward. No precise boundary marks the outer edge of the exosphere. It gradually blends into space.

The thermosphere and parts of the mesosphere and exosphere form a region called the ionosphere. This region absorbs a large amount of solar radiation. This radiation can change air particles to ions. These ions in the ionosphere aid radio communication. Some radio waves that enter the ionosphere are reflected back to earth. Reflection of radio waves from the ionosphere enables them to travel great distances across the earth.

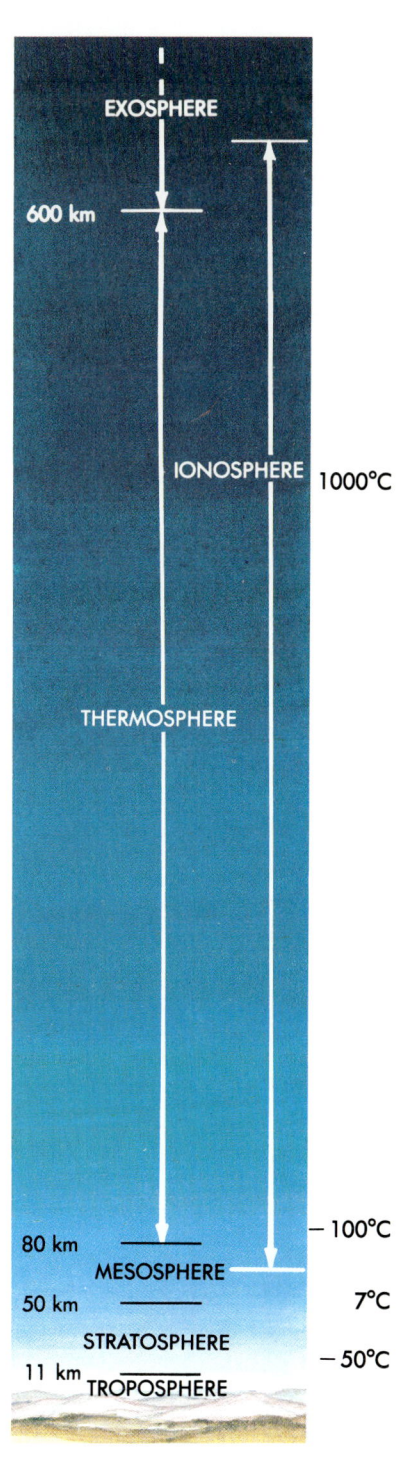

FIGURE 11–2. Each layer of the atmosphere has distinctive properties.

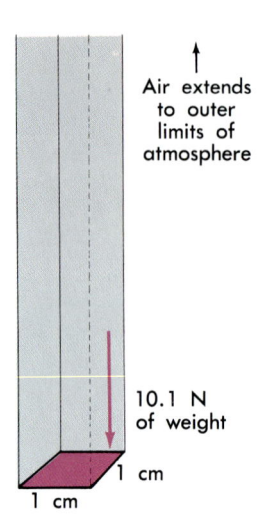

FIGURE 11-3. A force is exerted by the atmosphere on everything within and beneath it.

Why is air less dense at higher altitudes?

FIGURE 11-4. The column of mercury is maintained in the glass tube by the force of air exerted on the mercury in the dish.

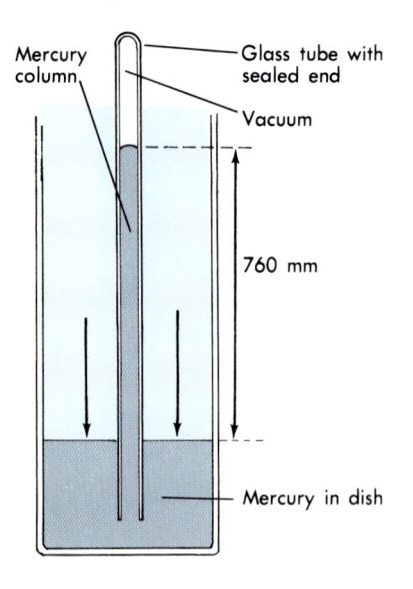

making sure

1. What temperature would you expect at 9000 m above the earth when ground temperature is 15°C? —43.5°C

11:3 Air Pressure

Review the definition of density, force/area.

Air is most dense near the surface of the earth. It becomes less dense at higher altitudes. Why? Gravity is the force that keeps the atmosphere from drifting into space. Gravity is strongest near the earth's surface. Three fourths of the atmosphere is held within 11 km of the earth.

Air exerts pressure in every part of the atmosphere. The average air pressure at sea level is 10.1 N/cm^2 or 101 kPa. You can understand how this pressure exists if you think of a square centimeter of the earth's surface. Above this surface, a column of air extends from the ground up to the outer limits of the atmosphere. Air in the column exerts a force of 10.1 N on the square centimeter of surface.

The amount of air pressure depends on the altitude. Air pressure is less at high altitudes than at sea level. As altitude increases, the density of air decreases. With decreasing density, air pressure decreases.

Air pressure never stays the same for very long. It may change from hour to hour or day to day. Air pressure is measured with an instrument called a **barometer** (buh RAHM ut ur). There are two types of barometers—mercury and aneroid (AN uh royd). A mercury barometer is a glass tube containing liquid mercury. One end of the tube is sealed, and the other is placed in a dish of mercury. The space above the column of mercury is a vacuum.

The height of the mercury in a barometer depends on the force of the air on the mercury in the dish. As the air pressure increases, the mercury rises. As the air pressure decreases, the mercury falls. Average sea level pressure is 760 mm of mercury. See Figure 11-4.

An aneroid barometer contains a sealed can from which air has been removed. The can contracts or expands with changes in air pressure (Figure 11-5). An increase in air pressure pushes the sides of the can inward. When air pressure decreases, a spring inside the can pushes the sides out. The amount the can contracts or expands is proportional to the air pressure. The can is connected to a needle which travels along a dial and records the movements of the can. From the scale on the dial, the air pressure can be read.

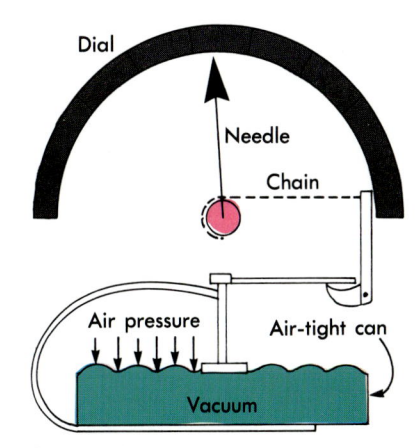

FIGURE 11-5. In an aneroid barometer, the movement of the needle depends on the force of air exerted on the air-tight can.

activity

EFFECTS OF AIR PRESSURE Activity 11-1

(1) Do this activity next to a sink. Fill an empty plastic milk container to the brim with water. (2) Obtain a 1-hole stopper to fit the opening in the container and a small piece of glass tubing to fit the stopper. You will also need a piece of rubber tubing 1 m in length. (3) Insert the glass tubing in the stopper. **CAUTION:** Moisten the end of the glass tubing with a lubricant such as glycerol. Hold the tubing with a rag or towel as you slide it into the stopper. (4) Attach the rubber tubing to the glass tubing. (5) Fill the rubber tubing with water. Hold a finger over each end so the water does not run out. (6) Insert the rubber stopper tightly into the container keeping the end of the hose elevated so the water does not leak out. See Figure 11-6. (7) Set the container on a stack of books and lower the hose into the sink. What do you observe? What causes the change in the container? Is the container still full of water after its shape changes? *See Teacher's Guide.*

FIGURE 11-6.

activity

See Teacher's Guide.

USING A BAROMETER Activity 11-2

Record the air pressure reading on a barometer at the same time each day for five days. Also record the general nature of the weather for each day. Note any change in the barometer if a storm approaches. If no stormy weather occurs during the five-day period, note if there is much change in the barometer reading. *Air pressure usually decreases when a storm approaches because most storms are associated with low pressure areas.*

11:3 Air Pressure

making sure

2. What keeps the atmosphere from floating off into space? Gravity attracts the atmosphere and holds it to the earth.
3. How does the density of air affect air pressure? Increased density causes increased air pressure.

11:4 Heating the Atmosphere

The sun gives off large amounts of energy each day. This energy from the sun is called radiation. Some of the radiation that travels to the earth is absorbed and scattered by the atmosphere. Most of the radiation, however, is absorbed by the earth's surface. Solar energy that reaches the surface is radiated back into the atmosphere. Carbon dioxide and water vapor in the air absorb the energy thereby warming the atmosphere.

Heating of the atmosphere is not uniform. Solar radiation that strikes the earth directly has a greater warming effect than radiation that strikes at an angle. For example, solar radiation strikes the equator at an angle close to 0°. The earth is warmed the most in this region. At the poles the angle is almost 90° and the warming effect is the lowest. Also, some of the sun's energy is reflected by the snow and ice, further reducing the warming effect at the poles.

When clouds are present during the day, much of the sun's energy does not reach the surface. Clouds block and reflect the energy. Cloud cover during the day reduces the sun's heating effect. On the other hand, clouds at night act as a blanket. Heat from the surface is prevented from escaping into outer space.

The temperature of the air affects its density. When air is heated, its density decreases. For example, air near the equator is less dense than air at the poles. Warm air rises due to its lower density. Cool air sinks because its density is greater. **Upward and downward movements of air are called convection currents.** Convection currents help to move heat quickly throughout the atmosphere. The difference in density between warm and cool air causes convection currents to form.

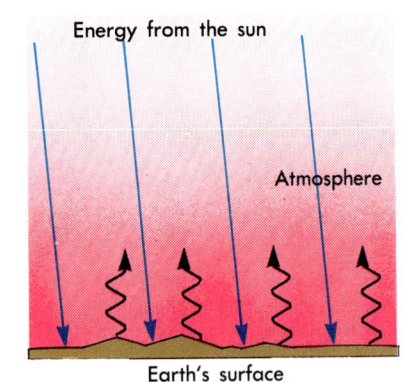

FIGURE 11–7. Energy from the sun is absorbed by the earth's surface. Reradiated energy from the surface is absorbed by the atmosphere, thereby warming the atmosphere.

What are convection currents?

FIGURE 11–8. Convective currents in the atmosphere are caused by density differences of the air.

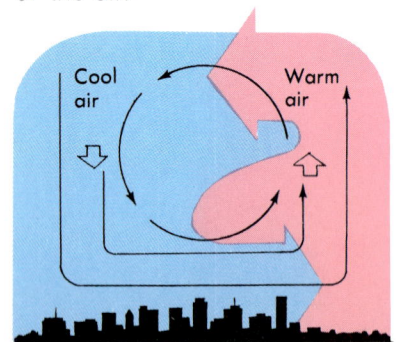

activity

MOVEMENT OF WARM AIR Activity 11-3

See Teacher's Guide.

Place a thermometer in an open cardboard box. Set the box in the sun for ten minutes. Record the temperature. Cover the box with clear plastic and repeat the procedure. Compare the two temperatures you recorded. If you remove the plastic at the end of ten minutes, does air move in or out of the box? Explain your observations.

The temperature increases when covered with clear plastic because light penetrates the plastic, is absorbed, and changed to heat. The plastic prevents the warm air from escaping.

This activity illustrates the "greenhouse effect."

What produces wind?

Define trade winds, westerlies, and polar easterlies.

FIGURE 11–9. The major wind systems of the earth. The curved flow of wind is due to Earth's rotation.

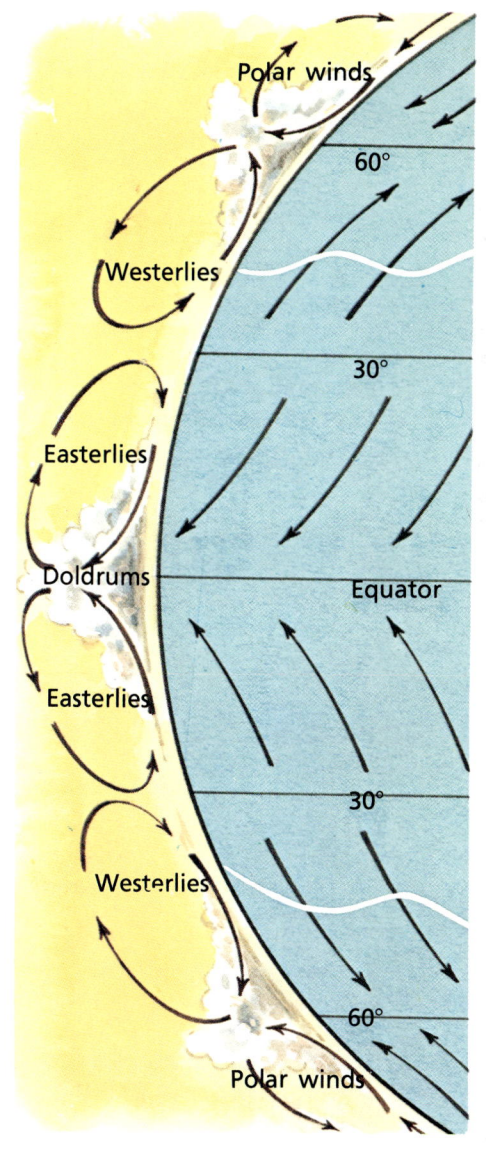

11:5 Winds

Warm air at the equator rises and moves out toward the poles. Cooler surface air flows towards the equator from both north and south to replace the rising air. The air coming from the equator gradually cools and sinks to the earth at the poles. The air then flows back to the equator again. The movement of air is called **wind**.

Rotation of the earth causes the winds blowing between the poles and the equator to break into large circular wind systems. One wind system forms between 30° latitude and the equator. The winds in this system are deflected towards the west because of Earth's rotation. These winds are called the trade winds. Winds flowing towards the poles from about 30° latitude curve to the east. These winds are known as the westerlies because they come from the west. Winds flowing from the poles are deflected to the west. They are called the polar easterlies. If the earth did not rotate, winds would flow north and south. Since the earth is rotating, however, the major wind systems flow from the east or west.

Swift, forceful winds blow from west to east in the upper troposphere. These strong winds are called **jet streams**. Jet stream winds are strong enough to reduce an airplane's speed by 100 km/h or more. Airplanes headed east at high altitudes may be speeded in their flight by jet streams.

Winds within the jet streams vary widely in speed. Some winds travel at about 80 km/h. Others speed

11:5 Winds 227

FIGURE 11-10. The jet stream winds are similar to the flow of water from a hose nozzle.

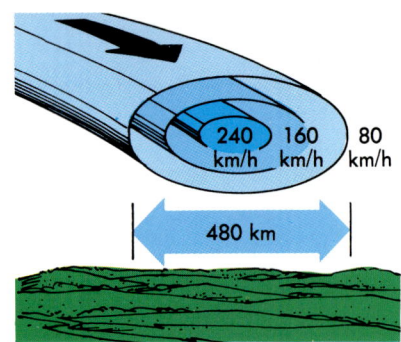

Scheduling of air traffic across the United States must take the effect of the jet streams into account.

How fast do winds travel in the jet stream?

along at as much as 600 km/h. The swiftest parts of the jet streams are the fastest winds on earth.

During the summer months a jet stream is located 10 to 12 km above the earth. In the winter months, it is about 6 to 9 km above the earth. The latitude of the jet stream also changes during the year. In summer, it is as far north as southern Canada. At other times, it is as far south as the Gulf of Mexico. The weather in North America is often associated with movement of the jet streams.

activity   See Teacher's Guide.

PATH OF A WATER DROP ON A SPINNING GLOBE   Activity 11-4

Cover the upper half of a globe with a layer of chalk dust. Spin the globe west to east, the way the earth rotates. While it is spinning, use a medicine dropper to place a drop of water near the North Pole. Wait a few seconds, then stop the globe and observe the path made by the water drop. Does the water follow a straight path **no** or an arc? Which way does it curve? **left or west**

The change in the direction of winds caused by the earth's rotation is called the Coriolis effect. The Coriolis force is dependent upon the rotational speed of the earth, wind speed, and latitude over which the moving air is located.

making sure

4. How might the trade winds affect sailboats headed toward or away from the equator?
5. How does the earth's rotation affect the winds blowing between the equator and the poles?
6. How does the position of jet streams change during the year?

4. increase speed going towards the equator and decrease speed going away from the equator
5. causes them to be deflected to the east or west
6. moves north and south and changes in altitude

11:6 Local Winds

The United States lies in the general flow of the westerlies. At times, different local winds may break this flow by changing the direction and speed of the winds in an area. Local winds exist in many areas and are created by the presence of mountains or bodies of water.

One kind of local wind, called a sea breeze, blows regularly in areas near an ocean or large lake. A sea breeze is caused by the uneven heating of water and land. When the sun shines during the day, the land heats up rapidly. Yet the temperature of the nearby body of water is changed very little. Air above the land is heated and rises. Air over the water flows in to replace the rising air, causing the sea breeze. In the evening if the land cools to a lower temperature than the water, the wind flow reverses. Now a land breeze blows out toward the water.

How are local winds created?

List three examples of local winds.

Remind students that the name of a wind indicates the direction from which it blows.

Sea breezes contribute to the cooling of seacoasts in the summer.

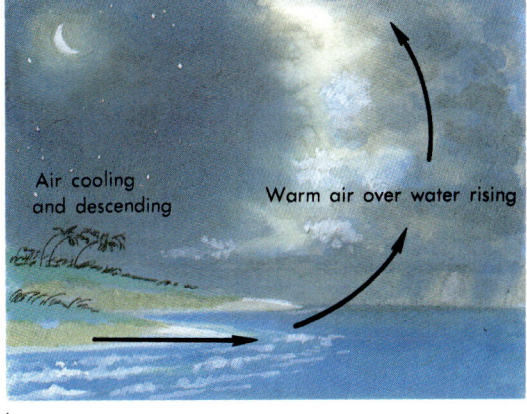

Many types of local winds occur in mountain regions. Valley breezes blow up the mountain slopes during sunny days. Mountain breezes blow down the slopes at night. Local winds near lakes or hilly areas make some places good for growing fruit. The steady breezes keep the air around the fruit trees moving and prevent frosts that might damage the fruit crop.

FIGURE 11–11. Sea breezes (a) and land breezes (b) occur because of different rates at which land and water surfaces absorb the sun's energy.

Warm Santa Ana winds in S. California that blow from high deserts to the coasts are downslope winds.

Just as rising air is cooled, descending air is warmed.

In some regions, strong downslope winds are common at certain times of the year. These downslope winds may have a warming effect. The warming by the wind is caused by the compression of air. The air is compressed because air pressure is greater at lower elevations. Compressed air increases in temperature because air molecules are forced closer together.

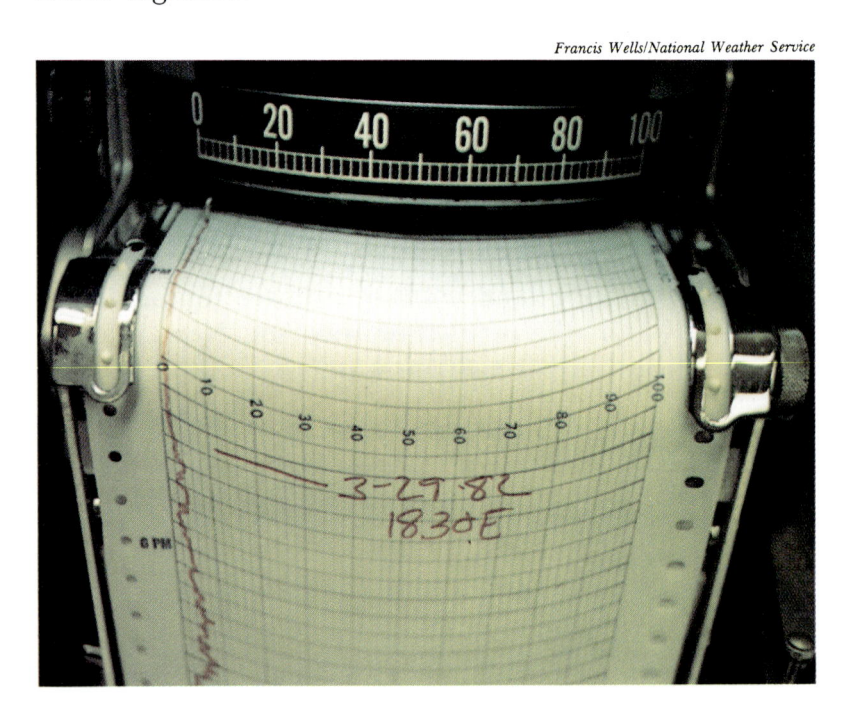

FIGURE 11–12. This equipment is used to record wind speed in knots.

activity   See Teacher's Guide.

WIND DIRECTION   Activity 11-5

Locate a wind vane near your home or school. Wind vanes are often placed on top of poles or buildings. Keep a record of the wind direction at the same time each day for five days. Does the wind blow in the same direction all the time? *no* Which way does it blow most of the time? *Answers will vary.*

In most parts of the U.S. it blows toward the east.

making sure   See Teacher's Guide.

7. What causes a sea breeze?
8. Why might a downslope wind have a warming effect?

11:7 Moisture in Air

Water is in the atmosphere in three states—solid, liquid, and gas. Water vapor is an invisible gas. Rain, clouds, and fog are all examples of water in the liquid state. Sleet, snow, and hail are solid forms of water.

Water enters the air by evaporation from soil, lakes, oceans, streams, rivers, plants, and animals. Plants release water vapor into the air through their leaves. You breathe out water vapor every time you exhale. Evaporation of sweat from your body also adds water vapor to the air.

When water vapor in the air is cooled to the dew point, it condenses into liquid water. **Dew point** is the temperature at which water vapor first begins to condense.

The liquid water takes the form of small droplets. These droplets may form clouds or fog. When conditions are right, the droplets join together and fall from the clouds as rain.

solid—ice crystals in clouds
liquid—water droplets in clouds
gas—water vapor in air

Define the dew point of air.

At night the ground may cool off to the temperature at which water condenses and forms dew. Dew is unlikely to form if the air is dry.

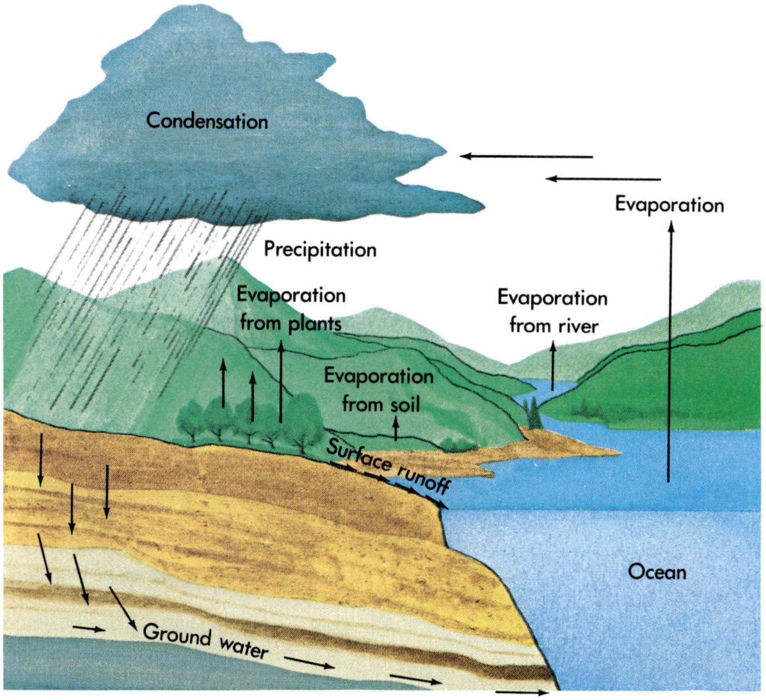

FIGURE 11-13. Evaporation, condensation, and precipitation are three parts of the water cycle.

Humidity is the amount of water vapor in air. Humidity is greatly affected by temperature. At higher temperatures, air can hold more water vapor. Air in tropical regions can hold about ten times as much water vapor as cold arctic air.

Call attention to the difference between humidity and relative humidity. Review grams of mass and the volume of a cubic meter.

Table 11–2.
Water Vapor in Saturated Air

Temperature (°C)	Humidity (g/m³)
−20	0.892
−10	2.154
0	4.835
10	9.330
20	17.118

Define humidity, relative humidity, and hygrometer.

Relative humidity is the percent of water vapor in air based on the amount the air can hold at a given temperature. When air is absolutely dry, the relative humidity is 0 percent. When air is saturated, the relative humidity is 100 percent. Saturated air contains all the moisture it can hold.

High relative humidity reduces the evaporation of perspiration and makes people feel uncomfortable.

Relative humidity is often measured with a **hygrometer** (hi GRAHM ut ur). One type of hygrometer contains both a dry bulb thermometer and a wet bulb thermometer. The dry bulb temperature is the actual air temperature. The wet bulb temperature depends on the relative humidity.

FIGURE 11–14. A sling psychrometer consists of two thermometers on a rotating frame. The difference in temperature between the wet bulb thermometer and the dry bulb thermometer indicates relative humidity.

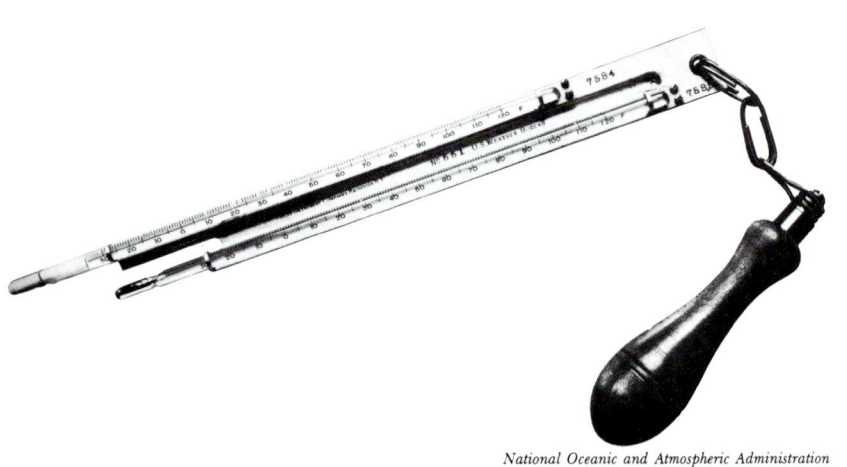

National Oceanic and Atmospheric Administration

These two temperatures are recorded and used to find the relative humidity on a special chart.

How does a wet bulb thermometer operate? It is moistened by a wick immersed in water. As water evaporates from the wet bulb, the bulb is cooled. The rate at which the water evaporates depends on the relative humidity. The lower the relative humidity, the faster the water evaporates, and the lower the wet bulb temperature.

Enrichment: Wrap absorbent cotton around the bulb of a thermometer and moisten it with tap water or rubbing alcohol at room temperature. Observe the decrease in temperature.

activity
See Teacher's Guide.

THE DEW POINT OF AIR INSIDE YOUR CLASSROOM   Activity 11-6

Fill a jar half full of water. Place a thermometer in the water and let the water stand until it reaches room temperature. Add several ice cubes and gently stir the water. Record the temperature at which water first begins to condense on the outside of the jar. This temperature is the dew point of the air in the room. Explain why dew may form on the ground on a cool humid evening.

Ground cools to the dew point of air and water condenses.

activity

USING A SLING PSYCHROMETER   Activity 11-7

Obtain a sling psychrometer. With water, saturate the cloth around the bulb of one thermometer. Swing the psychrometer around for a least 1 min. Record the temperature reading for each thermometer. Why are they different? Use Table C–1 in Appendix C to find the relative humidity. Dry bulb thermometer records the air temperature, and wet bulb thermometer is cooler because the water evaporates.

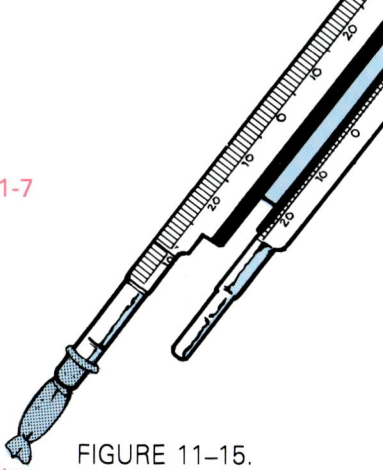

FIGURE 11–15.

making sure

9. What is the relative humidity when the dry bulb and wet bulb thermometer readings are the same? 100%

10. How does the relative humidity affect your comfort on a warm day? Higher relative humidity makes you uncomfortable.

11. Make a list of all the places in your home where water evaporates and enters the air. sink, plants, toilet, fish tank, water heated on stove

11:7 Moisture in Air

How do clouds formed at high altitudes differ from clouds formed at low altitudes?

11:8 Clouds

A cloud is a collection of water droplets or ice particles in the air. It is formed when water vapor is cooled and changed into water droplets or ice crystals. There are three basic cloud forms—cirrus, cumulus, and stratus. Cirrus clouds form at high altitudes and are composed of ice crystals. The ice crystals range in size from 0.2 mm to 0.6 mm. Cirrus clouds are white in color and often have a feathery appearance. They form at altitudes above 6 km and appear during all seasons.

FIGURE 11-16. Cirrus clouds which consist of ice crystals are classified as high clouds (a). Cumulus clouds form as rising columns of air are cooled to the dew point temperature (b).

Cumulus clouds are puffy, flat-based clouds, which often look like huge wads of cotton. Cumulus clouds form at low altitudes and are composed of water droplets. The base of a cumulus cloud is found at altitudes below 2 km. The top of the cloud, however, can extend upward to great heights.

FIGURE 11-17. A stratus cloud in contact with the earth's surface is called fog (a). Stratus clouds are layered and have a uniform gray appearance (b).

234 Atmosphere

Stratus clouds are flat layers of gray clouds composed of water droplets. Stratus clouds cover large portions of the sky. There are no individual cloud units. Stratus clouds are low clouds, found at altitudes below 2 km.

Certain clouds are combinations of the three basic cloud forms. For example, cirrocumulus clouds are white, puffy clouds found at high altitudes. Nimbus, the Latin word for rain, refers to a cloud associated with rain or snow. Cumulonimbus clouds are the well known "thunderheads" that produce thunderstorms.

Cirrus clouds are classified as high clouds. These clouds generally form with their bases over 6 kilometers above the surface. Other high cloud types include cirrostratus and cirrocumulus. Cumulus clouds are clouds of vertical development. Their bases usually form below 2 km but may extend to great heights. Stratus clouds are low clouds with bases below 2 km. Other forms of stratus clouds include stratocumulus and nimbostratus. Clouds with bases forming between 2 and 6 km are classified as middle clouds. Middle clouds include altostratus and altocumulus.

What does the term "nimbus" mean?

FIGURE 11–18. Thunderstorms produced by cumulonimbus clouds may only affect a small area of the earth's surface.

Edward Shay

activity

A CLOUD IN A BOTTLE   Activity 11-8

(1) Obtain one 4-L glass jar and a wooden match. (2) Hold the jar with the neck down. (3) Light a wooden match and hold it for about 7 s close to the opening of the jar. Smoke from the match should go up into the jar. Blow out the match. (4) Now *exhale* hard into the jar several times to fill it with as much of your breath as possible. (5) Quickly remove your mouth from the jar to release the pressure. What do you observe? How does water get into the jar? What causes the change in the air inside the jar?

The glass jar used in this activity must have a narrow neck.

A cloud forms due to condensation which occurs in the bottle.

water in exhaled breath

Water vapor condenses to water droplets

making sure

12. How do cumulus clouds differ from stratus clouds? *See Teacher's Guide.*

11:9 Precipitation

If the water droplets in a cloud become too large, the clouds may release precipitation. **Precipitation** is moisture that falls from the atmosphere. Rain is the most common type of precipitation. Raindrops range in diameter from 0.2 mm to 5 mm.

Snow is another form of precipitation. It consists of crystals of clear ice. Snow appears to be white because light is reflected from the crystals. A snowflake often has the shape of a flat, six-sided polygon or star (Figure 11–19b). Often it has a complex and beautiful design.

List forms of precipitation.

Call attention to the difference in formation of snow which forms by sublimation (water vapor changing to ice) and sleet which is liquid water that freezes.

FIGURE 11–19. Hailstones (a) are produced in cumulonimbus clouds and can be large enough to damage crops, homes, and automobiles. Snowflakes are flat six-sided crystals of ice (b).

Hail consists of hard pellets of ice that fall during a thunderstorm. It forms inside a cloud as water droplets change to ice crystals. The ice crystals grow in size by gathering added layers of ice. When they become too large to be supported by air currents within the cloud, the hail falls to the earth's surface. A lump of hail may be as large as 7.5 cm in diameter. If a hailstone is split open, it reveals a structure of alternate layers of clear and opaque white ice.

Sleet is rain that freezes as it falls through a layer of cold air near the ground. Sleet is commonly mixed with snow or rain.

main ideas — Chapter Review

1. You live in the atmosphere, which surrounds the earth.   11:1
2. Air is a mixture of nitrogen, oxygen, carbon dioxide, argon, water vapor, and small amounts of other gases.   11:1
3. Five layers of air make up the atmosphere.   11:2
4. Newtons per square centimeter, kilopascals, and millimeters of mercury are units of air pressure.   11:3
5. Winds and convection currents distribute heat through the atmosphere.   11:4, 11:5
6. The winds of highest speed are found in the jet streams.   11:5
7. Local winds are caused by the presence of mountains or bodies of water.   11:6
8. Water is in the air in three forms—solid, liquid, and gas.   11:7
9. Clouds are formed from water droplets or ice crystals.   11:8
10. Clouds may release precipitation such as rain, snow, or hail, which falls to the earth's surface.   11:9

vocabulary

Define each of the following words or terms.

atmosphere	exosphere	relative humidity
barometer	jet stream	stratosphere
convection current	mesosphere	thermosphere
dew point	precipitation	troposphere
		wind

study questions

DO NOT WRITE IN THIS BOOK.

A. True or False

Determine whether each of the following sentences is true or false. If the sentence is false, rewrite it to make it true.

F   1. Air is a ~~compound~~. mixture

F   2. There is ~~a~~ definite boundary at which the atmosphere ends. no

T   3. The density of air decreases with increased altitude.

F   4. The number of air particles per liter is ~~greater~~ at 3000 m altitude than at sea level. less

Atmosphere   237

T 5. You live in the troposphere.
F 6. Temperatures within the mesosphere ~~are fairly constant~~. decrease with height
F 7. Rain and snow occur often in the ~~stratosphere~~. troposphere
T 8. Sea level pressure is about 10 N/cm² or 100 kPa.
T 9. A mercury barometer measures air pressure in cm of mercury.
F 10. ~~All~~ clouds are made of water droplets. Some

B. Multiple Choice

Choose the word or phrase that completes correctly each of the following sentences.

1. About 78% of the air is (<u>nitrogen</u>, carbon dioxide, water vapor, argon).
2. Changing air (<u>pressure</u>, volume, temperature) causes a change in the reading of an aneroid barometer.
3. The density of warm air is (greater than, <u>less than</u>, the same as) the density of cold air.
4. As altitude increases, the reading on a barometer will (increase, <u>decrease</u>, remain the same).
5. A mercury barometer is used to measure (<u>air pressure</u>, temperature, humidity).
6. (Cirrus, <u>Cumulus</u>, Stratus) clouds look like large puffy pieces of cotton.
7. The (stratosphere, <u>ionosphere</u>, troposphere) is a region of the atmosphere that contains many electrically-charged particles.
8. (Rain, <u>Hail</u>, Sleet, Snow) is a type of precipitation that occurs only during thunderstorms.
9. When air is compressed its temperature (<u>increases</u>, decreases, remains the same).
10. (Nitrogen, <u>Oxygen</u>, Carbon dioxide, Argon) is removed from air when iron rusts.

C. Completion

Complete each of the following sentences with a word or phrase that will make the sentence correct.

1. Air travels from <u>high</u> pressure regions to <u>low</u> pressure regions.
2. A wind which blows from west to east is called a(n) <u>westerly</u> (west) wind.

3. Heat is distributed through the atmosphere by _____ and _____. convection current, winds
4. The rotation of the earth causes winds to be deflected from a straight course.
5. As moist air cools to the dew point, water is formed by a physical change called _____. condensation
6. Rain that freezes as it falls through a layer of cold air is called sleet.
7. Relative humidity is measured with a(n) _____. hygrometer
8. The lower the relative humidity, the faster the water evaporates on a wet bulb thermometer.
9. During the night a(n) cloud cover will reduce loss of heat from the earth.
10. At the equator warm air is rising.

D. How and Why See Teacher's Guide.
1. Why is air pressure lower at high altitudes than at sea level?
2. Explain how winds are produced in the atmosphere.
3. Why is the relative humidity 100% when water vapor in air condenses to form dew?
4. Why does warm air rise?
5. Rain drops that freeze on the surface are called freezing rain. Under what conditions does freezing rain form?

challenges

1. Locate the Beaufort wind scale in Appendix C. Use the scale to judge the velocity of winds you observe in your neighborhood. Keep a record of your observations.
2. Obtain a cloud atlas from a library and learn to identify clouds in your area.

interesting reading

Comptom, Grant, *What Does a Meteorologist Do?* New York, NY: Dodd, Mead, 1981.
Witty, Margot and Ken, *A Day in the Life of a Meteorologist.* Mahwah, NJ: Troll, 1981.

Weather affects many aspects of your daily life. When you awake in the morning, weather is one of the first things of which you become aware. The clothes you wear and the activities you take part in depend on the weather. What causes weather? What is climate? How are weather and climate related?

Dan McCoy/Rainbow

Weather and Climate

chapter 12

Introducing the chapter: Lead a discussion on how the weather affects our everyday lives. Ask students to list some of the benefits of more sophisticated weather forecasting.

12:1 Observing the Weather

What is the weather outside today? It may be clear or stormy, cold or hot. **Weather** is the condition of the atmosphere at a certain time and a certain place. Temperature, air pressure, relative humidity, wind speed, and precipitation are some terms used to describe the weather. Another term used to describe weather is cloud cover.

Cloudiness, or cloud cover, is measured by the amount or portion of sky covered by clouds. The scale of cloud cover is divided into tenths. For example five-tenths means that half the sky is covered by clouds. The terms clear, scattered, broken, and overcast are used to describe cloud cover.

The amount of cloud cover affects how far one can see in the sky. For instance, clouds may make it very difficult for an airplane pilot to see other airplanes. If there are too many clouds, air travel may be unsafe for planes without special equipment. Radar is used by pilots to find their way through clouds.

The **cloud ceiling** is the altitude at which the cloud cover becomes broken or overcast. A broken sky means that more than one half but not all of the sky is covered by clouds. An overcast sky is completely covered by clouds. The clouds are scattered if they cover one half or less of the sky.

GOAL: You will learn the factors that affect weather and how weather forecasts are made.

What are some terms used to describe weather?

Enrichment: Ask students to describe the clouds they are likely to see when the weather is fair and when it is raining or snowing.

241

Begin by having students make a table in their notebooks for recording their observations.

activity

Activity 12-1   OBSERVING THE WEATHER

Objective: To observe and record the weather for one week

An alternative is to make a large data table on the chalkboard or a bulletin board for recording data available to the whole class.

Materials   See Teacher's Guide.

anemometer
barometer
empty can
hygrometer or psychrometer
ruler
thermometer
wind vane

Procedure

1. Make each of the following observations and measurements each day for a week.
2. Record the outdoor temperature in your data table at least twice a day.
3. Use the barometer to observe the air pressure at least twice a day. Record the air pressure.
4. Observe and record the relative humidity at least twice a day. If you are using a psychrometer, the relative humidity may be found in Table C–1 in Appendix C.
5. Determine the percentage of the sky that is covered. Record it in your data table. Note: If the sky is broken or overcast, there is a ceiling.
6. Determine the wind speed and direction at least twice a day. If you do not have an anemometer, a Beaufort Scale, found in Appendix C, is helpful. Record the wind data.
7. Keep a record of the type and amount of precipitation. You can determine the amount of precipitation by collecting it in a container such as an empty can. Measure the height of water in the container with a ruler. Record the amount once each day.

Observations and Data

Date: 21 September 1983

Time	10:00 A.M.	
Temperature	18°C	
Pressure	1012.3 mb	or 76 cm
Relative Humidity	58%	
Clouds	Broken	
Wind Speed	18 km/h	
Wind Direction	NW	
Precipitation	Rain	

See Teacher's Guide

Questions and Conclusions

1. At what times did you record your highest and lowest temperatures? Why?
2. Did you notice a relationship between air pressure and cloud cover? Explain.
3. What was the total amount of precipitation you recorded for the week? How did your amount compare with amounts reported on television or in the local newspaper? Why might there be differences?
4. Did you notice any relationship between the type of precipitation and the type of clouds? Explain.

Explain the use of a Beaufort Scale before wind speeds are observed and recorded.

National Space Technology Laboratories, NOAA

Automated floating stations transmit surface weather conditions by radio from remote areas of the ocean.

FIGURE 12–1. Floating observing stations are used to obtain weather data over ocean surfaces.

making sure

1. A local weather station reported the following observations: air pressure—79 cm Hg; wind—from west, 16 km/h; temperature—32°C; sky—clear; relative humidity—14%; precipitation—none.
   (a) Describe the weather at the station on the day these observations were made. *hot, dry, clear, windy*
   (b) During what season of the year is weather of this kind most likely to occur? *summer*

12:2 Air Masses

An **air mass** is a huge body of air covering a land or ocean area. It may extend from the earth's surface to about 6.5 km above the ground. Temperature and humidity are fairly uniform at any given height in an air mass. But these conditions may be very different in nearby air masses.

An air mass forms as air gathers in one place and develops certain properties. These properties are determined by the surface beneath the air mass and the latitude. An air mass formed over an ocean has a high moisture content. An air mass formed within the Arctic Circle is extremely cold. Temperature and humidity are two major properties of an air mass.

The stability of an air mass is important in determining the weather of an area. Warm, moist air masses are said to be unstable. The warm air at the surface, because it is lighter, will tend to rise, producing clouds and precipitation.

Name two ways in which air masses differ.

12:2 Air Masses 243

Describe the worldwide system of air mass classification.

A worldwide system to classify air masses has been devised. Within this system, air masses are classified according to where they are formed. The small letter c (continental) is used for air masses that are formed over land. The small letter m (maritime) is used for those formed over oceans. Capital letter P (polar) means an air mass is cold because it forms at higher latitudes. Capital letter T (tropical) means an air mass is warm because it forms at lower latitudes. For example, the symbol mT stands for a maritime tropical air mass. This air mass is warm and moist.

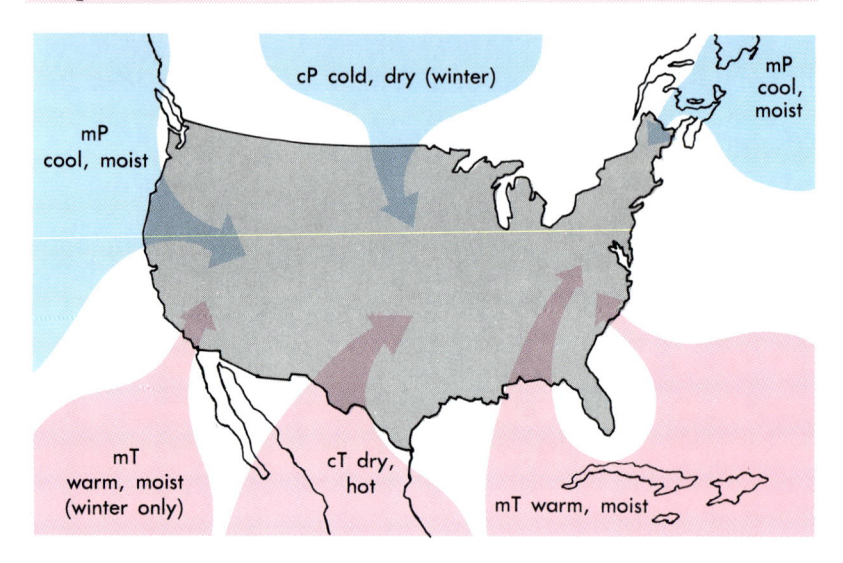

Within the temperate zone, where warm, moist air masses come in contact with cold, dry air masses, middle latitude lows or wave cyclones will form. These cyclones develop when different air masses flow side by side at different speeds and directions. The wave cyclone will form along the boundary (front) and is associated with clouds, precipitation, and the stormy weather that is characteristic of the temperate zone. These wave cyclones or lows are migrating systems and will move rapidly with the air masses.

FIGURE 12-2. Source regions of air masses that affect the United States.

Air masses tend to move. When air masses move, they pass over land or ocean surfaces. These surfaces may be warm or cold and may change the air masses. A cool air mass passing over a warm land surface becomes warmer. A warm air mass passing over a cold land surface becomes colder. An air mass passing over an ocean or large lake gains moisture.

Why does the temperature of an air mass change?

making sure

2. Over what type of surface might a dry warm air mass be formed? desert
3. What symbol would stand for each of the following air masses: continental polar, maritime polar, continental tropical, maritime tropical? Where might each of these air masses form?

3. cP—polar land,
mP—polar ocean,
cT—land near equator,
mT—ocean near equator

244 **Weather and Climate**

12:3 Fronts

Air masses do not mix. Like oil and water, they tend to remain separate. The region where two different air masses come together is called a front. A front is the boundary between two air masses. Clouds and precipitation are often associated with fronts.

Fronts move as the air masses move. One air mass usually moves into or invades a region occupied by another as the other air mass moves away. In this way, the front travels across the earth's surface.

When a warm air mass moves against a cold air mass, a **warm front** occurs. Warm air slides forward and above the cold air in a warm front. As the air rises, it cools. Cirrus clouds form about 1000 km ahead of the front. The clouds become lower and thicker as the front continues to move. A large area of snow or rain may fall from nimbostratus clouds and last for several hours. Temperatures often rise when a warm front passes.

Fronts tend to move from west to east across the U.S.

Mapping fronts and following their movements is a prime aspect of weather forecasting.

FIGURE 12-3. Clouds and precipitation associated with a warm front.

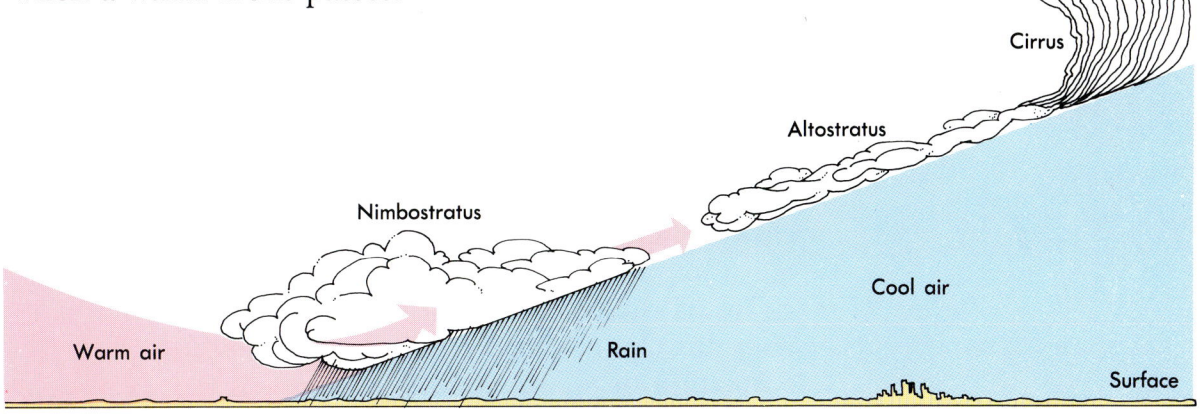

A **cold front** occurs when a cold air mass moves against a warm air mass. Cold fronts move almost twice as fast as warm fronts. As the cold air drives forward, the warm air is forced upward rapidly. Clouds that form along cold fronts are usually cumulus or cumulonimbus. Snowshowers or rainstorms may be severe, but they end quickly. When a cold front passes, temperatures drop and winds shift.

How is a cold front different from a warm front?

A squall line is an area of intense instability ahead of a fast moving cold front.

12:3 Fronts 245

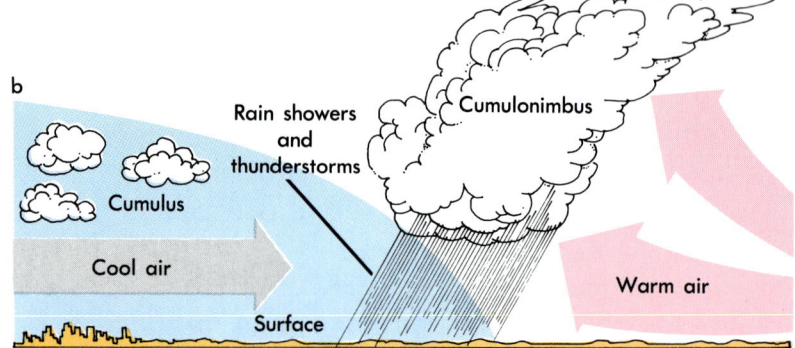

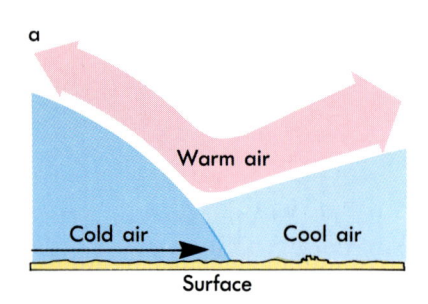

FIGURE 12-4. An occluded front has characteristics of both warm and cold fronts (a). Stormy weather may often accompany cold frontal passage (b).

How is an occluded front produced?

A cold front moves faster than a warm front. Therefore, a cold front may approach and overtake a warm front. When the two fronts meet, an **occluded front** forms. Weather in an occluded front has some of the characteristics of both warm and cold fronts. Periods of light rain or snow may be followed by heavy thunderstorms and snowshowers.

12:4 Thunderstorms

Thunderstorms may occur when warm, moist air rises rapidly through the atmosphere. For example, moist, warm air forced upward along a front or a mountain range may cause a thunderstorm. Thunderstorms also occur on hot summer afternoons. Why? Moist air is heated at the earth's surface. The warm air rises and cools. As it gains altitude, moisture in the air condenses to form cumulus clouds. Swiftly rising air currents within a cumulus cloud, called updrafts, can cause the formation of a cumulonimbus cloud. A cumulonimbus cloud is one that develops vertically into a tall, mountainlike white mass with a thick, dark base. Some cumulonimbus clouds may build as high as 18 km. The updrafts in these clouds may reach speeds of 100 km/h. Water droplets within the cloud join to become larger drops which fall as rain. Pieces of ice can also form in the cloud and fall as hail. When rain falls out of the cloud, it causes downdrafts that eventually stop the upward flow of air. As a result, the thunderstorm dies.

How do thunderstorms occur?

Most severe thunderstorms are associated with the interaction of low level and high level jet streams along the polar front.

246   Weather and Climate

FIGURE 12–5. Lightning and strong winds are usually associated with thunderstorms (a). Lightning is caused by a buildup of electrical charges within a cumulonimbus cloud (b).

Electric charges build within a cumulonimbus cloud. Some parts of the cloud have a positive charge. Other parts have a negative charge. When the charges become too great, lightning flashes. The lightning is caused by the discharge of electricity. Most lightning occurs within a cloud or from cloud to cloud. Occasionally lightning strikes the ground. A bolt of lightning heats the air to high temperatures. The heated air expands rapidly. The noise from this sudden expansion is called thunder.

What causes lightning?

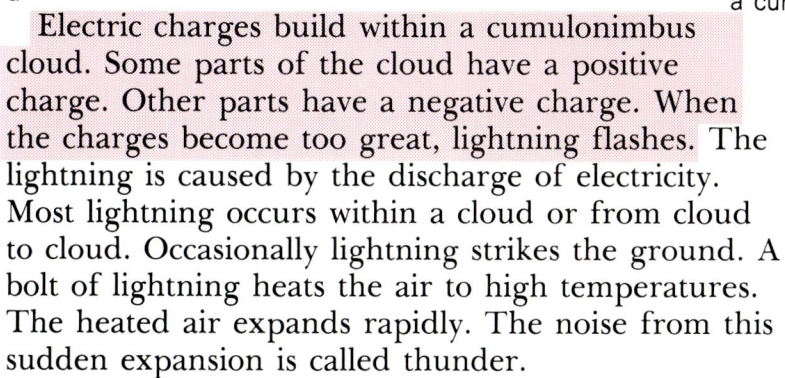

making sure   See Teacher's Guide.
4. How does a cumulonimbus cloud develop?
5. What causes a thunderstorm to "die?"

12:5 Highs and Lows

A **cyclone** is a region of low air pressure. The low pressure is caused by rising air at the center of the cyclone. In a cyclone, winds blow from the higher pressure region outside the cyclone toward the low pressure center. As they blow toward the low pressure region, the winds rotate. In the Northern Hemisphere, the winds rotate in a counterclockwise direction.

Define cyclone.

12:5 Highs and Lows 247

What kind of weather exists in cyclones?

Cyclones travel. In the Northern Hemisphere they usually move from west to east at a rate of 30 to 60 km/h. Some cyclones may travel half the distance around the earth before they break up.

Cyclones may occur every few days in a certain area. However, most of them are mild and go unnoticed. Bad weather conditions caused by condensation within rising air exist in some cyclones. The weather may range from cloudy skies and drizzle to intense storms with heavy rain or snow.

A high, or **anticyclone,** is a huge area of air in which the pressure is higher than the air around it. Winds blow out from the center of the high. In the Northern Hemisphere, the winds blow in a clockwise direction. Fair weather is generally associated with a high. As with cyclones, highs move from west to east in the Northern Hemisphere.

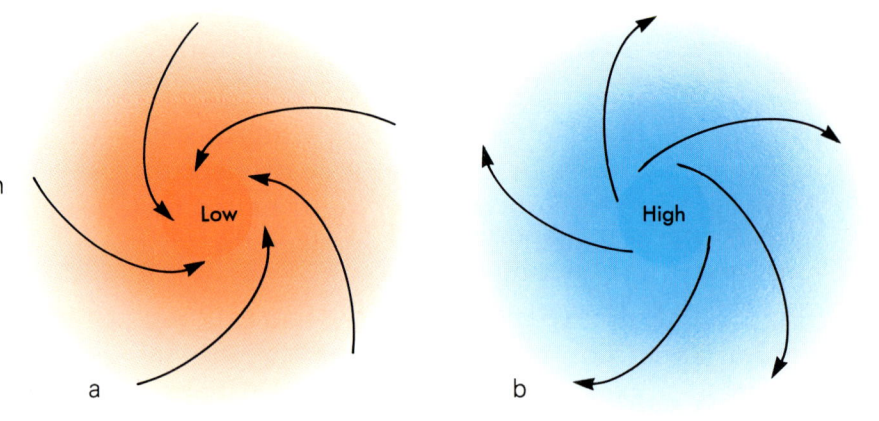

FIGURE 12–6. In the Northern Hemisphere, winds spiral counterclockwise into the center of a cyclone (a). Winds spiral in a clockwise direction as they flow from the high-pressure center of an anticyclone (b).

A **hurricane** is a stormy tropical cyclone. Hurricanes form in the trade winds, over regions of warm ocean water. In such regions a large low pressure area is created when warm, moist air rises. In the Northern Hemisphere, winds blow counterclockwise into the low pressure area. Wind speeds may be as high as 250 km/h. Huge amounts of water are released in heavy rainfall. The low-pressure center of the hurricane, called the eye, is the quiet part of the storm. Wind velocity in the eye is lower and there is much less rainfall. The sun may be shining and there may be little cloud cover in the eye.

Hurricanes are names given to tropical cyclones near North America. In Eastern Asia they are called typhoons, cyclones in India, willy-willies in Australia, and baguios in China.

FIGURE 12–7. A satellite photograph of hurricane Frederick in the Gulf of Mexico. Notice the absence of clouds within the eye.

A hurricane may travel hundreds of kilometers or it may remain in one area for several days. Many hurricanes form in the Gulf of Mexico and the Caribbean during the late summer and early fall. Sometimes these hurricanes travel up the east coast of the United States. A hurricane in the Pacific Ocean is called a typhoon.

A **tornado** is a special kind of cyclone. It is a whirling wind circling around a low pressure center about 60 m in diameter. A tornado is a funnel-shaped cloud and is associated with very strong thunderstorms. It extends from the base of a cumulonimbus cloud to the ground.

How do tornadoes form? This question has not yet been answered completely. One theory states that tornadoes develop from large thunderstorms that form along fast moving cold fronts. When warm and cold air meet along the front, the warm air is forced to rise rapidly in a spiral pattern. A sudden drop in pressure occurs in the center of the spiral. Violent winds begin to blow around and into the spiral producing the funnel shape of the tornado. Most tornadoes in the United States move at an average speed of 55 km/h and last 1 or 2 min. A tornado usually follows a path that is about 0.2 km wide and 1.5 km long.

Point out that a hurricane and tornado are both low pressure systems. Review the differences between a hurricane and a tornado.

Enrichment: Ask students who have experienced a hurricane or tornado to describe their observations.

12:5 *Highs and Lows*

Wind speeds within a tornado may reach 450 km/h. Extensive damage to houses and buildings can result when a tornado touches the ground. Strong winds can rip roofs from houses and hurl automobiles many meters into the air. It is wise to seek shelter from flying objects such as broken glass and wooden boards when a tornado is approaching. A cellar is a good place to be. If there is no cellar, you may gain protection in a closet or under a table or desk.

FIGURE 12–8. Tornadoes over bodies of water are called waterspouts (a). Tornadoes are associated with intense thunderstorms (b).

Although tornadoes appear in many parts of the world, the greatest number occur in the United States. Tornadoes are often found in the Great Plains region and the Mississippi Valley from early spring until late summer. About 700 tornadoes are reported each year in the United States.

making sure

6. Why is a gigantic amount of energy needed to create a hurricane? What is the source of this energy? *A tremendous amount of energy is used in evaporating the water and producing the high winds that make up a hurricane. The sun is the source of this energy.*

7. For a tornado and a hurricane, compare size, movement, total energy, and force. *A tornado is smaller, moves faster, has less total energy, and has greater wind force.*

12:6 Weather Maps

The National Weather Service has a network of weather stations throughout the United States. At certain times every day, each station reports the local weather conditions to the National Weather Service. This information is then plotted on national weather maps. Weather maps are important in **meteorology** (meet ee uh RAHL uh jee), the scientific study of weather.

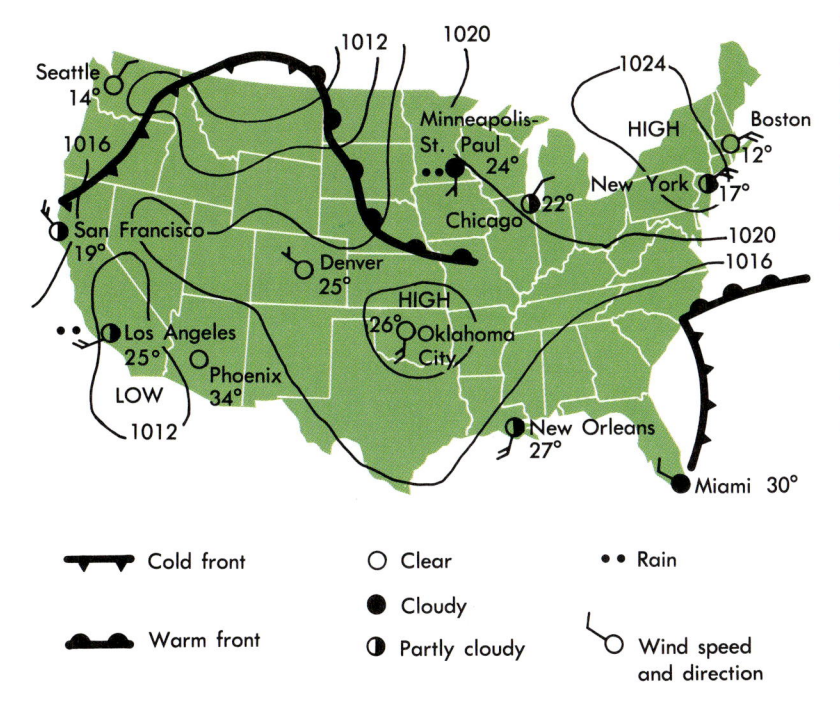

The spacing between isobars indicates the amount of pressure change occuring over a given distance—the pressure gradient. Closely spaced isobars indicate a steep pressure gradient and strong winds.

Feathers on the wind direction symbol point in the direction from which the wind is blowing. Number of feathers increases with increased force. Stress and practice the reading of map symbols.

FIGURE 12-9. Temperature, wind direction, pressure, and cloud cover are shown on a weather map.

A weather map is used to organize and display weather information. These maps show temperature, dew point, and amount of precipitation. They also show the location of major fronts and regions of high and low air pressure. The locations and movements of highs and lows are used by forecasters to predict weather. Highs and lows are indicated by isobars. An **isobar** is a line on a weather map that connects points having the same air pressure. The pressure is recorded in millibars. One thousand millibars is about average sea level pressure.

Why is a weather map important in meteorology?

Isotherms show the distribution of temperature and the direction in which either cold or warm air is moving.

12:6 Weather Maps 251

| activity | See Teacher's Guide. | Activity 12-2 |

LOCATING FRONTS ON A WEATHER MAP

Cut out the weather map in a newspaper and tape it to a page in your notebook. Locate the fronts shown on the map. What weather information is recorded for the largest city in your state? What is the forecast for your area?

Answers will depend on the data of the maps.

making sure   See Teacher's Guide.

8. Locate a front in Figure 12-9. Find two weather stations on opposite sides of the front. Compare the temperature at these stations. Are they alike or different?
9. Find a low pressure area and a high pressure area on the map in Figure 12-9. What is the weather in each of these areas?
10. Locate an isobar on the weather map and find the pressure in millibars.
11. On the weather map, what kind of front is near Miami, Florida?

12:7   Weather Forecasting

Enrichment: Ask students to list all the things they do that are affected by the weather.

A weather forecast is a prediction of the weather for some time in the future. As you know, weather forecasts are sometimes wrong. Forecasts are based on chance or probability rather than absolute certainty. For this reason, the forecasts specify percentages, such as "there is an 80 percent chance of rain tomorrow."

Weather forecasts are particularly important in aviation, ocean travel, and agriculture.

How is it possible to predict the weather? It is well known that weather travels from place to place. Generally, storms, fronts, and pressure systems move from west to east across the United States. Using weather maps and photographs made with weather satellites, forecasters keep track of these movements. For example, a forecaster observes an approaching snowstorm moving to the east at a rate of 10 km/h. The forecaster can use this speed to predict the arrival of the storm.

How is the movement of a storm forecasted?

Observing changes in temperature, humidity, and air pressure is important in forecasting. Hot summer days with high humidity may indicate thunderstorms in the afternoon. A steady or sudden drop in the air pressure may indicate that a cyclone with bad weather is approaching. Also, the kinds of clouds give clues to the coming weather. Distant clouds may show an approaching front. Clouds moving out of an area at the same time the barometer is rising indicates fair weather.

FIGURE 12-10. Forecasters at many National Weather Service offices use computers to help analyze weather maps and other weather information (a). Satellites provide valuable information about locations and movements of weather systems (b).

activity

See Teacher's Guide.

WEATHER FORECASTING Activity 12-3

Tune in an evening television weather forecast. Write down the forecast accurately. In addition, write down the facts or reasons the weather forecaster gives for the prediction. Observe the weather the next day. Was the weather forecast correct? *may be yes, no, or partially correct* If it was incorrect, what factors might account for the errors? *change in speed at which a front is moving, front may be stalled, clouds lack sufficient moisture for rainfall, wind direction changes*

making sure

12. What methods do forecasters use to predict the weather? *See Teacher's Guide.*

12:8 Climate

How is climate related to weather?

Enrichment: Ask students to describe the local climate. Then ask students who have lived in other areas to describe the climate where they lived previously. Ask students to describe the "ideal" climate they would prefer.

What are the three types of climate and where are they found?

Climate for any region depends upon its geographic location, the amount of radiant energy received from the sun, and the circulation pattern of the atmosphere. Local influences on climate include topography, plant life, and the presence of artificial structures.

Weather changes day by day. **Climate** is the average weather for a region over a period of years. Average temperature per year and average precipitation per year are two important aspects of climate.

The climate of a region is determined mainly by latitude. It is also affected by height above sea level, local winds, and the location of mountains, oceans, and large lakes.

Three general types of climate are temperate, tropical, and polar. Temperate climates are found between 66½° and 23½° latitude in both the Northern and Southern Hemispheres. Tropical climates are found from 23½° north latitude to 23½° south latitude. Polar climates are in the frigid zones above 66½° latitude around the poles. Within these major zones, climates may be modified by mountains or bodies of water. For example, marine climates occur along the seacoasts where moisture is high. The interior of continents have continental climates where there are hot summers and cold winters. The middle west of the United States has a continental climate.

Temperate climates are generally warmer than polar regions and cooler than tropical regions. The length of the growing season makes temperate

FIGURE 12–11. Cold temperatures in some polar climates allow snow and ice to accumulate many meters in depth.

James Westwater

254 **Weather and Climate**

regions ideal for agriculture where there is fertile soil and adequate rainfall. Rainy tropical areas are hot throughout the year with heavy rainfall. Some tropical areas have a dry season and a rainy season, when most of the rain falls. Deserts are present where little rain falls in both temperate and tropical regions.

Climates can cover vast ranges of land and water or just a small area. The climate in a valley surrounded by mountains and the climate in a city are examples of microclimates. A **microclimate** is a climate in a small area. For instance, it may be warmer and dryer in a valley than in surrounding mountains. Cities are often a few degrees warmer than nearby country areas. The blacktop streets and rooftops in a city absorb more of the sun's radiation than open fields absorb. Also, there is a warming effect caused by heated buildings. Microclimates are created inside houses by heating in winter and air conditioning in summer.

FIGURE 12–12. Rainfall in some moist tropical climates may exceed 1000 cm per year (a). Most temperate climates are characterized by a change in seasons (b).

Some plants such as certain grape varieties grow best in a microclimate that is particularly suited to the plant.

Most of the world's major deserts are located in the subtropical highs. The subsiding air currents within the high pressure area will absorb any available moisture.

making sure

13. Jet airliners fly at altitudes where the air is thin and very cold. What kind of microclimate is created inside the airliner? mild—temperature about 21°C and humidity about 40%
14. What kind of microclimate is created inside a terrarium? warm and humid

12:8 *Climate* 255

PERSPECTIVES
frontiers

Weather Watching

Have you ever wondered how your local weather forecaster can predict weather events? Obviously, weather forecasters use large amounts of data to determine tonight's or tomorrow's weather. However, despite all the available information the forecasts may be wrong. Studies are being done to help forecasters make better predictions. Scientists from the National Oceanographic and Atmospheric Administration (NOAA) are attempting to provide additional weather data through the use of a new satellite.

Since 1960, atmospheric scientists, or meteorologists, have been using satellites. Cloud pictures are relayed by the satellites to Earth, where the development and movement of storms and weather systems can be monitored. Using the pictures, forecasters have been able to observe major weather features, even in the most remote ocean areas.

Pictures of clouds, however, tell only part of the weather story. In other words, where there are no clouds, meteorologists can not see what is happening in the atmosphere. Therefore, NOAA scientists are developing another method of observing the atmosphere. A new satellite would enable meteorologists to "see" the wind, which produces the earth's weather.

The satellite will operate in low Earth orbit and use a laser system. The system will scan the atmosphere by shooting a laser beam twice each second. The beam would bounce off any dust particles carried by air currents and reflect back to the satellite. Wind direction could then be determined. Wind speed would be measured the same way as police radar detects speeders. The new laser satellite is appropriately named Windsat.

Most information about the wind is currently obtained by land stations. Twice each day, balloons are launched from 500 locations around the world. As the balloons drift into the stratosphere, wind direction and speed are relayed to Earth. It is hoped that Windsat, once in orbit, will provide additional wind data over remote locations. Meteorologists will then combine the data from the balloons and Windsat to construct maps of global wind patterns. The more complete maps will help forecasters make better long-range predictions.

The use of Windsat also has other advantages. For example, conditions for the formation of tropical cyclones could be detected before the storm actually develops. Today's satellites can only detect the storms after they have formed. Another benefit would be to airline pilots. By knowing the wind speed and direction provided by Windsat, pilots could avoid strong head winds and use tail winds. Windsat would help airlines save both time and money.

NOAA

main ideas

Chapter Review

1. Air pressure, temperature, relative humidity, wind direction and speed, cloud cover, and precipitation are major elements of weather.   12:1
2. The properties of an air mass are determined by the surface beneath the air mass.   12:2
3. Clouds and precipitation are usually associated with fronts.   12:3
4. Thunderstorms may occur when warm, moist air rises rapidly through the atmosphere.   12:4
5. Air moves from a region of high pressure to a region of low pressure.   12:5
6. Hurricanes and tornadoes are examples of intense cyclones.   12:5
7. Weather maps showing the location of fronts, highs and lows, temperatures, and relative humidity are used in predicting weather.   12:6, 12:7
8. Latitude, mountains, oceans, large lakes, and winds are factors that affect climate.   12:8

vocabulary

Define each of the following words or terms.

air mass	cyclone	occluded front
anticyclone	hurricane	tornado
climate	isobar	warm front
cloud ceiling	meteorology	weather
cold front	microclimate	

study questions

DO NOT WRITE IN THIS BOOK.

A. True or False

Determine whether each of the following sentences is true or false. If the sentence is false, rewrite it to make it true.

1. Weather is the condition of the atmosphere at a particular time.
2. Weather maps are used to forecast weather.
3. Fair weather is usually found in ~~low~~ pressure areas.   high

Weather and Climate   257

T   **4.** A rising barometer reading often means fair weather.

F   **5.** There is ~~little~~ difference in the weather on opposite sides of a front. *much*

F   **6.** Air masses travel from ~~east to west~~ across the United States. *west to east*

T   **7.** An air mass over an ocean gains moisture.

T   **8.** Precipitation often occurs along fronts.

F   **9.** Clouds are ~~never~~ found along fronts. *usually*

T   **10.** An occluded front occurs when warm and cold fronts meet.

B. Multiple Choice

Choose the word or phrase that completes correctly each of the following sentences.

1. A dry, warm air mass is most likely to form over *(the Arctic, a desert, an ocean, a seacoast)*.
2. A maritime polar air mass is represented by the symbols *(cT, mT, cP, mP)*.
3. Warm fronts travel *(more rapidly than, less rapidly than, at the same speed as)* cold fronts.
4. A sky completely covered by clouds is *(scattered, clear, overcast)*.
5. *(Temperate, Tropical, Polar)* climates are usually found between 23½° north latitude and 23½° south latitude.
6. A(n) *(cold, warm, occluded)* front forms when warm air moves against cold air.
7. Hurricanes are areas of *(low, high, equal)* pressure.
8. *(Hail, Rain, Lightning)* results when electric charges build inside cumulonimbus clouds.
9. A thunderstorm is associated with *(cirrus, cumulonimbus, stratus)* clouds.
10. Winds blow clockwise around a *(low, high, cold front)* in the Northern Hemisphere.

C. Completion

Complete each of the following sentences with a word or phrase that will make the sentence correct.

1. In the Northern Hemisphere, cyclones normally move in a(n) _____ direction. *easterly*
2. A(n) _____ front occurs where a warm air mass moves forward to take the place of a cold air mass. *warm*

3. _____ is the noise that is produced when air heated by lightning explodes. *thunder*
4. When clouds cover 8/10 of the sky, the sky is *broken*.
5. A tornado extends from the base of a(n) _____ cloud to the ground. *cumulonimbus*
6. Changes in temperature, humidity, and air pressure are important when observing the *weather*.
7. *climate* is the average weather for a region over many years.
8. A(n) _____ is a miniature climate. *microclimate*
9. Tropical climates occur between 23½° south and *23½° north* latitude.
10. The altitude at which cloud cover becomes broken or overcast is the *ceiling*.

D. How and Why *See Teacher's Guide at the front of this book.*
1. How is weather different from climate?
2. What is a weather map? How is it used?
3. What weather information is collected at a weather station?
4. Using the information in this chapter, explain how you would go about forecasting tomorrow's weather.
5. Why are tornadoes associated with cold fronts?

challenges

1. You can become a weather observer. Instructions are supplied in *Weather Forecasting* and *Instructions for Climatological Observers,* obtained at nominal cost from the Superintendent of Documents, Washington, D.C.
2. Build a weather station on your school grounds or in another suitable place.

interesting reading

Bosen, Victor, *Doing Something About the Weather.* New York, NY: Putnam's, 1975.
West, Susan, "It's Lonely at the Bottom." *Science '84.* Jan/Feb. 1984, pp 54-61.
Moyer, Robin, "The Great Wind." *Science 80.* November 1980, pp. 60–65.

Many ocean studies take place below the surface. A deep-sea diver must wear a heavy diving suit in order to withstand the pressures at great depths in the ocean. While a diver is beneath the ocean's surface, there are opportunities for observation and study. What forms of life are present in the ocean? What causes currents? What resources are found within the ocean?

Oceanography

chapter 13

Introducing the chapter: Display a globe or large map of the world to the class. Identify each ocean, and point out the vast area of Earth's surface covered by ocean waters. If it is feasible, arrange a field trip to an aquarium, the seashore, or a day trip on an oceanographic vessel as part of the study of this chapter. Also, show films pertaining to the ocean such as those listed for Unit 3, page 37T. Discuss similarities between the exploration of the oceans and the exploration of space.

13:1 Seawater

Almost three fourths of the earth's surface is covered by seawater. The oceans are a source of both food and recreation. In the future, they may also be an important source of energy, minerals, and fresh water. Scientists have much to learn about oceans. Many of the facts in this scientific field are yet to be discovered.

Oceanography is the scientific study of the oceans. Oceanographers study the composition of seawater and trace the path of ocean currents. Also, they study marine plant and animal life. Almost everything connected with the oceans, including their history, is of interest to oceanographers.

One important property of seawater is its composition. Seawater contains dissolved compounds known as salts. It also contains organic matter and nitrogen, oxygen, and carbon dioxide gas. More than half the known chemical elements are found in seawater.

Salts are carried into the oceans by rainwater draining from the continents. When water drains from soil it dissolves sodium chloride and other salts. Weathering and erosion dissolve mineral salts present in rocks. These dissolved salts are transported to the oceans by rivers.

GOAL: You will learn the major features of the oceans, ocean currents, tides, and sea life.

Define oceanography.

Salts are chemical compounds that contain a metal and a nonmetal combined together.

How are mineral salts added to seawater?

261

FIGURE 13–1. Seawater at different depths is collected using sampling bottles.

Remind students that table salt is sodium chloride and that this is only one kind of salt.

What is the most plentiful element in seawater?

Seawater is rich in five salts (Table 13–1). These salts contain the following elements: sodium, magnesium, calcium, potassium, chlorine, sulfur, and oxygen. Chlorine is the most plentiful element in seawater. About 55 percent of the total mass of all matter dissolved in seawater is chlorine. Sodium accounts for about 31 percent of the dissolved matter. The proportions of the elements dissolved in seawater are fairly constant throughout the oceans.

Table 13–1.
Substances Dissolved in Seawater

Salt	Chemical formula	Average composition (g salt/1000 g water)
Sodium chloride	$NaCl$	23
Magnesium chloride	$MgCl_2$	5
Sodium sulfate	Na_2SO_4	4
Calcium chloride	$CaCl_2$	1
Potassium chloride	KCl	0.7
With other minor substances to total		34.7

Oceanography

The total amount of dissolved salts in a sample of seawater varies from place to place. Some parts of the oceans are saltier than others. **Salinity** (say LIHN ut ee) is a measure of this saltiness. It is the amount of dissolved salts in a given mass of water.

Salinity of most ocean water ranges between 3.3 and 3.7 percent. In some regions, the salinity can be higher. In the waters of the Dead Sea, the salinity is 23 percent or more. The salinity is so high because only a small amount of fresh water flows into the Dead Sea. Also, the evaporation rate of water is very high. Thus, the amount of dissolved salts has built up. The Dead Sea has no fish and little plant life.

Define salinity.

What is the average ocean salinity?

Part per million (ppm) is another expression of concentration often used in science. Percent is changed to ppm by moving decimal point 4 places to the right.

FIGURE 13-2. When the water of the Dead Sea evaporates, large quantities of salt are deposited.

In the deep ocean salinity tends to be constant. It does not change much from day to day. However, the salinity of seawater near the coast may vary during the day. For example, the mixing of fresh river water with seawater causes the salinity to change slightly. Marine animals living in the deep ocean can stand only small changes in salinity. Most sea creatures living in coastal regions can stand a wide range of salinity. Oysters living in coastal areas are one kind of marine animal exposed to changes in salinity.

How does the salinity of seawater in the deep ocean compare with seawater near the coast?

13:1 Seawater

activity

MAKING A BRINE SOLUTION Activity 13-1

Make a solution similar to the average composition of seawater by dissolving salt compounds in 1000 g (one liter) of fresh water. Measure and dissolve the given amount of each compound as listed in Table 13–1. Test the solution with litmus paper or indicator paper to discover whether it is acidic, basic, or neutral. Save the solution for the activity on Page 266. **See Teacher's Guide.**

A 1-L sample of seawater, if available, may be used in place of the brine solution described here.

making sure

1. Which chlorine compound is most plentiful in seawater? **sodium chloride**
2. Using Table 13–1, determine the seven most plentiful elements in seawater.
3. Which metal is most plentiful in seawater? **sodium**

2. sodium, chlorine, magnesium, calcium, potassium, sulfur, oxygen, hydrogen

13:2 Temperature and Density

Average water temperatures at the surface of the ocean range from −1.3°C in the Arctic Ocean to 28°C in the tropics. Surface water temperatures vary from season to season by as much as 8 to 10 C° within the North Atlantic and North Pacific Oceans. In waters near the equator, the temperature changes throughout the year by only about 1°C.

Highest seawater temperatures have been recorded in the Persian Gulf where readings of 35°C are common.

Seawater temperature also varies with ocean depth. Deep ocean water is much colder than surface water. In the deepest parts of the world's oceans, water temperatures remain close to 0°C throughout the year. But they do not freeze because of the salt content of the water.

Relate the temperatures of seawater to solar energy reaching the earth.

Temperature is a major factor in the life of marine plants and animals. Living things in warm tropical waters tend to grow faster and have shorter life spans than those living in colder waters. Tropical marine life reproduces earlier and more frequently. Certain kinds of animals can live only in cold water. Others live only in warm water.

Why is temperature a major factor in marine life?

The density of seawater depends on its temperature and its salinity. Recall that density refers to the mass per unit volume of a substance. Density is usually expressed as grams per cubic centimeter (g/cm³).

Dissolved solids in seawater make its density slightly greater than that of pure water. Pure water has a density of 1.0 g/cm³ at 4°C. Seawater has an average density of 1.02 g/cm³ at 4°C. Water becomes more dense when its salinity increases and its temperature decreases. In general, cold water is more dense than warm water of the same salinity. Thus, the most dense seawater is found near the poles, where it is cold most of the year.

What affects seawater density?

Review the concepts of density and how it is recorded.

What is the average density of seawater at 4°C?

Density of seawater depends on its depth and temperature. These measurements also vary with area of the ocean and time of year. An average may be calculated from many observations.

FIGURE 13–3. The coldest and most dense seawater is found at the poles.

Ocean bottom water is more dense than surface water because it is colder and has more dissolved material. It does not mix easily with surface water. Seawater near the equator is not as dense as water at the poles. Why? When water is warmed by the sun it expands and becomes less dense.

13:2 Temperature and Density

FIGURE 13–4.

activity
See Teacher's Guide.

DENSITY OF SEAWATER   Activity 13-2

Use a laboratory balance to find the mass of 1 L of the solution prepared in the activity on Page 264. Calculate the density of the solution and record its temperature. Compare the density of the solution to the density of tap water. **Density is 1.034 g/cm³ which is more dense than tapwater.**

Gently place an egg in tap water. Does it float? **Place no** the egg in the salt solution. Based on your observation, do you think an egg will float in seawater? **Yes, egg floats in brine and seawater.**

making sure

4. Why is ocean water near the North Pole more dense than water near the equator? **It is colder.**
5. Where does the energy come from that warms the oceans? **sun**
6. Why is deep ocean water colder than water near the surface? **does not receive as much solar energy**
7. How does salinity affect the density of ocean water? **Increased salinity increases the density.**

13:3 The Seafloor

The oceans cover a vast area equal to about three fourths of the total surface of the earth. For many years, the floor of the sea was believed to be flat and smooth. But research has shown that the ocean basins contain mountains and valleys similar to those on land. For instance, the Mid-Atlantic Ridge stretches along the center of the Atlantic Ocean for a distance of more than 16 000 km.

Where is the Mid-Atlantic Ridge located?

Many deep trenches have been discovered in the seafloor. Ocean trenches commonly reach depths of 7 km. The deepest known point in the oceans is in a trench near the Mariana Islands. Here the water has been measured to a depth of nearly 11 km.

Where is the deepest known ocean trench?

How are the seafloors mapped? Knowledge of the profile of the seafloor has come largely from soundings. A profile of the seafloor is an outline of its surface. Sounding is a method used to measure

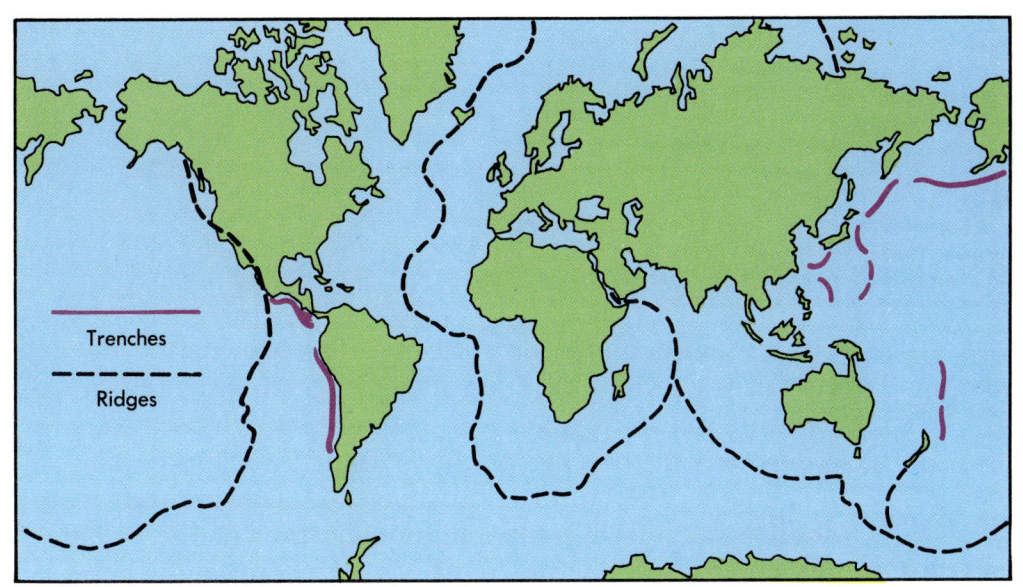

FIGURE 13–5. This map shows the location of the trenches and the ridges along the ocean floor.

the depth of water. Soundings are made in two ways. One way involves lowering a weight on a cable until it strikes the seafloor. The measured length of the cable is the water's depth. A second way to take soundings makes use of sonic (sound) depth recorders. Ships were first equipped with sonic depth recorders in 1940. Most of the facts about the seafloor have been discovered since that time.

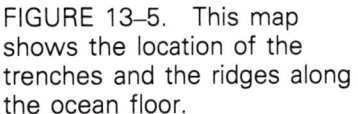

In what two ways are seafloors mapped?

National Oceanic and Atmospheric Administration

FIGURE 13–6. Electronic instruments receive echoes from the seafloor.

13:3 The Seafloor 267

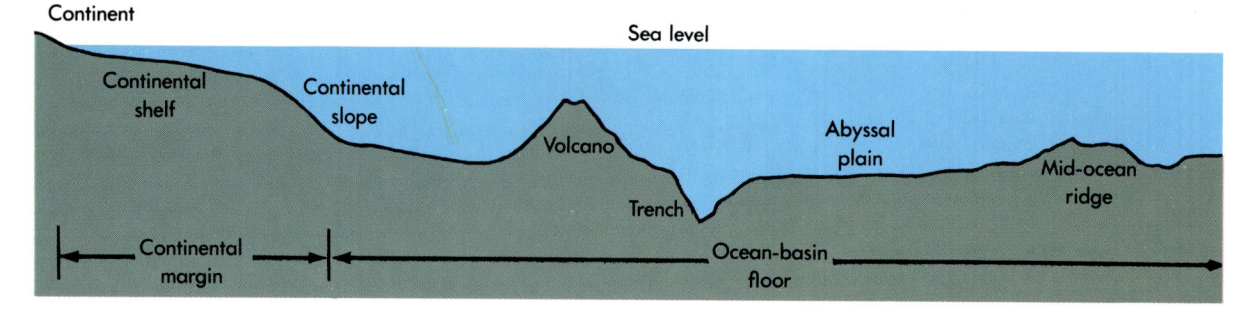

FIGURE 13-7. A profile of an ocean floor shows sloping shelves, volcanoes, trenches, and ridges.

Name the two main regions into which the seafloor may be classified.

Mountains, canyons, plateaus, and trenches have been discovered by instruments used to map the ocean bottom. Seismic reflections are returned from boundaries between contrasting rock materials such as the boundary between shale and sandstone, or between sandstone and igneous rocks. Seismic refraction waves travel along the ocean floor for some distance between the point where the wave enters the bottom material and where it emerges.

The seafloor may be classified into two main regions—continental margins and ocean-basin floors. **Continental margins** are the portions of the seafloor lying next to the continents. Within the continental margins lie the continental shelves. A **continental shelf** is a smooth, gently sloping portion of the ocean floor that borders the coastlines of a continent. They range in width from a few kilometers to more than 300 km. Continental shelves are less than 200 m below the ocean surface. **Ocean-basin floors** are mostly flat portions of the ocean floor. On the average, they are 4000 m beneath the ocean's surface. Mid-ocean ridges and ocean trenches are found in this region of the ocean.

FIGURE 13-8. Most of the islands on the earth were formed as a result of volcanic activity on the seafloor.

Dr. E.R. Degginger

making sure

See Teacher's Guide

8. What methods are used to map the ocean floor?
9. How are the features of ocean-basin floors similar to features on land? **Ocean ridges are like mountains and the trenches are like valleys.**

13:4 Sea Life

The ocean is the home of many kinds of organisms. Sponges, lobsters, whales, and sharks are but a few types of sea-living animals. One-celled plants called algae are the most plentiful ocean plant life. Algae make 80 to 90 percent of the food produced by the living organisms on Earth.

All animal sea life depends either directly or indirectly on plankton for food. **Plankton are a type of tiny animal and plant life that drift about in the sea.** Plankton provide food for very small fish. These small living creatures are eaten by larger fish. They in turn, are eaten by still larger fish. Plant plankton is the first link in a food chain that extends to the largest ocean animals.

What are plankton?

Explain the idea of an ocean food chain going from plankton to the largest fish-eating animals such as seals.

80–90% of the oxygen in the air is produced by photosynthesis in the oceans.

Photosynthesis is the basic chemical reaction in which food is made in green plants. All life, plant and animal, depends on this process for food.

a *Runk, Schoenberger/Grant Heilman Photography*   b   *Animals, Animals*

FIGURE 13–9. Plankton (a) and Portuguese men-of-war (b) are two links in ocean food chains.

The ocean contains a wealth of plankton. Tests made at the marine laboratory at Plymouth, England show the presence of as many as 4.5 million plankton in 1 L of seawater. The distribution of plankton in seawater varies from place to place. In the oceans, as on land, some regions are more fertile than others.

13:4 Sea Life

Name three physical factors that affect marine life.

Why is the ocean's food supply less at great depths than near the surface?

90% of marine species live on the continental shelf, around islands, or on rises, less than 200 meters below sealevel.

Plants and animals live in the parts of the ocean to which they are best suited. Sunlight, pressure, and temperature are important factors which affect marine life. Plant plankton live within the upper 60 m of seawater where light is available for photosynthesis. Deep within the ocean the absence of light prevents the growth of plant plankton. Thus, all fish that live in the ocean depths feed on organic matter that sinks from above. In general, the greater the depth, the smaller the food supply.

Most marine animals have gills which they use to obtain oxygen dissolved in seawater. Oxygen dissolves in seawater at the ocean's surface. It is also added by algae and other green plants as a product of photosynthesis. Some marine animals, such as whales, are mammals. They have lungs and obtain oxygen by coming to the surface and breathing air.

Sea life is most plentiful near coasts in tropical, subtropical, and temperate regions. Coastal waters are often shallow. Water depths are usually less than 50 m. The sunlight that penetrates these waters causes them to be rich in algae, seaweed, and other sea plants. Marine animals feed on these sea plants. Flounder, starfish, sea urchins, and crabs are a few of the animals that live within these shallow waters.

FIGURE 13–10. Coastal waters show a wide variety of life (a). Deep ocean animals must have special features to help them survive (b).

a

b

Joey Jacques

William H. Amos

270 **Oceanography**

13:5 Ocean Currents

Ocean currents are huge streams of water which flow within certain well-defined boundaries. The Gulf Stream, for instance, is a warm, swift current in the Atlantic Ocean. It flows out of the Gulf of Mexico between Florida and Cuba. It sweeps northward along the east coast of the United States. Then it moves across the Atlantic Ocean to Great Britain. It is about 80 km wide and travels as fast as 6 km/h off Florida. Its speed decreases to less than 1 km/h in the North Atlantic. It is one of the strongest ocean currents and moves huge amounts of water. Warm water is carried to the British Isles by the Gulf Stream. Thus the climate of Great Britain is warmer than would be expected from its location north of the equator.

The Gulf Stream is a surface current. Surface currents are caused by wind sweeping across broad expanses of water. As a wind exerts force on the ocean's surface, the surface water is set in motion. Most surface currents follow patterns similar to those of wind systems.

What is the Gulf Stream and where is it located?

Huge amounts of heat energy are moved by ocean currents.

Compare ocean currents to convection currents and winds in the atmosphere.

FIGURE 13–11. Cold surface currents flow from the poles toward the equator. Warm surface currents flow from the equator to the poles.

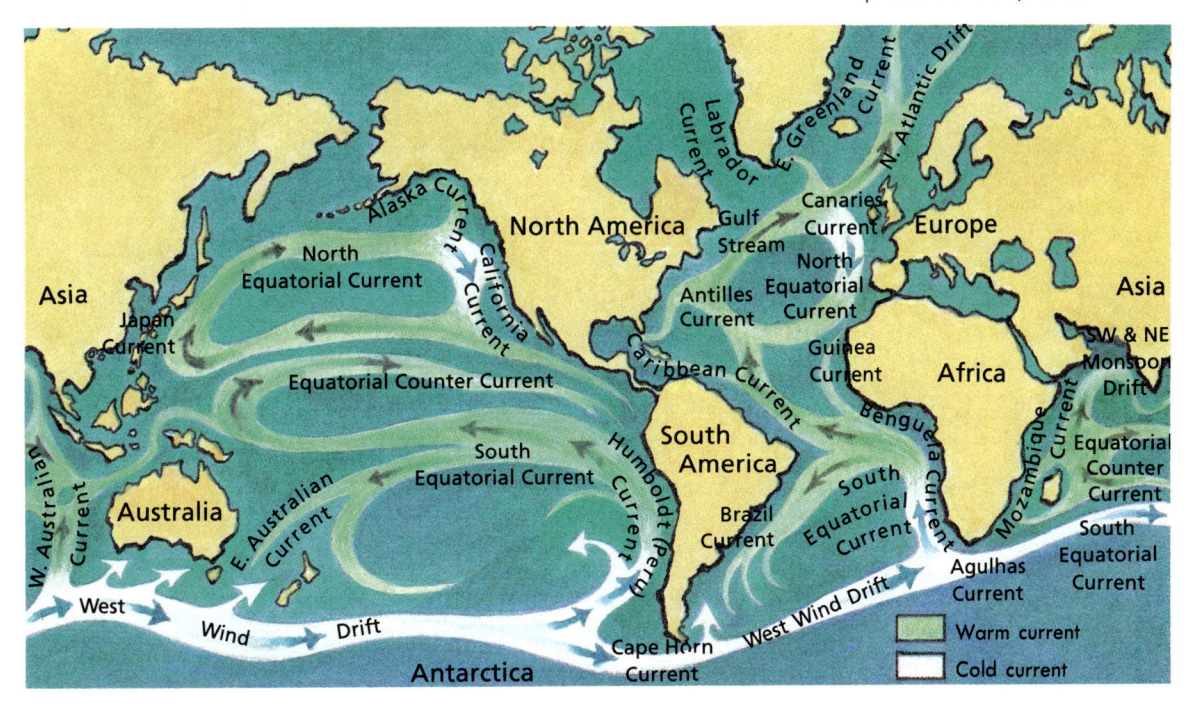

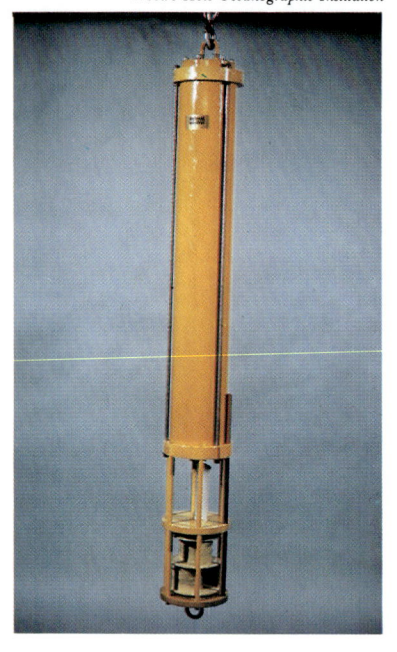

FIGURE 13–12. Oceanographers use a current meter to determine direction and speed of deep ocean currents.

What causes convection currents in the ocean?

Differences in water density result in convection currents within the ocean. A convection current in water is similar to a convection current in the atmosphere. Convection currents occur when water is cooled at the ocean's surface by loss of heat to the air. This cooled water increases in density and tends to sink because it is more dense than the warmer water beneath. At the same time, the warmer, less dense water beneath the surface water tends to rise.

Deep water currents are caused by differences in water density. Cold, dense water near the poles sinks toward the bottom of the ocean. Then this cold water flows toward the equator.

Oceanographers use sealed, floating drift bottles released at sea to check the direction of ocean currents. A drift bottle is weighted with enough sand to cause it to float nearly submerged. The bottle contains a card requesting the finder to record the date and the place in which the bottle is found. Those cards that are returned provide information about the speed and direction of ocean currents.

The speed and direction of an ocean current can also be measured with a water current meter. One of these meters is the Ekman current meter. It has a propeller and a tail fin. An Ekman current meter is suspended from a wire and lowered from a ship to the desired depth.

activity See Teacher's Guide.   Activity 13-3

CONVECTION CURRENTS IN WATER

Put two dozen ice cubes in the center of a large pan and fill it three-fourths full with water. Place the pan on a tripod and heat one side with a Bunsen burner. Add a few drops of ink or food coloring to the water on the side you are heating. Draw a diagram to show the movement of the ink. What causes the movement? *The ice cubes cool the water at the surface and increase its density. Heating decreases density at the bottom.*

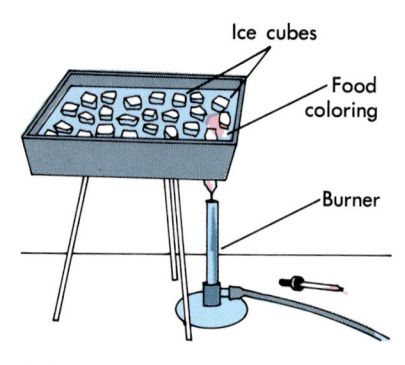

FIGURE 13–13.

making sure

10. How are convection currents in the ocean similar to convection currents in the atmosphere?
Both are caused by differences in density due to solar heating.

11. How is the climate of Great Britain affected by the Gulf Stream? It is warmer than would be expected if there were no Gulf Stream.

13:6 Waves

Where and how do ocean waves begin? Ocean waves are produced by winds. They possess kinetic energy gained from the winds. High waves are produced when winds sweep rapidly for great distances across a broad section of an ocean.

How are ocean waves produced?

W.C. Bradley Collection

A water wave is a transverse wave, that is, the movement of water particles in the wave is at right angles to the direction in which the wave is traveling.

FIGURE 13–14. Ocean surface waves may travel hundreds of kilometers before they reach shore.

An object floating on the ocean's surface rises and falls as a wave passes. It is not carried forward by the wave because water is not pushed forward by wave motion. The particles of water in a wave move up and down in a circular motion (Figure 13–15). Each water particle returns to the point from which it started.

Describe the motion of an object floating on the ocean as a wave passes.

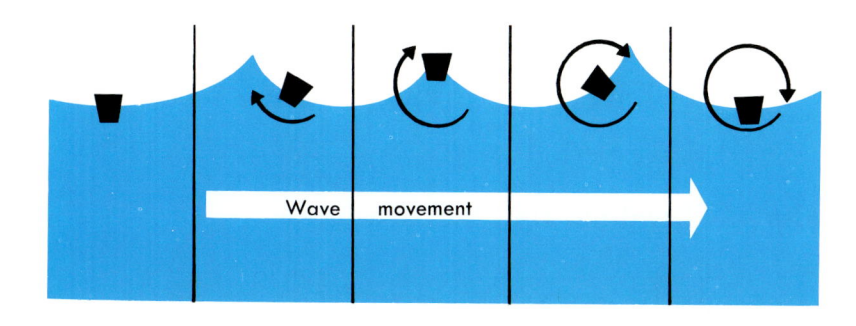

FIGURE 13–15. A cork rides a wave in an up and down circular motion.

13:6 Waves 273

Enrichment: Some students in class may have been ocean surfing for recreation. Ask them to describe how a surfer catches a wave and the kinds of waves best suited to surfing.

Three properties that describe a wave are amplitude, wavelength, and wave period (Figure 13-16). The **amplitude** of a wave is its height. It is one-half the distance between the top of a crest and the bottom of a trough. **Wavelength** is the horizontal distance from one crest to another. The height and wavelength of a water wave are commonly expressed in meters.

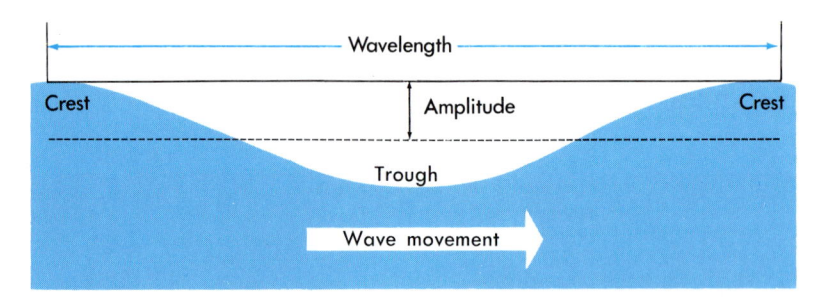

FIGURE 13-16. The crest is the highest part of a wave while the trough is the lowest.

A **wave period** is the time between the passage of two successive wave crests. Wave period is expressed in seconds. The shorter the period the greater the number of waves that pass a given point during a certain amount of time. The following is the equation for finding the period of a wave.

$$\text{wave period} = \frac{\text{time}}{\text{number of waves}}$$

How is the wave period determined?

Example:

Ten waves pass an anchored ship in 30 s. Find the wave period.

Solution

Work out the steps in this example on the chalkboard.

Step 1: Write the equation for finding the wave period.

$$\text{wave period} = \frac{\text{time}}{\text{number of waves}}$$

Step 2: Substitute the values for time and number of waves given in the problem.

$$\text{wave period} = \frac{30 \text{ s}}{10 \text{ waves}}$$

Step 3: Divide to find the answer.

$$\frac{30 \text{ s}}{10 \text{ waves}} = 3 \text{s} \quad \text{period} = 3 \text{ s}$$

Oceanography

As a wave approaches the shore, it touches bottom and slows. Other waves crowd in behind it. The ocean bottom exerts friction force (drag) on the wave. This force changes the form and motion of the wave. The drag force causes the circular paths of the water particles to be squeezed together (Figure 13–17). Both wave speed and wavelength decrease as a wave approaches the shore.

Explain how a wave becomes a breaker.

When a wave breaks, its forward speed at the top exceeds its forward speed of the wave as a whole.

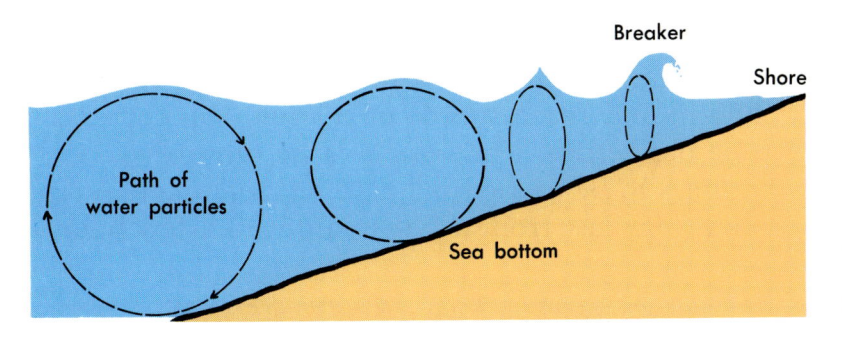

FIGURE 13–17. When water in a wave touches bottom, wavelength decreases and the height of the wave increases.

The height of a wave increases as it approaches the shore (Figure 13–18). Its crest becomes steeper and rises to a peak. Suddenly, the whole wave breaks into a tumbling mass of foaming water. At this point, the wave becomes a breaker. After the water spills onto the shore from the breaker, it slides back into the ocean underneath the incoming waves. This flow of water back into the ocean is called an **undertow**.

What is an undertow?

FIGURE 13–18. A breaker forms when a wave topples over, spilling water onto the shore.

13:6 Waves 275

Why do waves tend to bend and approach the shore head-on?

Waves approach a shoreline from many angles. As they reach the land, they are refracted or bent. <mark>They bend because the part of the wave entering shallow water first is slowed by the ocean bottom. The part in deeper water races ahead.</mark> Thus, waves tend to bend so that they approach the shore head-on.

Some waves do hit the shore at an angle and produce longshore currents. A **longshore current** is a flow of water along the shore.

Rip currents occur when longshore currents go back to sea. Rip currents range in length from 170 to 800 m and reach speeds of 8 km/h. Rip currents can be dangerous. They often cause unwary swimmers to be carried out into deeper water. A rip current may last from a few minutes to a few hours.

<mark>Rip currents widen and their strength lessens seaward from the breaker zone. A rip current usually can be observed from the beach because once the current has established a channel it flows seaward through this zone and causes a disruption in the wave front approaching shore. The wave front breaks into a large number of short steep waves.</mark>

Steve Lissau

FIGURE 13–19. A rip current can carry swimmers out into deep water quickly.

activity   See Teacher's Guide.

WAVE MOTION   Activity 13-4

Float a cork on water in a pan. Using a spoon, strike the water in the center of the container until waves are produced. Observe the motion of the cork. What do you conclude about the effect of a wave on the cork? [*makes it go up and down*] Is the cork carried forward by the motion of the wave? *no*

276   Oceanography

activity

See Teacher's Guide.   Activity 13-5

WAVES ALONG A SHORELINE

(1) Obtain a clear plastic container and fill it half full with water. **(2)** Place a thin, flat plate of metal or glass on some metal washers at one end of the container and adjust the plate so that its upper surface is about 1 mm below the water surface. This makes the water above the plate very shallow. The plate should be parallel to waves generated at the opposite end of the container.
(3) Generate waves with a wooden dowel (Figure 13–20a). Observe the waves as they reach the plate. Does the wavelength increase or decrease when the waves pass into shallow water? **decrease** **(4)** Now set the plate so that its edge is at a 45° angle to the waves (Figure 13–20b). Repeat the wave generation and observe the wave path. What change occurs? **Waves bend toward the glass plate.**

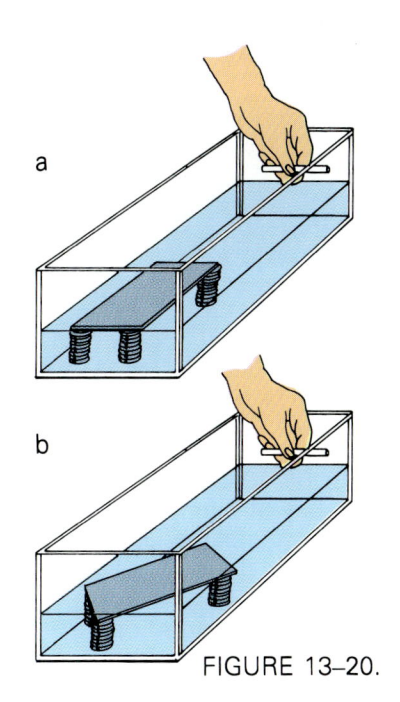

FIGURE 13–20.

making sure

12. From an ocean pier, you count 15 waves passing during 60 s. What is the wave period? **4 s**

13. An average of 24 waves passes a point on a dock in 2 min. Find the wave period. **5 s**

13:7 Tides

The ocean level at a seacoast is constantly changing because of tides. **A tide is a periodic rise and fall of the ocean.** Each day the ocean level along a coast rises to high tide, drops to low tide, rises to high tide again, and then drops to low tide. **Tides are caused by gravitational forces of the sun and moon.** In some areas the tides have little or no effect on the water level at the shore. In other areas, such as the Bay of Fundy near Nova Scotia, the water level changes by as much as 15 m. At low tide, the Bay of Fundy is almost completely drained of water. The height of the tides in a region is influenced by the depth and shape of the sea floor and the shape of the coast.

Define tide.

What causes tides?

13:7 Tides   277

FIGURE 13–21. High tide at the Bay of Fundy (a). Low tide at the same location leaves boats stranded (b).

Coastal configuration probably is the most important influence on the height and character of the tide. Tides are high on the coast of England but the Mediterranean has almost no tides.

Tides represent a vast source of kinetic energy. One project to harness the power of the tides has been completed in the Rance River in France. High tides cause the water in the river to reach heights of 11 to 13 m. Dams upriver trap the water at high tides. The return flow at low tide is directed through water turbines. The water turbines turn electric generators. As a result, the kinetic energy of the tides is used to produce electric energy.

activity

OCEAN TIDES   Activity 13-6

Locate daily tide listings for a port city such as New York, Miami, Galveston, San Francisco, or Los Angeles. Keep a record of the time of daily tides for one week. If such listings are not readily available, use the following data:

Day	Low	High	Low	High
1	2:38 A.M.	8:56 A.M.	3:09 P.M.	10:32 P.M.
2	3:08 A.M.	9:30 A.M.	5:00 P.M.	11:34 P.M.
3	3:43 A.M.	10:11 A.M.	5:57 P.M.	12:56 A.M.

How much time is there between high tides? Compare the time from high tide to high tide for three days. Is it always the same? How does the time between high tides compare to the time between low tides? See Teacher's Guide.

Answers will vary depending on the data used.

13.8 Seismic Sea Waves and Surges

Giant ocean waves are produced by earthquakes in the sea floor. Waves produced by an earthquake are called **seismic sea waves.** Some of these waves travel at speeds of more than 600 km/h. They cross the ocean in the form of low waves. They are so low in fact that ships at sea often do not know that a seismic sea wave is passing. In deep water, some are only a meter or two high. But when they reach the shore they may exceed a height of 30 m. Although they have nothing to do with the tides, seismic sea waves are often called "tidal waves." A seismic sea wave is also referred to as a tsunami (soo NAHM eeh), a Japanese word.

Seismic sea waves can be very destructive. Some caused great disasters when they smashed against coastlines. In 1883, a seismic sea wave killed 36 380 people living on small islands in the Dutch East Indies. In April, 1946, a seismic sea wave was caused by an earthquake in the sea floor near Alaska. This wave traveled toward Hawaii at a speed of 790 km/h. Reaching Hawaii, it caused extensive damage and death to 159 people.

What kind of waves are produced by an earthquake? What are they commonly called?

FIGURE 13–22 Tsunamis can do extensive damage when they strike the coastline (a). Seismic sea waves are caused by earthquakes on the ocean floor and may rise to great heights along the shore (b).

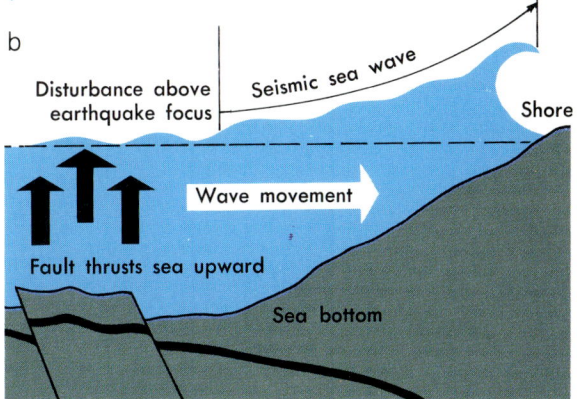

How is a storm surge formed?

A large cyclonic storm over an ocean may produce a storm surge. A storm **surge** is a mass of water that produces an abnormal rise in sea level along a seacoast. The surge forms when water, driven by storm winds, builds up along coastlines. The hill of water produced by the swelling may raise the water level by 10 m. When the storm reaches a seacoast, the surge of water may be carried onto land. Storm surges can be destructive to low-lying coastal areas. They are extremely dangerous when they coincide with high tides, often causing property damage and loss of life.

making sure

14. How are seismic sea waves different from ordinary wind-blown waves? Seismic sea waves travel low in the water and are faster than surface waves.

13:9 Ocean Resources

Natural resources are abundant in the oceans. For example, the fishing industry harvests millions of tons of fish each year. Deep sea drilling rigs sink wells thousands of meters into the ocean floor. These wells pump out oil and natural gas that are used for fuel. Elements used in industry, such as sodium and chlorine, are obtained from the salt in seawater. In some places, seawater is even a source of fresh drinking water. The salt is removed from the water in a distillation plant.

Oil spills from tankers, untreated sewage from some large cities, and the dumping of garbage and other wastes into the ocean cause pollution.

Continued pollution of the oceans may have a negative effect on ocean sea life.

More than 800 000 dry metric tons of seaweed are harvested from the oceans each year. The harvested seaweed plants are species of red and brown algae. They include kelp that grows to a length of 70 m. Among the many products obtained from seaweed is a substance called carrageenan (ker uh GEE nun). This material is used to thicken milkshakes, smooth skin cream, and to weave surgical thread. Agar, a gelatin substance used to grow bacteria in laboratory tests, comes from seaweed. Algin, another substance obtained from seaweed, is used to make paints, dyes, and paper. Experiments are being done to produce natural gas by decomposing seaweed. The natural gas could be used for heating homes.

What products are made from seaweed?

Latent Image

FIGURE 13–23. Many products are derived from seaweed.

In the future, it is likely that valuable minerals will be mined from the oceans. Round-shaped mineral pieces, called nodules, have been discovered on the deep ocean floors. These nodules are rich in manganese and iron. Manganese is mixed with iron to make steel hard. Other important minerals present in the nodules, in lesser amounts, are copper, cobalt, and nickel.

The oceans have a tremendous amount of energy that may someday be put to use on a large scale. Have you ever watched huge waves pounding on a beach? If so, you know the vast amount of power in waves and the work they can do. The rising and falling of the tides may someday become an important source of energy. Experiments are being done to use the kinetic energy of ocean waves and tides to generate electricity. The oceans are very large and are heated by the sun every day. Therefore they contain a tremendous amount of heat. Experiments are being done to use this heat to produce convection currents in the ocean water. The flow of water in the currents would be used to generate electric power.

The oceans may someday become a major source of hydrogen which will be obtained by electrolysis, using electricity generated by solar cells.

How can the ocean be used as a source of energy?

The hydrogen would be burned as a fuel with very little pollution.

making sure

15. In what ways can the ocean provide sources of energy?
16. How would drilling for oil in the ocean be more difficult than drilling for oil on land?

15. Waves have kinetic energy, tides have kinetic energy, the ocean is a huge source of heat energy

16. Drilling rig must be on a platform over the ocean. Waves can damage equipment; need more protection against oil spills

13:9 Ocean Resources

PERSPECTIVES

people

Commercial Diving

You have probably seen movies of SCUBA divers swimming in the ocean, exploring sunken wrecks or studying underwater life. People in other professions also use diving equipment to perform their jobs. For example, a large underground pipeline carrying water may spring a leak and require a diver to repair it. A city might hire a diver to determine if rust is damaging the inside of one of its water towers. These types of problems are encountered by Tony Kiefer, and he uses diving equipment to solve them.

Tony Kiefer is a commercial diver. He is a partner in a company that builds, inspects, repairs, or demolishes underwater structures. Such structures may include bridges, piers, dams, and pipelines. His company has been hired to raise barges and inspect a gasoline tank. In the latter job, only their special cameras entered the tank.

A typical day for Tony Kiefer involves driving up to three hours to a location, and evaluating the job. If the job is to make repairs, he will call for his employees to bring the necessary equipment. Video cameras may be used to determine the extent of the work involved. Divers will then determine what procedure should be used to do the job. Finally, the necessary materials and equipment are gathered.

Divers usually wear a type of helmet that has air pumped into it. A built-in radio enables communication with the surface. One or more people will stay at the surface to monitor the air pump and radio. Divers work at depths between 12 and 15 meters during most jobs. Water temperature ranges from $-2°C$ to $48°C$, depending on the season. A 12 hour work day may include 6 to 8 hours of underwater time. Usually the water is not clear and the work must be done using the sense of touch. Tony Kiefer and his divers have welded metal beams underwater, cut pipelines with special torches, and used explosives to reshape a river bottom.

Tony likes the variety of activities that his job offers. He especially likes to see new places and meet interesting people. He stresses that a person who is interested in a career of commercial diving should be especially interested in construction and building. Tony has a background in geology and his employees each have special skills. Besides diving and first aid, employees are trained in engineering, welding, and surveying. Each employee contributes to solving the special problems each new job presents.

Tony Kiefer

main ideas

Chapter Review

1. Seawater is a brine solution containing at least one half of the known chemical elements.   13:1
2. Differences in seawater density are caused by differences in temperature and differences in salinity.   13:2
3. Ocean basins contain mountains and valleys similar to those on land.   13:3
4. Sea life is most plentiful in the shallow coastal waters of temperate, subtropical, and tropical regions.   13:4
5. Ocean currents are produced by winds and by variations in the density of seawater.   13:5
6. Surface waves are produced by wind.   13:6
7. The heights of ocean tides are influenced by the depth and shape of the seafloor and the shape of the coast.   13:7
8. Tsunamis and surges are destructive to low-lying coastal areas.   13:8
9. The oceans can provide vast food, mineral, and energy resources.   13:9

vocabulary

Define each of the following words or terms.

amplitude	oceanography	surge
continental margin	plankton	tide
continental shelf	rip current	undertow
longshore current	salinity	wavelength
ocean-basin floor	seismic sea wave	wave period

study questions

DO NOT WRITE IN THIS BOOK.

A. True or False

Determine whether each of the following sentences is true or false. If the sentence is false, rewrite it to make it true.

T   1. Seawater contains sodium, magnesium, and chlorine.

T   2. About 55 percent of the dissolved matter in seawater is chlorine.

F   3. ~~Less~~ than one half of the earth is covered by water. *more*

F   4. The salinity of most ocean water is ~~over 4 percent.~~ *3.3-3.7*

F   5. When seawater is cooled, its density ~~decreases.~~ *increases*

T 6. The density of seawater is slightly greater than the density of fresh water.

F 7. Water within the Arctic Circle is ~~less~~ dense than surface water in the tropics. more

F 8. ~~One~~ high tide and ~~one~~ low tide occur each day. two, two

F 9. Water near the ocean surface is much ~~colder~~ than deep ocean water. warmer

T 10. The Gulf Stream in the Atlantic Ocean is one of the strongest ocean currents.

B. Multiple Choice

Choose the word or phrase that completes correctly each of the following sentences.

1. The ocean-basin floor is about *(1000, 4000, 8000, 10 000)* m below the ocean surface.
2. The deepest known point in the ocean lies off *(the Aleutians, Japan, Java, the Mariana Islands).*
3. Soundings of the ocean floor are made with *(radar, light, heat, sound).*
4. Continental shelves occur within the *(continental margins, ocean-basin floors, mid-ocean ridges).*
5. The water depth over a continental shelf is about *(100, 200, 300, 500)* m or less.
6. *(Rip currents, Undertows, Convection currents, Breakers)* occur when longshore currents go back to sea.
7. Plankton are microscopic *(plants, animals, plants and animals)* that float in water.
8. A flow of water back into the ocean underneath incoming waves is called a(n) *(rip current, breaker, undertow, current).*
9. A(n) *(rip current, undertow, surge, breaker)* is a wind driven mass of water produced by a storm.
10. Most marine animals obtain oxygen through *(lungs, gills, plankton, sunlight).*

C. Completion

Complete each of the following sentences with a word or phrase that will make the sentence correct.

1. As waves reach the shore, their wavelength decreases____.
2. ____ and iron are two metals found in nodules present on deep ocean floors. manganese

3. A seismic sea wave is produced by a(n) __earthquakes__.
4. The density of fresh water is __less__ than the density of seawater.
5. __Plankton__ are the first links in marine food chains.
6. Plankton are most plentiful in the upper 60 m of seawater because _____ is available. __sunlight__
7. The number of plankton in a liter of seawater is likely to be greater in the __tropics__ than near the poles.
8. Periodic changes in the ocean level along a coast are called __tides__.
9. Currents deep in the ocean are caused by differences in seawater _____. __density__
10. Ocean __waves__ and tides have kinetic energy that can be used to generate electricity.

D. How and Why   See Teacher's Guide.
1. How is the density of seawater affected by temperature?
2. Why do ocean currents travel in a direction different from the wind that causes them?
3. Describe what happens to a wave as it comes to the shore.
4. Why is sea life more plentiful in some parts of the ocean than in others?
5. How are surface currents produced in the oceans?

challenges

1. Construct a profile model of the Atlantic Ocean or Pacific Ocean seafloor with plaster of paris.
2. Build and maintain a saltwater aquarium.

interesting reading

Cook, Jan Leslie, *The Mysterious Undersea World*. Washington, D.C.: National Geographic Society, 1980.

Goldin, Augusta, *Oceans of Energy: Resources of Power for the Future*. New York, NY: Harcourt, 1980.

Olson, Steve, "The Contours Below." *Science '83*, July/August 1983, pp. 38-41

SIDE ROADS
The Sandy Seashore

Freda Leinwand

Unit 3 describes many physical conditions that are found on Earth. These conditions determine the manner in which plants and animals survive in a given area. In the marine environment, some factors include salinity, temperature, density, currents, waves, and tides. Each plant and animal reflects the physical conditions of its individual habitat. By examining a typical sandy beach in the Caribbean, we can observe the relationships between life forms and the physical conditions.

If one stands in shallow water and walks toward the shore, different zones of life are encountered. The first zone is called the subtidal zone. This area is always covered with water. It usually has a sandy bottom and very little plant growth. Most food in the subtidal zone comes in from outside the area. Predominant animals are burrowers that feed on plankton or small particles of food called detritus. Fan worms spread brightly colored tentacles into the water to collect plankton. When disturbed, these tentacles quickly retreat into tubes that are buried in the sand.

Sea urchins, protected by hard spines, move slowly across the bottom in search of food. Sand dollars also slide their thin bodies through the sand while feeding on detritus. Occasionally, a sting ray will swim in from outside the subtidal zone. Crabs are usually buried in the sand and are very difficult to see. When looking for food, a crab blends in with the color of the sand.

The next area encountered inland is the intertidal zone. It is covered with water during high tide and exposed during low tide. Intertidal animals must be able to breathe under water and also survive without water. Worms are the most common type of animal, living in burrows and crawling through the sand. Crabs, considered to be burrowers, can also be found in the intertidal zone. Most animal species of this zone are nocturnal, or night-dwelling. They avoid coming out during the day because of numerous predators.

At the upper region of the intertidal zone is a line of debris that was dropped by the high tide. This area is called the strand line, and is made up of pieces of dead algae, shells, and other items. It is a favorite area for beach combers.

Animals, Animals/Zig Leszczynski

Doug Wynn

Inland from the intertidal zone is the pioneer zone, named for the types of plants that grow there. They are the first plants to inhabit a beach, and face very harsh physical conditions. Very few nutrients are found in the soil of the pioneer zone, because erosion continually shifts the beach. Pioneer plants are low growing and have thick fibrous roots that are important for anchoring in the soil. Leaves are usually thick and sometimes hairy, two characteristics that minimize water loss. Most pioneer animals are not permanent, but move from one area to another.

Moving farther up the beach, one encounters the fixed dune zone. This area is characterized by large plants. The amount of nutrients in the soil is high and the sand is relatively stable. Many animals are found here, but most come in from outside the area.

The next inland area, sea grapes, derives it name from a plant species. These plants have thick leaves and can tolerate the occasional salt spray that reaches them. The amount of nutrients in the soil is very high, and much debris can be found on the ground. Ants scurry through the debris in search of food. Lizards can be seen flipping leaves with their heads, while looking for insects.

Beyond the dunes is the scrub-woodland zone, characterized by a large variety of plants and animals. The soil is rich, the trees are tall, and the ground is relatively cool. Most of the scrub-woodland animals are permanent residents, and few are burrowers. Of all the shore environments, this zone is least affected by the ocean.

Doug Martin

Brian Parker/Tom Stack & Associates

All living things have cells and reproduce. Yet, they have differences, too. A hummingbird is an animal; a cactus is a plant. How are they similar? How are they different? Plants and animals can be classified according to their differences. Why is a cactus classified as a plant? Why is a hummingbird classified as an animal?

Gunter Zeisler/Peter Arnold, Inc.

Living Things
unit 4

Just as all living organisms share many common features, they are also different from one another in many ways. Scientists study the features of living organisms. What features are common to all living organisms? What are the characteristics of the simplest life forms?

Sharon Kurgis

Life and the Cell

chapter 14

Introducing the chapter: Use Appendix D, page 510, to demonstrate the use of a microscope. Students should learn the parts of the microscope and the function of each part. Stress prudent care and proper use of the microscope. Show students how to cut a small letter "e" from a newspaper, make a wet mount, and observe the letter with a microscope. Have students record their observations.

14:1 Features of Living Things

Living things are found in hot climates, cold climates, oceans, mountains, and deserts. A study of life reveals that a great number of different plants and animals exist. Suppose you could find and identify all of the living things on your school grounds. You would name several hundred plants and animals. Many of them are so small you would need a microscope to see them.

Each living thing on the earth has certain features that make it different from all others. One of these features is size. An adult blue whale, the largest living animal, has a mass of 91 000 kg and is nearly 30 m long. The giant sequoia redwood trees of California grow to a height of 90 m and a diameter of 5.5 m. The smallest bacteria, however, are only 0.0001 cm long. Millions of them could swim in a spoonful of water.

How do you know if something is living or nonliving? Most living things have the following features.

(1) *Definite size and shape.* Most kinds of living things have certain shapes and grow to particular sizes.
(2) *Definite life span.* A living thing begins life, grows, matures, reproduces, and dies. The time from birth to death is called the life span.

GOAL: You will learn that all living organisms are alike in certain ways. However, every kind (species) of organism is unique and different from all the other species.

What are the six special features of most living things.

Edward Shay

FIGURE 14–1. The sequoia redwood is the largest kind of tree.

Enrichment: Discuss what would happen to a species that lost its ability to reproduce.

FIGURE 14–2. Response to the environment (a) and movement (b) are two features of living things.

a

b

(3) *Reproduction.* Reproduction is necessary if a species is to survive. Reproduction is the process of producing another living thing.

(4) *Metabolism.* All living things carry on a large number of chemical processes to stay alive. The sum of all these processes is called **metabolism.** One metabolic process is digestion. Digestion is the process of breaking down food into simpler substances which can be used by a living thing. Digestion provides the materials needed for living things to grow and produce energy.

(5) *Movement.* Another feature of most living things is the ability to move. Most animals run, swim, fly, squirm, or move in other obvious ways. Plants also move, although much more slowly.

(6) *Response to the environment.* Each living thing responds in some way to its environment. Anything in the environment that causes a change in behavior is called a **stimulus.** For example, different colors in traffic lights act as stimuli (plural of stimulus) for motorists and pedestrians. The behavior caused by a stimulus is a response. The response of most people to a flashing red light is caution because of possible danger.

Roger K. Burnard

292   *Life and the Cell*

activity

BURNING A CANDLE   Activity 14-1

Set a candle in a dish or pan and light it. Observe the flame. Suppose you wanted to convince a friend that the candle was alive. List on paper the reasons you would give. What reasons would you give to explain that the candle is not alive? See Teacher's Guide.

The purpose of this activity is to deepen students' insight into the nature and complexity of life and its characteristics.

making sure

1. List some of the features of each of the following living things: See Teacher's Guide.
    (a) human   (d) robin   (g) mushroom
    (b) grasshopper (e) cactus   (h) snake
    (c) spider   (f) frog   (i) octopus
2. List three stimuli that cause responses in humans. Answers will vary, sound of a bell, flash of a bright light, loud noise
3. How is plant movement different from movement of animals?
    Plants move slower and usually cannot move from place to place.

14:2 The Cell

Nearly all living things are made of cells. A **cell** is a basic unit of structure and function in living things. Most cells are very small—too small to be seen with the eye alone, but they can be studied with a microscope.

Define a cell and name its three basic parts.

A cell contains three basic parts. A thin layer called a cell membrane surrounds the cell. The cell membrane controls the movement of substances into and out of the cell. Near the center of the cell is a structure called the **nucleus** that controls the cell's life activities. The nucleus is enclosed by a membrane called the nuclear membrane. Some cells, such as muscle cells, may contain several nuclei. Between the cell membrane and the nucleus is a jellylike substance called **cytoplasm**. In most cells, the nucleus is darker than the cytoplasm because it is thicker and more dense.

Because different cells absorb different colors, a variety of stains are used to distinguish various structures.

Enrichment: Show students microscope slides of different kinds of cells. A microprojector may be used.

14:2   The Cell   293

FIGURE 14-3. There are some differences between plant and animal cells.

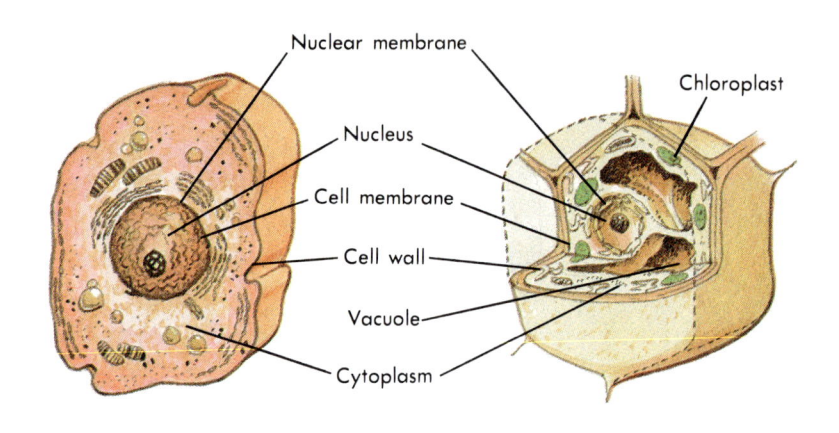

ANIMAL CELL   PLANT CELL

Review the proper use of the microscope.

How is a plant cell different from an animal cell?

What are protoplasm and cytoplasm?

In animal cells, the cell membrane is the outer layer of the cell (Figure 14-3). Plant cells, however, have an outer layer called a cell wall which surrounds the cell membrane. The cell wall forms a stiff case around the cell. It is made mostly of a material called cellulose (SEL yuh lohs). Cellulose gives strength to the cell wall.

Protoplasm is the "living material" of the cell. Each cell is a unit or mass of protoplasm. Protoplasm is 70 percent water. Proteins, fats, carbohydrates, and minerals make the remaining 30 percent. With the exception of the cell wall in plant cells, all parts of a cell are made of protoplasm. Cytoplasm, for example, is the protoplasm that lies outside the nucleus of a cell.

Living organisms are mostly water. Water has a high specific heat, resisting changes in temperature. It resists contraction and expansion at normal temperatures, dissolves many substances, and can flow, thereby transporting substances.

Table 14-1.
Elements in Human Protoplasm

Element	Percent
Oxygen	65
Carbon	18
Hydrogen	10
Nitrogen	3
Calcium	2
Phosphorus	1
Trace elements	1

Life and the Cell

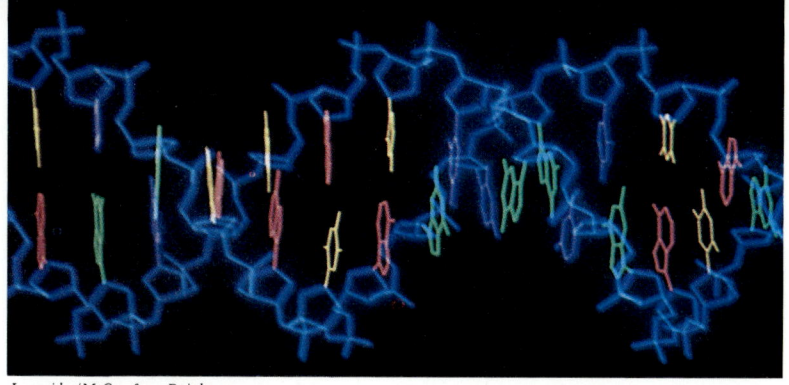

FIGURE 14-4. A DNA molecule has a spiral structure as shown in this computer model. Each molecule of DNA contains special "life codes," which are different for each living thing.

A chemical called DNA is found only in the nucleus. Another chemical called RNA is found throughout the cell. DNA and RNA molecules are some of the largest molecules in living things. Most DNA is made of thousands of smaller molecules linked to form one very large molecule. The specific order of these smaller molecules forms a "life code." This life code is used to control all activities of the cell. Every living thing has a special code of its own which makes it different from other species. Furthermore, each member of the species has a variation of the species code. Thus, all members of a species are somewhat different from each other.

What is the "life code" of DNA?

14:3 Cell Activities

Recall that the cell membrane controls the movement of substances through the cell. This membrane is selective, allowing only certain substances to pass through it. The process of diffusion occurs through cell membranes. **Diffusion is the movement of molecules from regions where they are more concentrated to regions where they are less concentrated.** Dissolved substances may move from one side of the membrane to the other. A cell gets food and oxygen it needs through diffusion.

In living things, some molecules can move in a direction opposite to the expected direction of diffusion. For example, sea plants may have a higher iodine concentration than exists in the seawater around them. Iodine moves into plants in a direction opposite to diffusion.

What is diffusion?

Cell activities keep an organism alive. Cell activities require the movement of materials into and out of the cell.

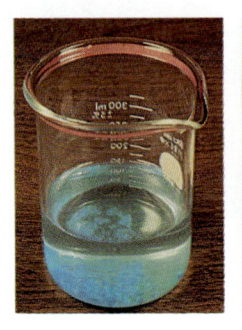

FIGURE 14–5. Molecules diffuse from where they are more concentrated to where they are less concentrated.

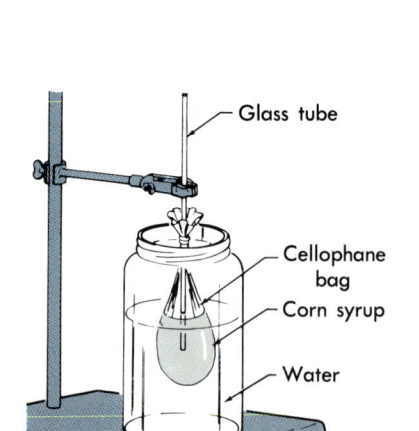

Diffusion of water through a membrane is called **osmosis** (ahs MOH sus). Suppose some pure water is separated from a sugar solution by a membrane as shown in Figure 14–7. Water molecules move from the pure water area into the sugar-water area by osmosis. Sugar molecules move into the pure water by diffusion. Water and sugar in this example diffuse in opposite directions.

Cells need energy for movement, growth, and other cell activities. Cells derive their energy from a metabolic process called respiration. **Respiration** is the release of energy from the breakdown of food within the cell.

Food enters the cell by diffusion and is digested within the cytoplasm. For respiration to occur, however, most cells require oxygen. Oxygen passes into the cell by diffusion. Digested food is then combined with the oxygen. During respiration, food and oxygen are chemically changed to produce energy.

The process of respiration also produces carbon dioxide and other waste materials. These waste products are not needed by the cell. Carbon dioxide along with materials not used by the cell are expelled from the cell by diffusion.

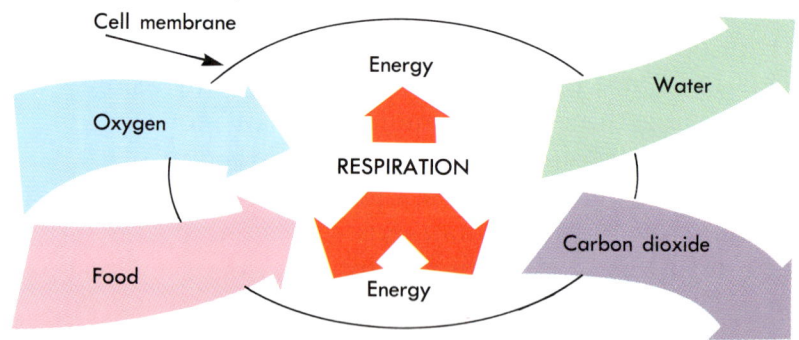

FIGURE 14–6. The solution rises in the tube as water enters the bag of corn syrup by osmosis.

Define osmosis.

How does a cell get the food and oxygen it needs for life?

Respiration and digestion are chemical changes.

FIGURE 14–7. Through respiration, a cell obtains energy by combining food with oxygen. Waste products of respiration include carbon dioxide and water.

296   *Life and the Cell*

activity

See Teacher's Guide.

OSMOSIS IN CARROT CELLS Activity 14-2

Hollow out a carrot and fill it with corn syrup. Corn syrup contains dissolved sugars. Set a one-hole stopper containing a glass tube into the top of the carrot. **CAUTION:** Before inserting the glass tube into the stopper moisten the end of the glass tube with a lubricant such as glycerol. Hold the tube with a rag or towel as you slide it into the stopper. When placed in the carrot, the tube must extend into the syrup. Use sealing wax to seal the stopper into the carrot. Place the carrot in a beaker of tap water. After several hours observe the liquid in the glass tube. What is the liquid? How did it get there? syrup and water mixture

Water moves into the carrot by osmosis.

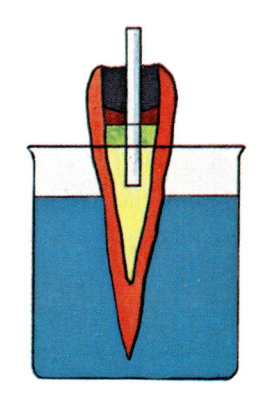

FIGURE 14–8.

making sure

4. As a solution freezes, does the rate at which substances diffuse through it increase or decrease? decrease
5. If a cell is placed in a concentrated salt solution, water diffuses out of the cell. Why?
6. How is cell digestion different from respiration?
7. What would happen if carbon dioxide and other waste materials did not diffuse out of a cell?
The cell would be poisoned and die.

5. Water is more concentrated inside the cell than outside.
6. Digestion breaks large food molecules into smaller molecules. Respiration combines food with oxygen to yield energy.

14:4 Classification

Many different kinds of organisms are alive on Earth. An **organism** is a whole and complete living thing. Organisms vary in size from those that have only a single cell to those that contain billions of cells. It would take many lifetimes to study all the organisms in the world. To make the study of living organisms easier, scientists have classified them into groups.

The naming and classification of organisms is a major problem for scientists. Why? Many plants and animals have names by which they are known to people in one part of a country. Yet, some of these very same plants and animals have different names in other regions of the same country. For example, the mountain lion is known in different areas of the

Define organism.

All classification is based on similarities and differences. Begin by going over examples of classification, e.g. clothes in a dresser, stamps in an album, books in a library.

Why are scientific names for living things used by all scientists?

United States as cougar, puma, and panther. To overcome this problem, a system has been devised by scientists to standardize the names. Each organism is given a name that is recognized and used by scientists throughout the world.

What are the five main kingdoms of living things?

Scientists classify living organisms into five main groups called **kingdoms.** The five kingdoms are monera, fungi, protist, plant, and animal. You probably know many plants and animals and their features. Protists, monerans, and fungi may be new to you. In studying these kingdoms, however, you will find familiar organisms in each group.

Most students are inclined to think of living things as plants or animals.

Others will be aware of the three-kingdom classification that adds a Protist Kingdom. Biologists use the updated five kingdom classification described here that most truly reflects the biological differences and similarities of organisms.

Organisms in each kingdom have been classified into smaller groups. Each group divides the previous group into smaller, more specific groups. There are six groups in each kingdom—phylum, class, order, family, genus, and species. Each smaller group contains organisms with similar features. At the species level there is one main type of organism.

Table 14-2.
The Five Kingdoms

Kingdom	Features	Examples
Monera	Single cell with scattered nuclear material	Bacteria
Protist	Single cell with the nucleus surrounded by a membrane	Amoeba, paramecium
Fungi	Many cells but cannot move, absorbs food from dead or living organisms	Yeasts, molds, mushrooms
Plant	Many specialized cells, uses chlorophyll to make food	Daffodils, cacti, palm trees
Animal	Many specialized cells moves, obtains food from outside sources	Whales, robins, bears

Life and the Cell

FIGURE 14–9. Ocelots (a) and mountain lions (b) belong to the genus Felis.

Each kind of organism is given two names—one for its genus and one for its species. Genus means group. A **genus** is a group of closely related species. A house cat, wildcat, ocelot, and cougar all belong to one genus, *Felis*. Although alike in many ways, each species in the genus *Felis* has distinct features. You might befriend a house cat and take it into your home. It is unlikely that you would have a wildcat or ocelot for a pet.

How are living things named?

Species means kind or type. A **species** is a single, distinct group of living things. *Felis sylvestris* is the scientific name for a wildcat. *Felis* is the genus and *sylvestris* is the species. Note that the genus name is spelled with a capital letter and the species name with a small letter. *Felis pardalis* is the scientific name for an ocelot, with *pardalis* being the species name. *Felis concolor* is the scientific name for a cougar. What is the species name for this animal?

Point out the diversity in structure and appearance within the same genus. Tulips belong to the lily family.

The scientific name for a species is either underlined or printed in italics.

More than one million species of animals are known today. At least 324 000 species of plants have been named. You belong to the human species. Dogs, horses, and white oak trees are examples of other species.

The scientific name often includes the discoverer's name.

making sure

8. How are organisms classified? into groups based on their similarities and differences
9. List the five kingdoms in which organisms are classified. Give one example of an organism in each kingdom. See Table 14–2.

14:4 Classification 299

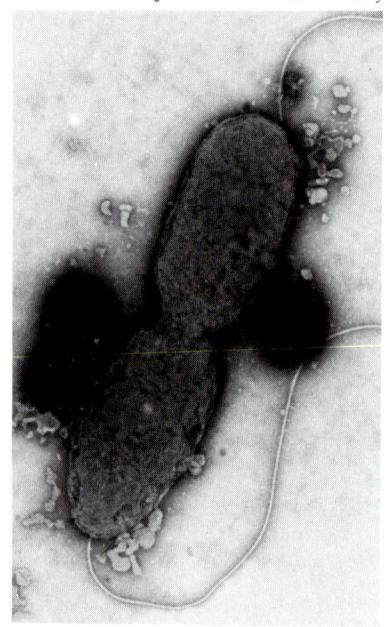

FIGURE 14–10. All bacteria contain cell walls. Some have flagella.

What are bacteria? Describe them. Where are bacteria found?

Do all bacteria need oxygen and heat to live?

FIGURE 14–11. Bacteria may be rods (a), spheres (b), and spirals (c).

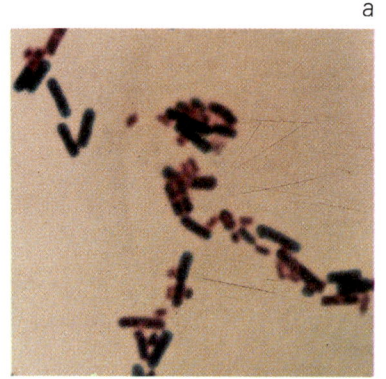

14:5 Bacteria

Bacteria are classified in the monera kingdom. The **monera** kingdom is composed of one-celled organisms with scattered nuclear material. Bacteria are one-celled organisms with cell walls. They are so small that 10 000 bacteria lined up in a row equals about 1 cm. Although bacteria do not have a nucleus, they do contain DNA. A few bacteria have whiplike structures called flagella (fluh JEL uh). The beating, whipping motion of the flagella moves the bacteria through a liquid. Organisms, such as bacteria that are so tiny we use a microscope to observe them, are called microorganisms (mi kroh OR guh nihz umz).

Bacteria are found almost everywhere. They are found deep in the oceans and high in the atmosphere. They cover your skin and are present in the air you breathe. Some bacteria live on dead plants and animals. Some live on other living plants and animals. Many live harmlessly inside your body.

Most bacteria need oxygen, a warm environment, food, and water. Some species of bacteria can live at temperatures below freezing. Others grow only in the absence of oxygen. A few species contain the green pigment chlorophyll and can make their own food.

Bacteria may be further classified by their shapes as spheres, rods, or spirals. From the three shapes, other forms are possible, such as pairs, groups, chains, or clumps.

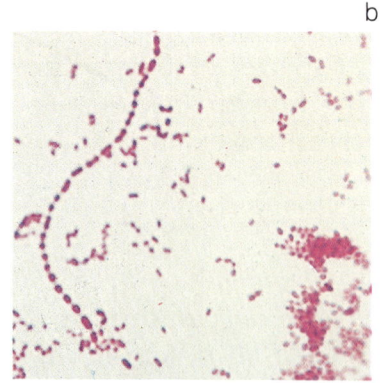

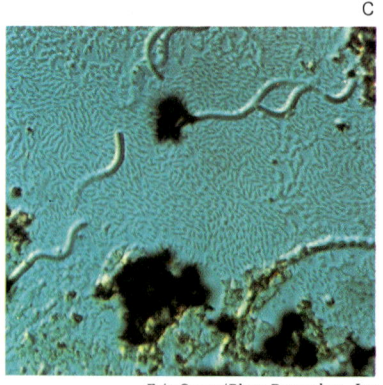

Bacteria reproduce by dividing into two parts. In this process, called fission, a cell wall forms in the middle of a bacterium cell. Two cells are produced as a result. A single bacterium cell is the parent in this kind of reproduction. When an offspring has only one parent it is called asexual reproduction. Asexual means that male and female organisms are not part of the reproductive process.

Under ideal conditions, bacteria can divide about every 20 mintues. At this rate, one bacterium could produce about 7000 metric tons of bacteria in three days. The rapid growth of bacteria accounts for the formation of large visible colonies as shown in Figure 14–12. A single colony contains billions of bacteria. The growth of a colony is limited because the bacteria poison themselves from the wastes they produce. Some bacteria in a large colony may not be able to get food. Usually, the growth rate of a bacteria colony is greatest on the first day. By the third day, growth often stops.

Bacteria can form endospores, which are structures containing a protective cell wall. Endospores can survive years of dryness, because the protective cell wall shields against harmful conditions. Some endospores survive boiling temperatures. Others are known to survive years at freezing temperatures. Many chemicals such as iodine kill bacteria, but not their endospores. When conditions improve, endospores grow into bacteria.

How do bacteria reproduce? Under what conditions do bacteria survive? What helps them survive these conditions?

Go over the concept of asexual reproduction, meaning without sex.

Bacteria can be helpful or harmful to people. Ask students to give examples of each.

Relate temperature to the growth of bacteria by discussing the storage of food in refrigerators.

FIGURE 14–12. Bacteria grow together in colonies. The colors and shape of a colony may depend on the species of bacteria.

A. Ottolenghi, Ohio State University

14:5 Bacteria   301

activity

See Teacher's Guide.

SEEING BACTERIA WITH A MICROSCOPE   Activity 14-3

Use a microscope to study prepared slides of different types of bacteria. The bacteria are made visible on the slide by using chemical stains which color the cells. Directions for using a microscope are given in Appendix D. First observe the slides with low power and then with high power. Draw the shape of each type and indicate the magnification used.

making sure

10. Why are the bodies of animals good places for the growth of certain kinds of bacteria?
11. How do bacteria respond to an unfavorable environment? **produce endospores**
12. How do bacteria reproduce? **fission (cell division)**
13. If bacteria reproduce so quickly, what controls the life span of a bacteria colony?
14. How are bacteria classified by shape? **rods, spheres, spirals**

Margin notes:
10. Bodies contain food, warmth, and moisture.
13. lack of food and wastes produced are poisonous

14:6 Fungi

Yeasts, molds, and mushrooms are classified as **fungi** (FUN ji). All species of **fungi** do not contain chlorophyll and therefore cannot make their own food. Fungi obtain their food from dead organic matter or living organisms on which they live.

Yeasts are one-celled organisms. Some yeasts obtain their food from sugar. Yeasts that live on sugar cause fermentation, a type of respiration. In fermentation, a yeast cell produces alcohol and carbon dioxide from sugar. The chemical change that occurs provides the energy the yeast needs to stay alive. Yeast is used in making raised dough products such as bread and rolls. Bubbles of carbon dioxide gas produced by the yeast cause the dough to rise. The fermentation of sugar by yeast is used in the commercial production of alcohol.

Margin notes:
Name three kinds of fungi.

Organic refers to compounds of carbon. Living or dead organisms are organic, meaning they contain carbon compounds.

What causes fermentation?

Life and the Cell

Asexual reproduction in yeast occurs through budding. In budding, a portion of the cell grows out from the rest of the cell. A new cell is formed when cell walls and cell membranes form between the cell and the "bud." The bud becomes a yeast cell offspring that has only one parent.

Molds grow well in warm, moist places where there is enough food. Mold can be grown on moist bread in a closed jar. Bread mold viewed with a hand lens appears as tiny threads. Under a microscope it can be seen that the bread mold threads are made of cells.

Molds reproduce by forming cells called spores. Tiny black spots visible on bread mold are spore cases. A spore case may contain millions of spores. Each spore is a single living cell. When the case matures, it opens and the spores spill out. They are often carried from place to place by wind. When a mold spore lands in a suitable place, it begins to grow. Many cell divisions result in a new growth of mold on the food.

Mushrooms are the most visible and well-known fungi. They grow in soil and decaying matter in damp areas of orchards, lawns, fields, and forests. The cap and stalk you see growing above the ground produce spores. A single mushroom may produce billions of spores. The spores are carried away by wind. If a spore lands in a suitable place, it begins forming a new mushroom.

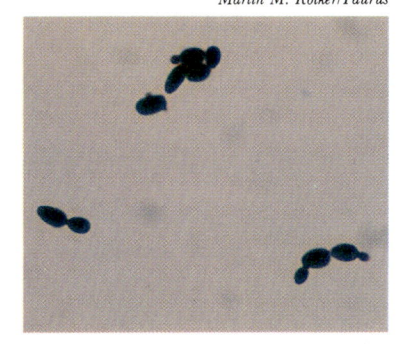

FIGURE 14–13. Yeast reproduces asexually through budding.

How does yeast reproduce?

How do molds reproduce?

Enrichment: Place a piece of bread made without preservatives in a plastic bag and seal it. Allow it to sit in a warm place for a few days and observe the growth of bread mold.

FIGURE 14–14. Mushrooms (a) and bread mold (b) are fungi.

14:6 Fungi

Why is it wise to avoid eating wild mushrooms?

Caution students not to eat wild mushrooms.

Although some species of mushroom are good to eat, many are poisonous. There is no simple way to tell a poisonous mushroom from one that can be eaten safely. Never eat wild mushrooms. You can be poisoned.

activity

OBSERVING YEAST   Activity 14-4

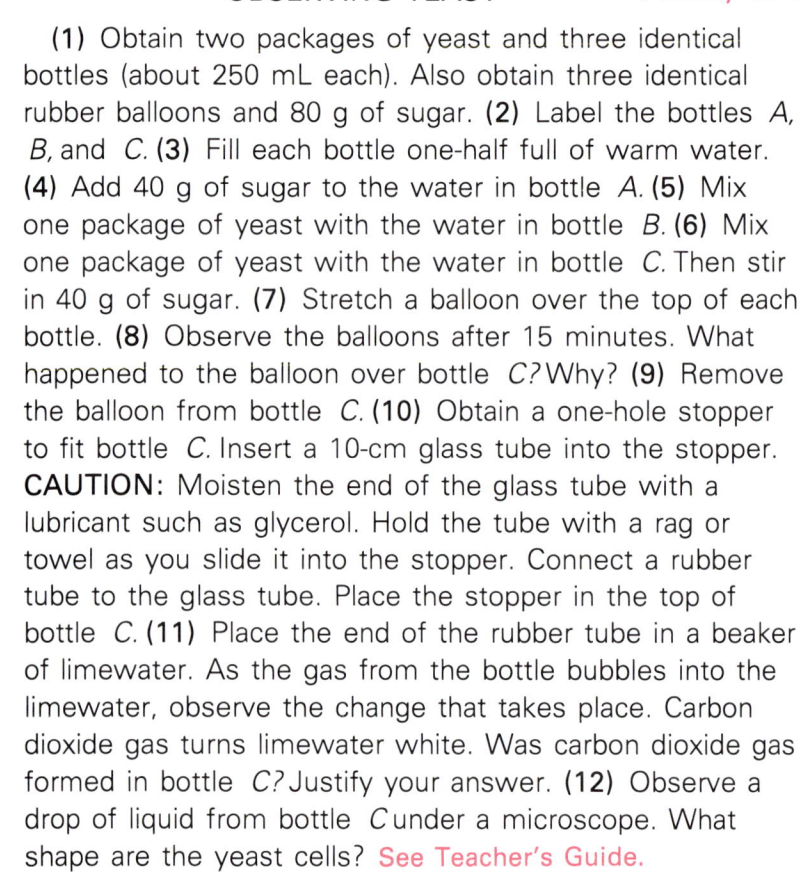

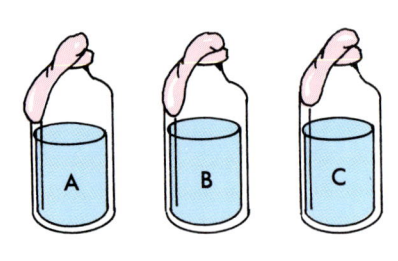

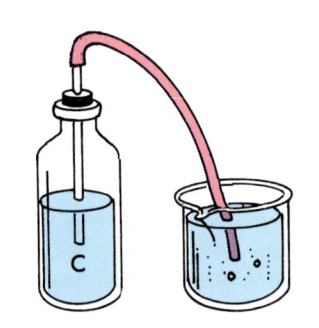

(1) Obtain two packages of yeast and three identical bottles (about 250 mL each). Also obtain three identical rubber balloons and 80 g of sugar. (2) Label the bottles *A, B,* and *C.* (3) Fill each bottle one-half full of warm water. (4) Add 40 g of sugar to the water in bottle *A.* (5) Mix one package of yeast with the water in bottle *B.* (6) Mix one package of yeast with the water in bottle *C.* Then stir in 40 g of sugar. (7) Stretch a balloon over the top of each bottle. (8) Observe the balloons after 15 minutes. What happened to the balloon over bottle *C?* Why? (9) Remove the balloon from bottle *C.* (10) Obtain a one-hole stopper to fit bottle *C.* Insert a 10-cm glass tube into the stopper. **CAUTION:** Moisten the end of the glass tube with a lubricant such as glycerol. Hold the tube with a rag or towel as you slide it into the stopper. Connect a rubber tube to the glass tube. Place the stopper in the top of bottle *C.* (11) Place the end of the rubber tube in a beaker of limewater. As the gas from the bottle bubbles into the limewater, observe the change that takes place. Carbon dioxide gas turns limewater white. Was carbon dioxide gas formed in bottle *C?* Justify your answer. (12) Observe a drop of liquid from bottle *C* under a microscope. What shape are the yeast cells? See Teacher's Guide.

FIGURE 14-15.

making sure

15. Why are yeasts, molds, and mushrooms classified as fungi? — no chlorophyll, live on other organisms or dead organic matter
16. How is reproduction in yeasts different from reproduction in molds? Yeast reproduces by budding.
17. What kind of respiration takes place in yeast? — fermentation
18. Why is budding considered a form of asexual reproduction? only one parent
19. How do spores move from place to place? wind blows them, carried on bodies of people and animals

Life and the Cell

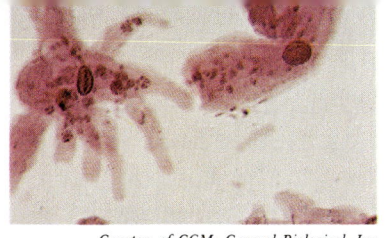

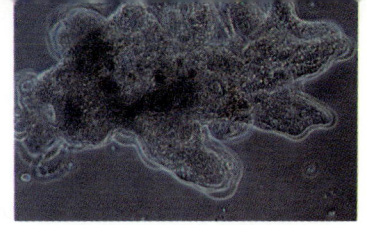

FIGURE 14–16. Two species of Amoeba. Amoebas move by using fingerlike projections of cytoplasm.

14:7 Amoebas and Paramecia

Amoebas (uh MEE buz) are one-celled organisms that move by changing their shape. They are members of the protist kingdom. Members of the **protist** kingdom are one-celled with a nucleus surrounded by a nuclear membrane. Amoebas are found in pond water on the underside of plant leaves. They are also found in the bodies of animals and in moist soil. An amoeba moves by pushing its cytoplasm against the cell membrane to form slender, fingerlike projections. These projections stick out in the direction the amoeba is moving. The rest of the cell mass flows along behind as the amoeba moves forward.

Amoebas eat tiny protists and bits of dead matter. When an amoeba comes in contact with food, the amoeba changes shape to surround the food. The food is trapped in a bubble-like structure called a food vacuole (VAK yuh wohl). Within the vacuole, the food is digested. In digestion, food is broken down into simple substances that can be used by the amoeba. The digested food passes from the food vacuole into the cytoplasm. Here it is used in respiration to produce energy or make more protoplasm. An amoeba obtains oxygen from the water around it by diffusion.

Amoebas and paramecia illustrate basic life activities. The life of these organisms is more complex than that of fungi, just as fungi are more complex than the bacteria.

How does an amoeba move?

How does an amoeba get food and oxygen?

FIGURE 14–17. An amoeba gets food by surrounding it with cytoplasm.

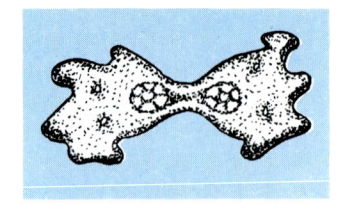

FIGURE 14–18. Amoebas reproduce asexually through fission.

What is a cyst?

How does a *Paramecium* move?

How does a *Paramecium* get food?

Stress the similarities and the differences of the amoeba and the paramecium.

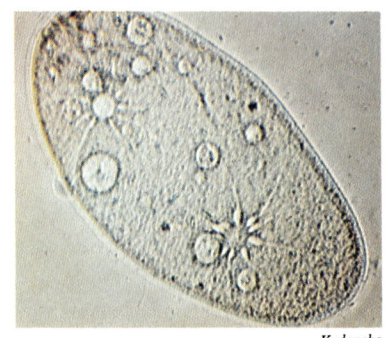

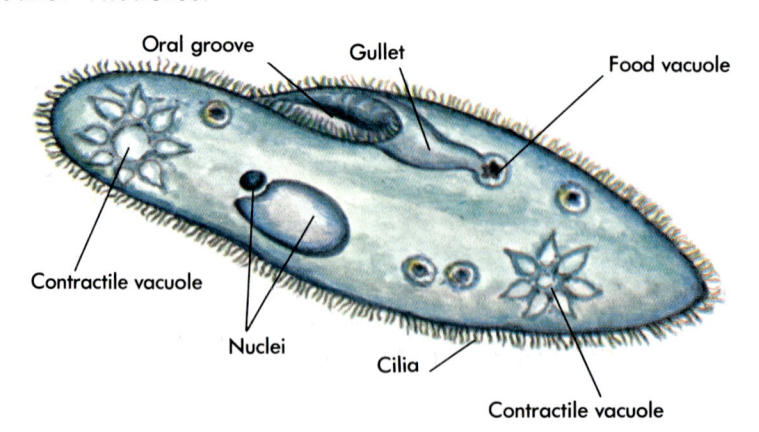

FIGURE 14–19. Paramecia have many structures or features.

Waste products are expelled by the amoeba through its cell membrane. Another vacuole collects excess water from the cytoplasm. The vacuole then forces the water out through the cell membrane.

An amoeba reproduces by fission. **Fission** is a reproductive process in which the nucleus of a cell splits in two. The two parts of the nucleus move to opposite ends of the cell. The cell separates slowly between the two nuclei until it splits, forming two cells (Figure 14-18). During unfavorable conditions, an amoeba may form a cyst (SIHST). A cyst is a cell enclosed in a thick, protective covering. Cysts enable amoebas to survive until conditions improve.

A *Paramecium* is another kind of protist. A *Paramecium* is one-celled and shaped like a shoe (Figure 14–19). The cell membrane of a *Paramecium* is covered with short, hairlike structures called cilia (SIHL ee uh). By beating the cilia, a *Paramecium* moves through water. Paramecia live in fresh water, salt water, and in animals. Paramecia move slowly. By reversing the motion of the cilia, a *Paramecium* can move backward.

The funnel-shaped area along one side of a *Paramecium* is called an oral groove. Cilia lining the oral groove beat inward and move food particles into the cell. Here the food is collected in the food vacuole. When filled, the vacuole breaks away and moves about inside the cell. Food in the vacuole is digested and passes into the cytoplasm. Wastes are expelled by paramecia through the anal spot and other vacuoles.

306   Life and the Cell

A *Paramecium* has two nuclei. The larger nucleus controls the activities of the paramecium. The smaller nucleus functions only during reproduction. In paramecia, reproduction by fission occurs in much the same way as it does in amoebas. Under ideal conditions, paramecia divide several times a day.

When conditions are unfavorable for fission, paramecia undergo a process called conjugation (kahn juh GAY shun). In conjugation, two paramecia join at their oral grooves. The larger nuclei of both cells break and slowly disappear. Inside each cell the smaller nucleus divides. Then part of the smaller nucleus in one cell joins part of the smaller nucleus in the other cell, and vice versa (Figure 14–20). The paramecia then move apart and each organizes a new large nucleus and a new small nucleus. Since there are still only two paramecia after conjugation, there is no increase in the number of cells. However, conjugation makes the paramecia hardier and better able to continue their life.

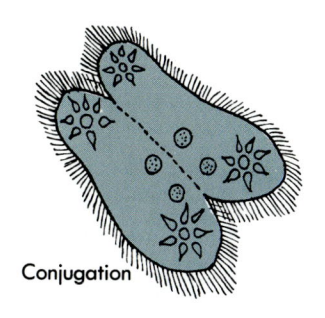

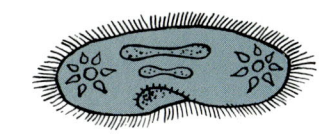

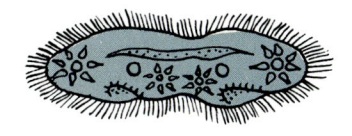

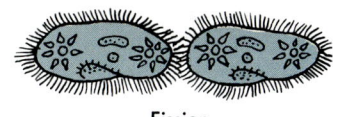

FIGURE 14–20. Paramecia reproduce by conjugation and by fission.

activity

See Teacher's Guide.   Activity 14-5

EXAMINING AMOEBA MOVEMENT

Make a ring on a clean glass slide by dipping the mouth of a test tube in petroleum jelly and pressing it against the slide. Place a drop of amoeba culture inside the ring. Place a coverslip over the slide. Observe the amoebas under a microscope using low power and dim light. Why is it necessary to seal the culture with petroleum jelly? so it does not dry out

How does a *Paramecium* reproduce?

activity

Activity 14-6

PROTISTS IN A HAY INFUSION

Make a hay infusion culture by boiling a handful of hay or dry grass in a liter of water. Place the culture in a jar and add a few grains of boiled rice. Let the mixture stand for 2 days; then add some pond water. Let the mixture stand for a few more days. Study the culture with a microscope. Describe the characteristics you see for any paramecia, amoebas, or other protists in the culture. Answers will vary.

14:7 Amoebas and Paramecia

making sure

20. What would happen to an amoeba if it were not able to expel wastes? *It would poison itself and die.*
21. A single amoeba was placed in a large tank. Within an hour it divided to form two amoebas. Assume that offspring continued to divide once every hour. If they all survive, how many amoebas would be in the tank after 12 hours? *4096*
22. What are the similarities in the way amoebas and paramecia obtain food? What are the differences? *Both take in food through the cell membrane into a food vacuole. An amoeba uses pseudopods. Paramecia take in food through the oral groove.*
23. What is the function of the smaller nucleus in the paramecium? *reproduction*
24. How does an amoeba respond to unfavorable conditions? *forms a cyst*

14:8 Flagellates

What is a flagellate?

A *Euglena* (yoo GLEE nuh) is an example of a flagellate, a one-celled protist with one or more flagella. Recall that a flagellum is a long, whiplike extension of a cell. By beating its flagella, a *Euglena* moves itself through water. A *Euglena* contains the green pigment chlorophyll which it uses to make food when light is present.

At one time botanists classified Euglena as a plant because it contains chlorophyll. Zoologists classified it as an animal because it is motile.

Some species of flagellates cause disease. African sleeping sickness is caused by a flagellate carried by the tsetse (SEET see) fly. People and animals can become infected by a bite from the tsetse fly. A person with the disease suffers from high fever, headache, and skin rash. Eventually, a person experiences mental dullness, paralysis, coma, and convulsions. If treatment with arsenic compounds is not given, death follows.

Some flagellates live together in colonies. For example, *Volvox* (VAHL vahks) is a flagellate which lives in a colony shaped like a ball about 1.5 mm in diameter. The walls of the ball contain thousands of *Volvox* cells joined by threads of cytoplasm. Flagella beat together to move the colony through the water.

Life and the Cell

a b

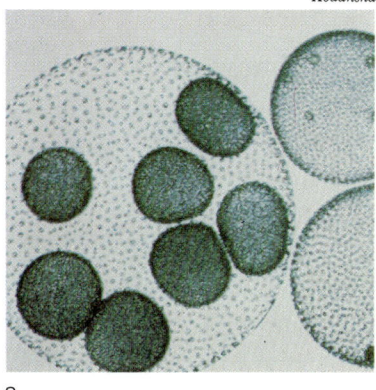

c
FIGURE 14–21. *Volvox* (a) is a colonial flagellate. *Euglena* (b) is a flagellate which can make its own food. A flagellate responsible for African sleeping sickness, is carried by tsetse fly (c).

making sure

25. How is the movement of paramecia, amoebas, and euglena different? paramecia—cilia, amoebas—pseudopods, *Euglena*—flagella
26. Why might some people classify *Euglena* in the plant kingdom? has chlorophyll

14:9 Sporozoans

The protist kingdom also includes sporozoans (spor uh ZOH unz). Unlike amoebas and paramecia, sporozoans cannot move to obtain food. All sporozoans obtain food from the organisms in which they live. During one stage in the life cycle, sporozoans produce spores for reproduction. Sporozoans cause disease. Malaria (muh LER ee uh) and certain fevers in cattle are diseases produced by sporozoans. Sporozoans live in muscles, kidneys, and other parts of an animal.

A sporozoan goes through several different stages in its life cycle. Sometimes different parts of its life cycle occur in different animals. The life cycle of the sporozoan that causes malaria in humans involves a human and a mosquito.

How are sporozoans harmful?

making sure

27. How are sporozoans different from flagellates and other protists? Sporozoans cannot move to obtain food
28. How are sporozoans similar to certain fungi in the method they use to obtain food? obtain food from other organisms

14:10 Viruses

List the main characteristics of viruses.

Viruses are smaller than cells. They are so small that they cannot be seen with an ordinary microscope. A virus has features of both living and nonliving things. It is not classified in any of the five kingdoms. Viruses reproduce inside the living cells of organisms and can be grown on living tissue. For example, certain viruses can be grown on chick embryos. Viruses can also be changed to a crystal form that does not reproduce. In the crystal form they are like nonliving substances. A virus is made of a core of either DNA or RNA surrounded by protein. Pictures of viruses enlarged more than 100 000 times have been made with electron microscopes. These pictures show that viruses can be rodlike, or hexagonal in shape. A hexagonal-shaped virus has a tail or stalk (Figure 14-22a).

A virus grows and reproduces in a living organism. Viruses may be removed from a plant or animal, crystallized, and stored. The stored virus may be later injected into a plant or animal where it will resume growth.

Viruses can cause disease in plants, animals, and people. The common cold, flu, smallpox, and polio are spread by air, food, water, and living organisms. When viruses infect a living organism, they invade and take over some of its cells. The virus becomes the "master" and the cell becomes a "slave." Normal activities of the cell change. It begins producing only more viruses. Eventually the cell bursts and releases new viruses. Once released, the new viruses can infect other cells.

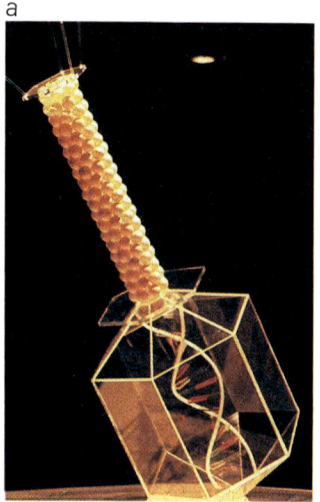

Scientists search for vaccines to immunize people against the common cold which is believed to be caused by a number of different viruses.

FIGURE 14–22. A model of a virus (a). Smallpox virus magnified 47 500 times (b).

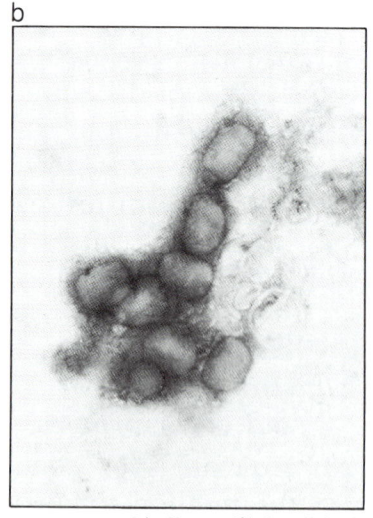

Jim Elliot

John Hanson, Ohio State University

310   **Life and the Cell**

Viruses have been described as genes in search of a cell. A gene is a section of a DNA molecule which determines hereditary traits. When a virus enters a cell, the RNA or DNA of the virus controls the cell's chemical reactions.

One theory states that the first life on Earth was a virus of some kind. But this theory does not fit all the facts. Viruses can only grow in the cells of living organisms. Therefore, viruses as we know them could not have been the first form of life.

Another theory states that viruses are genes that got "out of hand." Viruses may be DNA or RNA molecules that escaped from cells long ago in some unexplained way. These "out of hand" DNA molecules multiplied by taking control of a cell's DNA. Today we call them viruses.

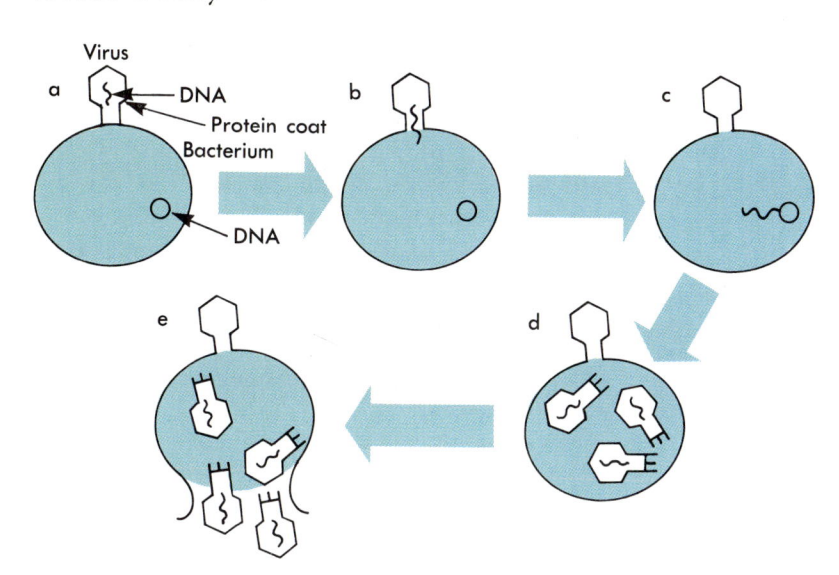

FIGURE 14-23. Some viruses attack bacteria. The virus attaches to the cell wall (a). The DNA from the virus is injected into the bacteria (b). The DNA from the virus attaches to the DNA of the bacteria (c). The DNA from the virus takes over and new viruses are formed within the bacteria (d). Within a short time bacteria burst open and the new viruses are released (e).

making sure

29. What features of a virus are similar to those of a living organism? reproduction
30. What feature of a virus makes it appear to be nonliving? can be crystallized and stored for many years as with any nonliving substance
31. If we accept the theory that the first form of life on Earth were viruses why would those viruses have to have been different than present-day viruses? Present-day viruses can only grow inside other organisms.

PERSPECTIVES

people

A Dedicated Microbiologist

Microbiology is the study of small organisms, also called microorganisms. Bacteria, viruses, fungi, and some algae are common examples of microorganisms. Natalie Jones graduated from college with a degree in microbiology. Numerous careers that were open to her included work in hospitals, private laboratories, food industries, and environmental agencies.

Natalie Jones decided to work with more than just microorganisms. Her job at the Ohio State University microbiology laboratories is to help over 1000 students perform experiments. The students that she works with are involved in pre-medical studies, pharmacy, veterinary medicine, optometry, dentistry, nursing, and other sciences. Microbiology is an important aspect of each of these sciences.

As an instructional assistant, her job is to make sure that each student receives the necessary equipment and supplies to perform a required experiment. The job is not easy, as Ms. Jones must know every aspect of the experiment. Many of the microorganisms that she works with are pathogenic, or disease-causing. She has to be very organized to make necessary materials available to students and researchers. All supplies, equipment, and specimens must be ordered well in advance of an experiment.

Cultures, or containers with living organisms, are maintained under the supervision of Natalie Jones. Such cultures cannot be contaminated with other microorganisms that are floating in the air. Special techniques are used when working with these microorganisms, and are monitored by Ms. Jones.

Natalie Jones also schedules all laboratory experiments, and supervises the maintenance of all equipment. She helps write laboratory manuals for the experiments, and helps prepare the examinations. If any experiments fail, she is involved in modifying procedures until they work.

Ms. Jones does not work alone. She supervises 40 teaching associates and four lab assistants. The success of the microbiology laboratory programs at the university depends on Natalie Jones and the individuals that she supervises.

Not all of Ms. Jones' efforts are spent with microbiology courses, however. Frequently, high school teachers seek advice from her about projects. Sometimes they need equipment, supplies, or information. She also helps teachers dispose of potentially dangerous cultures. Natalie Jones emphasizes that the most enjoyable part of her job is working with people who are interested in learning about microbiology.

Doug Wynn

main ideas — Chapter Review

1. The features of living things include reproduction, metabolism, movement, ability to respond to a stimulus, and a definite size, shape, and lifespan.   14:1
2. A cell contains a cell membrane, nucleus, and cytoplasm.   14:2
3. Cells derive energy from a process called respiration.   14:3
4. The scientific name of an organism contains the genus and species names.   14:4
5. Bacteria and blue-green algae are classified in the monera kingdom. All bacteria lack nuclei.   14:5
6. Yeasts, molds, and mushrooms are classified in the fungi kingdom.   14:6
7. Amoebas and paramecia are one-celled organisms in the protist kingdom.   14:7
8. A *Euglena* is a flagellate that contains chlorophyll.   14:8
9. Sporozoans obtain food from the organisms in which they live.   14:9
10. Viruses have living and nonliving features. They are made of a DNA or RNA core covered with protein.   14:10

vocabulary

Define the following words or terms.

cell	genus	nucleus	respiration
cytoplasm	kingdom	organism	species
diffusion	metabolism	osmosis	stimulus
fission	microorganism	protist	
fungi	monera	protoplasm	

study questions

DO NOT WRITE IN THIS BOOK.

A. True or False

Determine whether each of the sentences is true or false. If the sentence is false, rewrite it to make it true.

T   1. People belong to the species known as *Homo sapiens*.

Life and the Cell

F   2. The number of plant species is ~~greater~~ than the number of animal species. less

T   3. A scientific name includes the genus and species name of an organism.

F   4. One ~~species~~ name is *Canis*. genus

T   5. DNA contains a "life code."

F   6. Yeasts and molds ^ contain chlorophyll. do not

F   7. Amoebas move by ~~whipping their flagella~~. using fingerlike projections

T   8. *Euglena* contains chlorophyll.

F   9. One type of ~~paramecium~~ causes sleeping sickness. flagellate

F   10. Most viruses are ~~larger~~ than bacteria. smaller

B. Multiple Choice

Choose the word or phrase that correctly completes each of the following sentences.

1. The (<u>cell membrane</u>, nucleus, cytoplasm, vacoule) controls the movement of material into and out of the cell.
2. The lion, house cat, leopard, and tiger belong to different (phylum, class, genus, <u>species</u>).
3. Yeasts and molds are classified in the (monera, <u>fungi</u>, plant, animal) kingdom.
4. Diffusion of water through a membrane is called (respiration, fermentation, <u>osmosis</u>, metabolism).
5. A (<u>kingdom</u>, class, phylum, genus) is the largest group in the classification of organisms.
6. The (<u>monera</u>, plant, animal) kingdom includes bacteria.
7. Paramecia move by the use of (<u>cilia</u>, false feet, flagella, spores).
8. Yeast produces alcohol and carbon dioxide during (photosynthesis, <u>fermentation</u>, fission, osmosis).
9. Malaria is caused by a species of (mosquito, <u>sporozoan</u>, amoeba, bacteria).
10. Molds reproduce by (osmosis, <u>spores</u>, fermentation).

C. Completion

Complete each of the following sentences with a word or phrase that will make the sentence correct.

1. In the scientific name *Felis domestica*, *domestica* is the <u>species</u> name.

2. Molecules diffuse from a region of __high__ concentration to regions of __low__ concentration.
3. A *Paramecium* has __two__ nuclei.
4. A complete and entire living thing is called a(n) _____. organism
5. A plant cell __wall__ is rich in cellulose.
6. A virus can be seen with a(n) __electron__ microscope.
7. Every living organism is composed of one or more __cells__.
8. _____ form a cap and stalk above ground to produce spores. mushrooms
9. __Volvox__ is a species which lives in a ball-shaped colony.
10. Amoebas reproduce asexually by a process called __fission__.

D. How and Why See Teacher's Guide in front of this book.
1. How are living things different from nonliving things?
2. Draw a diagram of a cell and label the parts.
3. What advantages do scientific names have over common names for organisms?
4. Why is reproduction necessary for the survival of a species?
5. Why are bacteria able to survive in so many different environments?

challenges

1. Write a report on malaria, African sleeping sickness, or some other disease caused by a protist. Include in your report a discussion of the controls used to curb the spread of the disease and the effectiveness of the controls.
2. Obtain permission to visit a hospital or clinic laboratory to learn how bacteria are identified and cultured. Make a report to your class.

interesting reading

Simpson, Lance L. "Deadly Botulism," *Natural History.* January 1980, pp. 12–24.

Hall, Stephens, "The Fly." *Science '83*, November 1983, pp. 56-64.

The butterfly pea is a flowering plant. It grows in many areas and thrives in sandy soil. Many plants, however, grow only in water and do not have flowers. Some do not have stems or leaves. What plants consist only of single cells? How are desert plants different from tropical plants? What processes are common to all plants?

Gwen Fidler

Plants

chapter 15

Introducing the chapter: Arrange a display of plants from school or home. Briefly, identify and discuss the various features of each plant. Have students compare and contrast the features of each plant discussed.

15:1 Algae

Organisms in the plant kingdom come in a wide variety of sizes, shapes, and forms. Palm trees, seaweed, corn, grass, and dandelions all belong to the plant kingdom. Many useful products come from plants. Wood, charcoal, cotton, and paper are just a few of these products. All animals, including people, depend on plants for food.

One group of organisms that belong to the plant kingdom is algae. Algae are mostly one-celled plants which contain chlorophyll. **Chlorophyll** is a green substance in plant cells that is used to make food. It is chlorophyll that makes plants green.

Some species of algae are many-celled and very large. Algae are found from the poles to the equator. Certain species live in hot springs at temperatures as high as 89°C. Most algae live in fresh water or the ocean.

Protococcus is one of the many species of green algae. *Protococcus* grows on tree trunks. It forms a green coating on the bark. The cells are so small that it takes millions of them to cover a few square centimeters. *Protococcus* is a single, round or oval cell somewhat like yeast. Each cell contains a cell wall, membrane, cytoplasm, and nucleus. *Protococcus* differs from yeast in that it contains chlorophyll and can make its own food. Protococcus reproduces by fission.

GOAL: You will learn the main features of different kinds of plants and their life activities.

Remind students that blue-green algae are monerans. The algae included in the plant kingdom are listed in Appendix E on page 511.

What are algae?

The green growth on the inside of aquarium tanks and the green slime in ponds are examples of algae. Algae may someday be grown on a large scale for use as animal and human food. Also it may have use as a fuel.

Protococcus is sometimes mistaken for moss.

What is *Protococcus* and how does it differ from a yeast cell?

317

FIGURE 15–1. Kelp is a large brown algae (a). Diatoms are single celled algae which occur in thousands of different shapes (b).

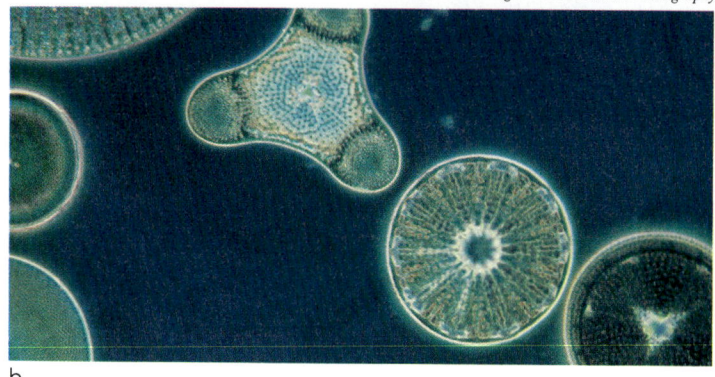

Protococcus cells live together in colonies of two, four, or six cells. Each cell contains many green chloroplasts (KLOR uh plasts), the disc-shaped bodies that contain chlorophyll. It is in the chloroplast that food is made.

making sure dry out and die when not moist
1. Why do most algae live in water environments?
2. How does *Protococcus* reproduce? fission

Spirogyra can be viewed with an overhead projector.

15:2 Mosses and Liverworts

Mosses and liverworts are small, many-celled plants that contain chlorophyll. They live close to the ground and rarely grow to more than a few centimeters in height. They do not have true roots, stems, and leaves. Mosses and liverworts lack the kind of transport system for moving water that is found in higher plants. Water and food move through moss and liverwort plants by diffusion.

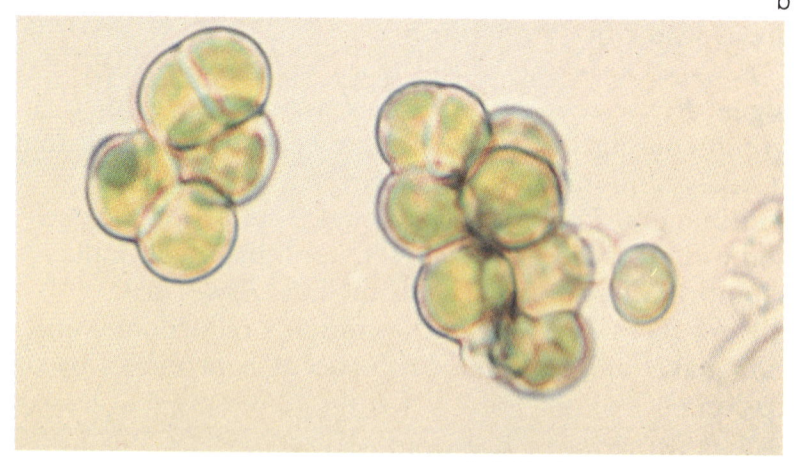

How do water and food move through mosses and liverworts?

FIGURE 15–2. *Spirogyra* (a) and *Protoccus* (b) are two species of algae.

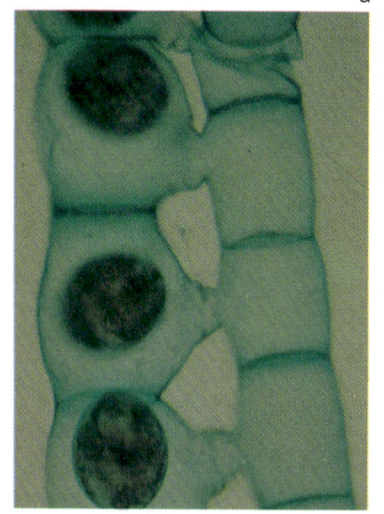

318 Plants

Liverworts grow only in wet places such as the banks of a stream or spring. A liverwort looks like a leathery "leaf" lying flat against the ground. Hairlike structures called rhizoids (RI zoydz) anchor the plant and absorb water from the soil. *Marchantia*, one species of liverwort, produces umbrella-shaped reproductive structures about 2.5 cm long.

Moss grows in shaded, moist areas such as cracks in shaded sidewalks or moist soil under trees. Moss plants appear to have "stems" with many "leaves." Rootlike rhizoids grow from the base of a moss "stem."

Liverworts and mosses do not have true roots, stems, and leaves as do seed plants. Note the quotation marks for these words in the text.

List the main characteristics of mosses and liverworts.

Alvin E. Staffan   *Roger K. Burnard*

a   b

FIGURE 15–3. *Marchantia* has umbrella-shaped reproductive structures (a). Mosses do not have true roots, stems, or leaves (b).

activity

See Teacher's Guide.

GROWING LIVERWORTS   Activity 15-1

Obtain some *Marchantia* from a plant store. Grow the Marchantia in a terrarium containing a layer of gravel and a layer of topsoil and humus. Cover the terrarium with a piece of glass. Observe the plants and watch for the growth of reproductive structures. Why should the relative humidity in the terrarium be kept at a high level? so plants do not dry out

making sure

3. Explain why mosses and liverworts grow close to the ground. They do not have a system to carry water and food very far.
4. Why do mosses and liverworts grow in damp areas? do not have roots to obtain water, dry out and die if not kept moist
5. How are mosses and liverworts similar to algae? contain chlorophyll, need moisture, made of cells How are they different? Most algae are one-celled organisms. Mosses and liverworts are many celled.

15:2 Mosses and Liverworts

15:3 Ferns

What conditions are best for ferns to grow and reproduce?

Students may be aware of ferns used by florists. Different kinds of ferns may be obtained from a local florist or nursery for study in class. Review asexual reproduction.

Ferns are larger than mosses and liverworts. They have true roots, stems, and leaves, and tubelike structures to carry water and food to all parts of the plant. Ferns grow in woods, swamps, and gardens where much moisture is present. Millions of years ago large fern forests covered the earth. At that time most of the land was wet and marshy, and the climate was warm. Tree ferns 9 to 12 m high were common. Today's ferns are much smaller.

Ferns have a life cycle with two stages. In one stage, small patches of spore cases form on the underside of a fern. When ripe, the spore cases open and spores are released. In a suitable environment, a spore grows to form a small plant about 0.5 to 1 m in diameter. When an organism is produced from a spore, it is by asexual reproduction. Asexual means there is only one parent.

In the second stage of reproduction the new plant produces sex cells—eggs and sperm. Sperm swim to an egg through rainwater or dew and one unites with the egg. The fused egg and sperm grow to form a new fern plant. This second stage is sexual reproduction. In sexual reproduction, two different sex cells unite to produce the offspring.

The sperm cells need water to move to the egg cells for fertilization.

FIGURE 15–4. Ferns (a) reproduce asexually by forming spores on the underside of the leaves (b).

making sure

6. Explain why ferns are able to grow several meters in height while mosses rarely exceed several centimeters. *Ferns have a system to transport water and food. Mosses do not.*
7. Why must ferns live in a moist environment in order to reproduce?

a

b

Roger K. Burnard

Roger K. Burnard

15:4 Seed Plants

Most of the plants you see are seed plants. There are two groups of seed plants—gymnosperms (JIHM nuh spurmz) and angiosperms (AN jee uh spurmz). **Gymnosperms** produce seeds that are not enclosed in a fruit. Pine, spruce, redwood, and fir trees are gymnosperms. Some gymnosperms, called conifers, produce seeds in cones.

Angiosperms have flowers and produce seeds inside a fruit. Maple trees, grass, corn, daisies, orchids, and tomatoes are angiosperms. More than half the known species of plants are angiosperms.

Enrichment: Explore the school grounds or a nearby park. Identify the kinds of seed plants growing there.

Name the two groups of seed plants.

Enrichment: Have students name several products we obtain from seed plants; for example, food, cotton, wood, and maple syrup.

FIGURE 15–5. Spruce trees (a) and other conifers produce seeds in cones (b).

Seed plants have true roots, stems, and leaves. Each of these parts is made of cells which are grouped together to form tissues. A **tissue** is a group of cells performing the same function. For example, a stem contains tissue whose function is to carry water up the stem. An **organ** is a group of tissues working together to perform a function. Organs in a flower produce cells that form seeds allowing a seed plant to reproduce.

Define tissue and organ.

FIGURE 15–6. Cherry trees have colorful flowers (a) and produce seeds inside a fruit (b).

15:5 Roots

What are some functions of roots?

Enrichment: Have students draw a full-page diagram of a plant showing the roots, stem, and leaves. On the diagram the student should list the functions of each part.

Roots anchor a plant. They also absorb, store, and transport water and dissolved substances. Have you ever tried to pull a plant out of the ground? A root system holds the plant firmly in the soil. Otherwise, wind and rain would easily uproot it. The root system of a plant spreads out from the plant. It sends many tiny roots throughout the soil in a complex pattern.

The size of a root system varies with different plants. Most land plants have as much or more growth below ground as above. The roots of some plants, such as alfalfa, are long and slender. Alfalfa roots may grow 5 m or more down into the soil.

Describe the structure and function of a root hair.

Small root hairs spread outward into the soil from the roots. Each root hair is a single cell that moves water and dissolved minerals through its cell membrane by osmosis.

Water and dissolved minerals travel through the root hairs into the xylem (XI lum) tissue of the root. **Xylem** tissue contains specialized xylem cells that transport water and dissolved minerals upward to all parts of the plant.

In many plants, the root is the main organ of food storage. The potato is an example of a root that is filled with stored food. Potatoes and most other roots in which food is stored are rich in starch. The stored starch allows the turnip plant to survive during the winter and to grow again in the spring.

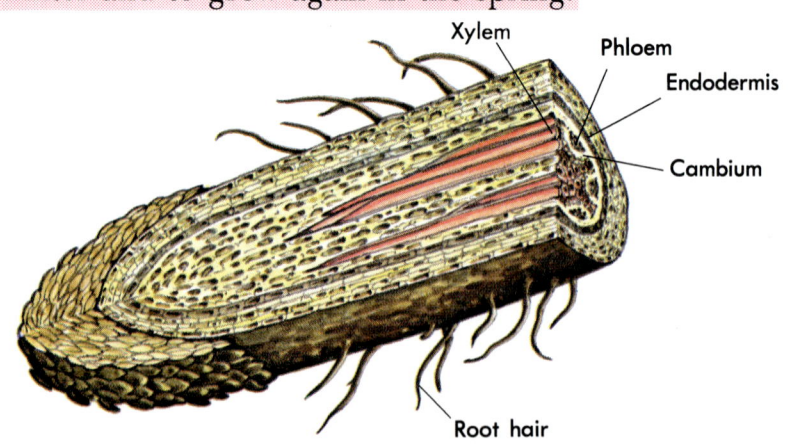

How are some plants like turnips able to survive winter and grow again in the spring?

Enrichment: Study the cross section of a root, longitudinal section of a root tip, and root hair. A carrot is suitable for preparing microscope slides of root structures.

FIGURE 15–7. Tissues in a root.

The growth of roots is affected by many factors in the environment. Gravity and water are two factors which have a great effect on roots. Roots respond to gravity by growing downward. They respond to water by growing toward the water source. The response of roots to water explains why tree roots often clog sewer pipes. Water leaking from a pipe stimulates the roots to grow toward it.

William D. Popejoy

What two factors affect the growth of roots?

The growth of the root is the response. Gravity and water are the stimuli.

a   See Teacher's Guide.   b

FIGURE 15-8. The root system of some plants consists of many roots and root hairs (a). The root system of a carrot consists of one main root (b).

activity

ROOTS AND ROOT HAIRS   Activity 15-2

Soak some radish seeds in water overnight. Place them in a petri dish on a piece of a moist paper towel. Cover and allow to stand in a warm place (28°C) for several days. Keep the paper towel moist. Use a binocular microscope or hand lens to study the roots. Be certain students identify the root hairs.

Soaking softens the seed coat.

Seeds must be kept moist for germination.

activity

HOW DO PLANT ROOTS RESPOND TO WATER?   Activity 15-3

Make a ball of sphagnum moss and hold it together with some cord. Soak seeds, such as radish or oats, in water. Place the seeds at the top, sides, and bottom in the outer portion of the moss. Hang up the ball and keep it moist for two weeks by spraying it daily with water. Watch the seeds sprout and begin to grow. Observe how the roots grow. Is the water a greater stimulus to root growth than gravity? yes   See Teacher's Guide.

The ball must be kept moist at all times. The ball can be placed in a sealed plastic bag and taken home for watering on weekends.

15:5 Roots   323

making sure Water increases the growth rate of roots.

8. How is the growth of roots affected by the amount of water available?
9. How does stored food aid the survival of a turnip plant? Food is available for use by the plant as needed.
10. Name at least five functions of roots.
    anchor plant, take in water, take in minerals, store food, absorb, store, and transport water

15:6 Stems

What is the main function of the stem?

What three tissues are found in stems?

Name the two types of plant stems. How are they different?

The transporting of water, minerals, and food is the main function of the stem. The stem also supports the leaves, allowing them to be exposed to sunlight. Food is made in the stems of some plants. Geranium and cactus are plants in which stem cells make food.

Stems contain several types of tissues. These include xylem, phloem (FLO em), and cambium (KAM be um). Xylem tissue in the stem, as in the root, transports water and dissolved minerals upward. **Phloem** tissue transports food substances downward. **Cambium** is the growth tissue of the stem. It produces new xylem and phloem cells.

There are two types of plant stems—herbaceous (hur BAY shus) and woody. Herbaceous stems are soft

FIGURE 15-9. Annuals such as corn have herbaceous stems and must be replanted each year (a). The apple trees have woody stems and live to produce fruit for several growing seasons (b).

a USDA

b David M. Dennis

and green. Usually they lack woody tissue and can be bent. Tomatoes, beans, peas, and corn are plants with herbaceous stems. Most plants with herbaceous stems live only one growing season. These plants are called annuals. Annual means once a year. Thus, an annual is a plant that grows, reproduces, and dies during one growing season.

Plants with woody stems are called perennials. A perennial is a plant that lives for several growing seasons. In fact, the age of a woody-stem plant may be found by counting the annual growth rings. During the spring and summer months, a woody stem increases in length and thickness and forms branches. Its woody tissue makes it rigid and stiff. Trees and shrubs have woody stems. The annual rings in a woody stem are new xylem cells produced by the cambium. The order and arrangement of xylem, phloem, and cambium tissue in herbaceous and woody plants is shown in Figure 15-11.

You probably think of a stem as the part of a plant that grows above the ground. Some plants, however, have underground stems. An underground stem is called a **rhizome.** Many grasses produce rhizomes, which spread the grass plants.

FIGURE 15-10. Plants with wood stems exhibit annual growth rings.

What is the difference between an annual and a perennial?

Define rhizome.

FIGURE 15-11. Tissues in a stem.

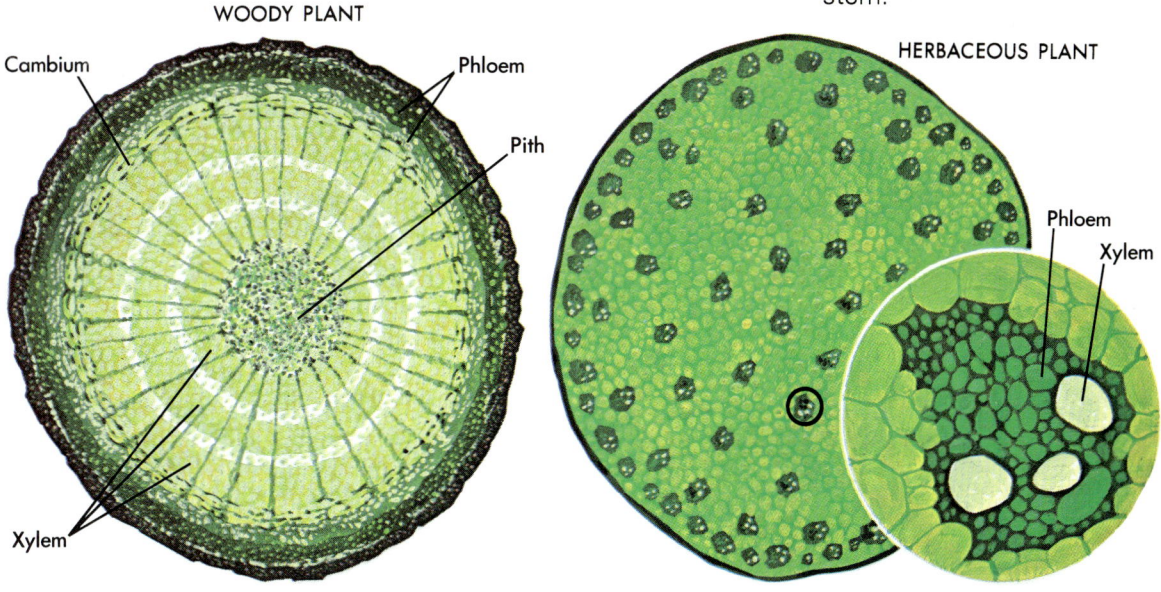

FIGURE 15–12. Different kinds of underground stems.

Many plants can be reproduced from a piece of stem, a method called vegetative propogation. Students may wish to propogate geraniums by cutting stems and placing them in moist sand. Transplant to soil when there is good root growth.

Define tuber and bulb.

Enrichment: Peel a potato and then put drops of iodine on it. The iodine turns blue-black, indicating the presence of starch stored in the potato.

The ink moves up the celery stalk causing the stalk to turn red. As the leaves on the stalk release water by transpiration, ink is drawn into and through the stalk.

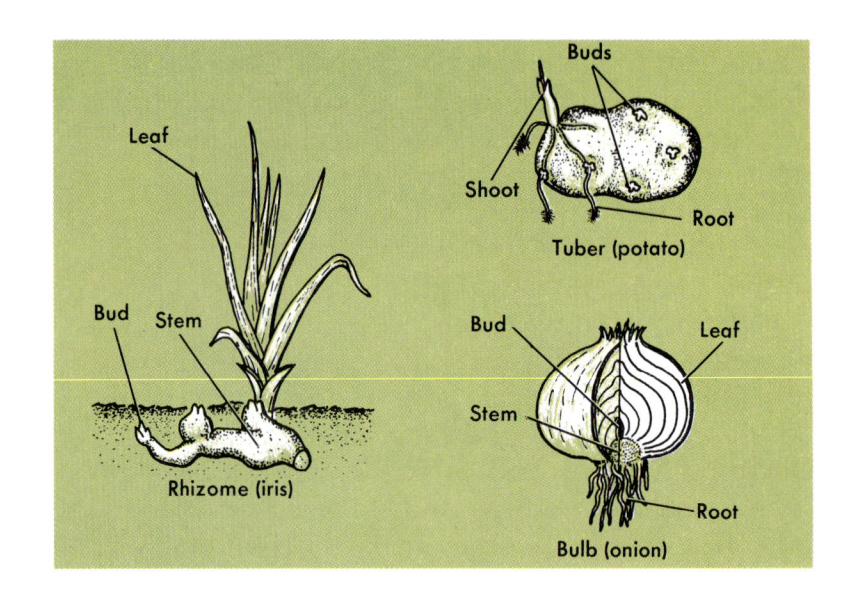

White potatoes and a few other plants produce a special kind of stem called a tuber. A tuber is the tip of a rhizome that is enlarged because of the storage of food in its cells. The white potato is the tuber of the potato plant. The "eyes" of the potato are buds. These eyes or buds in the white potato tuber may be used to grow new potato plants. One potato is cut into several sections. Each section must contain at least one bud and a piece of tuber. These sections are then planted. If all conditions are favorable, each bud grows to form a new potato plant. The bud gets its energy for growth from food stored in the portion of the tuber to which it is attached.

A plant bulb is a portion of the stem that grows underground. It consists of thick leaves that grow together in compact layers around a small bud. When separated from its parent plant and planted, the bud grows to form a new plant. Can you name any plants that reproduce by bulbs?

activity　　　　See Teacher's Guide.　　　Activity 15-4

TRANSPORT IN A PLANT STEM

Cut one end of a celery stalk and place it in a bottle of red ink. Observe the celery after a few hours. What happens? Explain.

activity

See Teacher's Guide.   Activity 15-5

COUNTING THE AGE OF A TREE

Obtain a cross-section piece of a tree trunk (stem). Observe the annual rings. Determine the age of the tree by counting the rings.

Use a hand lens. Copies of the ring patterns may be made with a photocopy machine.

activity

GROWING SWEET POTATOES   Activity 15-6

Suspend a fresh sweet potato, tapered end down, in a smallmouthed jar filled with water. You may have to insert 3 toothpicks to suspend the potato. Put the jar in a sunny window. Add water as needed to keep the water level constant. Record the number of days it takes the leaves and roots to grow. From which part of the sweet potato did new leaves and roots develop? See Teacher's Guide.

Water is lost from the jar by evaporation and it is essential that this water be replaced to prevent drying out.

15:7 Leaves

In most plants, leaves are the "food factories." Leaves produce the food the plant uses to build new tissues and to carry on life activities. Leaves vary in size, shape, structure, and arrangement. The shape of a leaf can be used to identify a plant.

The outer protective layer of a leaf is the **epidermis** (ep uh DUR mus). The epidermis contains many small porelike structures. These structures, each called a **stoma,** admit carbon dioxide and oxygen and control the release of water vapor. A leaf may contain 9000 to 70 000 stomata in each square centimeter of surface area. Most often, more stomata are found in the lower epidermis of a leaf.

Each stoma consists of an opening or pore about 1/20 as wide as the thickness of this page. The pore opens into air spaces within the leaf. Each pore is surrounded by two guard cells. Guard cells control the opening and closing of a stoma. When the guard cells absorb water, they swell and the stoma opens. When the guard cells lose water, they relax and the stoma closes. Stomata are most often closed at night and open during the day.

The broad flat surface of leaves make them suited to absorbing sunshine.

What is the epidermis?

FIGURE 15–13. A stoma on a leaf opens (a) and closes (b) to regulate the amount of gases that move into or out of a leaf.

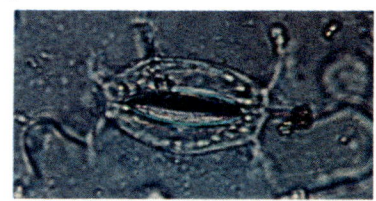

a

b

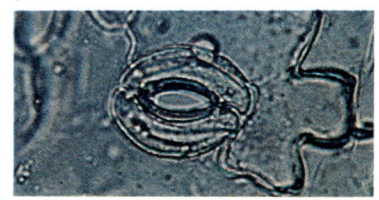

Kodansha

FIGURE 15–14. Structures in a leaf.

Compare this aspect of leaves to solar panels used in solar heating and on spacecraft for the production of electricity.

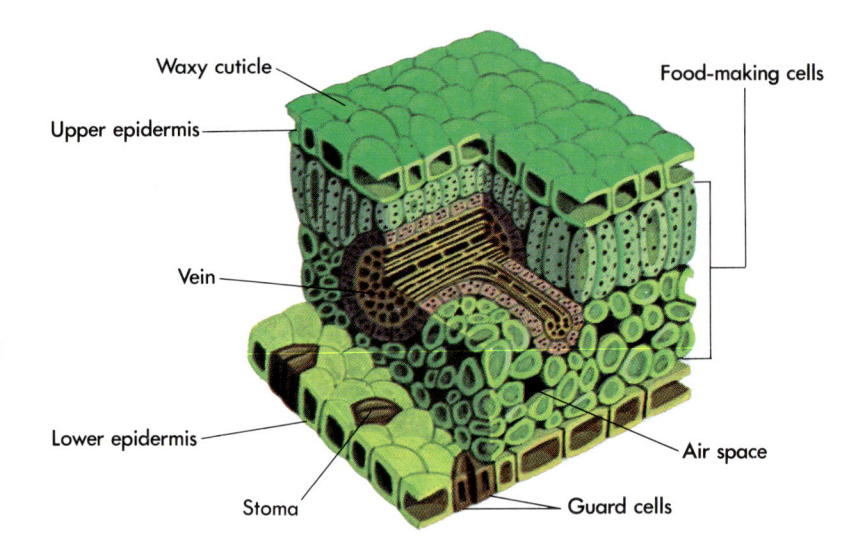

Enrichment: Observe stomates on a plant by preparing a microscope slide of a geranium leaf.

Some trees lose as much as a metric ton of water through their leaves in a single day.

Define transpiration.

FIGURE 15–15.

Stomata admit carbon dioxide to the leaf. Carbon dioxide diffuses from the air outside the leaf through the stomata into the air spaces within the leaf. The carbon dioxide is used to make food.

Stomata regulate the loss of water through the leaves of a plant. This water loss is called **transpiration**. Much of the water absorbed by the roots leaves the plant by transpiration through open stomata. In very dry weather, stomata often remain closed all day, which reduces the loss of water.

Some leaves also have a waxy covering that reduces water loss from the leaves. This covering prevents wilting and drying out during dry weather. Inside a leaf are soft, thin-walled cells arranged in two layers. Many of these cells have large amounts of chlorophyll. Veins within the leaf are part of the plant's transport system.

activity

WATER LOSS FROM PLANTS. Activity 15-7

Obtain a geranium plant in a flowerpot. Put a small plastic bag over the pot and tie the bag tightly around the stem. Use a large plastic bag to cover the entire plant and pot. Observe any changes that occur after 24 hours. Why did the changes occur? *Answers will vary. See Teacher's Guide.*

activity

See Teacher's Guide.

MAKING AND OBSERVING A LEAF SKELETON
Activity 15-8

(1) Obtain a leaf that is not damaged from insect-eating or breakage. Young leaves from early spring or old leaves from late fall are best. Magnolia leaves work well. **(2)** Fill a low tray or pan with pond water containing infusoria. Place the leaf in the water. Put the pan where it is warm, but not in direct sunlight. Why? **(3)** After two weeks, observe the extent of decomposition of the leaf tissue. Place the leaf under a fast-flowing stream of water from a faucet. The entire mass of tissue other than the vein structure will probably wash away. If the tissue does not wash away from the veins, place the leaf in the pond water for a few more days. **(4)** When the tissue is removed, press the skeletonized leaf between blotters or towels. The leaf can be preserved by pressing it between two sheets of plastic. What caused the tissue of the leaf to decompose? *microorganisms*

Infusoria can be obtained by taking a sample of pond water near decaying vegetation.

Infusoria are the microorganisms that cause decay. Infusoria grow faster when warm. Sunlight kills infusoria microorganisms.

15:8 Photosynthesis

If a plant is to grow, it must have light. Green plants absorb solar energy and use it in **photosynthesis** to make food. The energy stored in the food is used for the growth and repair of tissues. Plants grow most rapidly where there is plenty of light, water, and fertile soil.

Glucose, oxygen, and water are the products of photosynthesis. Glucose, a simple sugar, is the food used by the plant. Many glucose molecules join to make starch, which is stored in the plant. Oxygen gas and water vapor leave the plant through the stomata.

The equation for photosynthesis is written as follows:

$$6CO_2 + 12H_2O \xrightarrow[\text{light}]{\text{chlorophyll}} C_6H_{12}O_6 + 6O_2 + 6H_2O$$

carbon dioxide — water — glucose — oxygen — water

FIGURE 15–16. The small structures within a plant cell are chloroplasts.

Ward's Natural Science Establishment, Inc.

Discussion: Why is photosynthesis necessary for all life on earth?

activity

PHOTOSYNTHESIS

Activity 15-9

Objective: To determine if sunlight and chlorophyll are needed for photosynthesis

Materials See Teacher's Guide.

alcohol
baby food jar
beaker
coleus leaf
construction paper, black (or aluminum foil)
electric lamp (if direct sunlight is not available)
geranium plant
hot plate or ring stand and burner
iodine solution

Procedure

1. Keep the geranium plant in darkness for a few days to "starve" it. The purpose of this procedure is to stop the production of food.
2. Remove the plant from the darkness and cover one-half the surface area of 3 separate leaves with the black construction paper, Figure 15–17.
3. Place the plant in direct sunlight or under the electric lamp for 3 days.
4. After 3 days, pluck the leaves that were partially covered with paper and remove the paper.
5. Fill a small beaker with 200 mL of water. Heat the water to boiling with a hot plate or burner.
6. Place the leaves in boiling water for about 2 min.
7. Remove the leaves from the beaker and place them in a baby food jar containing 20 mL of alcohol. Place the lid tightly on the jar.
8. Carefully shake the jar containing the leaves and alcohol for 3 or 4 min to dissolve the chlorophyll that is present.
9. Remove the leaves from the alcohol and place them on a paper towel.
10. With the iodine solution, test for food in the leaves. Pour the iodine over the surface of each of the 3 leaves. An iodine-alcohol solution, turns blue when it comes in contact with starch. Record your results.
11. Obtain a coleus leaf. A coleus leaf has chlorophyll in certain sections.
12. Test the coleus leaf for starch by repeating Steps 5–8.

FIGURE 15–17.

Observations and Data

Record your observations.

Questions and Conclusions

1. Why was the geranium placed in the sunlight for 3 days? to start food production in the leaves
2. What process must have occurred in the leaf if the iodine solution changed colors when poured on the leaf? photosynthesis
3. According to this activity, what was produced in the leaf. starch

A large transparent plastic sack may be used in place of the bell jar.

activity

CARBON DIOXIDE AND PHOTOSYNTHESIS

See Teacher's Guide. Activity 15-10

Place one "starved" geranium plant on a shelf by a sunny window. Place another under an inverted bell jar, together with two small beakers one-half full of sodium hydroxide pellets. **CAUTION:** Sodium hydroxide can cause severe burns. Do not touch the chemical with your hands. Use petroleum jelly to form a seal around the base of the jar. An atmosphere with reduced carbon dioxide is within the bell jar. After three days, test a leaf from each plant for starch. Use the iodine test from the previous activity. What were the results? no starch produced in the leaf of the plant under the bell jar

FIGURE 15–18.

starch produced in the leaf in the air containing carbon dioxide

15:9 The Flower

A flower is a modified stem that contains organs for reproduction. These organs are usually surrounded by specialized leaves called petals. Often the petals of a flower, such as the rose, are brightly colored. Other plants, such as grass, have small flowers that often are not seen.

Other flower organs are sepals, stamens, and pistils. The **sepals** are green, leaflike structures on the underside of a flower. They enclose the parts of a flower before the flower bud opens and support it after it opens. Stamens and pistils are the reproductive parts of the flower. The **stamen** produces pollen. **Pollen** is tiny grains that contain nuclei for reproduction. The **pistil** is the flower part that contains the ovary (Figure 15–19).

Flowers which contain all four organs—sepals, petals, stamen, and pistil—are called complete flowers. Flowers which are missing one or more organs are incomplete flowers. In some plants, stamens and pistils are formed in different flowers. Each incomplete flower contains either stamens or pistils but not both. In the corn plant, pistil-bearing flowers form on the ear. Stamen-bearing flowers form on the tassel. In species such as the willow and meadow rue, pistil-bearing flowers grow on one

Name the parts of a flower.

If a model of a flower is available, use it to show the parts of a flower. You may be able to obtain live flowers to use.

Why are some flowers incomplete?

FIGURE 15–19. Parts of a complete flower.

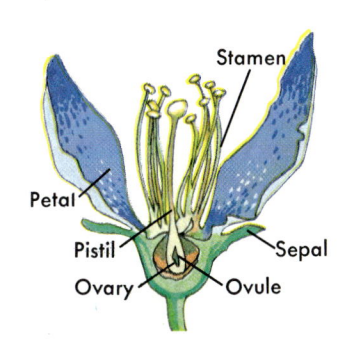

FIGURE 15–20. The reproductive structures of some flowers are easily identified.

Ruth Dixon

Enrichment: Have students make a drawing showing the parts of a complete flower using different colored pens for the different structures.

Define pollination. In what two ways can pollination occur?

Review pollination and fertilization to be certain that students understand the difference.

Define fertilization.

Which part of a flower becomes a seed? Which part becomes a fruit?

plant, and stamen-bearing flowers grow on another plant. Both types of plants must be present for the species to reproduce. One plant produces the pollen, and the other plant produces the eggs.

Pollen produced on the stamen is carried by wind, water, or a flying insect to the pistil. The transfer of pollen from stamen to pistil is called **pollination.** The transfer may be from a stamen to a pistil within the same flower. Pollination can also occur when pollen is carried from one flower to another.

When a pollen grain containing sperm cells lands on the pistil, it grows a tube downward through the pistil to the ovary. Inside the ovary are ovules, each containing an egg nucleus. When the tube reaches the ovary it grows into an ovule. Sperm nuclei are released. One sperm nucleus fuses with each egg nucleus. The joining of sperm and egg is called **fertilization.**

After fertilization, a small plant called an embryo begins to grow inside the ovule. Meanwhile, the walls of the ovule toughen, and starch is stored inside. The fertilized ovule is now a seed. The seed has a hard coat surrounding an embryo plant and stored food.

The ovary part of the pistil surrounds the growing seeds. The ovary enlarges and becomes a fruit. A pea pod, for example, is the ovary of a pea plant. For this reason, a pea pod is actually a fruit. Fruits such as oranges, grapes, tomatoes, blueberries, and plums are enlarged ovaries of plants. Each ovary enlarged with stored food after its ovules were fertilized.

making sure

11. Classify each of the following as a fruit or seed.
    (a) apple f   (c) bean s   (e) lemon f
    (b) pea s   (d) peach f
12. Which of the above is formed from an ovary? apple, peach, lemon
    Ovule? pea, bean

15:10 Seeds

Seeds provide food for people and animals. Every seed consists of two main parts—the embryo and the seed coat. The seed coat is a protective covering for the developing embryo. It develops from the wall of the ovule. In some seeds, a material called endosperm is present. Endosperm is a tissue that contains stored food. Both the endosperm and the embryo are enclosed within the seed coats. Cotyledons (kaht ul EED unz) are the seed leaves that are attached to the plant embryo. They store food. When the seed begins to grow, the one part of the embryo becomes the root of the new plant. Another part becomes the lower stem and the rest becomes the upper stem and leaves.

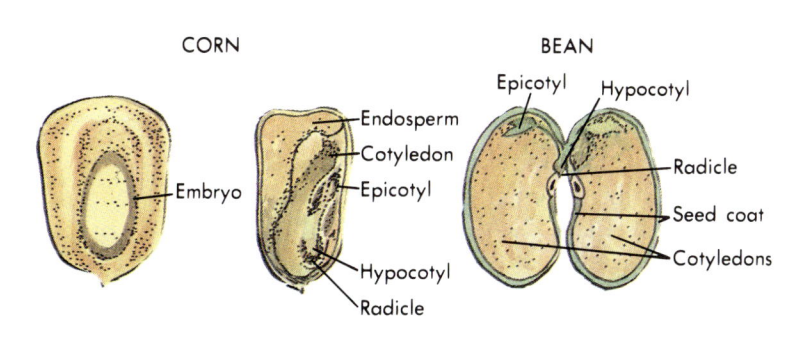

Name the two main parts of a seed.

Enrichment: Assign each student a different kind of seed to bring to class.

What important function do endosperm and cotyledons perform?

Enrichment: Dissect a variety of fruits (apple, strawberry, grape, etc.) and identify the seeds. If possible, examine the seed parts and their specialized function.

FIGURE 15-21. Parts of a bean seed.

FIGURE 15–22. New roots and stems can be observed growing from germinating seeds.

What conditions are favorable for germination? After germination, what conditions determine life?

FIGURE 15-23.

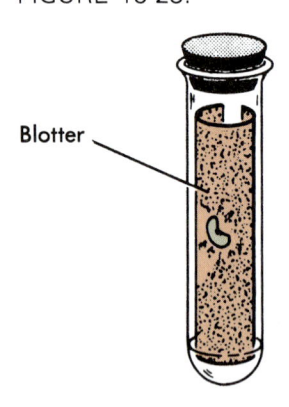

Blotter

Endosperm containing stored food is present in the seeds of corn, wheat, and other cereals. In plants suchs as peas, beans, clover, and alfalfa, food is stored in the cotyledons. The food stored in seeds is absorbed by the developing embryo as it grows into a new plant.

The early growth of an embryo plant is called **germination** (jur muh NAY shun). Germination will take place only if conditions are favorable. There must be a favorable temperature and enough moisture and oxygen. Some plants require a certain amount of light for germination to occur.

After germination is completed, a plant requires air, water, light, and fertile soil. The fertility of the soil is a very important factor. It may mean the difference between life and death for a plant.

activity
See Teacher's Guide.

FACTORS AFFECTING SEED GERMINATION
Activity 15-11

(1) Soak 54 bean seeds in water overnight. This process will soften the hard coat covering each. (2) The next day, place 3 beans and a piece of blotter in each of 18 test tubes. Press the seeds between the blotter and the wall of the test tube (Figure 15–23).

(3) In 9 of the test tubes, moisten the blotters with water. Seal the test tubes with stoppers. (4) Arrange the 18 tubes into 6 groups—3 groups with wet blotters and 3 groups with dry blotters. Each group will contain 3 test tubes. (5) Place group *A* (3 dry tubes) and group *B* (3 wet tubes) under the lamp. (6) Place group *C* (3 dry tubes) and group *D* (3 wet tubes) in a refrigerator set at 10°C. Be sure the light in the refrigerator is off. (7) Place group *E* (3 dry tubes) and group *F* (3 wet tubes) in a drawer where the temperature is about 25°C. (8) Observe all 6 groups each day. Record the number of seeds in each group that germinate after 1 week. Which group or groups germinated in the least amount of time? *B, F* Which group or groups took the most time to germinate? *C, D* What were the best conditions for seed germination? *warmth, moisture*

activity

SEED GERMINATION Activity 15-12

Fill a small tray or wooden flat with soil. Plant 20 bean seeds (soaked overnight in water) in the soil. Water regularly. Be careful not to overwater. Dig up 2 seeds every other day. Observe changes above and below the soil. Diagram what you see. *See Teacher's Guide.*

What special seed structures adapt the seed for wide dispersal?

15:11 Seed Dispersal

Few seeds would grow into new plants if they all grew in the same place. There would be too much competition for water, minerals, and sunlight. Only a few plants would survive. As a result, many seeds have features that aid seed dispersal. Seed dispersal is the carrying away of seeds from the plant on which they grow. Wind, water, and animals are the chief agents of seed dispersal.

Fruits and seeds often have special structures that allow them to be carried by the wind. For example, the fruit of the maple, ash, and elm have wings. Cotton, milkweed, and willow have long, silky hairs. Dandelion seeds are attached to hairlike tufts that form a parachute-like structure. A gentle breeze can carry a dandelion seed far from where it began its life.

Fleshy, edible fruits aid in seed dispersal. They are eaten by animals, often at a distance from where they grew. When the fruit is eaten, the seeds are exposed and may be dropped to the ground. If conditions are favorable, the seeds may grow.

Seeds of wild barley and most grasses have bristles. The bristles cling to the hair of animals or to clothing. Some fruits, such as cocklebur, have hooks or barbs that aid in their dispersal. You may have noticed after a walk in a park, field or forest, that your clothing has picked up some wild seeds or fruits. If so, you were an agent in the dispersal of plant seeds.

FIGURE 15-24. The seeds of milkweed (a) and maple (b) are dispersed by the wind.

a *J.W. Thompson*

b *General Biological Supply House*

PERSPECTIVES
people

A Marine Botanist

Dr. Llewellya Hillis-Colinvaux is a botanist, or a scientist who studies plants. *Halimeda,* the plant that she studies, may not be recognized as a plant by many people. However, Dr. Hillis-Colinvaux and other botanists have concluded that *Halimeda* plays a very important role in ocean environments.

Halimeda is like other plants. It carries on photosynthesis, using carbon dioxide from the water to make oxygen. It also produces organic compounds, which are then available to other marine organisms. Some fish, sea urchins, and snails feed on *Halimeda,* and thus obtain their energy from the plant.

Halimeda belongs to one of the five species of green algae that are capable of extracting calcium from sea water. The calcium is used to produce a very rigid rocklike skeleton. When *Halimeda* dies, its hard body parts fall to the ocean floor. Samples of the ocean bottom indicate that a large portion of sand from the ocean bottom is made up of *Halimeda* skeleton parts. Sand is very important to several marine organisms that live only on the seafloors.

Dr. Hillis-Colinvaux is involved in all aspects of *Halimeda* research. She studies the different types of the algae to determine a basis for classification. Growth rates, distribution, and relationships with other organisms are also examined. Her work has taken her to most of Earth's tropical oceans. Considerable laboratory work and library research are necessary before Dr. Hillis-Colinvaux goes into the water. She usually spends one month per year in the oceans, but wishes that there was more time for field work.

When Dr. Hillis-Colinvaux is in the field, her work is hard and long. Some days she spends up to eighteen hours marking areas on the ocean floor, collecting specimens, and carefully identifying and counting marine organisms. She frequently uses SCUBA diving equipment and submersibles, or small submarines, for this field work. In the evening, all observations from the day are recorded, and specimens are cleaned and prepared. Any living oganisms are also cared for.

The ocean, like all communities, consists of organisms that interact in numerous ways. The role that each plays is important. We obtain much of our food, many medicines, and most of our oxygen from the ocean. Humans can easily disrupt the relationships of marine life, if they do not have a good understanding of ocean ecosystems. Work by scientists like Dr. Hillis-Colinvaux help to increase our knowledge of the very important marine environment.

Doug Wynn

main ideas

Chapter Review

1. Algae are mostly one-celled green plants. 15:1
2. Mosses and liverworts grow close to the ground. They lack true roots, stems, and leaves. 15:2
3. Ferns are larger than mosses and liverworts. They have roots, stems, and leaves. 15:3
4. Seed plants dominate the plant life on Earth. 15:4
5. Roots provide anchorage, water and mineral absorption, and food storage for plants. 15:5
6. A plant stem conducts water and dissolved minerals upward to the leaves. It also conducts food substances downward from the leaves. 15:6
7. Leaves are the "food factories" for most plants. 15:7
8. Glucose, a simple sugar, is the food used by plants. 15:8
9. A flower is a modified stem that contains organs for reproduction. 15:9
10. A seed requires moisture, the proper temperature, and oxygen to germinate. 15:10
11. Seeds are dispersed by wind, water, and animals including people. 15:11

vocabulary

Define each of the following words or terms.

angiosperm	gymnosperm	pollen	stoma
cambium	organ	pollination	tissue
chlorophyll	phloem	rhizomes	transpiration
epidermis	photosynthesis	sepals	tuber
fertilization	pistil	stamens	xylem
germination			

study questions

DO NOT WRITE IN THIS BOOK.
A. True or False

Determine whether each of the following sentences is true or false. If the sentence is false, rewrite it to make it true.

F 1. Most of the land plants are ~~algae~~. seed plants
T 2. In most plants, food is stored in the leaves.

Plants 337

T 3. *Protococcus* is a green algae.
F 4. ~~All~~ stems grow ~~above~~ the ground. some, below
F 5. A moss has˄roots, stems, and leaves. no true
T 6. Tubers and bulbs are special roots that store food.
T 7. A fern produces spores.
F 8. Transpiration occurs most rapidly when the stomata of a plant are ~~closed~~. open
F 9. Ferns reproduce through ~~seeds~~. spores
T 10. Seed plants have roots, stems and leaves.

B. Multiple Choice

Choose the word or phrase that completes correctly each of the following sentences.

1. A *(stoma, pistil, root hair, stamen)* absorbs water and minerals from the soil.
2. A *(tomato, corn grain, bulb, white potato)* is a tuber.
3. Water enters a root through the process of *(pollination, osmosis, fertilization, transpiration)*.
4. *(Xylem, Cambium, Epidermis)* tissue conducts water upward in a plant stem.
5. The *(tomato, elm, maple, palm)* is an example of an herbaceous plant.
6. *(Xylem, Cambium, Phloem, Epidermis)* is the growth tissue of a stem.
7. Pollen is produced in a *(flower, leaf, stem, root)*.
8. A maple seed will most likely be dispersed by *(wind, water, animals, people)*.
9. Plants lose water from their leaves by *(transpiration, photosynthesis, osmosis)*.
10. *(Chlorophyll, Water, Glucose, Carbon dioxide)* is produced in photosynthesis.

C. Completion

Complete each of the following sentences with a word or phrase that will make the sentence correct.

1. Plants use _____ energy in photosynthesis. solar (light)
2. _____ flowers do not contain all the parts found in a complete flower. incomplete

3. Leaves are called the __food__ factories of plants.
4. Photosynthesis occurs primarily in the _____ of plants. chloroplasts
5. __Oxygen__ gas is formed in photosynthesis.
6. Plants are green because they contain a green pigment called _____. chlorophyll
7. In a flower, the reproductive parts are usually surrounded by very colorful, modified leaves called __petals__.
8. The fertilized __ovule__ of a flower becomes a seed.
9. The union of a nucleus in a pollen cell and the egg nucleus in the ovary is called _____. fertilization
10. _____, _____, and _____ are three conditions necessary for germination. moisture, oxygen, warmth

D. How and Why See Teacher's Guide in front of this book.
1. Why do many desert plants have small leaves or no leaves at all?
2. Why is seed dispersal important to a plant species' survival?
3. Why is a bee a good agent for pollination?
4. Draw a diagram of a *Protococcus* cell and label its parts.
5. Why is photosynthesis necessary for a green plant to stay alive?

challenges

1. Make a collection of tree leaves of 15 common species. Learn to identify these trees by their leaves.
2. Green leaves use only a portion of the light spectrum in photosynthesis. Plan and carry out, under the supervision of your teacher, an experiment to discover what colors of light are used in food making.

interesting reading

Kavaler, Lucy, *Green Magic: Algae Rediscovered.* New York: Cromwell, 1983.

Lerner, Carol, *Pitcher Plants: The Elegant Insect Traps.* New York: William Morrow Co., 1983.

Wexler, Jerome, *Secrets of the Venus Fly Trap.* New York: Dodd, Mead, 1981.

When you think of an animal, you may think of a furry creature like a dog or rabbit. However, a wide variety of living organisms without fur are animals. Many people do not consider insects like this praying mantis to be animals. What are the characteristics of animals? What are cells, tissues, and organs? What are the main animal groups?

Roger K. Burnard

Animals

chapter 16

Introducing the chapter: As an introduction to the study of animals, arrange with the school librarian for a class lesson that focuses on books, magazines, encyclopedias, and other media related to animals. Have students select a book or magazine containing information pertaining to animals. You may wish to have each student prepare a short oral or written report on what they have read.

16:1 Tissues, Organs, and Systems

More than one million species belong to the animal kingdom. Most animals move about, take in oxygen, and obtain food from plants or other animals. Some animals, such as sponges and jellyfish, live in water. Others, such as honeybees, can fly through air. Every animal species has special traits and structures that makes it different from other species.

Animals have cells that are specialized. That is, each cell has a certain kind of job to do. For example, a red blood cell carries oxygen. A muscle cell helps to move a part of an animal's body. These specialized cells are grouped together to form tissues. Blood is a tissue containing different kinds of blood cells. Muscle tissue is made of muscle cells. Bone tissue contains bone cells.

Groups of tissues working together make up an organ. Examples of organs in an animal are the heart, lungs, eyes, and ears. Name the function of each of these organs. Organs are grouped together in **systems**. Two systems in animals are the circulatory system and the digestive system. The circulatory system moves blood through the animal.

GOAL: You will learn the characteristics and life activities of invertebrate animals.

What are specialized cells?

How are a tissue, organ, and system different?

Although there is a wide diversity among animal species, all animal life is based on certain basic functions such as obtaining food and obtaining energy from the food. Also, every animal must rid itself of wastes.

341

FIGURE 16–1. A snake (a) and other complex organisms are made up of systems (b), organs (c), tissues, and cells (d).

The digestive system breaks down food into simpler substances that cells can use. Systems work together in doing all the activities needed to keep an animal alive.

Call attention to the basic life activities and how each animal carries on these activities.

16:2 Sponges

One of the simplest animals is the sponge. A sponge's body is a sack of cells containing pores, canals, and chambers. Most sponges are attached to the rocks at the ocean bottom near the coast. Ocean water constantly flows through the body of the sponge. From this water, the sponge obtains food and oxygen. As it uses food and oxygen, the sponge, like every animal, produces wastes which must be expelled from its body. The flow of ocean water through a sponge removes these wastes.

There are about 5000 different species of sponges. A natural sponge used for washing or cleaning is the

How does a sponge obtain food?

mineral matter or skeleton of a sponge's body. It is the material left after the cells have died and decayed. Artificial sponges for cleaning are made from cellulose, rubber, or plastic.

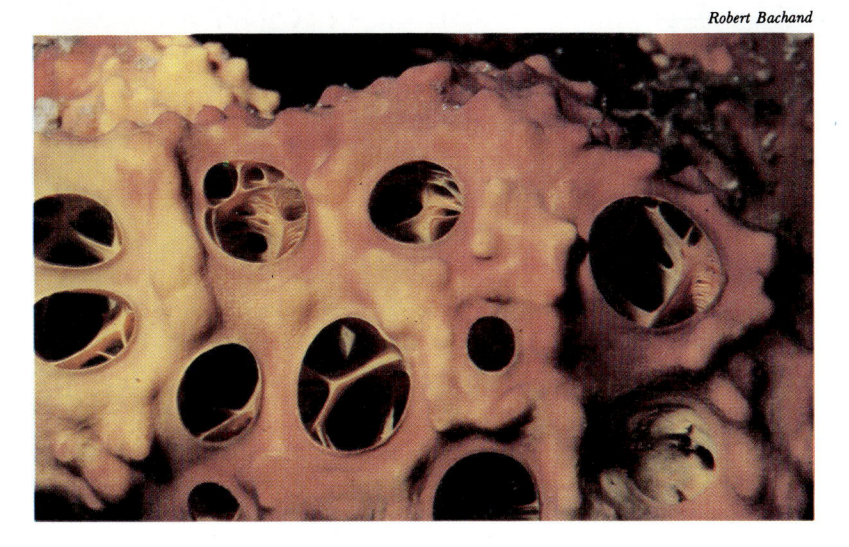

FIGURE 16–2. Adult sponges live along shallow coastlines, attached to rocks along the bottom.

Some students may have observed jellyfish in an aquarium or while swimming. Ask them to describe their observations.

How are jellyfish, sea anemones, corals, and hydra alike?

FIGURE 16–3. A jellyfish has a soft body that contains a central body cavity.

16:3 Jellyfish and Their Relatives

Jellyfish also live in water but are more complex than sponges. Their relatives include sea anemones (uh NEM uh neez), corals, and hydra. These animals are alike in that they all have a cavity in the center of their bodies. Unlike a sponge, jellyfish can move slowly through water. It swims by contracting and releasing fibers in its body.

Most jellyfish have tentacles. The tentacles are ropelike organs containing special stinging structures. Each stinger contains a slender coiled thread having spines attached to it and is used to sting other animals.

How does a jellyfish catch its food? It grasps a nearby animal with its tentacles. The coiled threads shoot out from the tentacles and stick into the animal. Poison from the threads is injected into the animal. The poison paralyzes it. The tentacles then pull the paralyzed creature into the jellyfish's central body cavity. Here the animal is digested and absorbed by the body cells.

FIGURE 16–4. Small fish are paralyzed by the poison injected from the tentacles of a jellyfish.

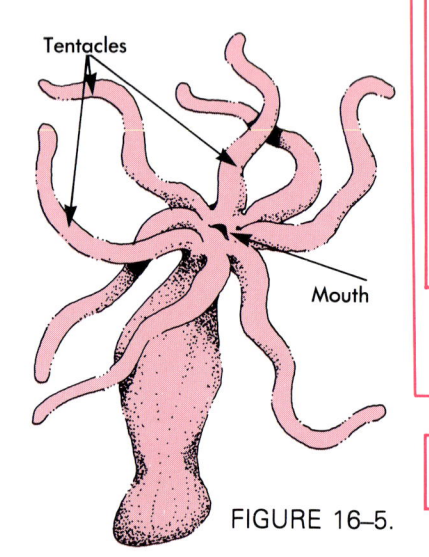

Animals, Animals

Tentacles

around the mouth

moves away

Mouth

FIGURE 16–5.

activity   See Teacher's Guide.

OBSERVING HYDRA   Activity 16-1

Obtain a watch glass (small jar) and add 10 mL of aquarium water. Use a dropper to pick up hydra from the surface of a hydra culture. Transfer the hydra to the water in the watch glass. Use a hand lens to observe the hydra. Draw a diagram showing the shape of a hydra's body. Where are the tentacles located? Describe any motion you observe. Touch a hydra gently with a dissecting needle. What is the animal's reaction? How do you think a hydra obtains food? After you have completed your observation, empty the liquid in the watch glass back into the culture container.   pulls it into its mouth with tentacles

making sure   has a central cavity, can swim

1. How is a jellyfish different from a sponge?
2. What special structures in a jellyfish enable it to catch food? tentacles and poison threads

Explain that free-living means an animal is not a parasite, e.g. planaria.

Define parasite.

16:4 Flatworms and Roundworms

Tapeworms and flukes are two kinds of flatworms. Both of these animals are parasites. A **parasite** is an organism that lives on or in another organism called the **host**.

344   *Animals*

Tapeworms may live in the bodies of animals. A tapeworm is a long flat worm with a small head at one end of its body. Hooks and suckers on the head are used by the tapeworm to attach itself to the inside of its host. The tapeworm obtains food, water, and a place to live from the host.

A planarian (pluh NER ee un) is an example of a flatworm that is not a parasite. A planarian obtains food from the water in which it lives. Different planaria species range from microscopic size to 7 or 8 cm in length. Planaria are found in streams and ponds. They live on the bottom surfaces of underwater plants, rocks, and logs. One method for collecting planaria is to place bits of meat in shallow water. The worms are drawn to the meat by juices that diffuse through the water.

As their name indicates, roundworms have rounded, tubelike bodies that taper to a point at each end. There are over 3000 species of roundworms. Hookworms, hairworms, and eelworms belong to the roundworm phylum. Many microscopic roundworms live in soil and water.

Trichina (trihk I nuh) is a parasitic roundworm that can cause a disease called trichinosis (trihk uh NOH sus). People may contract the disease if they eat pork that is not thoroughly cooked. Symptoms of the disease are severe muscle pains, fever, and weakness. The disease may even cause death. Trichinosis is prevented by freezing pork or cooking it until it is well done. Animals can become infected with trichina by eating garbage containing these roundworms. Farmers can stop the spread of the trichina worms by cooking garbage before it is fed to pigs.

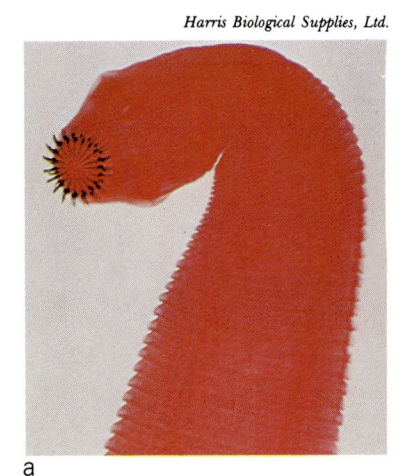

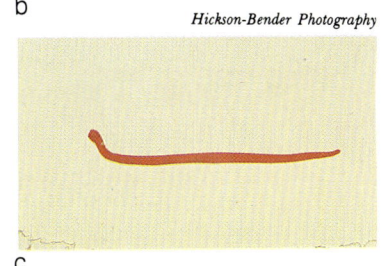

FIGURE 16-6. A tapeworm (a) and a planarian (b) are flatworms. A hookworm (c) is a roundworm.

Why are *Trichina* harmful?

making sure

3. How is a parasite different from a host? Parasite lives off the host.
4. Why is a tapeworm called a parasite? needs a host animal to live
5. How is a planarian different from a tapeworm? A tapeworm is a parasite. A planarian is free-living.
6. What features of a tapeworm allow it to live inside the bodies of animals? can absorb food through its body wall, special structures on the head to attach to host

16:5 Earthworms and Their Relatives

Earthworms, sandworms, and leeches are members of the same phylum. Animals in this group have bodies composed of ringlike segments. Most of the members of this group live in damp soil or fresh water. Some are marine worms that live along the seashore. Some species of segmented worms are free-living and others, such as leeches, are parasites.

Earthworms burrow underground. They feed near the surface on decaying leaves and other organic matter. In areas where the soil freezes in the winter, earthworms burrow deep below the frost line. They may burrow as much as 2 m below the soil surface.

An earthworm's body is long, cylindrical, and tapered at one end. Its underside is flat and light-colored. The body of the earthworm is divided into 100 to 180 segments. Tiny bristles on each of the body segments help the earthworm pull itself along the ground or through its burrow. If you have ever tried to pull an earthworm out of the ground, you know how tightly these bristles cling to the soil.

An earthworm's body has several systems that carry on the life activities of the animal. Food is digested and absorbed in a group of organs that compose the digestive system (Figure 16–7). Leaves, grass, and other plant materials are drawn in through the mouth. Rock particles mixed with the food aid the digestive system in grinding it up. Chemicals called enzymes produced by the digestive organs dissolve the food. The digested food is then absorbed into the earthworm's blood through the walls of the intestine.

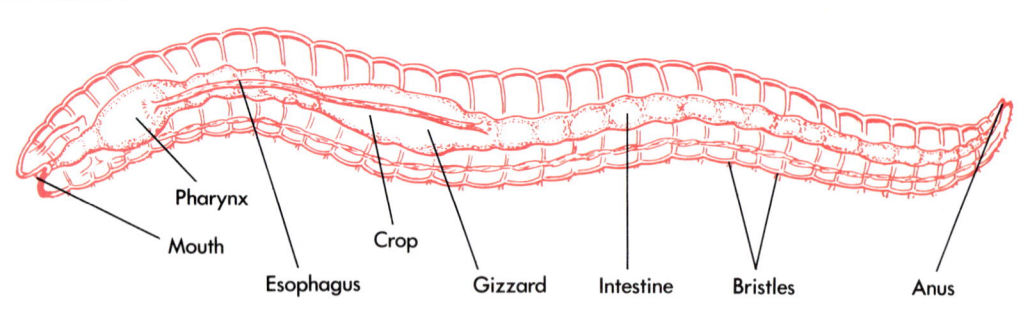

Earthworms increase soil fertility and contribute to the building of topsoil.

Identify the life activities of an earthworm and relate each activity to the structures identified here.

Relate the text description of an earthworm to the diagrams shown here.

FIGURE 16–7. Digestive system of the earthworm.

Mouth, Pharynx, Esophagus, Crop, Gizzard, Intestine, Bristles, Anus

The circulatory system of an earthworm has two blood vessels and several hearts. Blood moves digested food and oxygen to all of the cells in the earthworm's body. Oxygen absorbed through the skin is needed by the cells for respiration. The blood also picks up waste material from the cells and carries it away. Special coiled tubes filter the blood and remove the wastes. The tubes move the wastes to the surface of the earthworm. Food wastes are excreted through the open end of the digestive tract. **Excretion** (ihk SKREE shun) is the process by which wastes are removed from an organism.

Describe the circulatory system of the earthworm.

How does an earthworm obtain oxygen?

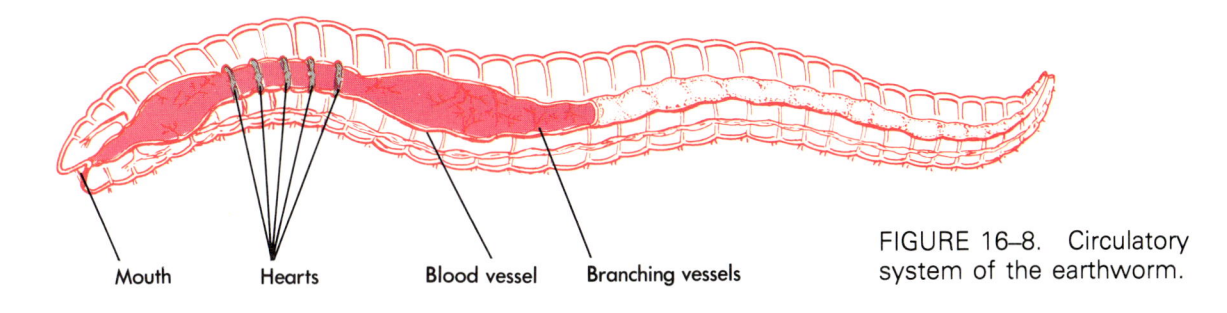

FIGURE 16–8. Circulatory system of the earthworm.

An earthworm has a simple nervous system. A nerve center is located just above the pharynx of the digestive system. Nerves extend from the nerve center through the length of the earthworm's body and connect to each of the body organs. These nerves control the operation of muscles and body organs.

Where is the nerve center of the earthworm located?

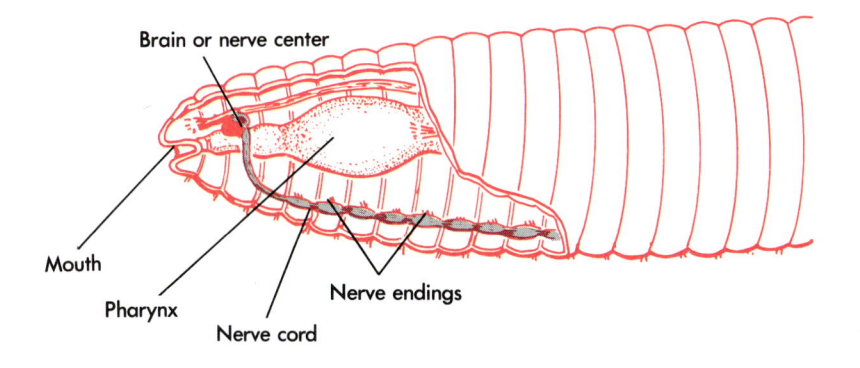

FIGURE 16–9. Nervous system of the earthworm.

16:5 *Earthworms and Their Relatives*

Point out that earthworms reproduce by sexual reproduction even though a worm is neither male or female.

Earthworms are neither male nor female. Each earthworm has both male and female sex organs. The male sex organs are called testes and the female sex organs are called ovaries. The testes produce the sperm cells. Sperm cells move with the aid of a long whiplike tail. Ovaries produce the egg cells.

Earthworms reproduce by sexual reproduction. Sexual reproduction occurs when sperm cells and eggs are produced and fertilization takes place. Earthworms reproduce mainly during warm, moist weather. Mating occurs at night and requires two to three hours. Two worms stretch out and lie parallel to one another. Sperm cells from each worm are exchanged. Union of sperm and egg cells, called fertilization, occurs after the worms move apart. The eggs are fertilized in a band of slime the earthworm secretes around itself. The slime band slips off, dries, and forms a cocoon around the fertilized eggs. Cocoons of the common earthworm measure about 5 mm by 7 mm. In each cocoon there are many eggs. The eggs grow by cell division and form young earthworms. The young soon break out of the cocoon and crawl away.

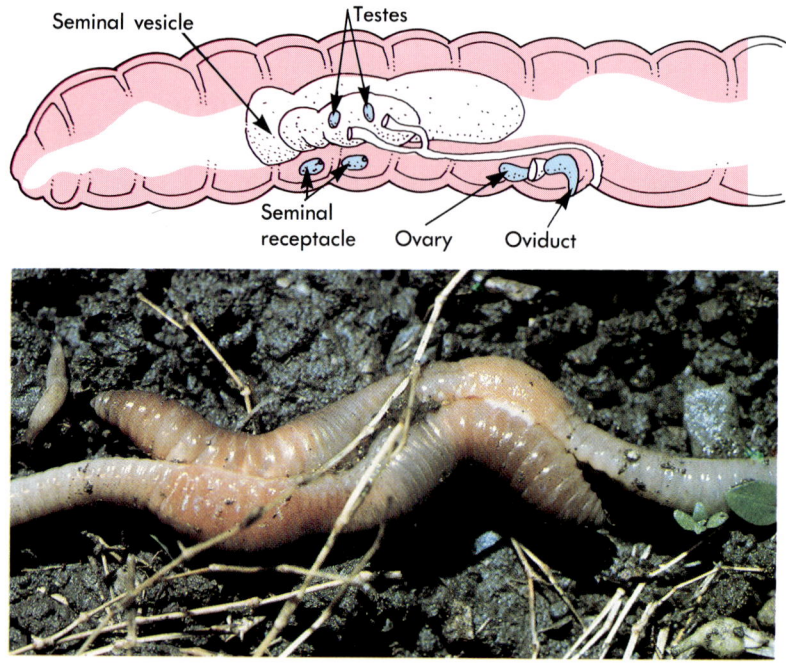

FIGURE 16–10. Reproductive system of the earthworm (a). During mating, earthworms exchange sperm through collarlike structures (b).

activity

DISSECTION OF AN EARTHWORM

Activity 16-2

Objective: To identify the major parts of an earthworm See Teacher's Guide.

Materials Use a freshly killed or preserved earthworm for this activity.
earthworm
dissecting needle
dissecting pan or board of soft wood
scalpel or sharp knife
20 straight pins

Procedure

1. Place the worm in a dissecting pan or on a board. See Figure 16–11.
2. Observe that the worm's body is made of ringed segments. Identify the mouth, clitellum, and anus.
3. With a scalpel or knife, cut lengthwise through the body wall from the anus to the mouth. Pin back the body wall on each side to hold it down.
4. Identify the parts that make up the digestive tract (Figure 16–7).
5. Locate the large blood vessels that run along the digestive tract. Find the blood vessels (hearts) that circle around the esophagus.
6. Remove the esophagus with a dissecting needle and identify the seminal vesicles that are part of the reproductive system. They appear as three lobes on each side of the esophagus. Try to locate the small testes at segments 10 and 11.
7. In segment 13, look for a small ovary. Also, look for the yellowish seminal receptacles in segments 9 and 10.
8. Find the cerebral nerves in segment 3 and the long nerve cord to which they are attached.

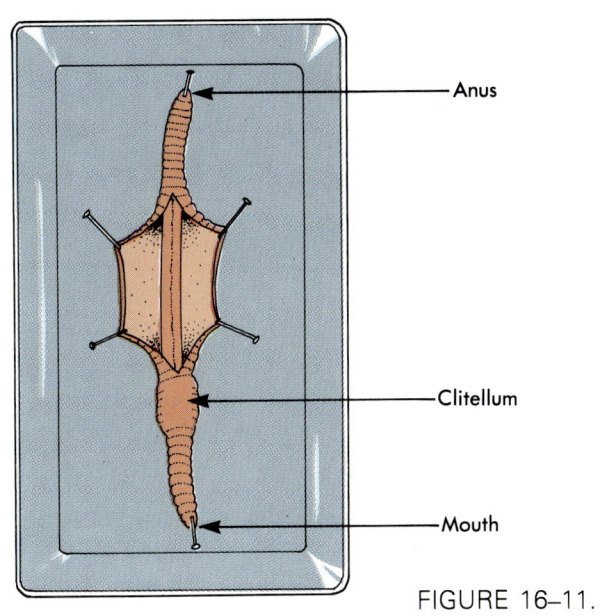

FIGURE 16–11.

Observations and Data See Teacher's Guide.

Draw a diagram of an earthworm dissection and label the parts you identified.

Questions and Conclusions

1. What are the parts of an earthworm's digestive tract?
2. How does the length of the digestive tract benefit the earthworm?
3. How are the large blood vessels along the digestive tract of benefit to the earthworm?
4. Why do you think the blood vessels around the esophagus are called hearts?
5. What are the male parts of an earthworm? Female parts?
6. What body parts carry "messages" through an earthworm?

7. sexual reproduction Two earthworms produce eggs and sperms that are exchanged. Fertilized eggs develop into young earthworms.

making sure

7. How do earthworms reproduce?
8. Why is it necessary for earthworms to live in moist soil? For exchange of gases (O_2 and CO_2) through the skin

16:6 Spiny-Skinned Animals

Spiny-skinned animals belong to the phylum that includes starfish, sea urchins, and sand dollars. All of these animals live in the ocean. Their bodies have a tough outer skin covered with coarse spines. Spiny-skinned animals also have a body form and shape called radial symmetry. **Radial symmetry means similar body parts extend out from the center.** A wagon wheel with spokes extending out from a hub has radial symmetry. Most starfish have five arms that extend out from the center of their bodies in the form of a star.

Define radial symmetry.

Compare radical symmetry with bilateral symmetry.

Describe the systems of the starfish.

Inside a starfish's body is a network of canals through which water is pumped. **A starfish has a simple digestive system with a mouth and stomach. Its circulatory system includes blood vessels surrounding the mouth. Each arm also contains a blood vessel. The nervous system has a central nerve ring with nerves that extend into each arm.**

FIGURE 16–12. A sea urchin (a) and a starfish (b) are spiny-skinned animals.

a

b

Geri Murphy

MJW

350   Animals

A starfish uses its arms to move and cling to objects. The underside of each arm has rows of small structures which work like suction cups. To move, a starfish slowly reaches out and attaches an arm to the seafloor. Then it uses the arm to pull the rest of its body along. The arms are also used to grasp and pull open oysters and clams, which provide food for a starfish.

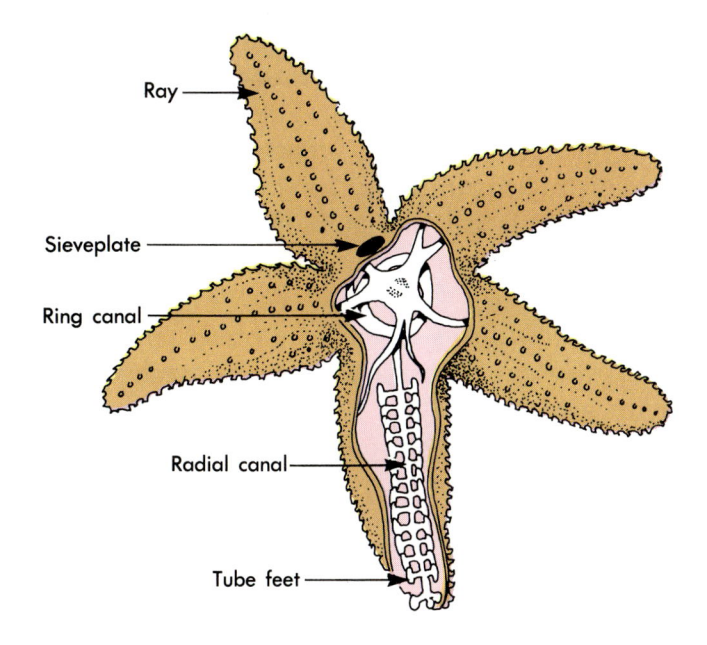

FIGURE 16–13. Internal features of a starfish.

making sure

9. How do starfish move? pulls itself with its arms
10. What feature makes spiny-skinned animals different from other animals? radial symmetry, tough outer skin with coarse spines

16:7 Mollusks

All mollusks have soft bodies; many have hard shells or shell-like coverings. Snails, oysters, and clams are mollusks. This group also includes the octopus and squid. There are about 40 000 different species of mollusks. Most of the species live in the ocean in shallow water near the shore. Some snails live in fresh water and others live on land.

What do all mollusks have in common?

Students who have collections of shells at home may want to bring them to class for viewing by other students.

16:7 Mollusks 351

Name two uses for mollusks.

Many mollusks are used for food. Some have beautiful shells and are collected and used for decorations. Oyster shells are crushed and fed to poultry as a source of calcium in their diet. Pearls are removed from oysters and used in jewelry. A pearl is made when an oyster secretes material that surrounds a foreign object lodged inside the shell.

A common garden snail is another example of the animals in the mollusk phylum. The snail has a round, looped shell which is used to protect the snail's soft body. Attached to the body are two pairs of tentacles and a pair of eyes. The garden snail is most active at night in damp weather. It moves slowly by contracting muscles in the bottom of its foot. Garden snails eat plants. When present in large numbers, the snails may severely damage gardens and fields.

Squids and octopuses live in the ocean and have a large head, two eyes, and a mouth. Tentacles or fleshy arms attached to the body are used to obtain food. Cuplike suckers that aid in grasping and holding things are located on the tentacles. Both squids and octopuses can swim. An octopus has eight

FIGURE 16-14. A snail has a single, coiled shell that protects the snail's soft body (a). Octopuses have well developed eyesight (b).

a

b

David M. Dennis

Zig Leszczynski/Animals, Animals

arms which it uses to crawl about rocks and tidepools along seacoasts. Squids and octopuses feed on small marine animals such as fish and shrimp. Small squids are eaten by fish and other marine animals. They are used as bait for fishing and in some places are used for food by people.

Although squids and octopuses do not have hard shells, their internal body parts are similar to other mollusks.

16:8 Arthropods

Members of the **arthropod** (AR thruh pahd) phylum have segmented bodies, jointed legs, and exoskeletons. Insects, spiders, crayfish, crabs, shrimps, and lobsters belong to the arthropod phylum. An **exoskeleton** is a skeleton on the outside of an animal's body. It protects the soft, internal body parts. There are more than 800 000 species of arthropods, making it the largest phylum in the animal kingdom.

Stress the fact that this group contains more species than any other. Most of the species are insects.

Name the characteristics of the arthropod phylum.

A group within the phylum, the crustaceans (krus TAY shunz), includes fairy shrimp, water fleas, barnacles, crayfish, and crabs. Although most of these animals live in the sea, many live in inland waters. Various species of crayfish live in freshwater streams, ponds, and lakes all over the world.

Other groups of arthropods include the millipedes and centipedes. Both have wormlike bodies. Centipedes have one pair of legs for each body segment. Millipedes have two pair of legs for each segment.

FIGURE 16–15. A crayfish (a) and a crab (b) are examples of crustaceans.

Name three animals with exoskeletons.

a b

FIGURE 16–16. The sting of a scorpion (a) may be harmful to humans. Most spiders inject a poison to paralyze their prey (b).

Spiders and scorpions are two closely related types of arthropods. These animals live on land with more living in warm, dry regions than elsewhere. Many species of spiders and scorpions have poison glands and poison claws or "fangs." They use the poison to kill insects and other small animals for food. Some spiders catch their food, while others spin webs in which they trap their prey. A few spiders and scorpions have bites or stings that can cause people serious illness and even death. Two body segments, head and thorax, and four pair of legs are distinct features of spiders.

making sure   to protect soft internal body parts
11. What is the purpose of an exoskeleton?
12. What features make arthropods different from other animals? external skeleton, segmented bodies, jointed legs

Call attention to the differences between insects and spiders in the number of body parts and legs.

List the body parts common to all insects.

16:9 Insects

Insects are the largest group of arthropods with about 700 000 different species. Grasshoppers, flies, lice, butterflies, beetles, and bees are but a few of the many different kinds of insect. An insect's body is divided into three parts, the head, thorax, and abdomen. Insects have three pair of legs and usually two pair of wings. Insects are the most plentiful and widespread of all land animals.

a b

FIGURE 16–17. Insects such as a grasshopper (a) and a stag beetle (b) have three pairs of legs and usually two pairs of wings.

Insects go through a series of changes in their lives called metamorphosis (met uh MOR fuh sus). Some insects, like the grasshopper, undergo an incomplete metamorphosis in which young grasshoppers develop gradually into adults. Other insects undergo a complete metamorphosis in a series of four different stages. **The stages are egg, larva, pupa, and adult.** For example, a caterpillar emerges from a butterfly egg. A caterpillar is the larva stage of a butterfly. The pupa stage occurs when a caterpillar spins a cocoon around itself. The caterpillar changes in the cocoon, and an adult butterfly hatches out of the cocoon.

What is metamorphosis?

Name the four stages of complete metamorphosis.

Complete metamorphosis is a loop or cycle with each stage always repeated again.

a b c

FIGURE 16–18. The larva stage (a), the pupa stage (b), and the adult stage (c) of a Monarch butterfly.

A grasshopper has all the body parts of a typical insect. The hind legs are unusually strong and its wings fold in a narrow line along the body. The wings unfold and spread when the grasshopper flies.

16:9 Insects

FIGURE 16-19. External body parts of a grasshopper (a). Digestive system of a grasshopper (b).

The grasshopper is used here to illustrate basic insect structures and functions.

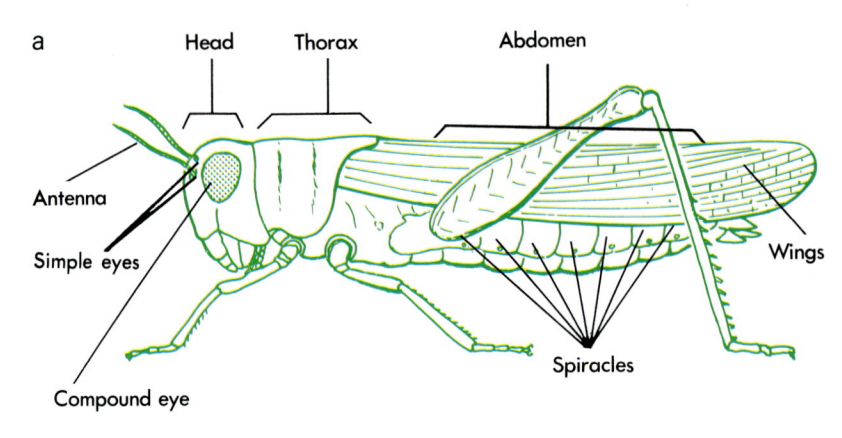

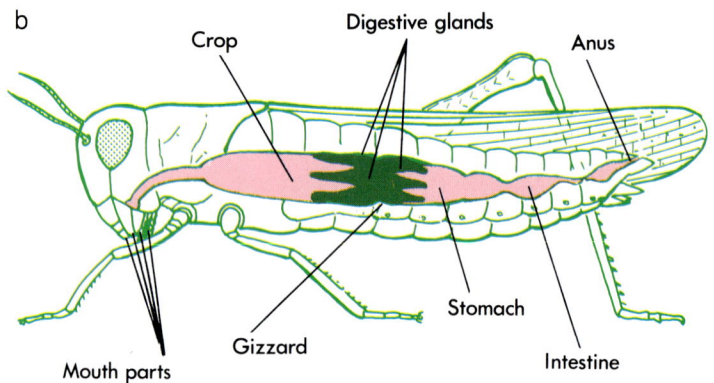

As the food a grasshopper eats passes through its digestive system, it is digested and absorbed into the blood. An insect's blood fills its entire body cavity. A long tube on top of the digestive tract acts as a heart and pumps blood. The blood is pumped forward through the tube to the head. Blood bathes all the organs and is circulated by muscular movements. It then reenters the heart through tiny openings. This cycle is repeated over and over. A grasshopper's blood does not carry oxygen. It is used solely to carry food and wastes.

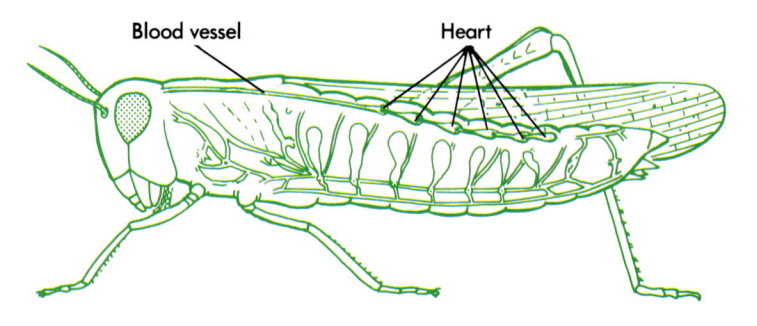

FIGURE 16-20. Circulatory system of a grasshopper.

356   Animals

A grasshopper's respiratory system is a complex network of tubes inside the body. These tubes open to the outside through tiny openings called spiracles (SPIHR ih kulz). There are ten pair of spiracles, eight pair on the abdomen and two pair on the thorax. Air is pumped into and out of the spiracles by the movement of the abdomen and wings. The pumping allows an exchange of gases between the cells and the air in the tubes. Oxygen diffuses into the cells and carbon dioxide diffuses out.

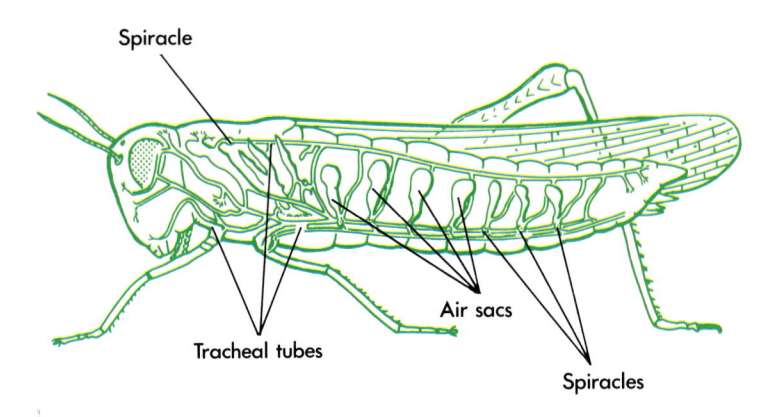

FIGURE 16-21. Respiratory system of a grasshopper.

The nervous system of the grasshopper is located in the bottom of the body cavity. The system is made of nerves and a nerve center. Antennas on the grasshopper's head provide a sense of touch. Three simple eyes are located on the front of the grasshopper's head. These eyes detect darkness and light. Two large eyes on the sides of the head detect light, color, and motion.

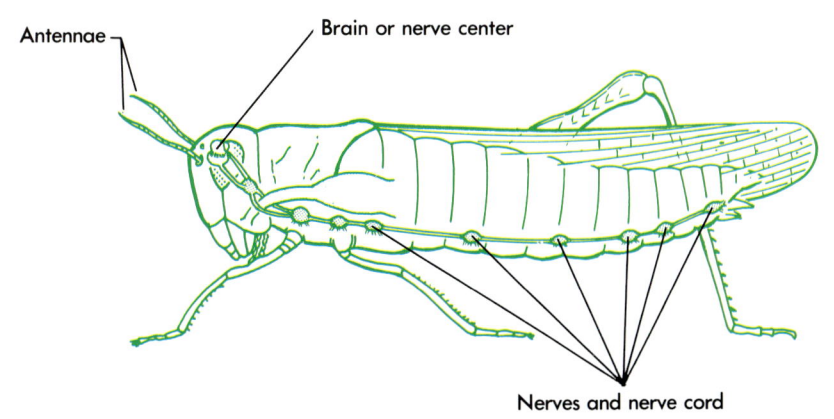

FIGURE 16-22. Nervous system of a grasshopper.

16:9 Insects

A grasshopper also has a sense of hearing. Earlike structures are located in the abdomen next to the thorax. They are large, membrane-covered cavities.

Grasshoppers have separate sexes. The male deposits sperm inside the female. Fertilization of the eggs takes place in the body of the female.

A female grasshopper lays eggs in a hole in the ground. She uses a special structure called an ovipositor (OH vuh pahz ut ur) to dig the hole. A female grasshopper may deposit 100 or more eggs during late autumn. These eggs hatch in the spring.

Young grasshoppers are called nymphs. They resemble the adult grasshopper but are wingless. The nymph grows until its exoskeleton becomes too small. Then it sheds, or molts, its exoskeleton. Molting occurs at different times until the grasshopper reaches its full size.

activity   See Teacher's Guide.   Activity 16-3

START AN INSECT COLLECTION

Catch insects with a net or jar. To kill them, put them in a jar with a piece of cotton soaked in fingernail polish remover. Cover the jar with a lid. **CAUTION:** The fumes of fingernail polish remover are poisonous. When the insects are dead, mount them with pins on cardboard or Styrofoam. Label the insects in your collection with their scientific names.

making sure   See Teacher's Guide.

13. How are insects different from lobsters and spiders? How are they similar?
14. Describe the four stages of metamorphosis of a Monarch butterfly.
15. Describe the external features of a grasshopper.
16. How does the circulatory system of a grasshopper differ from the circulatory system of an earthworm?
17. How do grasshoppers reproduce?

Chapter Review

main ideas

1. The animal kingdom consists of more than one million species. — 16:1
2. Examples of systems in animals are the digestive, circulatory, respiratory, excretory, and nervous systems. — 16:1
3. A sponge's body is a sack of cells combining pores, canals, and chambers. — 16:2
4. Members of the jellyfish family include sea anemones, corals, and hydra. — 16:3
5. Some animals, such as the tapeworm, are parasites. — 16:4
6. Most animals reproduce sexually through the fertilization of eggs by sperm. — 16:5
7. Spiny-skinned animals have tough outer skins covered with spines. — 16:6
8. Mollusks have soft bodies and many have hard shells or shell-like coverings. — 16:7
9. The arthropod phylum has the largest number of species of any animal phylum. Most of these species are insects. — 16:8
10. As they grow and develop, insects go through a series of changes called metamorphosis. — 16:9

vocabulary

Define each of the following words or terms.

arthropod	mollusk	roundworm
excretion	metamorphosis	specialized cell
exoskeleton	organ	spiracle
flatworm	parasite	system
host	radial symmetry	

study questions

DO NOT WRITE IN THIS BOOK.

A. True or False

Determine whether each of the following sentences is true or false. If the sentence is false, rewrite it to make it true.

F 1. A ~~muscle~~ cell is a specialized cell that carries oxygen to all parts of the body. [blood]

T 2. A system is made of organs.

Animals 359

F 3. One example of an arthropod is a ~~sponge~~. grasshopper   Answers will vary.
T 4. The tentacles of a jellyfish are used to obtain food.
F 5. An earthworm is ~~a parasite~~. free-living
T 6. Trichina is a parasitic roundworm.
F 7. Food is digested in an earthworm's ~~circulatory~~ system. digestive
T 8. Earthworms produce offspring through sexual reproduction.
T 9. Every earthworm has a nervous system.
T 10. A starfish has radial symmetry.

B. Multiple Choice

Choose the word or phrase that completes correctly each of the following sentences.

1. A(n) (*sponge, earthworm, ~~clam~~, fly*) is a mollusk.
2. An insect is a(n) (*roundworm, mollusk, ~~arthropod~~*).
3. (*Flatworms, Mollusks, ~~Arthropods~~*) have the greatest number of species.
4. (*Antennae, ~~Spiracles~~, Eyes*) are openings through which a grasshopper takes air into its body.
5. Grasshoppers reproduce by (*~~sexual~~, asexual*) reproduction.
6. In complete metamorphosis, the adult stage comes after the (*egg, larva, ~~pupa~~*) stage.
7. Jellyfish and their relatives have a(n) (*lung, ~~cavity~~, heart, brain*) in the center of their bodies.
8. Animals with jointed legs and segmented bodies are (*flatworms, mollusks, ~~arthropods~~*).
9. (*~~Blood~~, Water, Air*) moves food and oxygen through the body of an earthworm.
10. In (*~~sexual~~, asexual*) reproduction, there is a union of a sperm cell and an egg cell.

C. Completion

Complete each of the following sentences with a word or phrase that will make the sentence correct.

1. A group of tissues working together is called a(n) __organ__.
2. __Trichinosis__ is a disease people may contract by eating uncooked pork.
3. A crayfish belongs to the crustacean group in the __arthropod__ phylum.

4. _Insects_ are the largest group of animals in the arthropod phylum.
5. _Larva_ stage of a butterfly is a caterpillar.
6. The two _eyes_ of a grasshopper detect colors.
7. _Insects_ have three pairs of legs and their bodies are divided into three parts.
8. An earthworm contains both _male_ and _female_ sex organs.
9. There are over one million species in the _animal_ kingdom.
10. A group of _organs_ working together is called a system.

D. How and Why See Teacher's Guide in the front of this book.
1. What happens to parasites if they kill their host animal?
2. Name four animal body systems. What is the function of each system?
3. What are the general body features of an insect?
4. Describe the stages in the complete metamorphosis of an insect.
5. Why do you think insects are the most plentiful and widespread of all land animals?

challenges

1. Obtain a library book on entomology, the study of insects, and learn how to make an insect collection. Visit a natural history museum to see insect collections prepared by entomologists.
2. Obtain information on a career in entomology. Find out what kinds of jobs exist and how a person prepares for a career in this field.

interesting reading

Gutman, Bill, *Women Who Work With Animals.* New York: Dodd, Mead, and Co., 1982.

Kohl, Judith and Herbert Kohl, *Pack, Band, and Colony: The World of Social Animals.* New York: Farrar, Strauss, and Giroux, 1983.

Animals can be divided into two major groups—those with backbones and those without backbones. Animals such as birds, reptiles, and amphibians have backbones. Why is this animal classified as a bird? What are the differences between warm-blooded and cold-blooded animals? What characteristics are unique to each of the groups of animals with backbones?

Tom Stack & Assoc.

Animals with Backbones

chapter 17

Introducing the chapter: List the vertebrate animal phyla on the chalkboard: fish, amphibians, reptiles, birds, mammals. Review briefly the main characteristics of each phylum and examples of each as an overview of the chapter's content. Discuss careers related to animals, such as zoo keeper, veterinarian, game warden, and biologist.

17:1 Backbones and Skeletons

A catfish, a toad, and a squirrel are three very different kinds of animals. Yet these animals are alike in a very important way. They all have a brain and a skeleton inside their bodies. How many animals can you name that have a brain and an internal skeleton made of bones?

Animals can be divided into two large groups called the vertebrates (VURT uh brayts) and invertebrates (ihn VURT uh brayts). A **vertebrate** is an animal that has an internal skeleton. Part of the skeleton is a case that surrounds and protects the animal's brain. Vertebrates are part of a large phylum called chordates (KOR dayts). This phylum includes fish, amphibians, reptiles, birds, and mammals. The skeleton of a vertebrate supports and protects body organs. Muscles attached to the skeleton cover and provide protection for the brain and the spinal cord. A **spinal cord** is a long bundle of nerves that connects the brain to other parts of the body. The backbone of an animal consists of bones or vertebrae that protect the spinal cord. In this chapter, you will study the features of vertebrate animals.

GOAL: You will learn the characteristics of species classified in the chordate animal phylum.

Remind students that human beings are vertebrates.

Name the animals in the chordate phylum.

Enrichment: Have students reach around to their backs and feel their vertebrae. Discuss the importance of the skull and backbone in protecting the brain and spinal cord. Relate the discussion to the use of safety helmets in jobs and sports.

FIGURE 17–1. A vertebrate has an internal skeleton which protects the brain, spinal cord, and other internal organs.

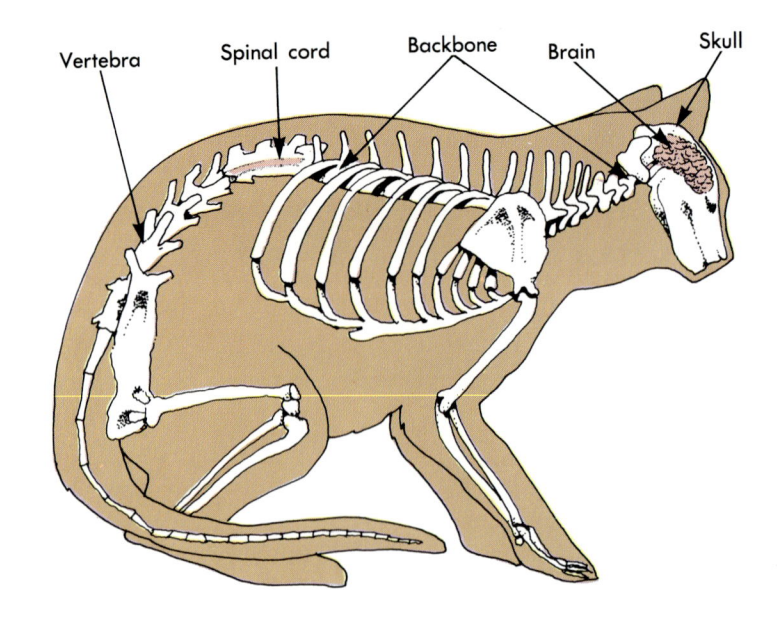

In Chapter 16, you learned about the animals that are invertebrates. **Invertebrates** do not have a backbone or an internal skeleton. Many of these invertebrate animals, such as worms and jellyfish, have soft bodies. Others, such as insects, lobsters, and clams, have a hard exoskeleton that protects the soft internal body parts.

making sure

1. How do vertebrates differ from invertebrates?
2. Why are fish considered vertebrates when jellyfish are not?

1. vertebrates—backbone, internal skeleton
   invertebrates—no backbone, soft bodies, exoskeleton
2. fish—backbone, jellyfish—no backbone

The oxygen fish obtain from water is dissolved oxygen. It is not the oxygen in water molecules.

How is cartilage different from bone?

17:2 Fish

What kind of animal do you think of when you see the word fish? Fish are vertebrates that live in water. They obtain oxygen from water and have fins which are used for swimming. Sharks, skates, and rays belong to one group of fish. The fish in this group have skeletons made of cartilage (KART ul ihj). **Cartilage** is a dense, strong, rubbery tissue. It is different from bone in that it lacks the minerals that make bone hard. Cartilage is present in your body in your nose, ears, and parts of your backbone.

364   Animals with Backbones

FIGURE 17–2. A blue shark is an example of an animal with a skeleton made of cartilage.

Fish with skeletons made of bones are classified in a group called bony fish. Most of the fish you know belong to this group. Trout, bass, salmon, catfish, and goldfish are examples of bony fish. All fish are cold-blooded. Cold-blooded means an animal's body temperature changes as the temperature of its environment changes. The temperature of a fish depends on the temperature of the water in which it lives. As the water temperature changes, so does the body temperature of the fish.

Define cold-blooded.

There are many different sizes and kinds of bony fish. Some minnows are about 1 cm in length. Swordfish grow as long as 4 m. Mackerel have streamlined bodies for very fast swimming. In contrast, the seahorse has an unusual shape and swims slowly in an erect position. Flounders can change color to blend with the sea bottom. Most fish species live either in salt or in fresh water only. Sticklebacks, however, are at home in either kind of water. Salmon live in salt water but return to freshwater rivers to lay their eggs.

Give an example of a bony fish.

FIGURE 17–3. A flounder can change color to match the ocean bottom.

17:2 Fish 365

Enrichment: Dissect a perch or other bony fish. Label the major parts using tags and pins. Display the labeled perch in class.

A yellow perch provides a good example of the structure of a bony fish. It has a large mouth and small, thin teeth. One eye is located on each side of the head. The eyes do not have eyelids. Behind each eye is a thin bony gill cover. Under each gill cover are four comblike gills. Two fins are attached to the back of the perch and there is a large tail fin. One fin is located behind each gill and two fins are located on the bottom of the perch.

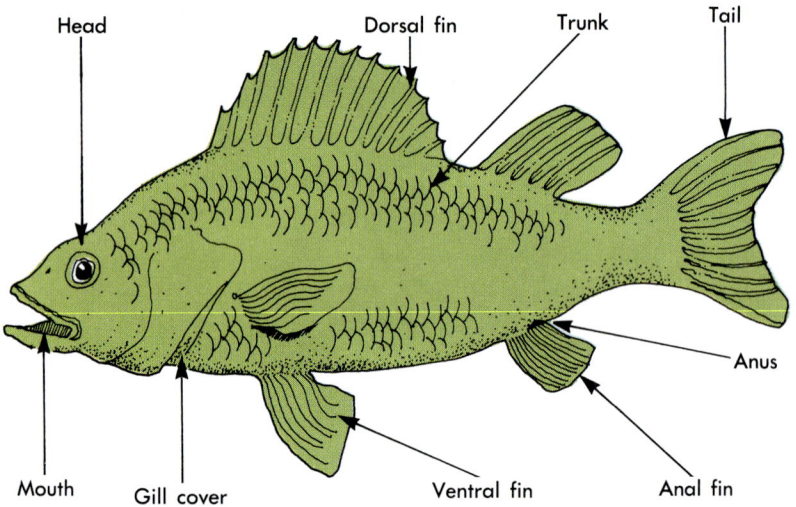

FIGURE 17-4. External body parts of a yellow perch.

Enrichment: A live goldfish in an aquarium tank may be used to observe the external features of a fish.

The skeleton of a yellow perch contains about 40 bones. A skull and a spinal column made of bones called vertebrae are part of the skeleton. Muscles attached to the bones are used by the fish to swim and turn in the water. The digestive system includes a stomach and an intestine in which food is digested and absorbed. Bile is produced in an organ called the liver and is stored in the gall bladder. Bile, which aids in the digestion of fat, is carried through a duct, or small tube, to the intestine. Food that is not digested passes out of the fish through an opening called the anus.

Like other fish, a yellow perch has a heart with two sections or chambers. The pumping action of the chambers forces blood through the circulatory system. Blood carries food and oxygen to cells throughout the fish. The blood also has white cells and substances that fight disease. Arteries carry

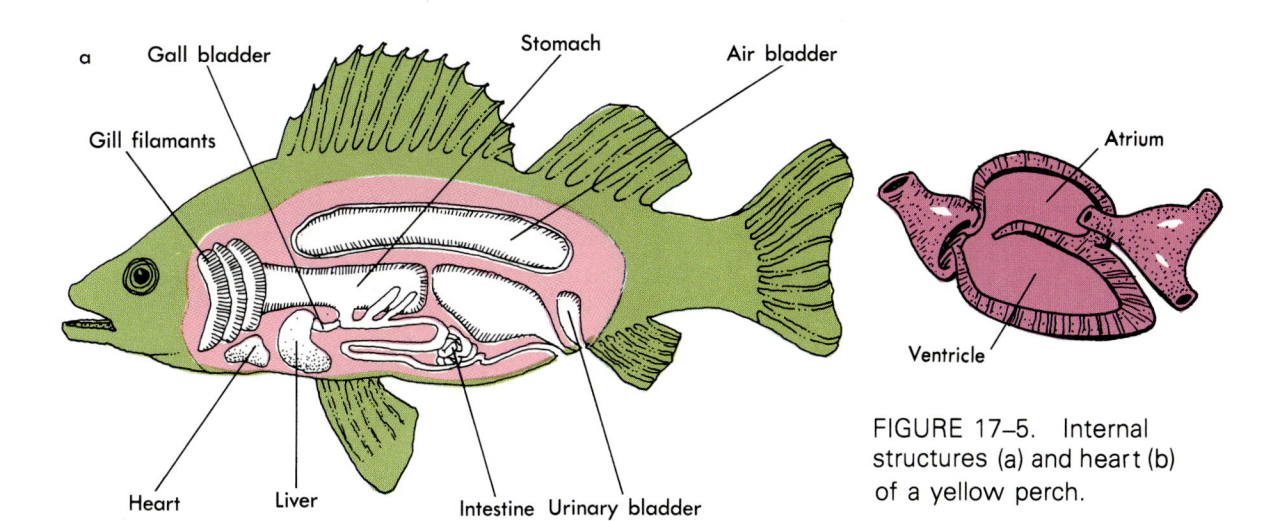

FIGURE 17-5. Internal structures (a) and heart (b) of a yellow perch.

blood away from the heart. Veins carry blood back to the heart. Tiny capillaries connect the arteries and veins. The food and oxygen from the blood passes through the capillary walls to reach the cells.

To obtain oxygen, a fish pumps water over its gills. Oxygen dissolved in the water diffuses through the thin membranes of the gills into the fish's blood. Carbon dioxide, a waste product, moves out of the blood into the water. A fish soon dies when removed from water because it cannot obtain oxygen.

activity
Keep the fish moist at all times.

CIRCULATION OF BLOOD   Activity 17-1

(1) Place a goldfish in a beaker of water at room temperature. (2) Take some cotton and wet it in the water. (3) Remove the goldfish from the beaker and wrap it with the moist cotton. Leave the tail exposed. (4) Place the fish in a petri dish and lay a microscope slide over the tail. (5) Place the dish on a microscope stage. Focus on the tail so you can observe the circulation of blood. (6). Add water to the cotton with a dropper to keep it moist. (7) After you have observed the circulation quickly return the goldfish to the beaker of water. Does the blood flow in blood vessels? Does the blood always flow in the same direction? yes yes   See Teacher's Guide.

FIGURE 17-6.

17:2 Fish

activity   See Teacher's Guide.

FEATURES OF A FISH　　　　Activity 17-2

Observe some fish in an aquarium tank. Select one fish and carefully study its size, shape, and other features. Make a drawing of the fish that shows its shape, fins, eyes, gills, and other parts. Label all the parts you can identify. Repeat this procedure with another kind of fish in the tank. Compare your two drawings. How are the two fish similar? How are they different? How are the two fish different from the perch shown in Figure 17–4?

making sure

3. List three features common to all fish. *fins, gills, scales, internal skeleton*
4. How do fish obtain the oxygen they need to live? *from water through gills*
5. How do the cells in a fish obtain food and oxygen? *Blood carries food and oxygen to the cells.*
6. What structures do fish use to move through the water? *fins*

17:3 Amphibians

What is an amphibian?

Frogs, toads, and salamanders are amphibians. **Amphibians** are animals that live part of their lives in water and part on land. They are vertebrates that have moist skin and no scales. Adult amphibians have two eyes, ears, and breathe through lungs. Amphibians, like fish, are cold-blooded.

Frogs live in or near bodies of fresh water while toads live on land. Some species of toads live in moist areas such as the floor of a forest, while others live in desert regions. Frogs and toads are usually 5 to 12 cm long, through the head and back. They eat live insects and worms. Some species, such as the bullfrog, also feed on small fish.

How does a frog reproduce?

Frogs and toads reproduce during the spring. **Fertilization of the eggs by the sperm takes place outside the female. After a female lays its eggs in water, the male fertilizes them with sperm. The fertilized eggs develop into tadpoles. A tadpole has gills, a tail, and lives in water.** As it develops into an

FIGURE 17–7. A toad is an amphibian that lives mostly on land.

Courtesy CCM: General Biological, Inc.

Animals with Backbones

FIGURE 17-8. A green salamander is an amphibian.

adult, the tadpole loses its tail and gills and develops lungs and legs. Some species of toads lay their eggs in streams. Others lay their eggs in temporary pools of water formed after spring or summer rains. Tadpoles feed on algae and small plants.

In regions with severe winters, frogs burrow into the muddy bottom of a pond during the cold season. Toads burrow down into the ground below the frost line. During the cold winter months, these frogs and toads are in a winter sleep called hibernation. During hibernation an animal's body metabolism decreases and its heartbeat is very slow. The animal lives on materials stored within its body.

FIGURE 17-9. A tadpole has gills, a tail, and lives in water (a). Most frogs lay their eggs in large masses in the water (b).

a

b

17:3 Amphibians

FIGURE 17-10. Frogs have a long, sticky tongue for catching insects.

In what two ways does a frog obtain oxygen?

Detailed discussion of the frog is given to illustrate the main features of vertebrates.

Digestion changes large food molecules into smaller food molecules that can be used by cells.

FIGURE 17-11. The brain of a frog (a) and major internal organs (b).

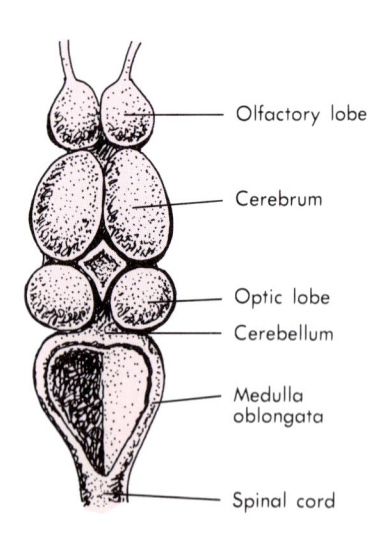

17:4 The Frog

A frog obtains oxygen through its moist skin. It also obtains oxygen by breathing through a pair of lungs. An exchange of gases occurs in the lungs of the frogs. Oxygen diffuses from the lungs into the blood. Carbon dioxide and water vapor diffuse out of the blood into the lungs. Thus, a frog inhales air rich in oxygen. It exhales air rich in carbon dioxide and water vapor. In water, a frog must come to the surface to breathe air.

The digestive system of a frog is shown in Figure 17-11. Muscles in the walls of these organs cause them to contract and push the food through the system. It is the same kind of motion that occurs in your own digestive system.

Digestive juices containing enzymes are made in the walls of the stomach and intestines. An enzyme is a chemical that aids in the digestion of food. Two organs, the **liver** and **pancreas** (PAN kree us), also produce digestive juices. Bile from the liver aids the digestion of fat in the small intestine. The pancreas produces digestive enzymes that pass through a duct to the small intestine.

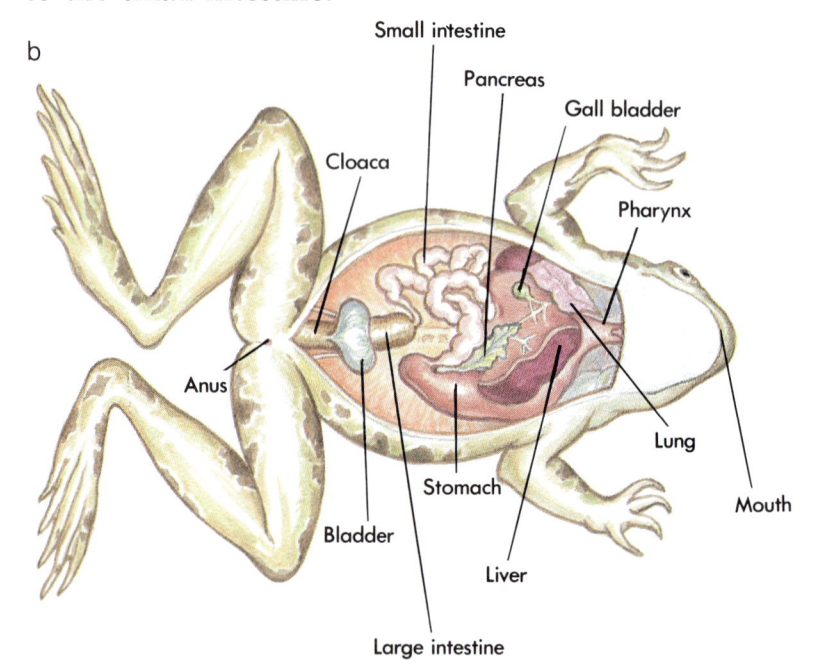

370 *Animals with Backbones*

Most of the digestion and absorption of food occurs in the **small intestine** of the frog. Waste products pass through the **large intestine** into the cloaca, a storage area. Wastes are then excreted through the anus.

A frog's circulatory system contains a heart and a network of blood vessels. The heart has three chambers. They are the right atrium, left atrium, and ventricle. Blood from the frog's body enters the right atrium. Blood returning from the lungs enters the left atrium. Both the right atrium and the left atrium contract and push blood into the ventricle. Then the ventricle contracts and pushes blood through vessels to the lungs and other body organs.

The blood is pumped through a circulatory system of arteries, veins, and capillaries. Red blood cells contain a compound called hemoglobin (HEE muh glo bun) which carries oxygen. White blood cells, which fight disease, are also present in the blood.

A frog's brain and spinal cord compose its central nervous system. Many nerves go from the brain and spinal cord to other organs in the body of the frog. Also, nerves return from these organs to the central nervous system. Through these nerves, the brain controls the life activities of the frog.

The eyes of a frog are much like your eyes. Six muscles are attached to the eyeball and rotate the eye in its socket. The eye is connected by nerves to the brain. A frog does not have an external fleshy ear. Its eardrum is on the surface of the head. The eardrum can receive sound through either water or air. Sound reaching the eardrum causes it to vibrate. These vibrations are carried to the inner ear. Here the vibrations are changed to nerve messages that travel to the brain.

All animals with backbones have compounds called hormones in their blood. They help control various organs such as muscles, stomach, and heart. Hormones are produced by glands and sent directly into the bloodstream. In the bloodstream, hormones are transported to all parts of the body.

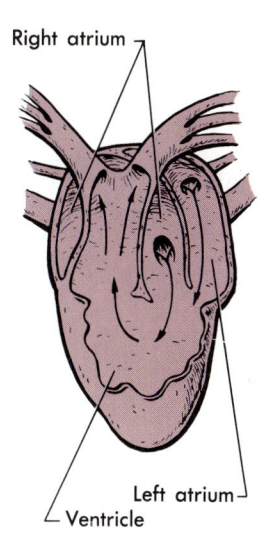

FIGURE 17–12. Circulation through the heart of a frog.

Describe the circulatory system of the frog.

What makes up the frog's nervous system?

Trace the flow of a drop of blood from the right atrium through the frog's body and back to the heart.

Hormones are produced by ductless glands called endocrine glands.

Supervise students during the activity to be certain they follow directions and observe the structures that are listed here.

activity

PARTS OF A FROG

Activity 17-3

Objective: To dissect and learn the parts of a frog

Introduce this activity with a preview of the work to be done.

Materials See Teacher's Guide.

dissecting needle
dissecting pan or soft wooden board
frog, preserved — Use a dead or preserved frog for this dissection.
paper towel
pointed scissors, small and sharp
scalpel or small knife
6 straight pins
tweezers

Procedure

1. Lay the frog, backside up, on a paper towel in the dissecting pan.
2. Find the nostrils, mouth, tympanic membrane, forelegs, hind legs, and feet.
3. Make a cut with the scissors on both sides of the mouth so it can be opened wide. Locate the internal nostrils and teeth. Find the gullet, glottis, and tongue. Observe how the tongue is attached inside the mouth.
4. Turn the frog over on its back. Use the scalpel to cut through the abdomen in a line from the lower jaw to the anus. Make a cut across one end of the first cut at the lower jaw. Make a second cut across the body just forward of the anus. Locate the muscles in the chest inside the skin. Pull the body wall and pin it down with straight pins.
5. Use the dissecting needle to move and separate body parts as needed. Locate the heart. Observe the blood vessels leading from the heart. Find the lungs, liver, and gall bladder.
6. If your frog is female, it will have large masses of small dark eggs in the body cavity. Remove the egg masses if they are present.
7. Examine the parts of the digestive tract. Find the pancreas and spleen. The pancreas is in the first loop of the small intestine. The spleen is a round, red organ under the intestines.
8. Lift up and cut out the intestines. Locate the kidneys and bladder.
9. If you have a female frog, you removed the ovaries with the eggs. The two long, coiled white tubes are the oviducts. In a male frog, the round white organs near the kidneys are testes.
10. Remove the pins and turn the frog over. Cut through the skin, muscle, and skull to expose the brain. Do not cut too deeply. Find the cerebrum, cerebellum, medulla, and spinal cord.

Observations and Data

Draw a diagram that shows the organs inside the body cavity.

Questions and Conclusions

1. What parts of the frog protect the internal body organs? skeleton, muscles, skin
2. Name one organ you observed in each body system. Answers will vary.
3. What organs are part of the digestive tract? mouth, tongue, pharynx, stomach, small intestine, cloaca, anus

Animals with Backbones

activity

See Teacher's Guide.

OBSERVING METAMORPHOSIS IN FROGS
Activity 17-4

During the spring, some tadpoles may be obtained from a pond. Keep them in an aquarium with an air pump and filter. Maintain the water at room temperature. Feed the tadpoles every day with chopped lettuce, spinach, or cooked egg yolk. Remove any excess food so it does not decay in the water. Observe the tadpoles until they grow into mature frogs.

making sure

7. How are amphibians similar to fish?
8. Give an example of how an amphibian reproduces.
9. How do frogs respond to the change in seasons from fall to winter?
10. How is the body temperature of a cold-blooded animal affected by a change in the temperature around it?
11. What compound in a frog is used to transport oxygen through the body?

7. One life stage has gills and lives in water.
8. Frog—female deposits eggs in water which are fertilized by sperm from males.
Fertilized egg develops into a tadpole.
9. hibernate in mud at bottom of a pond
10. Body temperature changes as the temperature around it changes.
11. hemoglobin

17:5 Reptiles

Reptiles are vertebrate animals that are cold-blooded, breathe air, and live mainly on land. Lizards, snakes, turtles, crocodiles, and alligators are reptiles. The outer skin of a reptile consists of dead cells that form a layer of dry scales. These dry scales resist water loss from the body and protect the animal from the rough surfaces of sand and rocks.

What is a reptile?

Name five reptiles.

FIGURE 17–13. The iguana (a), turtle (b), and alligator (c) are reptiles.

a

b

c

Carl W. Rettenmeyer/University of Connecticut Biology Dept.

Alvin E. Staffan

Warren Garst/Tom Stack & Assoc.

17:5 Reptiles 373

Turtles are usually found near water. Lizards and snakes are often found out in the open "sunning" themselves on warm days. Some snakes are located underneath rocks.

Enrichment: Have students describe experiences they may have had with wild or pet snakes. Some students may have observed snakes in a zoo.

Except for snakes and some lizards, reptiles have four legs which are suited for running, crawling, or climbing. Most species of reptiles live in tropical or subtropical regions.

Snakes have long round bodies and no legs. The skeleton of a snake includes the skull, backbone, and ribs. Very long snakes like boas and pythons may have 200 to 400 bones in their backbones. Snakes have teeth, each slanted backwards, on the roofs of their mouths. The teeth hold food while it is swallowed. When snakes eat, they swallow their food whole without tearing or chewing it. The bones of a snake's jaw are arranged so the jaws can open very wide. This feature allows a snake to swallow prey greater than the diameter of the snake's body. Some snakes have special teeth called fangs that are used to inject venom. Venom is a poison used to paralyze and kill animals for food and as a defense against enemies. Rattlesnakes have fangs which are folded back when not in use. A snake has a ribbonlike tongue with a forked tip that provides the animal with a sense of smell.

Describe the reproduction of reptiles.

During reproduction, the fertilization of eggs in reptiles occurs inside the female's body. Most species deposit the fertilized eggs outside the body where they develop into young reptiles. Often the eggs are deposited in spaces between rocks and logs, or in

FIGURE 17–14. A female sea turtle lays eggs in a sandy beach at night.

C. Allan Morgan/Peter Arnold, Inc.

374 **Animals with Backbones**

mud or sand. In some species, such as rattlesnakes and garter snakes, the eggs develop into young inside the female reptile. The eggs of many reptile species are somewhat like bird's eggs having a leathery or hard shell. Sea turtles produce about 400 eggs each year. Small snakes and lizards produce 10 to 20 eggs and the American alligator produces 30 to 60 eggs each year.

making sure

12. How is a reptile different from an amphibian?
13. Amphibians and reptiles lay many eggs each year. Why is there no large increase in the number of these animals?
14. How are many reptiles suited to life in tropical or subtropical regions?
15. How is reproduction in rattlesnakes different from reproduction in other species of snakes?

12. A reptile has scales and does not have a stage that lives in water.
13. Predators eat most of the eggs and offspring.
14. The dry scales on the skin resist water loss and protect the animal from rough surfaces.
15. Eggs develop into young inside the female.

17:6 Birds

Birds are warm-blooded. Warm-blooded means an animal's body temperature is maintained at a certain level. For most birds, the daytime body temperature is 40 to 43°C. They have wings, two legs, feathers, and reproduce by laying eggs. Feathers insulate a bird's body and help regulate its body temperature. Birds are noted for their colors. Using these colors, different species can be identified. A blue jay, for example, has a distinctive blue color. One way to tell the difference in the sexes of some bird species is to notice the difference in their colors. Usually the males are more brightly colored than the females. Some birds, such as pheasants, are hunted as game, and others, such as chickens, are raised for meat and eggs.

Birds have a four-chambered heart, a delicate, strong skeleton, no teeth, and a projecting beak or bill. The bones of a bird have many hollow air spaces. Between the internal organs are air sacks.

Define warm-blooded.

What is the function of feathers?

Introduction: Birds rank high in students' interest. Ask students to name and describe birds they know. Then ask students to list the ways in which birds are alike.

Traits that distinguish males from females are called secondary sex characteristics.

Compare the four-chambered heart of a bird with the three-chambered heart of a frog and the two-chambered heart of a fish.

FIGURE 17-15. A bald eagle can be easily identified because of the white feathers on its head.

Birds are noted for their loud voices and characteristic calls. They use their calls to warn of danger, attract mates, and keep other birds out of their territory.

Some birds, such as the bobwhite, live in the same area throughout the year. Other species, such as the robin, migrate to another region with the change of seasons. Most migration occurs north and south. Migratory birds go north to feed and nest in the spring and summer. They return south for the winter. Some birds migrate to the mountains for the summer and return to the lowlands in winter.

FIGURE 17-16. Cardinals live in the same area throughout the year (a). Canada geese are migratory birds (b).

a

b

Birds that migrate use the same routes every year. They arrive and depart from a region at about the same time each year. Some birds migrate at altitudes up to 1700 m. Even with stopping occasionally to feed, a bird travels an average of 50 km each day. Species migrating long distances fly at night and use the stars to find their way.

During reproduction, fertilization of the eggs takes place inside the female bird. The female then lays the fertilized eggs, which have hard shells. One or more eggs may be laid. A California quail hen will lay about 14 eggs. Either parent may sit on the eggs and keep them warm until they hatch. The length of time before the eggs hatch varies with the species. It is about 18 days for pigeons, 28 days for hawks, and 42 to 60 days for ostriches. Young birds require great care or they will die. Their parents feed and guard them and keep them warm.

activity See Teacher's Guide.   Activity 17-5

INTERNAL ORGANS OF A CHICKEN

Obtain the internal body parts of a chicken from a butcher shop. These are the parts that are removed when a chicken is cleaned before it is sold for food. Wash the organs with cold water, then place them in a flat pan. Use Figure 17–17 to identify the heart, liver, lungs, and kidneys. Identify each part of the digestive system shown in Figure 17–17. Make a drawing of the digestive system, labeling each of the organs. Slice the gizzard in half and examine its contents. Why is a gizzard called the "hen's teeth"?

It contains small stones and grinds up the chicken's food.

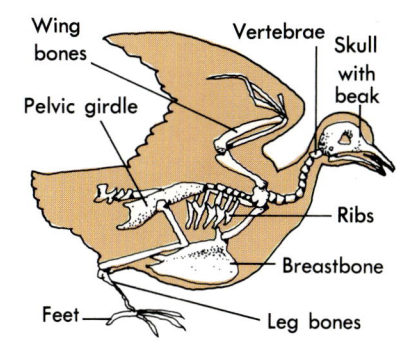

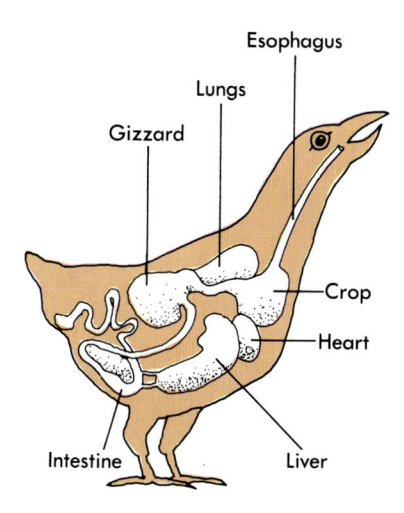

FIGURE 17–17. The internal body parts of a bird.

making sure

16. How does the sight and hearing of birds help to keep them alive?
17. Why do birds have bones with hollow spaces and air sacs between internal organs?
18. Why are female birds usually not as brightly colored as the male?
19. Why do most species of birds have their own distinct "birdcall"?

16. They can react quickly to escape predators and to catch food.
They are colorful and many species fly in the air where they can be seen.
17. decreases density and makes flight possible.
18. Bright coloring of males attracts females for mating.

The call enables one member of a species to locate another member of the same species.

17:6 Birds

17:7 Mammals

Whales, porpoises, bats, kangaroos, bears, and rats are all mammals. **Mammals are warm-blooded vertebrates that have hair and feed their young with milk. Compared to other kinds of animals, mammals have a large brain.** Many domestic animals, such as cats, dogs, cows, and horses, are also mammals. Domestic means the animals are usually tame and are raised and cared for by people.

There is a wide variety of mammal species. Shrews and mice are the smallest mammals, measuring less than 5 cm in length. Elephants are almost 4 m tall. The largest mammals are whales, which reach a length of 35 m and have a mass of about 100 metric tons. A bat is an unusual mammal because it has wings and can fly.

Name some characteristics of mammals.

Mammal parents provide more care for their young than do other animals.

FIGURE 17-18. A killer whale has lungs and cannot breathe underwater (a). A shrew is the smallest mammal (b).

a

b

A mammal generally has a large head due to the size of its brain. Many mammal species have external ears that aid in hearing. Deer and other animals that graze on grass have long necks. Tails of some mammals, such as monkeys, are used to hold tree branches and vines.

Human beings are mammals. Their large brains and intelligence make them distinct from other animals.

Different species of mammals have many different sources of food. Hoofed mammals, such as cows and sheep, eat grass and other plants. Some mammals, such as the weasel and seal, eat only the flesh of other animals. Rats, bears, and pigs are examples of mammals that eat both plants and animals.

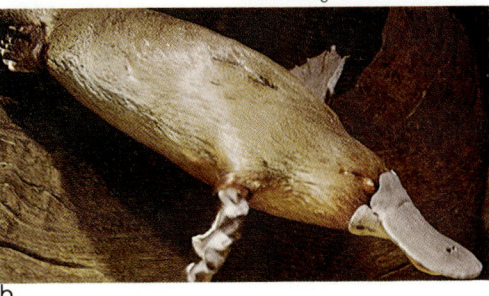

FIGURE 17–19. Most bats are active only at night (a). A platypus is a mammal that has birdlike features (b).

Some mammals that eat flesh are **predators**, meaning they must kill other animals, prey, for their food. For example, cattle are eaten by tigers, deer are eaten by wolves, and mice are eaten by weasels.

Most animals hatch from eggs. However, the young of most mammals develop inside the mother's body. A kangaroo is an exception. The young kangaroo develops inside the mother for only a few days and then crawls out of the mother's body and into a pouch. Here the development of the new animal continues. Female mammals feed their young on milk produced by mammary glands in the chest or abdomen. A newborn mammal requires more care for a longer period of time than the young of other kinds of animals.

Do the young of most mammals develop inside or outside the mother's body?

A duck-billed platypus is a very unusual mammal. This animal lives in Australia and is 45 to 50 cm long. It has a hard bill like a duck, webbed feet, a flat tail, and lives in or near water. The duck-billed platypus reproduces by building a nest in which it lays one to three small eggs. When hatched, the young are about 2.5 cm long and nurse on the mother's mammary glands.

Discussion: Ask students to explain why the duck-billed platypus is unusual.

People do not always see themselves as animals because they think, speak, read, and write. Also, people make and use all kinds of tools. People are mammals and they belong to the animal kingdom. People have hair, an internal skeleton, feed their young on milk, and are warm-blooded.

Why is the duck-billed platypus an unusual animal?

making sure

20. Why are mammals considered the most advanced of all animals? They have the largest brains and the most complex bodies.

PERSPECTIVES
frontiers

Why Rattlesnakes Have Rattles

Have you ever observed a plant or animal structure and wondered what function it served? For example, what is the purpose of a bald eagle having white head feathers, or a rat having a long tail? Many structures of living organisms serve a purpose that is often difficult for us to determine. To ask questions, however, is an important aspect of science. It is also one of the first steps toward solving a scientific problem.

Rattlesnakes are well known for the rattles found on the end of their tails. However, the purpose of the rattle has been questioned by biologists. The most widely accepted hypothesis is that the rattle serves as a warning to other animals to stay away. The rattle may also be used to keep large animals such as bison, deer, and antelope from stepping on the snake. This theory, like other scientific theories, cannot be proven directly. Instead, scientists evaluate the existing evidence, and develop the best theory from this evidence.

In order to understand the function of rattles on rattlesnakes, it may be helpful to examine a situation where the rattles are absent. On Santa Catalina Island, off the coast of California, a variety of rattlesnake is found that does not have rattles on the end of its tail. Since no large animals are found on Santa Catalina, some scientists assumed that the snakes did not need rattles to defend themselves.

Recently, some scientists have challenged the explanation for the absence of rattles. One hypothesis states that the Santa Catalina rattlesnakes climb trees more frequently than other rattlesnakes in search of birds for food. If the snake had a long chain of rattles brushing against branches, the noise would scare away its prey. Therefore, the snake has an advantage without the rattles.

A second function of rattles has been demonstrated by other biologists. The Massasauga rattlesnake is a small snake that feeds on a variety of organisms, including frogs. Biologists observed that when a frog was placed in a cage with the snake, the tail rattled. The frog then leaped at the rattles as if they were an insect and bit the snake's tail. The snake proceeded to devour the frog.

It should be emphasized that the rattles can serve a variety of functions. Also, that one particular use is not more important than another use. However, the origin of the rattle may be difficult to determine. Studying the rattle on a rattlesnake illustrates how scientists support theories that are difficult to prove by direct experiments.

Joseph T. Collins

main ideas   Chapter Review

1. Animals can be divided into two groups, vertebrates and invertebrates. 17:1
2. Fish are vertebrates that live in water, obtain oxygen through gills, and are covered with scales. 17:2
3. Amphibians are vertebrates with moist skin and no scales. They live part of their lives in water and part on land. 17:3
4. A frog's body is a good illustration of the systems of organs present in vertebrate animals. 17:4
5. Fish, amphibians, and reptiles are cold-blooded animals. Birds and mammals are warm-blooded. 17:2–17:7
6. Most species of reptiles live in tropical or subtropical regions 17:5
7. Birds are vertebrates that have wings, two legs, feathers, and reproduce by laying eggs. 17:6
8. Some birds stay in the same area all year and others migrate when the seasons change. 17:6
9. Mammals are vertebrates that have hair, are warm-blooded, and feed their young on milk. 17:7
10. Most animals reproduce by laying eggs. 17:7
11. Fertilization and development in most species of mammals occur inside the mother's body. 17:7

vocabulary

Define each of the following words or terms.

mammal   reptile   amphibian   invertebrate

study questions

DO NOT WRITE IN THIS BOOK.
A. True or False

Determine whether each of the following sentences is true or false. If the sentence is false, rewrite it to make it true.

F 1. The fertilization of eggs in a reptile occurs ~~outside~~ *inside* the female's body.
T 2. Bones protect internal body organs.

Animals with Backbones 381

F 3. A fish's heart has ~~four~~ chambers. two
T 4. Fish obtain oxygen through gills.
T 5. In mammals, blood circulates through blood vessels.
F 6. Tadpoles have ~~lungs~~. gills
T 7. Hormones help regulate the body functions of an animal.
F 8. Reptiles are ~~warm~~-blooded animals. cold
F 9. ~~All~~ reptiles have four legs. Most
T 10. The bones of a bird contain space filled with air.

B. Multiple Choice

Choose the word or phrase that completes correctly each of the following sentences.

1. A(n) (*crayfish*, *perch*, *alligator*, *snake*) is an invertebrate animal.
2. Oxygen is taken into an adult frog through its (*gills*, *lungs*, *right atrium*, *ventricle*).
3. (*Arteries*, *Veins*, *Capillaries*) are tiny blood vessels.
4. A toad is a(n) (*reptile*, *amphibian*, *sponge*, *mammal*).
5. Food is absorbed into the blood mostly through a frog's (*mouth*, *stomach*, *intestine*, *pancreas*).
6. An animal with a backbone is classified in the (*arthropod*, *mollusk*, *chordate*) phylum.
7. The heart of a bird has (*one*, *two*, *three*, *four*) chambers.
8. (*Reptiles*, *Fish*, *Mammals*, *Birds*) are the most advanced of all animals.
9. (*Reptiles*, *Fish*, *Mammals*, *Birds*) feed their young on milk.
10. In most mammals, the young develop (*inside*, *outside*) the mother's body.

C. Completion

Complete each of the following sentences with a word or phrase that will make the sentence correct.

1. A spinal column of __bone__ surrounds and protects the spinal cord.
2. __Muscles__ and bones are used by vertebrates to move their body parts.
3. __Capillaries__ are blood vessels that connect arteries and veins.
4. Birds and __mammals__ are two groups of warm-blooded animals.

5. The central nervous system of a frog includes the <u>brain</u> and spinal cord.
6. A turtle belongs to the <u>reptile</u> group of vertebrates.
7. <u>Feathers</u> insulate a bird's body.
8. Many migratory birds fly at night and use the <u>stars</u> to find their way.
9. A(n) _____ is an example of a domestic animal. chicken (Answers will vary.)
10. A wolf is a <u>predator</u> because it kills prey for food.

D. How and Why See Teacher's Guide in the front of this book.
1. How are vertebrates different from invertebrates?
2. How are fish different from amphibians?
3. Name three systems in a frog and indicate the functions of each system.
4. What features are used to identify a species of bird?
5. Name three species of mammals. List the ways in which these three species are alike.

challenges

1. Plan a field trip to a zoo. Take along a camera and take pictures of different animals. Find out the scientific name for each animal. Mount your photos on a poster board and label each with the animal's common name and scientific name.
2. Investigate raising tropical fish as a hobby. Visit a pet shop or aquarium store to obtain information. Find out what equipment is needed and what care is needed to keep the fish alive and healthy.

interesting reading

Ryden, Hope, *Bobcat*. New York: Putnam, 1983.

Roland-Entwistle, Theodore, *Illustrated Facts and Records Book of Animals*. New York: Arco, 1983.

McLoughin, John C., *The Canine Clan: A New Look at Man's Best Friend*. New York: Viking Press, 1983.

SIDE ROADS
Reading the Environment

Joseph T. Collins *Carolina Biological Supply Co.*

Organisms use a variety of senses to receive information from their environment. For example, birds hear other birds singing and learn that they may be trespassing in another bird's territory. A bear tastes a berry and learns that it is edible. A zebra smells a lion and retreats to a safer location. A bee sees the bright color of a flower and feeds on the nectar. Most animals use their five major senses to gain information about their surroundings. Some animals are exceptions and do not use all five senses, such as blind cave salamanders and blind cave fish.

Bats, some birds, and several marine animals use high frequency sounds called sonar, to pick up information they cannot see. These sonar waves echo off obstacles and are received by the animal. Bats use this technique to move through dark caves without hitting the walls. The bat emits a sound that bounces off the cave wall and is picked up by its sensitive ears. Bats also use this technique to feed on insects that live in the caves. There are some insects that are capable of hearing the bat's high frequency sounds, and quickly diving when the bat approaches.

Bats are not the only animals to use sonar. A bird in South America called the oilbird lives in caves and uses sonar to find its way in and out. Dolphins, seals, and sea lions also use sonar to locate food and avoid obstacles.

Several species of fish, tadpoles, salamanders, and frogs have special structures called lateral lines, that enable them to detect when something is approaching. A lateral line picks up the slightest vibrations in water. It enables many fish to swim in schools and coordinate their movements. Some salamanders use it to locate food.

Some types of blind fish produce an electrical field that surrounds their bodies. As they swim and approach obstacles, the electric field enables the blind fish to change course. Sharks, and several species of skates, rays, and catfish are capable of detecting weak electric fields that are given off by all living organisms. Sharks are sensitive to distances of about three meters. They are able to detect their food even when it is hidden under the sand.

Doug Wynn

Boa constrictors, pythons, and pit vipers feed on birds and small mammals. These snakes are able to find their prey in total darkness by detecting the slightly higher body temperature. The pit vipers get their name from small pits on their bodies that detect heat. Boas and pythons do not have visible pits, but do possess several under the scales of their upper and lower lips. (See Figures 3 and 4.)

Animals often reflect their lifestyles. When examined by scientists most animals give clues about their means of interacting with the environment. This interaction is often a result of the animals reading their environments through the senses.

Doug Wynn

Environment is all the surroundings of an organism. The environment includes the plants, animals, climate, and nonliving parts of an area. Only the plants and animals that are suited to a particular environment live in it. How are these big horn sheep suited to their environment? Why is a water buffalo unsuited to this environment?

Rick McIntyre for Tom Stack & Assoc.

Environment and Heredity
unit 5

A group of living things makes up a community. Many environmental factors determine the type and size of a community. What are those factors? How does a community differ from a population? What cycles occur in a community? What is an organism's habitat?

Populations and Communities chapter 18

Introducing the chapter: Do a demonstration in which a student exhales through a straw into a beaker of limewater. Explain that the limewater turns cloudy due to the presence of carbon dioxide (CO_2) in the exhaled air. (The CO_2 combines with limewater, $Ca(OH)_2$ to form insoluble CaO.) Point out that carbon dioxide is part of the environment, comprising 0.03 percent of the atmosphere. Ask students to identify ways in which carbon dioxide is added to the atmosphere (animals, burning, volcanic eruptions). Define and discuss the meaning of the term environment.

18:1 Communities

Living organisms are found in many different places from the equator to the poles. Some animals spend much of their lives underground while others spend their time flying through the air. Living creatures inhabit the deepest oceans and the highest mountains.

A **community** consists of all the living things in a certain area. For example, fields, forests, swamps, and lakes are communities. Each community contains populations of plants and animals that live together. A pond is an example of a community. Living in a pond are populations of fish, different species of plants, and various kinds of insects. The pond community may also include frogs and turtles. Within a community, each different species affects each other in many ways. For example, in a pond frogs eat insects and fish eat tadpoles.

GOAL: You will learn about the interrelationships among living things and their environment and the communities in which they live.

Define community. List five examples.

Introduce this topic with observations of a terrarium or aquarium. Relate these observations to life in a student's community.

FIGURE 18–1. Although a pond and a tidepool are both water communities, life in a pond (a) is quite different from the life in a tidepool (b).

1. A pond is smaller in size and the water is not moving as in a river or stream. A pond will have different species of fish.
2. Where organisms go, what they eat, how they move, and how they relate to other organisms can be learned by studying organisms in their community.

Many fascinating facts can be learned about organisms by studying them in their own community. Where do they go? What do they eat? How do they move? Other living things in the community must be observed also. How do they relate to other living things? These are a few questions a scientist asks when studying a plant or animal in its community.

making sure

1. How does a pond community differ from a community in a river or stream?
2. What facts can be learned by studying organisms in their community?

Define population. List five examples.

The human population in a city or town is a concept familiar to students. Discuss this idea then go on to consider the bird population in the local area, for example, sparrows or starlings.

18:2 Populations

A **population** is the total number of any one species living in a community. For instance, all of the bass in a lake is a population of bass. All of the mice in a city block make up the mouse population of the block. Other examples of populations are the total number of ants in an anthill or dandelions in a lawn. What are some other plant or animal populations?

goldenrod in a field, redwood trees in a forest, algae in a pond

390 **Populations and Communities**

To describe a community, you need to know the types and sizes of the populations present. Any one community contains hundreds of different plant and animal populations. Among these hundreds, there are often two or three **dominant** (DAHM uh nunt) **species.** Dominant, in this case, means present in the greatest numbers. Dominant plant species may make up from 30 to 90 percent of the living things in a community.

Define dominant population.

A census is used to count the number of people that make up a population. Populations of wildlife are determined by counting representative samples in selected areas.

FIGURE 18–2. A coral reef community supports a number of fish and invertebrate populations.

Oak and hickory trees are dominant in some forests in the United States. They are the major plants growing in these forest communities. In the same way, a small pond may be dominated by sunfish and catfish. Although other fish species are present, the sunfish and catfish are present in greatest numbers.

Stress the fact that within the community each population has a habitat that may be different than the other populations.

Each species occupies a certain place within a community. A **habitat** (HAB uh tat) is the place where a species of plant or animal normally lives. Where would you look for birds in a forest? Many birds live in trees and bushes. This is their habitat. Where would you expect to find deer or earthworms? Deer spend their lives among trees and bushes. Earthworms live in the topsoil. All of these animals have their own unique habitats. However, they may be part of the same community.

Define habitat. Describe the habitat of a deer and an earthworm.

FIGURE 18–3. The black buck, threatened by hunting and destruction of its habitat in India, now lives only on game ranches in Texas.

If you were to fish for catfish, you would plan to catch them in their habitat. Catfish are usually found near the bottom of a freshwater pond or stream. To catch them, baited hooks are tied at the end of a weighted line. These hooks then sink to the bottom of the pond, the catfish's habitat.

Every population and every living organism in a population has an environment. The **environment** is the surroundings of an organism or population. Environment includes all other living and nonliving things. For example, the environment of a rabbit includes grass, trees, and other animals. It also includes enemies, such as foxes. The rabbit breathes air. It also drinks water. Air and water are also part of its environment.

FIGURE 18–4. Some of the factors in a rabbit's environment.

392   Populations and Communities

activity

POPULATIONS IN A COMMUNITY

Activity 18-1

Objective: To study the life in a community

Materials See Teacher's Guide.

hand lens
meter stick
plastic dishpan
shovel
string
4 wooden stakes

Procedure

1. Obtain permission to remove a 900-cm² sample of soil from an area in your yard, an empty lot, or woods.
2. Measure a square area, 30 cm by 30 cm, to study. Mark the area using stakes at the four corners of the square.
3. Count and record the number of plants and the number of animals in the square. List the names of the plants and animals you can identify.
4. Dig up some soil within your square and fill the dishpan one-third full.
5. Gently break up the soil and look for organisms. Use the hand lens to observe any organisms you find.
6. Make a sketch of one organism you find.
7. Return the soil to its original place when you finish your observations.

Observations and Data

Number of plants	Number of animals	Plants identified	Animals identified
3	4	moss grass algae	centipede sowbug spider ants

Most of the organisms in the soil are microorganisms.

FIGURE 18–5.

Questions and Conclusions

1. What type of biological community did you observe—yard, forest, or empty lot? Answers will vary.
2. Which type of living organisms—plants or animals—were present in greater number? Why? Answers will vary.
3. What living organisms present in your soil sample were not observed? Why? microorganisms
4. How do organisms living in soil aid in the growth of plants? break down organic material into compounds plants can use
5. Why are plants necessary for the life of organisms living in the soil? provide food
6. What is the source of the energy used by the plants, animals, and microorganisms in the soil? sun

activity

See Teacher's Guide.

COMPARING ENVIRONMENTS Activity 18-2

Study the environment of an aquarium. Make a list of the things it contains. What populations are present? How is the aquarium different from the room around it? How is it similar to the room? Repeat this activity with a terrarium.

Margin annotations:
- gravel, water, plants, fish, snails, algae
- Answers will vary.
- has water, not air, contains fish or other aquatic life
- contains living things

making sure

3. Where would you be most likely to find cockroaches? Rats? Pigeons? How are their habitats similar? How are they different?

See Teacher's Guide.

18:3 Population Change

Over a year's time, the size of a population within a community may stay the same. The size of another population may change. A stable population remains about the same size from year to year. For example, the number of beech trees in a mature forest would be a stable population. This means that the total number of beech trees does not change much from year to year.

An unstable population changes in size from year to year. The dandelion population in a lawn can be unstable. When no one tends the lawn, the number of dandelions increases. When weed killer is sprayed on the lawn, the number of dandelions decreases.

Margin questions:
- Define the words stable and unstable.
- How is a stable population different from an unstable one?

FIGURE 18-6. Spraying weed killer on this lawn (a) caused the dandelion population to decline (b).

a

b

Courtesy Chemlawn Corporation

Courtesy Chemlawn Corporation

394 **Populations and Communities**

Why does the number of individuals in a population change from year to year? This important question is not yet completely answered. It is known that the size of any animal population depends on the food supply. The birthrate for the species is also a factor. Those species that reproduce quickly can increase their number in a short time. Those that have enough food will be able to reproduce. Disease can also affect population size. The number of natural enemies of a species is also a factor.

Changes in the size of an unstable population may occur in cycles. These cycles are up-and-down trends in the population size over the years (Figure 18–7).

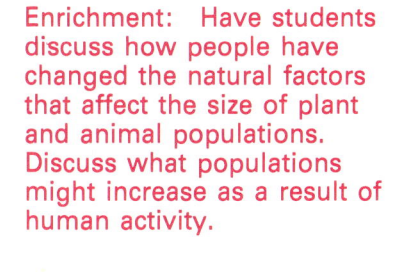

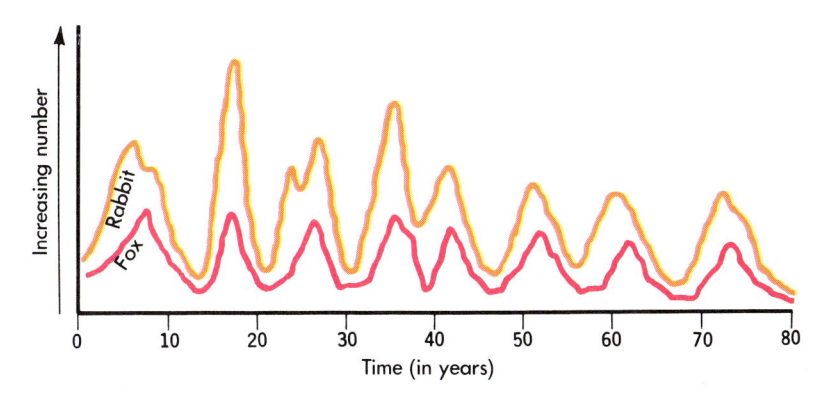

Enrichment: Have students discuss how people have changed the natural factors that affect the size of plant and animal populations. Discuss what populations might increase as a result of human activity.

FIGURE 18–7. Some predator and prey populations rise and fall in regular cycles.

Individuals can be gained or lost in several ways by an animal population. Births and immigration (ihm uh GRAY shun) increase the size of the population. **Immigration means movement into the community.** Deaths and emigration (em uh GRAY shun) decrease the size of a population. **Emigration means movement out of the community.**

How are immigration and emigration different?

Consider the case of field mice living in a grain field. When the grain ripens, it is harvested in late summer. Some of it is left as waste on the ground. Mice eat grain as their major food source. Mice from nearby areas move into the grain field. This immigration causes an increase in the mice population. Research shows that the birthrate for mice increases when food is plentiful. This also causes an increase in the population. Eventually, the increased number of mice is checked by deaths, predators, and emigration.

18:3 *Population Change* 395

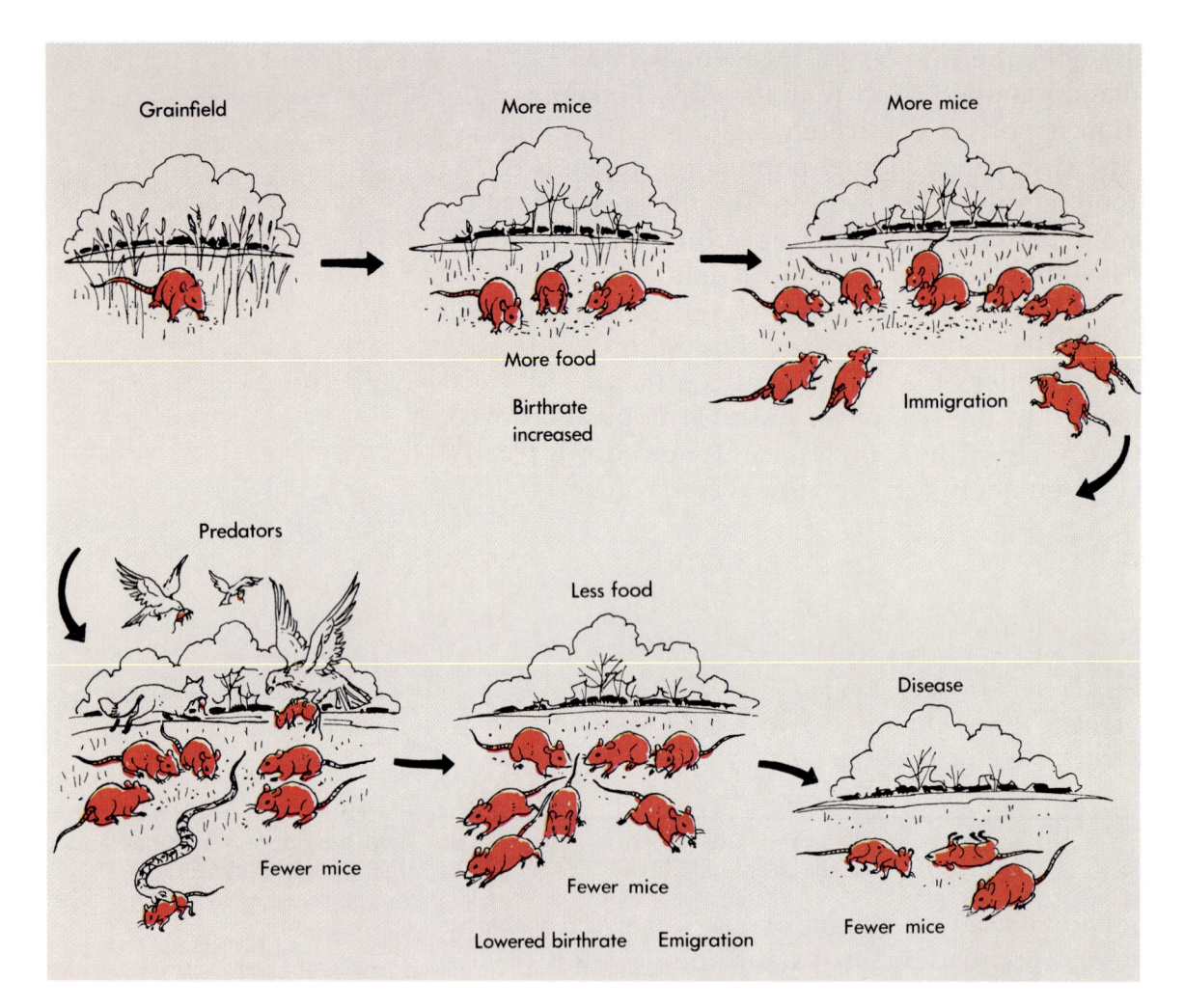

FIGURE 18–8. Changes in the size of a field mouse population.

When a population becomes large, certain factors tend to limit its size. Predators and disease are the most important of these factors. As the number of mice increases, the natural enemies of the mice also increase. Hawks, snakes, and other mouse-eaters are attracted by their growing food supply—the mice. These predators eat the mice and reduce the mice population.

A large mouse population also leads to a smaller food supply. Some mice will not get enough to eat. Many will then move from the field and go elsewhere to find food. A lack of food also causes the birthrate to drop.

It is known that when the number of mice increases, epidemic diseases may start. Disease-causing bacteria may also be a kind of natural enemy. An increase in the number of individuals of a species increases the chance for contagious diseases to spread. There are more mice in the community. Thus, there is a greater chance for disease to be spread from sick to healthy mice.

The size of one population affects another. Both the number of bacteria and the number of mouse-eaters can increase when the number of mice increases. An increase in the number of one species affects the numbers of some other species.

making sure

4. What factors can affect the size of a rat population within a small city?
5. What factors affect the growth of the human population?
6. How does a change in a mice population affect the population of the mice's natural enemies?
7. How does a change in seasons affect changes in some animal populations?

4. amount of food, poisoning by people, presence of predators such as cats, disease
5. food supply, disease
6. Decrease in mice reduces the food supply for the natural enemies.
7. It causes the populations to increase and decrease. For example, more insects are present in the summer than in winter.

18:4 Competition

The organisms within a community may compete for the things needed for survival. These include sunlight, food, water, and shelter. For example, trees in a forest compete for the available water. If a severe drought occurs, there may not be enough water for all. Some of the trees may die. Tall, mature trees limit the sunlight reaching the forest floor. Small, young trees compete for the available light. If a tree is in shade most of the time, it may grow very slowly or die.

Competition increases when many species have the same needs. Competition for the same food source has an effect on each species. One example of this has been observed in the laboratory. Two different

Enrichment: Discuss why an animal that could not compete would be at a disadvantage in its environment.

What may happen to plants and animals in competition for the things needed for survival?

FIGURE 18–9. Organisms in many communities compete for the same food source.

paramecia species were grown in separate dishes. Each was given the same food source. Both increased in number. Then, the two species were grown in the same dish. The same food source was used. Competition for the same food source became too great. One species did not survive.

A species may reduce or avoid competition with other species. One way is by living in different parts of a community. This fact has been observed in three related warbler species in a Pennsylvania forest. They each feed on the same food source—spruce trees. One species feeds mostly in the lower branches of spruce trees. Another feeds in the middle part. The third warbler species feeds on the outer tips of high branches.

Competition may be reduced by some birds and mammals in another way. Many of these animals set up territories. One or more animals defend a certain area or **territory**. If another animal comes near, it is usually chased out. As a result, there is little or no competition for the resources in the territory.

In what two ways may some birds and mammals reduce competition?

FIGURE 18–10. Bull elephant seals defend territories on their mating grounds.

Populations and Communities

activity

POPULATION CHANGES

Activity 18-3

Objective: To study the effects of competition on the size of a population

Materials   See Teacher's Guide.

beaker, 50-mL
dropper
dry yeast
microscope
1 slide and coverslip
spoon
sugar cube
water

FIGURE 18-11.

Procedure

1. Mix the yeast in 25 mL of water at room temperature.
2. Place a drop of the yeast-water mixture on the slide and add a coverslip.
3. Observe the yeast with the microscope. Count and record the number of yeast cells in your field of view.
4. Remove the slide and rinse it with water.
5. Add the sugar cube to the beaker and stir until it is dissolved.
6. Wait two minutes; then place a drop of the mixture on the slide and add a coverslip.
7. Observe the yeast under the microscope. Count and record the number of yeast cells in your field of view.
8. Repeat these observations every two minutes for 20 minutes. Remember to wash the slide and coverslip after each observation. Record all numbers.
9. Record any changes in appearance and odor of the solution in the beaker.

Observations and Data

No sugar, number of yeast cells: _____

Time (min)	Number of yeast cells
2	
4	
6	

Draw a line graph showing the relationship between the time and the number of yeast cells in the population.

Questions and Conclusions

1. What things were needed for the yeast to survive? **sugar, water**
2. From your graph, describe the changes in the size of the yeast population. **increases over time**
3. When did the yeast population reach its highest number? **Answers will vary.**
4. How might crowding, food, and waste products affect the size of a population? **food—increases, crowding and waste products—decrease**

18:4 Competition   399

FIGURE 18-12. In this field, succession is occuring as grasses are giving way to taller shrubs and trees.

Define succession.

Define climax community.

FIGURE 18-13. In the succession from lichens to maples and beeches, each stage has different dominant species.

making sure

8. Setting up territories tends to reduce the crowding of animals in a population. Why? *It reduces the competition for resources in an area.*
9. How does competition affect the number of living things in a community? *decreases the number*

18:5 The Climax Community

Since populations change, communities undergo slow, gradual change. The process of slow, gradual change in a community is known as **succession** (suk SESH un). During succession, dominant species of plants and animals slowly give way to new dominant species. The end result of succession is called a **climax** (KLI maks) **community.** Each region follows a certain pattern of succession and climax. This depends on the climate and soil conditions. A climax community remains as long as the climate of the region does not change.

Succession can be seen in many regions. One example is in abandoned farm fields of northeastern United States. The land of the Northeast was once covered with forests. Trees were cut down by the settlers to make way for farms. However, it became unprofitable to farm some areas. As the fields were abandoned, "nature slowly took over." Grass, shrubs, and trees finally returned. Succession continued until the climax community was reached.

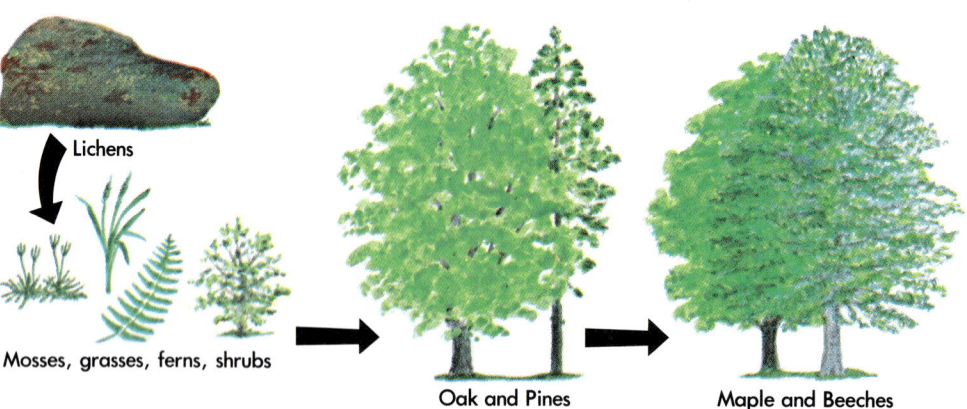

Lichens → Mosses, grasses, ferns, shrubs → Oak and Pines → Maple and Beeches

Populations and Communities

FIGURE 18–14. Succession of a pond to a forest.

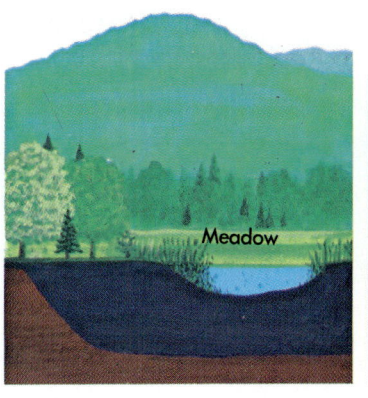

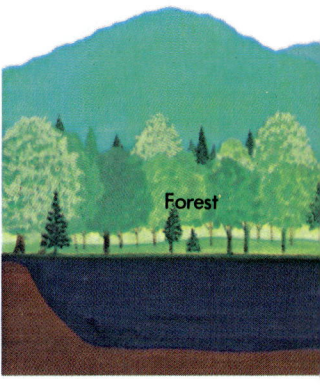

A freshwater pond community may be the first stage in a succession. Water, running into the pond, carries soil particles with it. With time, the pond may fill with soil and become a bog. A bog is an area of wet, spongy ground. Eventually, the bog may become a meadow. Finally, the meadow may become a forest. If the forest remains unchanged for a long time, it is the climax community.

As the plant life changes during succession, the kinds of animal life will change also. For example, woodchucks and rabbits are most plentiful in fields.

In succession, species succeed those that die off.

activity

COMMUNITY SUCCESSION   Activity 18-4

Prepare a hay infusion culture by boiling a handful of hay or dry grass in 1 L of water. Place the culture in a jar and add a few grains of boiled rice. Let the mixture stand for two days. Then add some pond water. Let the culture stand for 10 days. Each day, remove a drop of the culture liquid. Observe it under a microscope with low and high power. Record the changes in the species each day. What, if any, other species appear in the culture? See Teacher's Guide.

making sure

10. How does climate affect a climax community?
11. How does a succession of plants affect the succession of animals in a community?
12. Explain the succession that occurs when farmlands are abandoned.

10. The climax community is suited to the climate. When the climate changes, the community changes.
11. Dominant animal species change during succession.
12. grass–shrubs and weeds–trees

18:5 **The Climax Community**

18:6 Boundary Communities

The boundary between two communities is a most interesting place because a great variety of species live here. A boundary is a region between two different communities. One boundary community is the seashore. The land that borders on a forest is another example.

A boundary region has the advantages of two communities. Usually more species live in this area than in either of the communities it separates. Boundaries provide the best environment for many plants and animals. Species common only to the boundary region live here. Also, species from both adjacent communities live here.

Frogs and snapping turtles live in boundaries between freshwater ponds and grassy meadows. They spend part of their life in water and part on land. Frogs lay their eggs in the water but come on land to catch insects for food. Snapping turtles do the opposite. They lay their eggs on land but get their food from the water.

What name is given to the area that separates two communities?

Seagulls are plentiful in seacoast areas.

FIGURE 18–15. The place where land and ocean meet has a unique boundary community.

402 Populations and Communities

18:7 Cooperation in Populations

Members of the same population use the same resources. If the resources are limited, individuals must compete with one another. Sometimes, members of the same species cooperate with each other. For example, wolves hunt together. A flock of birds may scare off an attacker. A crow may "stand guard" in a tall tree while the rest of the flock feeds in a cornfield. Ants live and work together in colonies. Cooperation helps the entire ant colony.

Cooperation is greatest among animals that live in societies (suh SI ut eez). A **society** is a group of individuals of the same species that live together in an organized way. Bees, ants, and wasps are organized into societies. Within a society, there is a division of labor. There is cooperation for the welfare of all.

Survival of an individual is based on the survival of the group. This fact is well illustrated by a honeybee colony. There are three kinds of honeybees—female queen, male drone, and female worker. The roles of the queen, drone, and worker are determined in large part by heredity. Under ordinary conditions there is only one queen in each colony. The queen is the only bee that lays fertilized and unfertilized eggs. The queen's contribution to the colony is to produce new male and female offspring. Drones produce sperm to fertilize the queen's eggs. This is their only role. Worker bees carry on work tasks necessary for the society's survival.

Worker bees perform many different jobs. In early life, they serve as nurse bees that feed the young bees. Later, they become housebees. Their jobs now include cleaning, storing honey, producing wax, and guarding the hive. Adult workers become field bees. Field bees collect nectar and pollen from plants.

FIGURE 18–16. Each wasp in the society has a certain job to do. Cooperation helps all members in a society live and survive.

Explain an example of cooperation in a population.

Why is cooperation usually greatest among animals living in societies?

Scientists who study animal behavior are interested in the trait known as altruism, the aiding of another without benefit to self. Altruism is part of the cooperation within some animal populations.

making sure

13. How would the loss of a queen, drones, or workers threaten the survival of a hive?
    queen—no new bees produced
    drones—queen's eggs not fertilized
    worker—food (honey) not produced

18:8 Ecology

What is ecology? Why is it important?

Go over the definition of ecology given here. A more common definition of ecology with which students may be more familiar is the care or protection of the environment; as in the phrase "good ecology." Some students think it means fighting pollution.

The relationships between organisms and their environments is called **ecology** (ih KAHL uh jee). One of the most important relationships is the use of green plants by animals for food. Another is the decay of dead plant and animal matter. Decay returns the elements and compounds in dead matter to the soil. The decayed material is used as nutrients for growing plants. You have learned that disease, food supply, and many other factors can affect the size of a population. A change in the size of one population in an area can cause a change in the size of other populations. In other words, the life of any one species is related to the lives of other plants and animals in the environment. People who study the relationship between organisms and their environment are called ecologists.

Why is ecology important? Facts about living things can be used to save plants and animals that are in danger of becoming extinct. Extinct means all the members of a species die and the species is gone forever. For example, scientists study the life habits of the California condor, a species near extinction. They seek to learn how this bird raises its young, obtains food, and survives in the wild. If the scientists can find a way to increase the reproduction of the condor, they may save it from extinction.

Understanding ecology can also help preserve the environment. Scientists study the harmful effects of fertilizers, insecticides, and other chemicals. The knowledge gained may help us find ways to use these materials without causing harmful air and water pollution. Increasing the population of a species and preventing damage to forests by harmful insects are two problems attacked by ecologists.

Two important elements needed by living organisms are oxygen and carbon. In nature, these elements are part of the oxygen–carbon dioxide cycle. Here the word cycle means that each of these elements is used over and over again. Oxygen is a

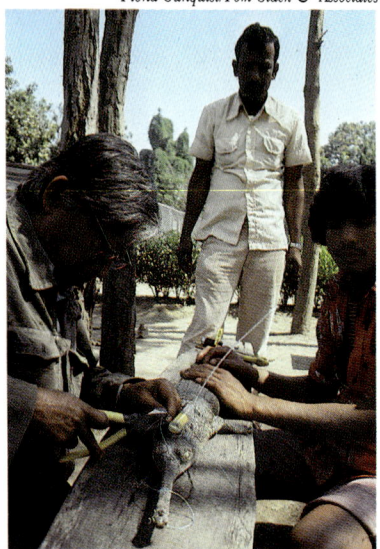

Fiona Sunquist/Tom Stack & Associates

FIGURE 18–17. Ecologists study relationships between living things and their environment.

One planned experiment is to capture several condors and attempt to breed and raise them under controlled conditions.

Explain the oxygen–carbon dioxide cycle.

404   *Populations and Communities*

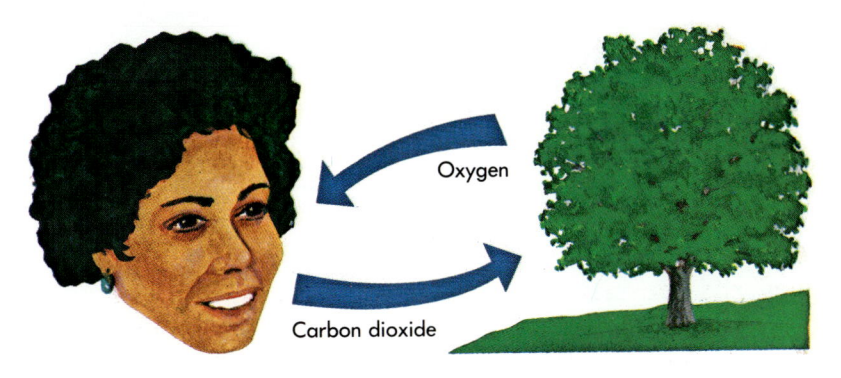

FIGURE 18–18. Oxygen and carbon dioxide are constantly being used and reused by animals and plants in the oxygen–carbon dioxide cycle.

Most of the oxygen is produced by plants in the oceans. Diagram the O_2–CO_2 cycle on the chalkboard.

gas present in air. Carbon is also present in air in the form of carbon dioxide gas. Animals remove oxygen from air and add carbon dioxide when they breathe. Plants take in carbon dioxide and give off oxygen during photosynthesis. The carbon in carbon dioxide is used along with water to make food. In the oxygen–carbon dioxide cycle, oxygen and carbon go from plants to animals and back to plants in a continuous cycle. Neither element is ever used up.

In the nitrogen cycle, nitrogen gas is removed from air to form compounds that are added to soil. Nitrogen-fixing bacteria take nitrogen gas and change it into nitrogen compounds. These bacteria live in colonies inside nodules, which are swellings on the roots of legumes. Legumes are a group of pod-bearing plants such as alfalfa, clover, and beans. Plants use nitrogen compounds from soil to make protein. Protein is needed for the growth and repair of tissues. Certain bacteria in soil break down organic matter and change the nitrogen in it into a compound called ammonia. Other bacteria change the ammonia to nitrate compounds. Another kind of soil bacteria converts nitrogen in compounds to nitrogen gas and returns it to the atmosphere.

FIGURE 18–19. Legume crops such as alfalfa return nitrogen to the soil.

making sure   See Teacher's Guide.

14. How would the world change if there was no decay of dead plants and animals?
15. Describe the flow of carbon in the oxygen-carbon dioxide cycle.

18:8 Ecology

PERSPECTIVES

people

Devotion to Wildlife

For three months Paula Koeneman continually watched over two rare jaguarundi kittens. She fed them every two hours—night and day—until they could feed themselves. As head keeper of birds and mammals at the Arizona-Sonora Desert Museum, Paula Koeneman works with wild animals that most people will never see. The jaguarundi kittens are only two of many animals under her care.

The Arizona-Sonora Desert Museum is a combination of a zoo and museum. Numerous plants and animals that are native to the desert and other communities of Arizona, southern California, and Mexico, are on display at the museum. These desert communities are very fragile and they have suffered greatly when humans destroyed many natural habitats.

Paula Koeneman's goal is to teach people the importance of wildlife. She supervises six zookeepers who care for hundreds of birds and mammals. They provide suitable habitats and proper diets for the animals to survive. For a beaver, for example, two aspen trees are cut every month, because beavers prefer to eat aspen bark. Some animals have been brought to the museum for recovery from injuries or because they were orphaned. Many are later released in their natural habitats, once they are able to care for themselves.

C. Koeneman

When new plants and animals arrive at the museum they are first quarantined, or kept apart from the others, to determine if they have any disease. Once it has been determined that they are healthy, the plants and animals are placed in with the others. One of Paula Koeneman's responsibilities involves the care of quarantined animals. Her past experience working with veterinarians is invaluable to her job.

Many of the animals that Paula Koeneman works with are endangered or threatened species. For example, the museum was the first institution to begin breeding the Mexican Gray Wolf. This wolf has been hunted and trapped throughout history, until only a few remain in remote area of Mexico. Working with the Mexican government, the museum hopes to build a larger population in captivity. The wolves would then be distributed to other zoos, as a precaution against population extinction through disease. Once a large enough population is established, individuals would be released in their natural habitats in Mexico, with the hope that the animals will survive.

Paula Koeneman would like to see a similar program started with the ocelot, a small cat that has been hunted for its beautiful spotted skin. Very few ocelots are known to exist in the wild. Both the ocelot and its plain colored cousin, the jaguarundi, have had their habitats reduced by humans.

The survival of many animal species depends upon people knowing the importance of wildlife, protecting their native habitats, and limiting human interaction. Paula Koeneman is making progress towards saving many species from extinction, and giving people the opportunity to see these live animals.

Chapter Review

main ideas

1. A community consists of all of the living things in a given area.   18:1
2. Each species normally lives in a certain place called a habitat.   18:2
3. All living and nonliving things surrounding an organism make up its environment.   18:2
4. Populations are either stable or unstable. Changes in many unstable populations occur in cycles.   18:3
5. Organisms within a community may compete for survival needs.   18:4
6. Through succession, dominant species slowly give way to new dominant species. This results in a climax community.   18:5
7. A society is an organized group of the same species which live together in cooperation.   18:7
8. Ecology is the science dealing with the study of relationships between organisms and their environment.   18:8
9. The oxygen, carbon, and nitrogen cycles are part of the environment.   18:8

vocabulary

Define each of the following words or terms.

climax community	emigration	population
community	environment	society
dominant species	habitat	succession
ecology	immigration	territory

study questions

DO NOT WRITE IN THIS BOOK.

A. True or False

Determine whether each of the following sentences is true or false. If the sentence is false, rewrite it to make it true.

T   1. Stable populations do not change much.
F   2. The environment of an organism includes ~~only~~ the ^living and^ nonliving things around it.
T   3. Ocean life is more abundant near coastlines.
T   4. Plants and animals are part of the oxygen–carbon dioxide cycle.

Populations and Communities   407

F **5.** Food supply is ~~the~~ only ˄ factor that affects population size. one
T **6.** By defending their territories, animals reduce the amount of competition in a region.
F **7.** The population of a species tends to ~~decrease~~ when there is a large food supply. increase
F **8.** Succession usually occurs in a ~~short~~ time period. long
T **9.** Ecology is the study of the relationships among living organisms.
T **10.** Most communities contain hundreds of populations.

B. Multiple Choice
Choose the word or phrase that completes correctly each of the following sentences.

1. All the oaks in a forest make up a (*community, habitat, population*).
2. Populations increase by (*death and immigration, birth and emigration, birth and immigration*).
3. A (*climax, succession, cycle*) is the slow and gradual change in a community.
4. When the grain in a field decreases, the mice population (*increases, decreases, remains the same*).
5. Plants and animals living together in a pond make up a (*habitat, community, population*).
6. An increase in predators causes (*an increase, a decrease, no change*) in the population of the species they eat.
7. The formation of a bog is a good illustration of (*competition, succession, emigration*).
8. Emigration causes the size of a population to (*increase, decrease, stay the same*).
9. Dominant plant species may make up (*less than 30%, 30% to 90%, 100%*) of all living things in a community.
10. The area between two communities is a (*climax, boundary, territorial*) community.

C. Completion
Complete each of the following sentences with a word or phrase that will make the sentence correct.

1. The total number of frogs in a pond is called a(n) __population__.
2. The population of an ant colony is an organized __society__.

3. The population of _foxes_ in a community would eventually be reduced by a decrease in rabbits.
4. An abandoned field in northeastern United States may someday become a(n) _forest_.
5. A climax community is the end result of _succession_.
6. Photosynthesis is part of the oxygen–_____ cycle. **carbon dioxide**
7. _____ and _____ tend to reduce the size of a population. **lack of food, predators, emigration, disease**
8. Slow changes in populations occur over many years during _succession_.
9. A(n) _climax_ community will remain as long as the climate stays the same.
10. Scientists who study the relationships among living things and their environment are _ecologists_.

D. How and Why See Teacher's Guide at the front of this book.
1. Explain why a change in the population size of one species may affect the size of another species.
2. Explain how a climax community is formed.
3. How does a population gain or lose individuals?
4. Give two examples of a community. Name the living things you might expect to find in these communities.
5. How does competition affect the size of populations? Give one example.

challenges

1. Obtain information from a library on an endangered species in your home state.
2. Obtain information on urban ecology, the relationship between people and their environment. Plan a study you can do to find out how the life of a city depends upon many different people doing many different kinds of jobs.

interesting reading

Harris, Susan, *Whales*. New York, NY: Watts, 1980.

Milne, Lorus and Margery, *Gadabouts and Stick-at-Homes: Wild Animals and Their Habitats*. San Francisco CA: Sierra Club, 1980.

Raeburn, Paul, "An Uncommon Chimp." *Science '83*. June 1983, pp. 40-48

The grasshopper balanced on the edge of a sundew plant is the food which gives the sundew energy. All living things need energy to carry on life processes. Food supplies this necessary energy. What is a food chain and how is it related to a food web? How are predators different from parasites?

Alvin E. Staffan

Food and Energy

chapter 19

Introducing the chapter: Exhibit an aquarium that contains fish and plants. Begin a class discussion by asking students to identify the organisms in the aquarium that produce food. Have students also identify the organisms that eat food. Point out that the carbon dioxide in the water is exhaled by the fish and used by the plants to make food. Ask students why light is needed for life in an aquarium. Have students identify other features of the life system in the aquarium.

19:1 Food Chains

Every living organism needs energy for its life activities. Energy is needed for breathing, digesting food, circulating blood, and reproducing. Running, flying, and swimming are activities that increase an animal's need for energy. All the energy that plants and animals use comes directly or indirectly from the sun.

Green plants have one advantage over most other living things. They make their own food. Plants use the energy from sunlight to make sugar. This sugar is changed to starch and other substances which are stored food. Because animals cannot make their own food, they depend on plants or other animals for food. Many animals, such as deer and cows, eat only plants. Other animals, such as lions and wolves, eat only other animals. Some animals, such as humans, bears, and rats eat both plants and animals.

GOAL: You will learn the food and energy relationships that exist among organisms in food chains and food webs.

Scientists define energy as the capacity to do work. Work is done when something is moved. Relate this definition to the activities described here.

Review photosynthesis, the reactants (CO_2 + H_2O), and products (O_2 + sugar). Sunlight is the source of chemical energy stored in food.

FIGURE 19–1. A food chain shows a simple food relationship. A plant is always the first link in a food chain.

Explain what is meant by a food chain and give one example.

In a community there are food links between different plants and animals. A plant may be eaten by an animal. Another animal eats the animal that ate the plant. In turn, another animal eats this animal. A series of animals feeding on plants and/or other animals makes a **food chain.** The first link in a food chain is always a plant. One example of a food chain found in many forests is acorn–mouse–owl. Another example is wood–termite–anteater–lion. Food and the energy it contains are passed through a food chain. This transfer of energy begins at the source, which is plants. It then continues as one animal eats another animal.

FIGURE 19–2. In this food chain the eucalyptus tree is the producer and the koala bear is the consumer.

Miriam Austerman/Animals, Animals

412 *Food and Energy*

Most animals eat more than one species as food. Thus, the term food chain does not always describe the food relationships. The term food web is used for the food relationships in a community. **A food web** is a complex feeding system containing more than one chain. In one food web, mice, rabbits, and squirrels eat plants. Owls eat both the mice and rabbits. The fox eats both rabbits and birds. The food web formed in this way is shown in Figure 19-3.

What is a food web? Describe an example.

FIGURE 19-3. A food web shows the food relationships in a community. An animal may occupy different levels within a food web.

19:1 Food Chains 413

1. Food web consists of more than one food chain.
2. Green plants use the sun's energy to make food. Animals depend on plants for food.
3. Plants make the food on which animals depend.
4. corn–chicken–person
   Answers will vary.

making sure

1. How is a food web different from a food chain?
2. Explain why all the energy animals use comes from the sun.
3. Why is the first link in the food chain always a plant?
4. Give an example of a food chain of which you are a part.

19:2 Producers and Consumers

State the difference between a food producer, consumer, and decomposer. Name examples of each.

plant eater—herbivore
flesh eater—carnivore
flesh and plant eater—omnivore

Organisms in a food web that make their own food are called **producers.** All green plants are producers. All animals in a food web are consumers. **Consumers** cannot make food and must eat plants or other animals. There are many different kinds of consumers. Some are small plant-eaters, such as mice. Others are large flesh-eaters, such as lions. Mice and lions are usually the second and third links in the food chains of which they are a part.

Bacteria and fungi are both **decomposers.** They cause the decay or break down of dead plants and animals. Decomposers are always the last link in the food chain.

Scavengers (SKAV un jurz) are animals that feed on dead animals. Vultures and jackals are two examples of scavengers. These animals "clean up" the environment. Both decomposers and scavengers are consumers. They are important and necessary links in every food chain.

Why are scavengers and decomposers important consumers?

FIGURE 19–4. Decomposers (a) and scavengers (b) are vital links in food webs. Without them materials could not be recycled and used again.

activity

OBSERVING POND WATER

Activity 19-1

Objective: To study some producers and consumers in this community

See Teacher's Guide.

Materials
dropper
4 jars, large (widemouth)/lids
microscope
4 samples of pond water (surface, under the surface, bottom, near shore)
slide and coverslip

FIGURE 19–5.

Procedure

1. Collect pond water in the jars from each location listed in the materials above. **CAUTION:** Be sure an adult accompanies you to obtain the pond water samples.
2. Place a drop of pond water from the surface sample on the slide.
3. Add a coverslip and examine the water using low power first and then using high power.
4. Make a sketch of several different organisms that you see. Record the color of each organism and the number of each organism present.
5. Remove the slide and wash it with water.
6. Repeat Steps 2 through 5 for each of the other samples of pond water. Be sure to sketch several organisms and record their colors. Also, record the number of each organism present.
7. Determine if each organism you observed is a producer or a consumer. Producers usually are yellow, green, or blue-green. Consumers usually are colorless. Record your findings.

Observations and Data

Source of water sample	Sketches of organisms	Color and number	Producer or consumer

Questions and Conclusions

1. Did you find more producers or consumers in your samples? State a reason for your observations. **producers**
2. How are producers different from consumers? **Producers make their own food. Consumers eat other organisms.**
3. In which sample did you find the most producers? **surface**
4. Where did most of the consumers live in the pond? **under surface**
5. How did the producers obtain their food? **photosynthesis**
6. Where would you expect to find decomposers in the pond? **bottom**
7. Draw a food web using your sketches of the organisms you observed in the pond water.

19:2 Producers and Consumers

making sure

5. Classify the following as producer, consumer, or decomposer.
   (a) grass   P
   (b) rabbit   C
   (c) tree   P
   (d) decay bacteria   D
   (e) bread mold   D
   (f) grasshopper   C

6. Why are decomposers and scavengers important in every food chain or web? They help clean up the environment and recycle nutrients.

19:3 Energy Pyramids

Every organism needs energy. Green plants trap light and store this captured energy in the food they make. Some animals eat plants to obtain energy and others eat the animals that feed on plants. As food passes through a food chain, energy is obtained for use by each of the organisms in the chain. The energy captured by the green plants is used by all of the consumers in the food chain.

The acorn–mouse–owl food chain is a good example of the transfer of energy from one organism to another. An oak tree uses sunlight to make food and stores some of the food in its acorns. A mouse eats acorns. Most of the energy in the acorns is used by the mouse for its life activities. Only about 10 to 15 percent of the energy in the acorns is stored in the mouse. When an owl eats a mouse, most of the energy the owl obtains is used for the owl's life activities. Only 10 to 15 percent is stored in the owl's body.

What happens to energy as it moves through a food chain? Most of it is used for the life activities of the organisms in the food chain. These activities include cell division, circulation of blood, breathing, running, and flying. The loss of energy used for life activities can be pictured with an energy pyramid. A pyramid is wide at the bottom and pointed at the top. Each level in the pyramid is smaller in size than the level below it. Green plants, the producers in a

Energy is lost by animals as they lose heat to their surroundings.

What does the shape of an energy pyramid show?

416   Food and Energy

food chain, make up the first level of an energy pyramid. Animals that eat only plants make up the second level. Flesh-eating animals are in the third and higher levels.

In each successive level of an energy pyramid, there is less energy. The loss of energy is shown by the decreasing size of the pyramid from bottom to top. Most of the energy is in producers, such as oak trees or other green plants at the bottom. The least amount of energy is in the consumers at the top. Further, the shape of the pyramid shows there is a large number of producers supplying energy for a smaller number of consumers.

In the energy pyramid shown in Figure 19-6, producers make up the first level. Consumers make up all the remaining levels. This arrangement is true for all energy pyramids. Animals that eat only plants make up the second level. Flesh-eating animals make up the third and higher levels.

In general, the smaller the animal, the larger the daily requirement per unit of body mass.

FIGURE 19-6. A general model of an energy pyramid (a) and a corn–mouse–owl energy pyramid (b).

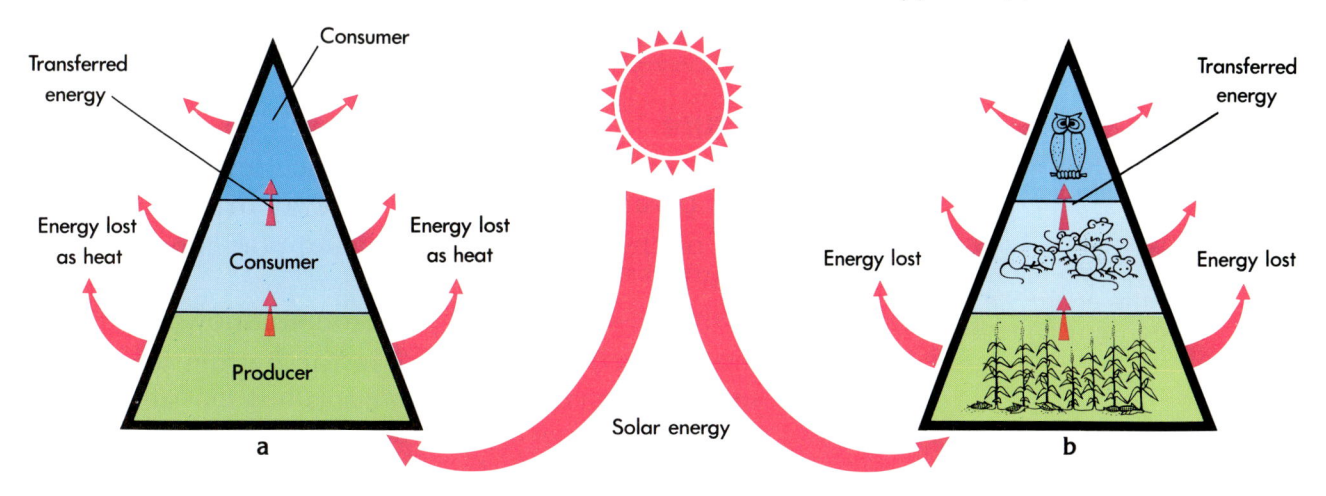

The pyramid's shape represents the energy lost in the food chain. With each successive level, there is less energy. For this reason, there is more energy available in a short food chain than in a long one. The decreasing size of the pyramid also represents the large number of producers needed to supply energy for a small number of consumers.

19:3 *Energy Pyramids*

activity

See Teacher's Guide.   Activity 19-2

CONSTRUCTING AN ENERGY PYRAMID

(1) Cut four triangles, 30 cm on each side, out of a piece of cardboard. Tape the sides of the triangles together to form a pyramid. (2) Use a pen and ruler to divide the pyramid into three levels by marking two horizontal lines on each side of the pyramid (Figure 19–6). (3) Make a list of foods you ate yesterday. (4) Divide the foods listed into two groups, those that come from producers and those that come from consumers. (5) From old magazines, obtain small pictures of the producers and consumers in your food list. Also, obtain some pictures of people. (6) Paste the pictures of the producers on the pyramid in the lowest level. Paste the pictures of the consumers in the middle level. Paste the pictures of people in the top level. (Note: If you are unable to obtain pictures, make colored drawings of producers, consumers, and people for each level.) How many producers are in your energy pyramid? How many consumers? Why are the producers in the lowest level?

The number of producers and consumers will vary depending on diet of students.

They produce the food that contains the energy for all the organisms in the pyramid.

making sure

7. Energy is lost as it is used for life activities.

7. Why is the total food energy in each level of an energy pyramid less than in the level below it?

8. (a)—There is no loss of energy in the minnows. All of the energy in the plankton go to the bass.

8. Which of these food chains will support more bass? Why?
   (a) plankton–bass   (b) plankton–minnows–bass

9. Why are green plants always in the bottom level of an energy pyramid? *Green plants trap the sun's energy. All the animals in levels above receive this stored energy.*

19:4 Predators

Define predators. Why are they important in the food chain?

Predators are at the top of all food chains. **A predator is an animal that feeds on other animals.** Most predators, such as foxes, kill their prey. **Prey** is an animal that a predator uses for food. Some, such as mosquitos, feed on only part of their prey. Their prey usually does not die from an attack.

Predators help control the population of the prey species. This prevents the prey population from growing too large for the food supply. For example, the deer population in an area may become so large that there is not enough food for all the deer. Many of the deer die or become sick. The killing of deer by wolves and other predators helps keep the deer population from becoming too large.

In a community, the number of predators depends on the number of prey. For example, if the number of mice increases in a community, the number of predators, such as snakes and hawks, increases. As more mice are eaten by the predators, the mouse population decreases. As a result, the number of hawks and snakes decreases in the community.

Predators also have another effect on the prey species. They tend to attack older, ill, and injured members of a population. As a result, weak animals are removed. The remaining animals are healthy and strong. These animals are the ones most likely to survive and reproduce.

Stress the relationship between predator and prey. Predators depend on prey for food and keep the population of prey in check.

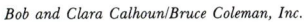

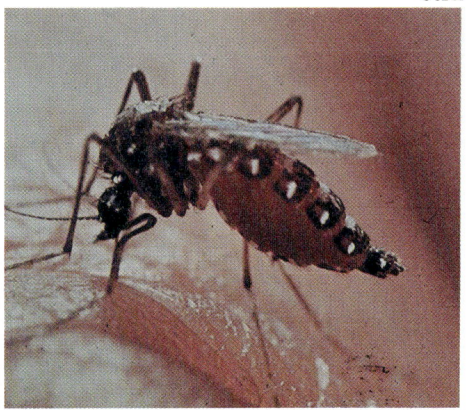

FIGURE 19–7. The timber wolf (a) kills its prey; a mosquito (b) does not. Both animals, however, are predators.

making sure

10. How might the population size of a predator species change if its prey began to increase in large numbers? increase in size
11. What would happen to a predator species if its prey species became diseased and began to die off? decrease in number

19:4 Predators

Contrast a predator and a parasite.

Describe the relationship between a parasite and its host.

19:5 Parasites

Another type of consumer lives and feeds in or on other living things. Tapeworms, fleas, and disease-causing bacteria are examples. These organisms are parasites. A parasite obtains its food and other needs from another organism called the host. Parasites spend most of their lives on or in the host. Most parasites harm their host organisms.

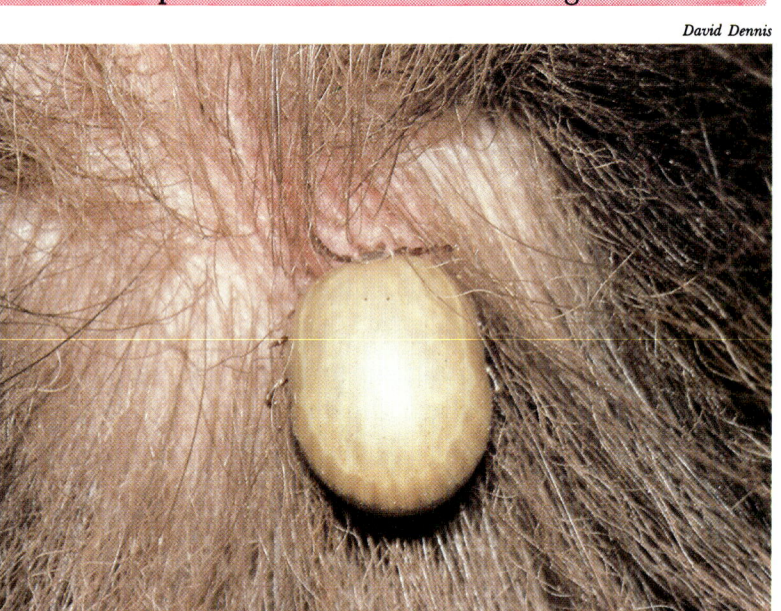

David Dennis

FIGURE 19–8. Ticks are parasites. They live on the blood sucked from the bodies of animals.

The life cycle of a parasite may include many stages. One or more stages may be spent on or in a host. Beef tapeworms have stages in both cattle and humans. A person becomes infected by eating raw or barely cooked beef that contains tapeworm larvae. These larvae become active in the person's intestine. Here, they grow into mature tapeworms. The heads of the tapeworms attach to the intestinal wall. The human intestine provides food, water, and a warm environment. Here the adult tapeworms produce eggs. The eggs are then shed through a person's feces. Tapeworms often weaken the human host because they take away food and cause bleeding in the intestine.

Food and Energy

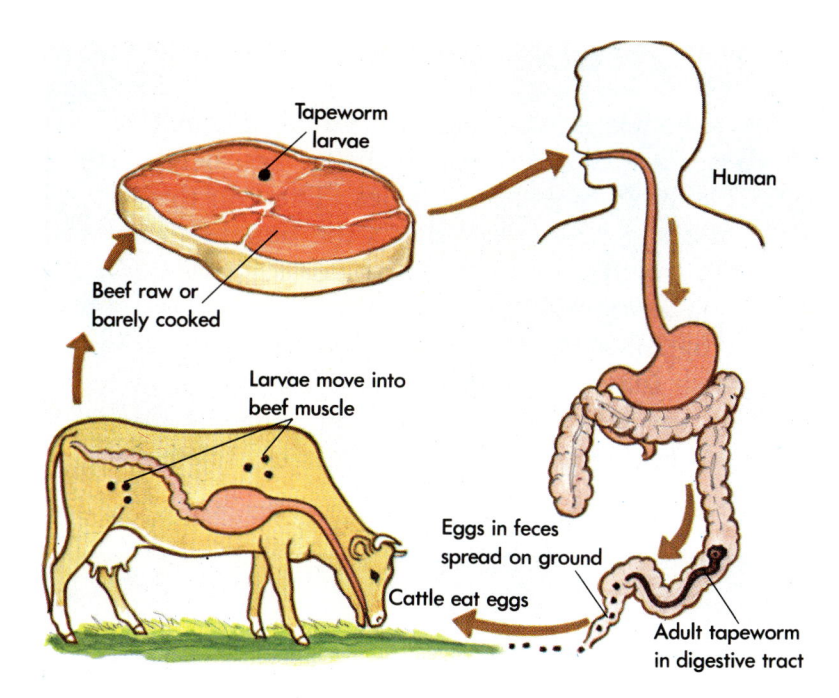

FIGURE 19-9. Life cycle of a tapeworm.

making sure

12. Parasites can make an animal ill. They can even cause death. Why does a parasite "lose" if it kills its host? It loses its food and habitat.
13. How can people protect themselves from becoming infected by beef tapeworms? Cook beef thoroughly.
14. Why might a mosquito be considered a parasite as well as a predator? A mosquito obtains its food from living organisms.

19:6 Symbiosis

Symbiosis (sihm bee OH sus) refers to two organisms living together. In one form of symbiosis, one species benefits and the other species receives neither benefit nor harm. The benefit to one species may be shelter, food, or a surface on which to grow. For example, tropical orchids grow on the branches of larger trees. The orchids are attached to the trees but they do not take food from the tree or harm it in any way.

Define symbiosis.

List two examples of symbiosis.

Another example of symbiosis is the sea anemone that lives in harmony with only one fish species, the decoy fish. Sea anemones have tentacles which paralyze most fish that touch them. Yet, decoy fish swim freely among the tentacles. They do this without being harmed. What is the benefit? The sea anemones provide protection for decoy fish. Although anemones feed on other fish species, they do not harm decoy fish. It has been observed that decoy fish may "help" the sea anemone. When feeding, the fish drop bits of food. The food drifts down into the sea anemone's tentacles and becomes food for the sea anemone.

FIGURE 19–10. Symbiosis occurs between decoy fish and sea anemones (a) and in lichens (b).

In some cases of symbiosis, both species receive benefits from each other. A honeybee obtains nectar from a flower on a plant. At the same time, the honey bee aids in pollinating the plant by carrying pollen from one flower to another. A termite can eat wood because of certain protozoans that live in the termite's intestine. The protozoans help by digesting the cellulose in the wood. In turn, food for the protozoan is provided by the termite.

Another example of symbiosis where both species benefit is a **lichen** (LI kun). A lichen is an alga and a fungus living together. The alga makes food that is used by the fungus. The fungus supplies water and minerals for the alga. The partnership between alga and fungus allows lichens to grow almost anywhere, even on bare rock. Lichens may also grow on trees, decaying wood, and soil. A lichen may look like a leathery leaf, a scaly patch of paint, or a clump of moss. The color of a lichen may be gray-green, bright yellow, or orange. In the arctic, lichens provide food for animals. The best known of the arctic lichens is reindeer moss.

FIGURE 19-11. In a type of symbiosis, the small remoras receive a "free" ride by attaching to a shark. Remoras also obtain food scraps left by the shark.

activity
See Teacher's Guide.　　Activity 19-3

THE STRUCTURE OF A LICHEN

Place a piece of lichen in a drop of water on a slide. Separate it with dissecting needles. Add a coverslip and observe the lichen with a microscope. Sketch and describe the appearance of the lichen. How are the algae different from the fungi? Algae are green.

making sure

Symbiosis is two organisms living together.
Parasitism is one organism living off another.

15. How is symbiosis different from parasitism?
16. How do the alga and the fungus in a lichen benefit each other? Alga makes food. Fungus supplies water and minerals.
17. In a lichen, which is the producer and which is the consumer? Alga—producer, fungus—consumer

19:6 Symbiosis　423

PERSPECTIVES

frontiers

A World Without Sunlight

Marine biologists recently discovered two deep sea communities that exist without the sun's energy. Small submersibles were used to search the floor of the Pacific Ocean west of Central America and South America in an expedition sponsored by the National Geographic Society. In both locations, heat and minerals such as sulfur are being released from water that percolates through molten rocks on the ocean floor. The heat, in an otherwise cold ocean floor environment, enables a special type of bacteria to thrive.

Chapter 19 describes the sun's importance in food chains where plants grow by photosynthesis and serve as producers. The bacteria found on the ocean floor use a different source of energy to survive. The energy is produced by a process called chemosynthesis. Through this process, naturally occuring elements and compounds are combined with oxygen and release energy. The organisms that use this energy are called chemotrophs. The common substances that are combined include ammonia, nitrite, sulfur, a form of iron, and hydrogen gas. The bacteria that use ammonia and nitrite are very important in the nitrogen cycle.

The chemosynthetic bacteria serve as a producer in a food chain. Larval animals feed directly on the bacteria. These larvae and the bacteria are then fed on by certain species of segmented worms, leeches, clams, and crabs. Some of these species have never been observed before and were not named prior to the expedition.

One type of worm, bright red in color, was observed to live in a tube and reach up to four meters in length. It survives by filtering chemosynthetic bacteria and some forms of plankton from the water. Another type of marine animal was nicknamed the "dandelion" because of its resemblance to the plant. It was discovered while examining photographs of the first expedition. When finally collected, it was identified by zoologists as a relative of the jellyfish. The chemosynthetic bacteria are the ultimate source of food and energy for the organisms that make up these deep sea communities.

The warm mineral-laden communities are important because they point out that it is possible for a community to exist at depths where sunlight cannot reach. Since these deep sea animals have been there for millions of years, they have developed features that enabled them to survive only in such a unique habitat.

Dudley Foster/Woods Hole Oceanographic Institute

main ideas

Chapter Review

1. A food chain is made up of a series of animals feeding on plants and/or other animals.   19:1
2. Food webs contain many food chains.   19:1
3. Producers, green plants and algae, make their own food. They are the first level of every food chain.   19:2
4. Consumers cannot make their own food.   19:2
5. Decomposers cause the decay of dead organisms.   19:2
6. Animals that feed on dead animals are scavengers.   19:2
7. Food energy is transferred from producer to each level of consumer in an energy pyramid.   19:3
8. Predators are consumers that are located at the top of food chains.   19:4
9. Parasites are consumers that live and feed on or in other living things.   19:5
10. Symbiosis is when two organisms live together. In some cases, the organisms benefit each other.   19:6

vocabulary

Define each of the following words or terms.

consumer	food web	producer
decomposer	lichen	scavenger
energy pyramid	prey	symbiosis
food chain		

study questions

DO NOT WRITE IN THIS BOOK.

A. True or False

Determine whether each of the following sentences is true or false. If the sentence is false, rewrite it to make it true.

F   1. Food energy is ~~gained~~ at each level of an energy pyramid.   lost

T   2. Only producers can make their own food.

T   3. Decomposers are always the last link of a food chain.

T   4. Energy used by living organisms comes from the sun.

F 5. Both organisms ~~always~~ benefit from their relationship in symbiosis. *sometimes*

F 6. ~~All~~ parasites kill their hosts. *Few*

T 7. Predators help control the population of their prey.

F 8. A parasite is ⌃ a predator. *not*

T 9. More producers are needed to support a smaller number of consumers.

T 10. A food web contains many food chains.

B. Multiple Choice

Choose the word or phrase that completes correctly each of the following sentences.

1. Fungi are (*producers, consumers, **decomposers***).
2. Energy loss in a food chain (***increases**, stays the same, decreases*) as the number of links increases.
3. Algae are (***producers**, consumers, decomposers*).
4. Transfer of food energy is shown best in a drawing of a(n) (*food chain, food web, **energy pyramid***).
5. A mosquito is an example of a (*producer, **predator**, scavenger*).
6. A tapeworm is an example of a (***parasite**, predator, scavenger*).
7. (***Producers**, Consumers, Decomposers*) are always the first link of any food chain.
8. An example of symbiosis is (*mold on bread, **algae on fungi**, ticks on dog*).
9. A person who eats only vegetables would be on the (*first, **second**, third*) level of an energy pyramid.
10. The adult beef tapeworm spends its life in (*cattle muscle, **human intestine**, soil*).

C. Completion

Complete each of the following sentences with a word or phrase that will make the sentence correct.

1. _____ cause the decay of dead plants and animals. *Decomposers*
2. A lichen is an example of _____. *symbiosis*
3. Most __energy__ is lost as it is transferred through a food chain.
4. Animals that eat other animals are _____ in a food web. *predators*

5. A(n) _scavenger_ cleans up the environment by eating dead animals.
6. The series of food links between plants and animals is called a(n) _____. _food chain_
7. A(n) _producer_ is always at the base of an energy pyramid.
8. Fungi growing on a rotten log is an example of _____. _decomposer_
9. The _____ in a food chain traps solar energy. _producer_
10. Vultures and jackals are _____ in a food web. _scavengers_

D. How and Why _See Teacher's Guide in the front of this book._
1. Explain how a food web is different from a food chain.
2. How are scavengers and decomposers alike? How are they different?
3. Describe a food chain that exists in a community near your school or home.
4. Describe two examples of symbiosis.
5. Why are decomposers always the last link in a food chain?

challenges

1. Prepare a bulletin board display or drawing of food webs. Show the many relationships between and among food chains.
2. Make a food chain mobile. Cut out pictures of different foods, plants, animals, and other organisms from old magazines. Paste the picture of each food or organism on an index card. Tie the cards together with string in a chain in the same order as the positions of the organisms in a food chain. Hang the mobile in your classroom.

interesting reading

Blaustein, Elliott H. and Rose, *Investigating Ecology.* New York: Arco, 1978.

Patent, Dorothy Hinshaw, *Hunters and Hunted.* New York: Holiday House, 1981.

Scott, Jack Denton, *Window on the Wild.* New York: Putnam's, 1980.

You can easily observe how features of a bobcat kitten were passed down from its parents. Likewise, your physical features were obtained from your parents and more distant ancestors. How are traits passed from parents to offspring? What is the importance of genes and chromosomes? What is a mutation?

Warren Garst/Tom Stack & Associates

Heredity

chapter 20

Introducing the chapter: Ask each student to write down the name of a good friend and then list as many of the person's features and traits as they can. Examples include eye color, hair color, height, weight, and personality traits. Discuss whether the traits identified are inherited, acquired, or a combination of both. Have students read Section 20:1. Relate the effects of environment to the development of inherited traits.

20:1 Inherited Traits

Every living organism is a product of its heredity and environment. **Heredity is the passing of traits from parents to offspring.** Through heredity, traits such as color, size, and shape are passed on from parents to offspring. The traits of an animal or plant often go back through many generations of ancestors. Why are some plants and animals very much like their parents while others are not? Scientists study heredity in search of answers to this important question.

A living organism has an inheritance (ihn HER ut unts) it receives from its parents. This inheritance consists of inborn traits or characteristics. Much has been learned about this subject by studying the heredity of plants, such as the pea, and animals, such as the fruit fly.

You may have seen fruit flies buzzing around some fruit or decaying plant tissue on a warm summer day. There are several reasons fruit flies are used for research on heredity. Hundreds of flies may be kept in a small container and fed on small amounts of food. Therefore, they are inexpensive to raise and require very little space. Another advantage of using fruit flies is that there are only about ten days between generations (jen uh RAY shunz). A

GOAL: You will learn how inherited traits are passed from parents to offspring.

Define heredity.

How do scientists study heredity?

You may want to begin this chapter with a discussion of the traits of the students' families, e.g., how they look like (or don't look like) their mother, father, sisters, brothers, etc. The students may give their explanations and observations.

FIGURE 20-1. Eye color is one of the traits of fruit flies that scientists study.

Some students may have had experience in breeding guinea pigs, mice, dogs, or other animals. Ask them to relate their experiences and to explain whether or not the offspring were like the parents.

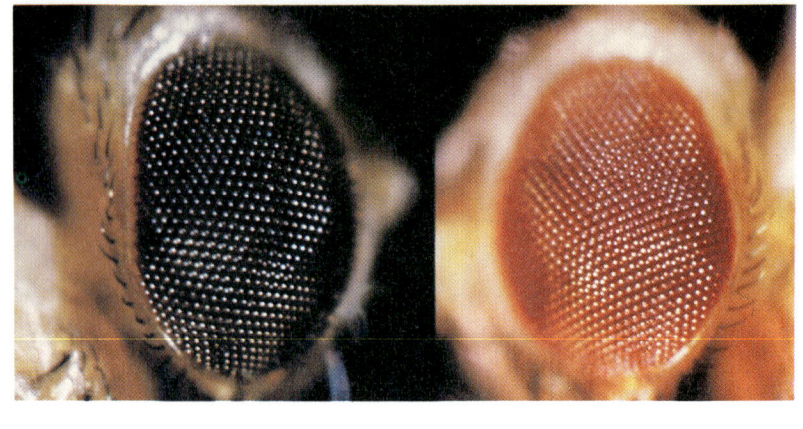

generation is the time it takes an offspring to mature and reproduce. Compare a generation of ten days for the fruit fly to six months for rabbits, four years for horses, and twelve years for elephants. The passage of traits can be followed much more quickly through many generations of fruit flies than any other animals. Fortunately, heredity works about the same for most plants and animals. Therefore, much of what has been learned about the fruit fly can be applied to other living things.

making sure

1. Why are elephants not used in heredity experiments? *They are expensive to raise and have a long time (12 years) between generations.*
2. What are the advantages of using fruit flies in heredity experiments? *They are cheap to raise and have a short time (10 days) between generations.*

20:2 Law of Dominance

Gregor Mendel (1822–1884), an Austrian monk, was one of the first persons to do research on heredity. Mendel studied heredity in pea plants. He observed that some pea plants were short and bushy, and others were tall and climbing. He saw that some pea plants produced yellow seeds and some produced green seeds. He also saw differences in the color of the flowers of the pea plants. Mendel experimented to find out how the traits of pea plants were passed from one generation to the next. He planted seeds from short pea plants and discovered they produced more short pea plants. Seeds from

Genetic problems may be solved by the Punnett square method.

TT—pure tall   ss—pure short

TT × ss

	T	T
s	Ts	Ts
s	Ts	Ts

100% hybrid tall

FIGURE 20-2. Smooth pea seeds are dominant over wrinkled seeds.

these short plants also produced short plants. Every time Mendel planted the seeds from short pea plants, new short plants were produced.

When a trait such as shortness is inherited through several generations the trait is said to be pure. Thus, shortness is a pure trait in certain pea plants. The seeds from short pea plants produce more short pea plants. Other pure traits that Mendel discovered in pea plants were tallness, yellow seeds, and green seeds.

Mendel wondered what would happen if he crossed tall and short pea plants. Would the resulting plants be tall or short?

To cross pea plants, the stamens in the flowers of one plant are removed. The pistil is then pollinated with pollen taken from the other pea plant in the cross. Mendel crossed pure short and pure tall pea plants by transferring pollen from tall plants to short plants. Mendel observed that every pea plant from this cross was tall. Neither short nor medium plants were among the offspring. The offspring from a cross between plants that have different traits is called a **hybrid.**

From his experimental results, Mendel theorized that a trait from one parent may cover or mask a trait from the other parent. This theory is now known as the **law of dominance.** In a hybrid, the trait that determines how the plant or animal will appear is said to be the **dominant trait.** The trait which is present, but hidden or masked, is said to be the **recessive trait.** In pea plants, tall is dominant and short is recessive. Pea plants are short only when tallness is not inherited.

Describe Mendel's experiments with pea plants.

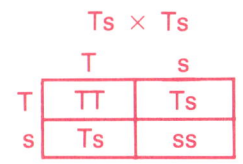

Ts × Ts

	T	s
T	TT	Ts
s	Ts	ss

25% pure tall
50% hybrid tall
25% pure short

Define: hybrid, dominant trait, recessive trait.

FIGURE 20-3. Developing a hybrid pea plant.

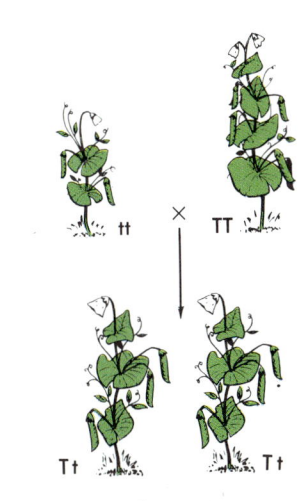

20:2 *Law of Dominance* 431

activity

OBSERVING FRUIT FLIES

Activity 20-1

Objective: To observe various traits of fruit flies See Teacher's Guide.

Materials
2 baby food jars/lids
dark cloth
Drosophila fruit fly culture
hammer
hand lens
masking tape
nail
straight pin

In a wild-type fly, the eyes will be red and the body and wings will be normal. If a laboratory culture containing different traits such as white eye, vestigial wing, or dark bodies are used, be certain students identify those traits.

Procedure

1. Label the jars A and B.
2. Punch three holes in each jar lid using the hammer and nail. The purpose of the holes is for air.
3. Cover the holes with masking tape.
4. Use a straight pin and make some small holes in the masking tape.
5. Transfer the fruit flies in the culture container to jar A using the following method. Tap the fruit flies to the bottom of the culture container. Remove the top of the container and immediately cover it with jar A. Cover the culture container with the dark cloth. The fruit flies should move into jar A. Remove jar A and cover the jar and the culture container immediately so the flies do not escape.
6. Observe the flies with the hand lens. Record the color of eyes, the shape and size of the wings, and the body color. Record the number of flies having each of the three traits in Table A.
7. After 24 hours, transfer the fruit flies to jar B. Use the transfer procedure you followed in Step 5. Return jar B to your teacher.
8. Store jar A at 20–25°C. Observe the jar daily and note the number of days it takes for new flies to appear.
9. Observe the new flies in jar A with a hand lens. Record in your data table the 3 traits you observe. Record the number of flies having each trait in Table B.

Observations and Data

A

	Traits	Number of Flies
Eye color		
Wing		
Body color		

B

	Traits	Number of Flies
Eye color		
Wing		
Body color		

Answers will vary depending on the cultures used.

Questions and Conclusions

1. What is the source of the flies that appeared in jar A. eggs from the parents
2. Were the traits of these flies similar to or different from their parents? How?
3. Why are fruit flies used to study heredity? cheap to raise and a short time between generations

432 **Heredity**

20:3 Crossing Hybrids

Pea plants resulting from a cross between pure tall and pure short plants are true hybrids. Although they grow tall, the plants are not pure for this trait. Mendel also produced other hybrid pea plants, such as hybrid yellow seed plants. These plants resulted from a cross between pure yellow seed plants and pure green seed plants. The offspring plants produced yellow seeds, but they were not pure for this trait.

Mendel's next step in his study was to cross the hybrid pea plants he had produced. For example, he crossed hybrid tall pea plants with hybrid tall pea plants. What do you think resulted?

When the seeds from hybrid tall cross were planted, some of the resulting pea plants were tall and some were short. A careful counting showed that about three of every four plants were tall. Only about one out of four was short.

Mendel also crossed hybrid yellow seed plants with hybrid yellow seed plants. As a result, he obtained some plants that produced yellow seeds and some

Diagram crosses on the chalkboard using Punnett square.

The breeding of hybrids illustrates how traits can skip a generation.

Some historians of science suspect that Mendel already knew the results of his crosses from observances he had made of pea plants over the years. His experiements and calculations confirmed these observations.

Describe two examples of a hybrid cross.

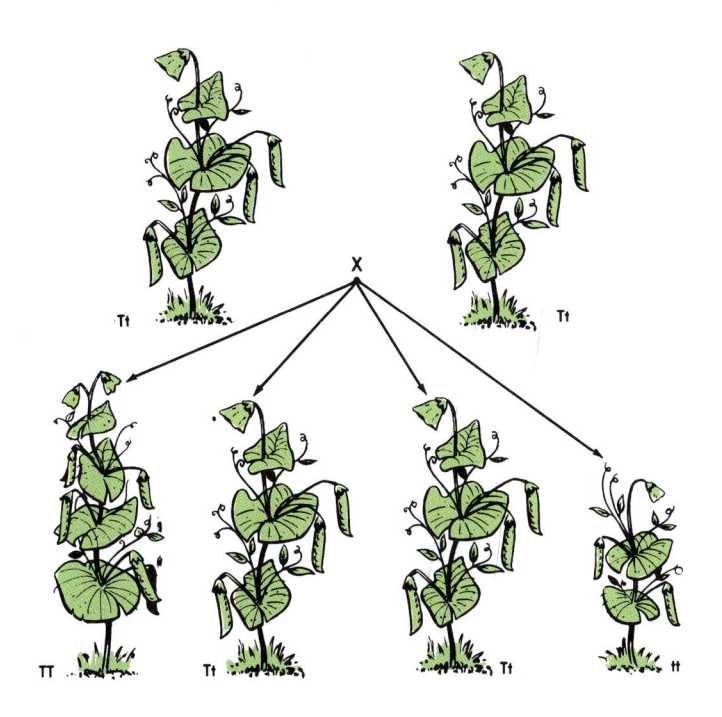

FIGURE 20-4. Cross of hybrid tall pea plants.

FIGURE 20–5. Cross of hybrid yellow seed pea plants.

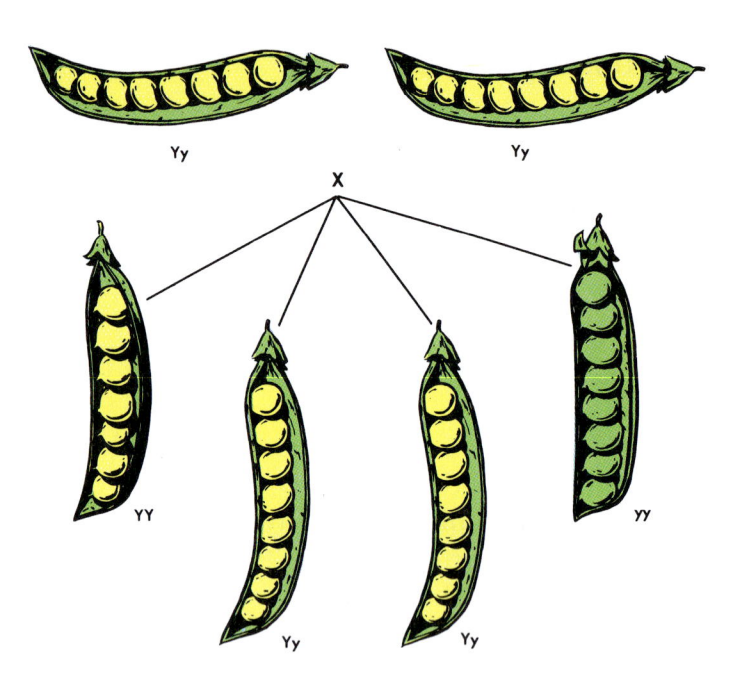

What did Mendel learn when he crossed hybrids?

FIGURE 20–6. In this hybrid cross, the purple kernels in the corn are dominant.

that produced green seeds. About three of every four plants had yellow seeds. Only about one out of four had green seeds.

The crossing of hybrid plants showed two things. First, hybrids do not produce only other hybrids. For example, hybrid tall plants can give rise to short plants. Second, inherited traits do not always appear in an individual. They may skip one or more generations. Mendel's conclusions can be summed up as follows:

(1) Some inherited traits are dominant. Others are recessive.

(2) When both dominant and recessive traits are present in an individual, only the dominant appear.

(3) Recessive traits may be passed from one generation to another without appearing. They may be found in a later generation.

Further experimentation has shown that the facts of inheritance in pea plants are also true in other plants and animals.

434 *Heredity*

activity

See Teacher's Guide.

PROBABILITY AND COINS Activity 20-2

Probability or chance affects the inheritance of traits. Some traits have a greater chance of appearing in the offspring resulting from a cross between two animals or plants. Toss a coin 50 times and keep track of the number of heads and tails. What is the chance of a head when the coin is tossed? What is the chance of a tail? Toss two coins together 50 times. Record the number of each combination: head–head, head–tail, and tail–tail. What is the chance of getting each combination? Combine your results with those of your classmates and determine the totals for the class. How does the chance of getting each combination change as the number of tosses increases? How is the tossing of two coins together similar to a hybrid cross? The head-tail combination represents the dominant-recessive gene combination.

Relate probability to the change for different traits to appear in offspring.

1 chance out of 2 for either head or tail.

1 out of 4, 2 out of 4, 1 out of 4

Actual results agree more closely with the probability ratios.

making sure

Two recessive genes must be present for the recessive trait to show up.

3. Why is a recessive trait pure when it shows up in an organism?
4. Explain how a white kitten may be born even though both parents are black.
   Both parents are hybrid black and contain a gene for white.

Bb × Bb

	B	b
B	BB	Bb
b	Bb	bb

75% black, 25% white

20:4 Blending

The law of dominance does not apply to all traits. Since Mendel's time, scientists have learned that some traits are inherited by blending. **Blending is the inheritance of a trait that is a combination of the traits present in the parents.** For example, white-flowered four o'clocks crossed with red-flowered four o'clocks produce all pink-flowered four o'clocks. Incomplete dominance is another term for blending.

Blending also occurs in shorthorn cattle. When red shorthorns are bred to white shorthorns, the offspring are roan. Roan is a blend of red and white hairs in the coat.

What is blending inheritance?

FIGURE 20–7. Roan cattle result from blending when red shorthorns are bred with white shorthorns.

Jean Wentworth

FIGURE 20–8. Flower color in four o'clocks is inherited through blending. Crossing red-flowered plants with white-flowered plants produced all pink-flowered offspring.

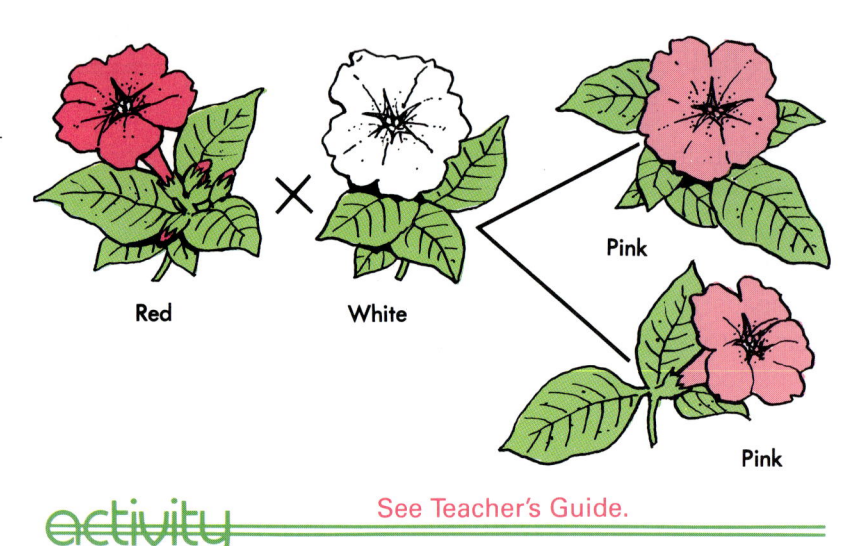

activity
See Teacher's Guide.

OBSERVING SOYBEAN PLANTS Activity 20-3

Obtain 20 hybrid soybean seeds. Soak the seeds in water for an hour. Then plant them in soil or sand. Keep the soil or sand moist. Plants should appear in 8 to 12 days. The leaves may be yellow, green, or yellow green. What is the number and percent of plants showing each color? green 25%, yellow green 50%, yellow 25%

making sure

5. What experiments would you do to find out if a plant trait was dominant, recessive, or caused by blending? Describe the results you would expect in your experiments.

5. Cross the plant with another plant that is pure recessive for the trait. If the trait appears in the offspring, it is dominant. If the recessive trait appears in all the offspring, then the trait is recessive. If the cross results in three traits appearing in the offspring, the trait is caused by blending.

20:5 Chromosomes and Genes

Mendel described heredity in pea plants, but he did not explain how traits are inherited. It took scientists many years to discover how traits are passed from parents to offspring. Even now, all the answers are not known.

We do know that the nucleus of a cell is important in heredity. Under high magnification, tiny, dark grains are visible in a stained nucleus. This grainy material is called chromatin (KROH mut un). When a

436 **Heredity**

FIGURE 20-9. Chromosomes.

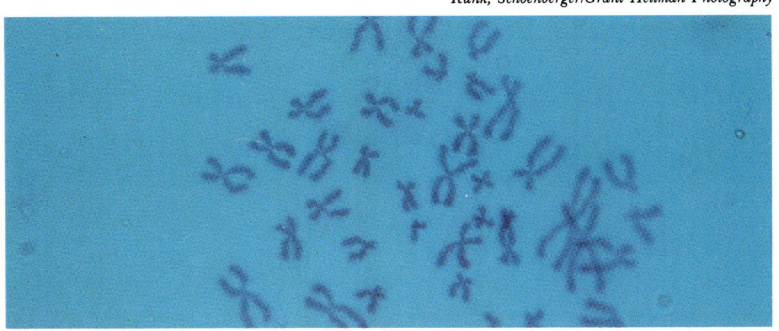

cell begins to divide, the chromatin forms thin, threadlike bodies. These threadlike bodies are called **chromosomes.** Chromosomes vary in size and shape, and almost always occur in pairs. Each chromosome in a cell nucleus has a mate in the same nucleus.

Each species has a certain number of chromosomes. The nucleus of every human body cell has 46 chromosomes that form 23 pairs. The number of chromosomes for various species is given in Table 20-1. Why are chromosomes important in studying heredity? Chromosomes contain **genes** that control the inheritance of different traits. Genes are arranged in a chromosome like beads on a string. Every inherited trait is controlled by at least one pair of genes. Like chromosomes, genes also occur in pairs. A gene in one chromosome has a mate in the other chromosome of the pair.

Define: chromosome, gene.

Where are chromosomes and genes located?

Since Mendel's time scientists have learned that heredity is more complicated than originally thought. Most traits are controlled by more than one pair of genes. A gene may be involved in determining more than one trait. Recessive genes may have an effect on the functioning of an organism that is not obvious. A person's inheritance is the net combination of their genes and not pairs of genes acting in isolation. DNA is said to contain a life code with genetic information for each trait.

Table 20-1. Species Number of Chromosomes

Species	Chromosomes
Fruit Fly	8
Housefly	12
Onion	16
Cabbage	18
Corn	20
Toad	22
Tobacco	24
Frog	26
McIntosh apple	34
Lizard	140

20:5 Chromosomes and Genes

How do genes control the inheritance of traits?

Height in pea plants is controlled by a single pair of genes. Seed color and flower color in pea plants are each controlled by a single pair of genes. Some traits, such as the size and shape of bones, eye color, and skin color are controlled by many pairs of genes. Also, a pair of genes may affect not one, but two or more traits.

It is not certain how a gene determines the traits of a plant or animal. It is known, however, that a gene is a part of a DNA molecule. Scientists believe that the DNA in a gene controls the chemical changes within a cell. By regulating chemical changes, the genes control traits such as height, seed color, eye color, and skin color.

DNA contains information for the production of enzymes that regulate the chemistry of a cell.

Dominant and recessive traits can be explained by genes. One gene is dominant for a trait and the other is recessive. For example, in hybrid tall pea plants, the gene for tallness is on one chromosome. A gene for shortness is on the other member of the chromosome pair. The effect of the gene for tallness is dominant over the gene for shortness. The gene for tallness controls the chemical changes that make the plant grow tall.

FIGURE 20–10. Dominant and recessive traits within this population are determined by genes.

Image Workshop

MAKING SURE
6. Why is a cell nucleus important in heredity? *It contains the genes that control inherited traits.*

438 *Heredity*

20:6 Reduction Division and Fertilization

How does an organism get its genes? In the reproduction of plants and animals, sperm and eggs unite. Sperm and eggs are formed in a special type of cell division called **reduction division.** Reduction division is unlike the division of a body cell in which the chromosome number of the cells remains constant. In reduction division, the chromosome number of the dividing cells is reduced by half.

Reduction division actually consists of two divisions. The chromosome number for frogs is 26. In the first division stage of a male frog sex cell, 26 chromosomes pair up and duplicate. This duplication results in 52 chromosomes. They separate when the cell divides. Each of the two new cells has 13 double-stranded chromosomes. In the second division, the two cells with the double-stranded chromosomes divide to form four cells. Each of the four new cells has 13 chromosomes. Thus, reduction division in the frog starts with one cell having 26 chromosomes and ends with four sperm having 13 chromosomes each.

Models of chromosomes may be made with clay, pipe cleaners, and electrical wire. Models may be made to show the chromosomes in cell division and reduction division.

Explain how reduction division changes the chromosome number.

FIGURE 20–11. Body cell division (a) and reduction division (b). Note that the egg and sperm cells contain half the number of chromosomes contained in the original cell.

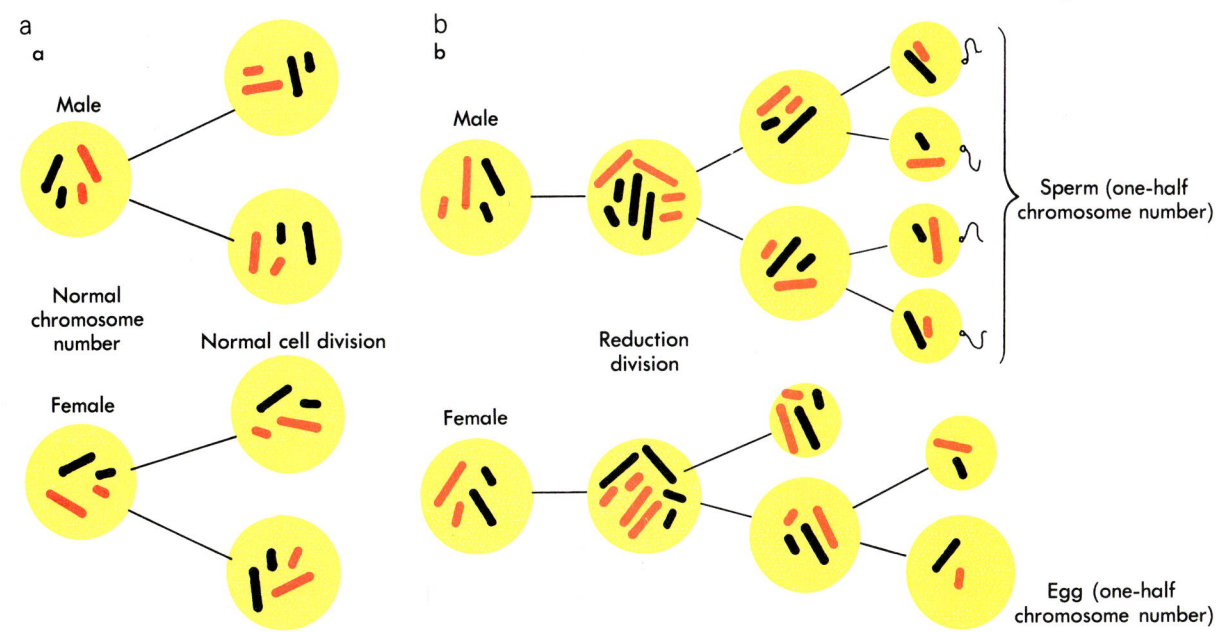

20:6 Reduction Division and Fertilization 439

How does fertilization restore the chromosome number of the species?

Review fertilization and relate it to the chromosome number of the egg, sperm, and zygote.

In the female frog, the first stage is the same as in the male. However, one of the two cells with 13 double-stranded chromosomes retains most of the cytoplasm. The other cell dies. The remaining cell divides to produce two cells, each with 13 chromosomes. Again, one cell retains most of the cytoplasm and the other one dies. The remaining cell becomes the egg cell containing 13 chromosomes and most of the cytoplasm. Because a sperm and an egg cell have half the species number of chromosomes, they have only half the number of genes.

Fertilization is the process by which a sperm cell and an egg cell unite to form a single cell. This process results in the chromosomes combining so that the species number of chromosomes is restored. Each frog egg and each frog sperm has 13 chromosomes. Thus, a fertilized egg has 26 chromosomes, the species number for frogs. Because half the chromosomes come from each parent, a new frog gets an equal number of genes from each parent. Through reduction division and fertilization, inherited traits are passed from parents to offspring.

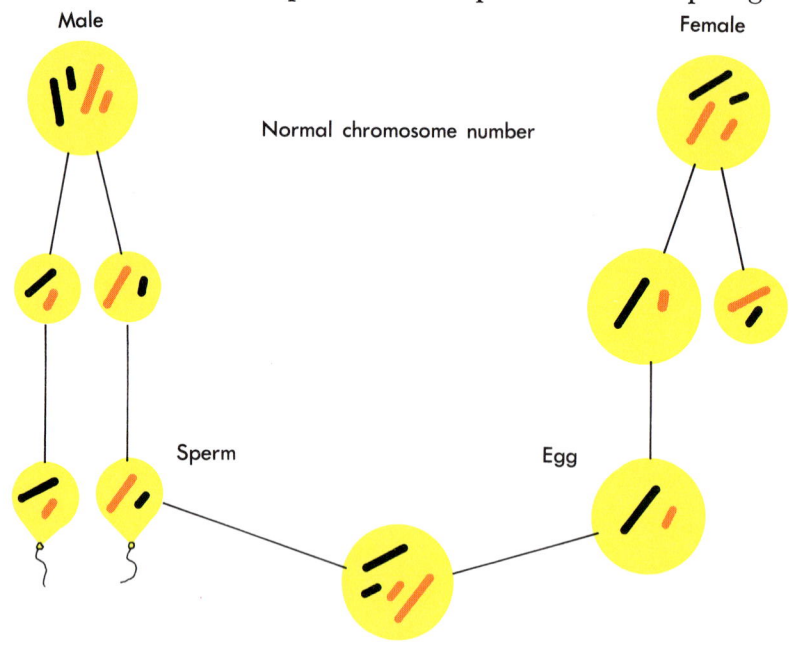

FIGURE 20–12. A zygote forms when a sperm cell fertilizes an egg cell. The zygote contains the same number of chromosomes found in each parent.

440   *Heredity*

activity

See Teacher's Guide.

COMPARING BODY CELL DIVISION AND REDUCTION DIVISION   Activity 20-4

Use a microscope to study prepared slides of onion root tip, whitefish eggs, and animal testis. Observe the differences between body cell division and reduction division.

Point out the difference in the way in which the chromosomes line up on the spindle.

making sure

7. Corn plants have 20 chromosomes in each cell. Pollen and egg nuclei are produced by the corn in reproduction.
   (a) How many chromosomes are in a pollen nucleus and an egg? **10 in each**
   (b) How does fertilization restore the complete number of 20 chromosomes? **Chromosomes in the pollen and egg nuclei are combined together in the egg.**

20:7 Sex Determination

The sex of most plants and animals is controlled by genes. These genes are usually on a certain pair of chromosomes called the sex chromosomes. Most species have two types of sex chromosomes, X and Y. The X chromosome is long and rodlike. The Y chromosome is short and shaped like a "J." In many animals, a female has a pair of X chromosomes (XX). A male has XY chromosomes. Genes on the X and Y chromosomes cause many of the features that make males and females different.

A female has two X chromosomes. Therefore, each of her eggs can have only one X chromosome. Since a male has both X and Y chromosomes, half the sperm will have an X chromosome, and half will have a Y. Since all eggs contain an X chromosome, an egg fertilized by an X-carrying sperm results in a female (XX). An egg fertilized by a Y-carrying sperm results in a male (XY).

How is sex in animals produced by chromosomes?

FIGURE 20–13. X and Y chromosomes determine the sex of an organism.

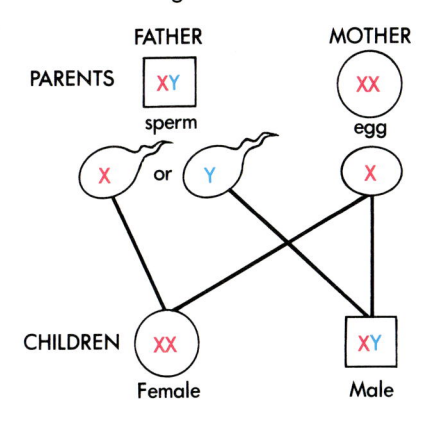

20:7 *Sex Determination*   441

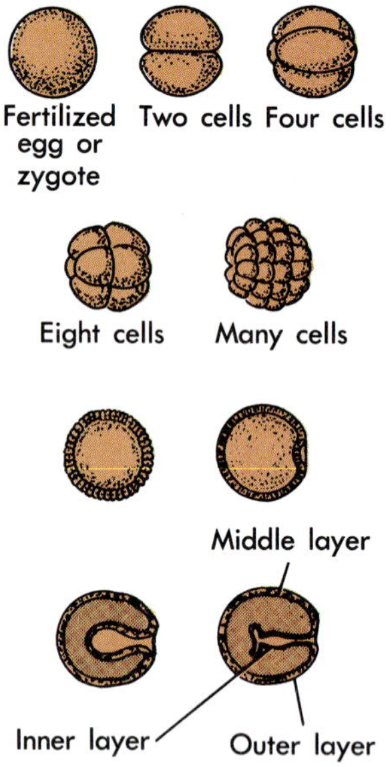

Fertilized egg or zygote
Two cells
Four cells
Eight cells
Many cells
Middle layer
Inner layer
Outer layer

FIGURE 20–14. Development of an embryo from a fertilized egg.

How does a zygote become an embryo?

FIGURE 20–15. Development of a chick embryo at 56 hours (a) and 96 hours (b).

making sure

8. How are the sex chromosomes in males different from those in females? XY—males, XX—females
9. What are the chances of an animal offspring being a female? Male? 1:2, 1:2
10. What sex cell determines the sex of the offspring? sperm cell

20:8 Reproduction

A fertilized egg is the beginning of a new member of a species. It is the first cell in a new animal or plant. The fertilized egg cell, called a **zygote** (ZI goht), divides to become four, eight, sixteen cells, and so on. The process continues until a new member of the species is formed. In these early stages of life, the organism is called an **embryo**. Figure 20–14 shows the early development of an embryo.

The early stages in the life of an embryo are basically the same in all backboned animals. First, many cell divisions occur. Then the cells form a hollow ball or sphere. One side of the sphere pushes inward to form a deep pocket. As a result, the embryo becomes shaped like a cup made of two layers of cells. More cells grow between these two layers to form a third layer of cells. These cell layers will form the various tissues, organs, and systems of the new animal (Figure 20–14).

a

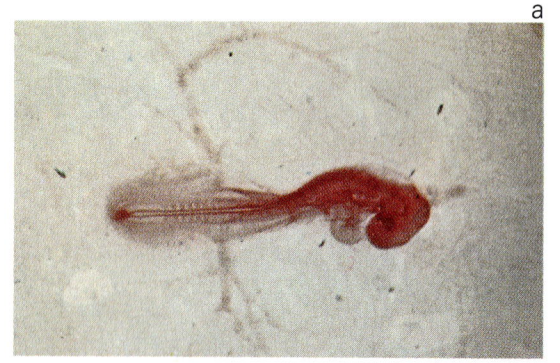

b

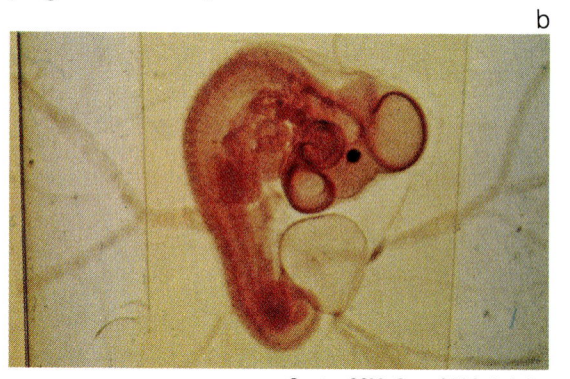

Courtesy CCM: General Biological, Inc.

Courtesy CCM: General Biological, Inc.

In species such as the horse, cow, and sheep, one offspring is usually produced at a time. Once in a while, twins are produced. There are two kinds of twins—fraternal and identical. **Fraternal twins result from two different fertilized eggs.** They are no more alike than any two offspring from the same parents.

Identical twins come from one embryo that was formed from a single fertilized egg. In the very early stages of its development, the embryo divides in two. Each part then develops further as a separate embryo. Two offspring, called twins, are produced from the embryos. Because they come from the same egg and sperm, identical twins have the same genes. This fact explains why they are so much alike.

As the embryo differentiates to form tissues of specialized cells it is called a fetus.

Explain how identical twins are different from fraternal twins.

One fertilized egg divides into two cells

Cells separate

Two nearly identical individuals

IDENTICAL TWINS

Two fertilized eggs divide but cells do not separate

Two very different individuals

FRATERNAL TWINS

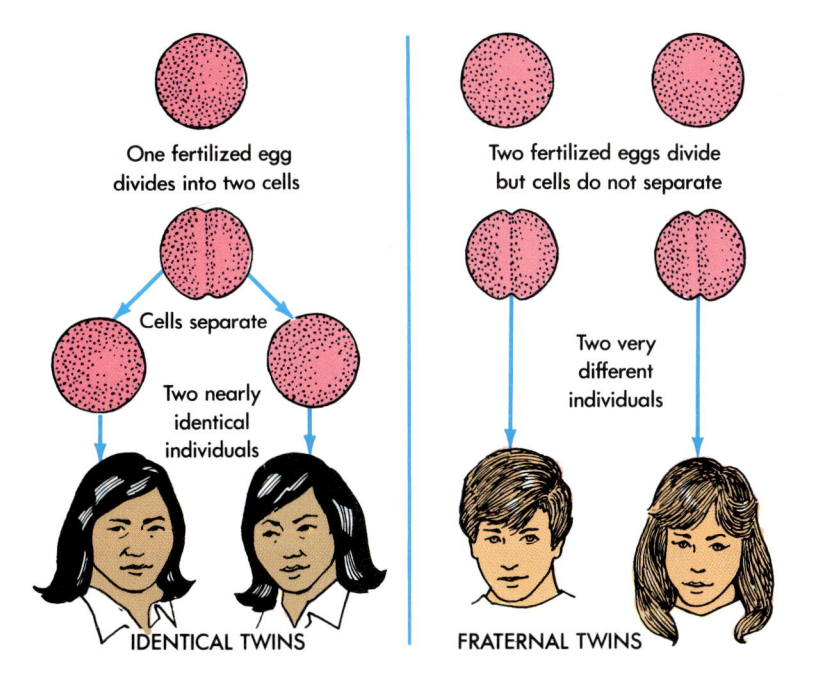

FIGURE 20–16. Identical twins come from a single fertilized egg. Fraternal twins come from two fertilized eggs.

making sure

11. Scientists often study twins to learn the effects of environment on development. What kinds of twins are used in this kind of research, identical or fraternal? Why are twins good subjects for these studies? identical—Identical twins are alike in inheritance. Differences are due to the effects of environment.

20:8 Reproduction 443

20:9 Mutations

What is a mutation?

A change in a gene or chromosome that causes a change in an inherited trait is called a **mutation** (myew TAY shun). A mutant is a living thing that has a mutation. Because a gene is made of DNA, a gene mutation is caused by a change in the DNA. DNA is a complex chemical that carries hereditary traits from one generation to the next. Hence, changes in the structure of DNA will result in changes in the traits passed to the offspring.

Many mutations are recessive and may remain hidden for generations.

Many mutations occur from unknown causes. Other mutations result from exposure to X rays, nuclear radiation, and certain chemicals.

Relate mutations to environmental hazards such as radiation and chemicals that can cause birth defects in animals.

Although many mutations go unnoticed, some have been observed in animals and plants. Mutations may produce unexpected traits, such as short-legged dogs and seedless grapes. These mutants are produced by a change in one or more genes.

Many mutations are harmful to an organism. These mutations make the organism less likely to survive. For example, insects with a mutation that produces short wings do not fly well. Some harmful mutations can even cause death.

Albinism is a color mutation. In albinism, the genes for color are changed so that no color is produced. Thus, albino animals are usually white with pink eyes. Albinism occurs in many animals and plants, such as squirrels and corn.

FIGURE 20–17. Mutations produced the yellow flowers (a) and the albino rattlesnake (b).

444 **Heredity**

activity

ALBINISM IN TOBACCO   Activity 20-5

Obtain tobacco seeds from a biological supply house. Place about 60 seeds on moist paper towels in the bottom of a glass baking dish. Keep the paper towels moist. After 6 or 8 days observe the seeds. What percent of the plants are albino? Can albino plants survive? Explain your answer.

Seedlings use the stored food in the seed for early growth.

Answers will vary.

No, they do not have chlorophyll needed to make food

making sure

12. Why might albinism be a harmful mutation? *A white animal would stand out and not be camouflaged. Thus, it might be easy prey.*

20:10 Plant and Animal Breeding

Many domestic plants and animals are very different from their ancestors. These new varieties include seedless oranges, cattle without horns, and hens that lay eggs almost every day. Each new variety has special traits and often these traits can be passed on to the offspring. New varieties are produced by scientists and other people who breed plants and animals.

One process that scientists use to improve plant and animal breeds is called selection. **Selection means to choose only certain individuals for reproduction.** For example, a dairy farmer uses calves from high-producing milk cows to replace older cows in the herd. Selection is also used to

A seed catalog provides information and pictures of specific plant varieties developed by plant breeders.

Define: selection, breed, purebred, crossbred.

FIGURE 20–18. The parents of these puppies were probably mixed breeds (a). This "beefalo" resulted from a cross between a bison and a species of beef cattle (b).

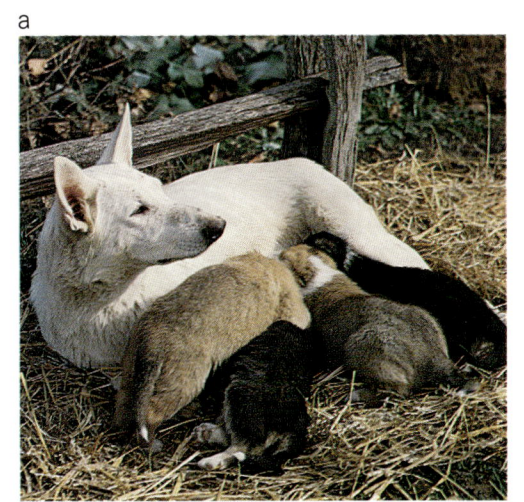

a

b

Grant Heilman Photography

Pat Lanza Field

20:10 *Plant and Animal Breeding*   445

What is hybrid vigor?

You could explain to students that trees that produce seedless oranges are grown by grafting.

How are new varieties of plants produced?

FIGURE 20-19. This woman is artificially pollinating a pine tree.

improve egg production. The fertilized eggs from high-producing hens are kept to raise more high-producing hens.

Through the process of selection, different breeds have been developed. A **breed** is a group of animals or plants that are alike in their traits and have a common ancestor. A member of a breed is called a **purebred** animal or plant. Purebreds breed true; that is, they pass their traits on to their offspring.

A cross between two different breeds results in a **crossbred,** or mixed breed. For example, a mongrel dog is a crossbred. A crossbred animal is often hardier and more disease-resistant than a purebred. This feature is called **hybrid vigor.** However, one disadvantage to crossing breeds of animals is that the traits of the offspring cannot be controlled. For example purebred pups resemble their parents, but there is a great variety among crossbred pups.

Disease-resistant plants have been produced by selection and crossing different varieties of breeds. In developing a disease-resistant variety, only the seeds from plants that are disease-resistant are planted. This selection is repeated generation after generation. Finally, after many selections and crosses, most of the plants will be disease-resistant.

In plants, artificial pollination is used to produce new varieties. Two breeds of plants with desirable characteristics are selected. Then pollen is transferred from the flower of one plant to the flower of the other plant. Many strains of hybrid corn have been produced this way. Hybrid corn has greatly increased corn production. The ears are larger, and the plants are more disease-resistant than the parents.

Many new plant breeds have come from mutations. These breeds include the California navel orange, pink grapefruit, and the nectarine. Mutations occur naturally in plants from time to time. X rays, nuclear radiation, and chemicals have been used to increase the mutation rate. Plant breeders select desirable mutants and use them to develop new varieties. New varieties of roses have come from mutations.

main ideas

Chapter Review

1. Heredity is the passing of traits from parents to offspring. **20:1**
2. The law of dominance states that a dominant trait will mask the recessive trait. **20:2**
3. Hybrids do not produce only hybrids; inherited traits do not always appear in an individual. **20:3**
4. Blending occurs when none of the traits inherited is dominant. **20:4**
5. Chromosomes contain the genes that control the inheritance of traits. **20:5**
6. Reduction division results in sperm and eggs which have only half the number of chromosomes of a body cell. **20:6**
7. Fertilization, the union of a sperm and egg cell, restores the species number of chromosomes. **20:6**
8. The sex of most living things is determined by the inheritance of X and Y sex chromosomes. **20:7**
9. A mutation is a change in a gene or chromosome which results in a change in one or more traits. **20:9**
10. Improved breeds of plants and animals are developed through selection and crossing. **20:10**

vocabulary

Define each of the following words or terms.

blending	gene	purebred
breed	heredity	recessive trait
chromosome	hybrid	reduction division
crossbred	hybrid vigor	selection
dominant trait	law of dominance	zygote
embryo	mutation	

study questions

DO NOT WRITE IN THIS BOOK.
A. True or False
   Determine whether each of the following sentences is true or false. If the sentence is false, rewrite it to make it true.

F 1. ~~Blending~~ is an example of dominance. Tallness in pea plants   Answers will vary.

F 2. The law of dominance explains the inheritance of ~~all~~ traits. many

T 3. Mendel experimented with pea plants.

T 4. Hybrids are the result of cross breeding.
T 5. A dominant trait may cover up a recessive trait.
F 6. The body cells and sex cells of a species have ~~the same~~ number of chromosomes. a different
T 7. A gene is made of DNA.
T 8. The union of a male sperm with a female egg restores the chromosome number for the species.
T 9. A fertilized egg develops into an embryo.
F 10. A purebred animal or plant has ~~more~~ hybrid vigor than a crossbred animal or plant. less

B. Multiple Choice

Choose the word or phrase that completes correctly each of the following sentences.

1. A (*purebred, crossbred, mixed breed*) animal comes from a cross between two animals of the same breed.
2. (*Dominance, Blending, Mutation*) best explains how pink-flowered plants result from white-flowered and red-flowered parents.
3. (*Purebred animals, Crossbred animals, Mongrels, Hybrids*) are likely to breed true.
4. (*Hybrid corn, A purebred, A cocker spaniel, A beagle*) results from a cross between two varieties.
5. (*Fruit flies, Elephants, Horses*) have been used in many heredity experiments.
6. Seeds from pure tall pea plants produce (*short, tall and short, tall, medium*) pea plants.
7. A fertilized egg is called a (*zygote, crossbreed, chromosome*).
8. If two pea plants, hybrid for yellow seeds, are crossed, (*3 of 4, 1 of 4, 2 of 4*) offspring will produce green seeds.
9. When a body cell divides the chromosome number of a cell (*increases, decreases, remains the same*).
10. The number of chromosomes in an egg cell is (*half, two times, the same as, three times*) the number of chromosomes in a body cell.

C. Completion

Complete each of the following sentences with a word or phrase that will make the sentence correct.

1. A cross between tall and short pea plants illustrate the law of _____. dominance

2. Roan color in cattle is produced from a type of inheritance called blending
3. XX in humans are the sex chromosomes of a female.
4. Inherited traits are controlled by pairs of genes on chromosomes.
5. Twins formed from the same egg are called identical twins.
6. Reduction division reduces the number of _____. chromosomes
7. A sudden change in an inherited trait is called a(n) _____. mutation
8. To choose only individuals with desirable traits for breeding is _____. selection
9. A trait that is masked by another trait is _____. recessive
10. The _____ is an example of a mutant. seedless grape (Answers will vary.)

D. How and Why See Teacher's Guide at the front of this book.
1. How is it possible for a trait to skip a generation?
2. Why are identical twins different from fraternal twins?
3. How does hybrid vigor help an animal or plant?
4. Select a desirable quality of a fruit that you would like to develop and propose a method of development.
5. Why does the sperm of a male rabbit determine the sex of its offspring?

challenges

1. Obtain the latest information on recombinant DNA (gene transfer). Find out how this technique may be used to produce new varieties.
2. Do you have a favorite kind of dog? Do library research to find out how this variety of dog was developed through selective breeding.

interesting reading

Chedd, Graham, "Genetic Gibberish in the Code of Life," *Science 81*. Nov. 1981, pp 50-55.

Stwertka, Eve, *Genetic Engineering*. New York: Watts, 1982.

Silverstein, Alvin and Virginia, *The Genetics Explosion*. New York: Four Winds, 1980.

These butterflies are all related to one another. The genetic changes that produce differences in the butterflies have taken place over a long period of time. What changes in the environment could produce these kinds of changes? How can changes be caused by natural selection? What evidence shows that species are changing?

Descent and Change

chapter 21

Introducing the chapter: Prior to introducing the chapter have students prepare a brief report on the life and work of Charles Darwin. Discuss Darwin's life and the reasons he is recognized as one of the great scientists of all time. Review the methods of science discussed in Section 21:3 and relate them to Darwin's work as a biologist.

21:1 Origin of Living Things

Scientists have proposed many theories to explain how life began. Most of these theories state that the first living things came from nonliving matter. How could nonliving matter form living organisms? Elements and compounds were the only matter present when the earth was formed. Perhaps some form of energy such as lightning caused simple molecules to combine. As a result, larger, more complex molecules formed. Some of the large molecules were the kinds present in living organisms.

In 1953, Stanley Miller, an American scientist, performed an important experiment. The experiment was based on the theory that the early atmosphere of the earth was much different than it is today. The early atmosphere probably was made of many gases, such as methane, ammonia, and hydrogen. Water was also present. In Miller's experiments, these gases combined to form amino (uh MEE noh) acids when an electric discharge passed through them. **Amino acids** are the building blocks of **protein** which is the basic material of living organisms. Amino acids contain carbon, hydrogen, nitrogen, and oxygen. According to scientific theory, a lightning bolt passed through a mixture of gases and caused the first amino acids to form.

GOAL: You will study the theories that scientists have proposed to explain the origin of life and changes in species.

Elements present in living organisms are found in the earth's crust. When an organism dies, the elements in its body are returned to the earth through natural decay.

Describe Stanley Miller's experiment.

Organic chemistry is the study of organic compounds such as amino acids and proteins.

451

FIGURE 21–1. Model of Miller's experiment.

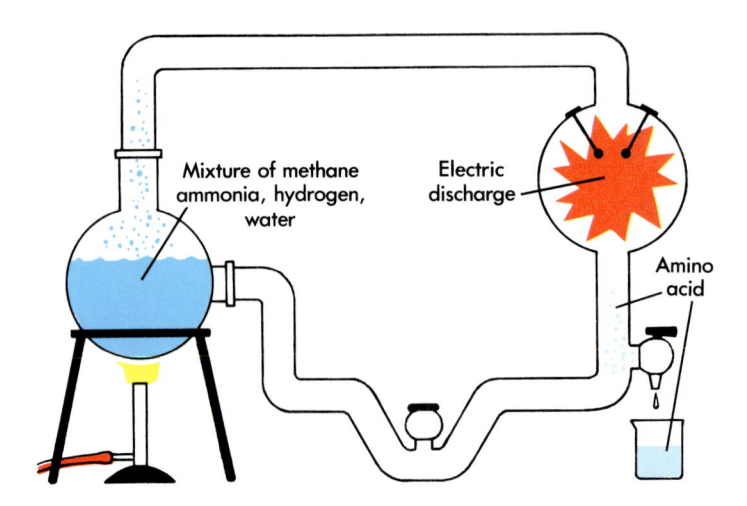

Mixture of methane ammonia, hydrogen, water

Electric discharge

Amino acid

The theory does not stop here. It states that, in some way, the first amino acids combined to form proteins. These proteins then combined to form larger, more complex compounds. These compounds had the ability to reproduce themselves in some unexplained way. Exact copies of the proteins resulted. In other words, a protein molecule became a blueprint for making new protein molecules.

The theory states that the first simple forms of life began in the oceans. The land surfaces of Earth were not suited for life. They were covered with rough rock and no soil. The first living substances may have developed into simple one-celled organisms. These organisms may have been somewhat like the one-celled protists described in Chapter 14. From these simple organisms, more complex forms of life developed.

No one can be sure if the forming of the beginning of life from nonliving matter happened only once. It may have taken place many times and in many places. At this time, there is no evidence to answer this question. Perhaps no answer will ever be found.

How do scientists think the first forms of life developed?

FIGURE 21–2. Bacteria (a) and blue-green algae (b) were probably some of the earliest forms of life on Earth.

a

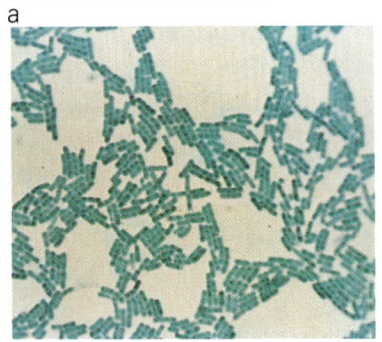

Phillip Harris Biological, Inc.

b

Robert Mitchell/Tom Stack & Assoc.

making sure

1. According to theory, how did the first amino acids form on Earth? — They were made from gases when an electric discharge passed through the gases.
2. Explain why the first simple forms of life probably lived in the ocean. — Land was bare rock and not suited to life. Seawater contained nutrients needed to support life.

452 **Descent and Change**

21:2 Darwin's Theory

Charles Darwin was a British scientist. He is given credit for developing a theory that explains why there is a great number and variety of plant and animal species. In 1831, the 22-year old Darwin sailed from England on a five year journey around the world aboard the British ship, the *Beagle*. Darwin was the ship's naturalist. Darwin's main job was to study the plants, animals, and rocks found along the coast of South America.

During his journey, Darwin collected and recorded as many species as he could find. After his return to England, he spent much time studying his samples, drawings and notes. Darwin was amazed at the number of different plants and animals he had observed. He wondered how so many different species had come about.

At the age of twenty-two Charles Darwin had a degree in theology and was about to enter the clergy. Because of his exceptional ability as an amateur naturalist, he was chosen by the British Admiralty as ship's naturalist for the H.M.S. Beagle. This voyage led to a change in career goals and Darwin's life vocation as a scientist.

How did Darwin obtain information for his theory?

FIGURE 21–3. Some of the plant and animal species studied by Darwin were found on the Galapagos Islands.

For the next twenty years, Darwin collected evidence to support a new theory that he was developing. He visited farms throughout England and spoke with plant and animal breeders. He found that different plants and animals had been developed through the careful selection of breeding stock. For example, the race horse and the work horse had been developed through careful selection. Horses that could run swiftly were bred to produce very fast race horses. Stocky, strong horses were bred to produce better work horses. From such evidence,

21:2 Darwin's Theory 453

Explain Darwin's theory of evolution by natural selection.

Darwin began to think that selection might take place in nature. He developed a theory of **natural selection** that explained how new species were formed.

In 1858, Alfred Wallace, a British biologist, was working in Malaya. He sent Darwin a copy of a scientific article he was going to publish. Darwin was surprised to find his own ideas in Wallace's article. Wallace had come to the same conclusions as Darwin. This often takes place in science. Two scientists working alone make the same discovery or conclusions.

Simultaneous discovery of a new idea in science is not uncommon. Scientists working in different places may come up with the same discoveries even though they have not communicated with each other.

Both Darwin and Wallace concluded that different species of plants and animals were related. They also believed that new species were appearing. At the same time, other species were disappearing. Any new plant and animal species came from the old. However, the new species had different traits.

Review the definition of a theory as an explanation based on available facts and observations.

Darwin and Wallace agreed to present their findings to the world. Neither sought the major credit. Both men were interested in making their ideas known to people everywhere. Their theory is known as the theory of evolution by natural selection. It states that slow and gradual changes in living organisms occur over a period of many years. Changes result from a selection of those organisms that are best suited to the environment.

FIGURE 21-4. Charles Darwin (a) and Alfred Wallace (b).

a 
Historical Pictures Service, Chicago

b 
Historical Pictures Service, Chicago

In 1859, Charles Darwin published his theory in a book called *The Origin of Species*. Darwin stated that among all living things there is a surplus of offspring. For example, a maple tree may release thousands of seeds each year. A female insect may lay thousands of eggs in one lifetime. All fish, birds, and reptiles produce many more offspring than are needed to maintain their number. However, only a small fraction of these live to become adults. In their natural environment, the food supply for animals is limited. There is not enough food for all. Thus, only those animals that get enough to eat will live. The plants and animals most likely to survive are those best suited to their surroundings. They are the fittest.

Descent and Change

FIGURE 21-5. The horned lizard (a) and the cactus (b) are adapted to survive desert conditions.

To survive, an organism must be able to avoid its enemies. It must remain healthy. It must also be suited to the environment. If the environment changes, many species may be destroyed. Suppose a humid area becomes a hot, dry desert. Many animal and plant species will die. Which ones will survive? The species that survive on the desert are those that need little water to live. They must also be capable of adjusting to very hot weather.

Point out that the fittest are those organisms whose bodies and physiology are best suited to their environment. Suited to the environment is called adaptation.

activity   See Teacher's Guide.

SEEDS AND SURVIVAL   Activity 21-1

Find a small tree that has pine cones on it. Count the number of seeds in one pine cone. Then, count the number of cones on the pine tree. Estimate the total number of seeds produced by the tree. Repeat this procedure for a milkweed pod and its plant. What is the advantage in producing many seeds? At least a few seeds will land where they can grow.

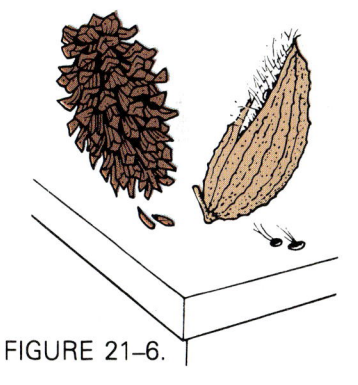

FIGURE 21-6.

making sure   long hair, keep animals warm

3. When the climate of a region becomes very cold, a change in species may occur. Which kinds of animals will be more likely to survive, those with short hair or those with long hair? Explain.

4. How is the careful selection of plants for breeding related to Darwin's theory of natural selection?

4. Artificial selection is used to develop new varieties. According to Darwin's theory natural selection causes species to change.

21:2 Darwin's Theory

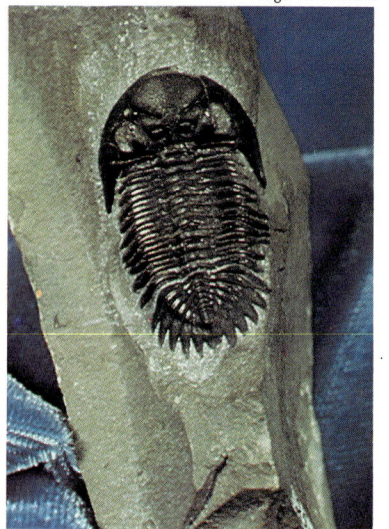

FIGURE 21-7. Trilobites became extinct millions of years ago.

What do fossils show about past life?

Fossils may form as casts, molds, carbon layers, or through petrification.

FIGURE 21-8. Fossil records show that tropical vegetation (a) once covered Greenland (b).

21:3 Fossil Evidence

Fossils provide evidence that organisms of the past were much different from the organisms of today. They also indicate that the organisms of today have arisen from the organisms of the past by evolution. At one time there were huge forests of giant ferns, and dinosaurs roamed the earth. Fossils of one reptile species show it had wings and could fly. Woolly mammoths have been discovered preserved in ice in Siberia.

Fossil records show that species have changed during millions of years. The changes in species over time caused most species to die off and new species to take their place. Scientists believe that the new species were descended from the old species. According to one theory, birds and reptiles are descended from the same ancestors.

Trilobites are an example of an animal for which there are many fossils. Trilobites lived about 500 million years ago. Trilobites no longer exist on the earth. They became **extinct** many years ago. However, species that resemble trilobites are alive today. Among these species are crabs and lobsters. It is possible that the trilobites that lived long ago are the distant ancestors of crabs and lobsters.

Descent and Change

The number of species alive today is much less than the number of species that have lived in the past and left fossils. Many species that lived in the past are now extinct. For example, dinosaurs and saber-toothed tigers no longer exist. Changes in the world that have taken place over millions of years may be responsible for the changes in species. Most noted of these changes are the changes in climate. The fossil record shows that at one time palm trees and tropical plants grew in Greenland. Now it is a cold arctic area. As changes in climate took place, only those living things with traits suited to the new environment survived. These species produced the new and different generations of plants and animals.

If a fossil collection is available, exhibit it in class and discuss the kinds of fossils it contains. Students may have fossils which they can share with others.

21:4 Natural Selection

In nature, selection occurs in many ways. In one experiment a field was seeded with grass seed. One section of grass was fenced off and allowed to grow to full height. The grass was allowed to grow so that it could be cut and dried to make hay. Another section was used as a pasture. This pattern was continued for three years—one half of the field was grown for hay and the other half was grazed. During the fourth year, seeds were taken from both sections of the field and planted. What do you think the grass grown from these seeds looked like?

Grass from the section used for pasture was short and rambling. Grass from the section used for hay was tall and straight. Form a scientific theory to explain the difference between plants from the pasture field and the hay field. For three years, the tall grass in the pasture was eaten by animals. This grass stood tall and erect and could be easily eaten. Most of the short rambling grass escaped being eaten because it grew close to the ground. Thus, the short grass in the pasture area lived to produce seeds for the next crop of grass. The next generation of grass in the pasture resembled its parents. It was short and rambling.

Have students organize the information for this experiment under the headings Problem, Procedure, Observations, Conclusion. Ask them to state the hypothesis and explain how the hypothesis is supported by the observations.

FIGURE 21-9. Within a few years, only short grass will grow on the side of the field used for pasture.

How does the experiment with grass illustrate natural selection?

Explain how natural selection affects species.

Seeds from the hay field produced plants that were tall and straight. Tall plants received the sunlight needed to make food. They grew until they reached maturity. Short plants in the hay field were shaded by the tall plants. They did not get enough sunlight. The tall grass survived each year in the hay field. The short grass died.

There were two different environments in the field. In one, tall plants had the advantage, and they survived. In the other, short plants had the advantage, and they survived. The surviving plants determined the type of grass that grew in each area in the next generation.

The life and death struggle among plants and animals is the basis for natural selection. There are changes between generations in each plant and animal species. Some of the changes are advantages for the new generation. Others are disadvantages.

According to the theory of natural selection, the species that survive are those best suited to the environment. For instance, an animal may have a color that blends with the environment. It is less likely to be seen and killed by its enemies, therefore, it has a good chance to survive. An animal that does not need much food is able to stay alive when a food shortage occurs. Such an animal is better suited for survival than an animal that needs a large food supply.

Descent and Change

FIGURE 21-10. Can you find the golden plover chick among the plants in the photo? Because of its protective coloration, predators also have a difficult time finding it.

making sure

5. Rats have lived in cities for centuries. Why is a rat most fit to survive in a large city?

   A rat is suited to the environment in a large city. It eats garbage and reproduces at a rapid rate. There are many secluded places to hide from enemies.

6. Why is it unlikely the housefly will become extinct as long as there are people on Earth?

   A housefly is adapted to conditions under which many people live.

21:5 Changes in Species

The theory of evolution proposes that living things can change from generation to generation. Over time, these changes may be great enough to produce new species. Some species alive today will someday become extinct. New species will take their place. Evolution in living things goes on now just as in the past.

Evolution is a change in the hereditary features shown by a population of organisms from one generation to the next. Figure 21-11 shows the evolution of the horse. Fossils of the first known horse show that it was about the size of a fox. Through many generations, the horse has increased in size. It developed a larger head and lost all but one toe on each foot. The modern horse is very different from the early horse, its ancestor.

What does evolution mean?

Wide gaps in the fossil record suggest that there were sudden, rapid changes in species at certain times in the earth's past geologic history.

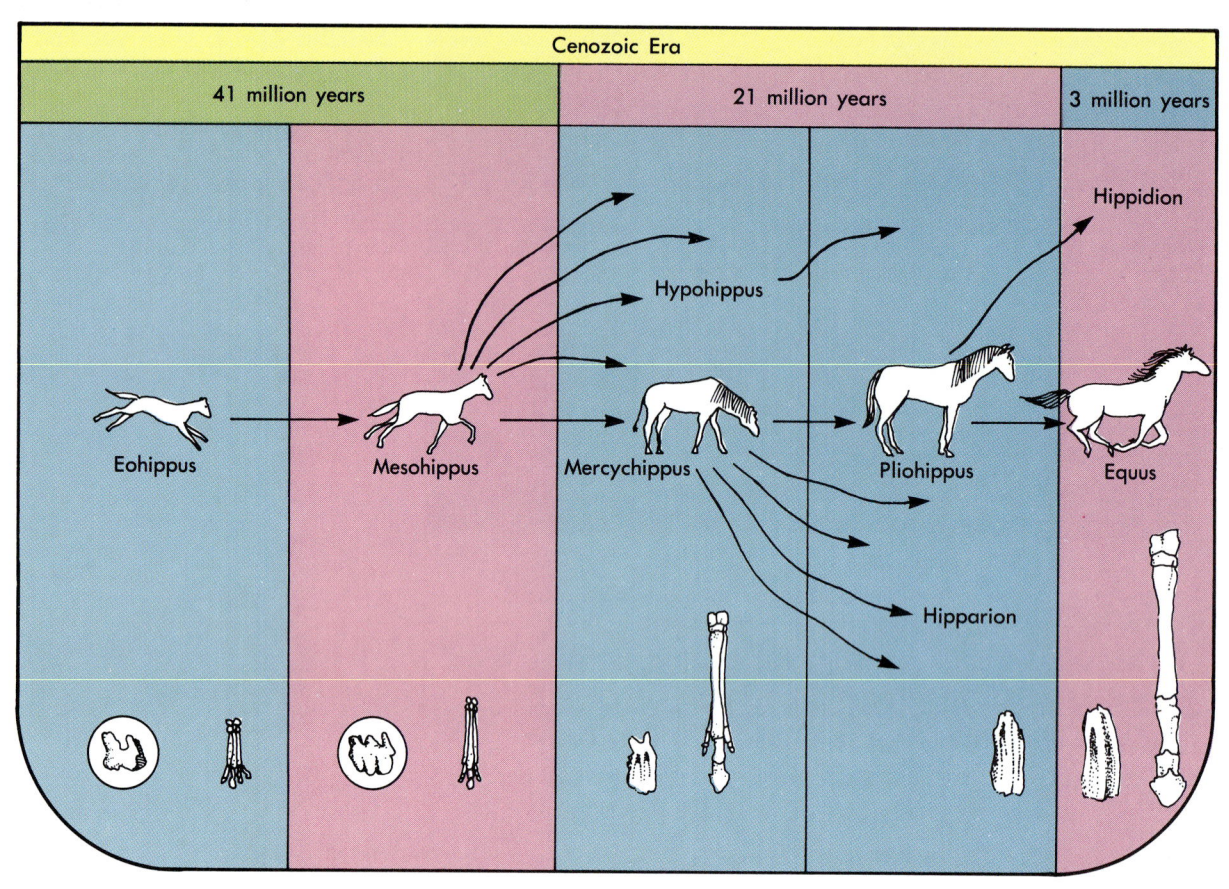

[Figure: Evolution of the horse across the Cenozoic Era — 41 million years: Eohippus → Mesohippus; 21 million years: Mercychippus → Pliohippus, with branches to Hypohippus and Hipparion; 3 million years: Equus, with branch to Hippidion. Tooth and foot bone illustrations shown below each stage.]

FIGURE 21–11. In the evolution of the horse, the number of toes decreased from four to only one. Teeth have also changed.

Much evidence shows that living things are still changing. About 1850, dark-colored peppered moths were observed near Manchester, England. Earlier these moths had been very light in color. Where did the dark moth come from? In the early 1800s in England, many factories were built. Their chimneys gave off large amounts of soot. The soot settled all across the country. In many areas, the soot collected on the trunks and limbs of trees.

Before the factories were built, the light-colored moths blended with the color of the tree trunks. But, when the light-colored moths landed on the dark soot-covered trees they could be seen by birds. They were soon eaten. The dark moths could not easily be seen against the dark tree trunks and were not eaten by birds. Dark moths survived to reproduce more dark-colored moths.

Evidence was needed to explain the change in moth color. A scientist set up several high speed cameras near some soot-covered trees. Many moths were released. Half of the moths were dark. Half of the moths were light. The cameras recorded the birds eating moths. The photos showed that the birds ate more light-colored moths than dark-colored moths.

Only those plants and animals that survive and reproduce pass their traits to a new generation. Who is most likely in each generation to survive and reproduce? It is those individuals that are best suited or adapted to their environment. For example, some bacteria are resistant to antibiotics. (an ti bi AHT ihks). They are more likely to survive and reproduce than bacteria that are not resistant.

How do scientists determine whether or not a scientific theory is acceptable? They look for evidence to support, change, or reject the theory. Scientists have been testing Darwin's theory since he first presented it in 1858. Many of the questions that they have raised still remain unanswered. The main points of Darwin's theory are listed below.

(1) Every organism comes from another organism.
(2) In all species of plants and animals more offspring are produced than are needed to replace the parents.
(3) There is a variety of traits among individuals in a species.
(4) Conditions for life on earth have changed in the past and will continue to change in the future.
(5) In each generation, the organisms best suited to their environment survive and produce the new generation. This process is called natural selection.
(6) Natural selection causes some species to become extinct.
(7) Natural selection causes some species to develop into new and different species.

FIGURE 21–12. The ability of a species to survive may depend on the environment. Which moth in each photo is most likely to escape being eaten?

Describe the peppered moth experiment.

State the main points in Darwin's theory.

21:5 Changes in Species

7. More light-colored moths appear. As pollution decreases the trees become lighter in color and the light-colored moths are less visible against the bark.
8. Through many generations it has increased in size and lost all but one toe on each foot.
9. Climate changes and the species is not suited to the new climate.
10. whether or not it agrees with known facts, whether or not the theory can be used to predict events

How can mutations be helpful and harmful?

FIGURE 21–13. These short legs on a basset hound are caused by a mutation. The legs of most dogs are usually longer.

Grant Heilman Photography

making sure

7. Air pollution control devices are used to reduce the amount of soot in the air. How might this change affect the color of moths in the future?
8. How is the horse an example of descent and change?
9. How might natural selection cause a species to become extinct?
10. How do scientists determine whether a theory is acceptable?

21:6 Mutations and Change

It is known that organisms best suited to the environment are most likely to survive. Suited means that an organism has the traits needed for survival in the place where the organism lives. A fish survives in water because it has gills, but it cannot live on land. The fish is not suited to life on land. A green plant survives because it uses sunlight to make food. However, the green plant will not live long in total darkness. It is not suited to an environment without light.

How do organisms acquire the traits that increase their chance of survival? One way is through mutations. A change in color, a loss of toes, and a resistance to an insecticide may result from mutations. Suppose a change in color causes a reptile to blend better with the surface on which it lives. The color change would provide protection and increase the reptile's chance of survival. The loss of one or more toes may enable an animal to run faster. If so, there is a greater chance of the animal surviving by escaping its predators. If a fly becomes resistant to a poisonous insecticide, the fly increases its chance of survival in an environment where people spray insecticides.

When a harmful mutation occurs, the ability of an organism to survive will decrease. A mutation that prevents a plant from making chlorophyll would result in the death of the plant. If death occurs at an

Descent and Change

early enough age, the organism has not lived long enough to reproduce. In this case the mutant trait is not inherited.

Mutations are inherited. Therefore, mutations that help organisms survive are passed on to the offspring of the organisms. As a result, over millions of years, mutations may have caused many changes in species. Some mutations may have caused some species to become extinct. However, many mutations may have caused new species to appear. The new species survive because they are better suited to the environment.

What causes mutations? Nuclear radiation, certain chemicals, and high temperature's can cause mutations. Radiation levels on earth may have been very high in the past. If so, many mutations may have occurred in plants and animals that lived long ago. These changes may have taken place at a high rate. As the landscape and climate of earth changed, natural selection may have taken place. Organisms with mutations suited to the new environment would have survived. Eventually, new species would have developed that would be much different from their ancestors.

Stress the fact that evolution refers to changes in a species over time and not to changes in a given number of a species.

How might mutations change a species?

For changes in an organism to be inherited they must be due to changes in the genes. Changes resulting from exercise or injury would not be inherited since they do not affect the genetic material of the organism.

Relate the material in this section to principles of heredity discussed in Chapter 20.

FIGURE 21-14. Some chemicals may cause mutations. These chemicals must be prevented from entering the environment.

Gilbert Dupoy/Black Star

making sure

11. How is mutation related to the processes of natural selection? Mutations cause changes that produce the traits of new species.

21:6 Mutations and Change

advances in science

Evolutionary Clues

Evolution is supported by fossil evidence and the existence of mutations. Scientists also have found other sources of evidence for evolution. One area of scientific evidence suggesting that organisms on Earth have changed over a period of time is comparative biochemistry. Red blood cells, for example, contain a complex protein called hemoglobin. If the hemoglobin of a chimpanzee is compared to that of a human, a similarity is found. However, if human hemoglobin is compared to the hemoglobin of a dog, much less similarity is found.

According to scientists, this similarity in hemoglobin shows that humans and chimpanzees are more closely related than humans and dogs. The reason may be that they share a closer evolutionary path than humans and dogs. They might have a more recent common ancestor from which they evolved in their own different ways. Other chemical similarities between chimpanzees and humans would suggest an even greater relationship between the two organisms.

Comparative anatomy also offers evidence for evolution. In the study of comparative anatomy, scientists look for similarities between two organs or other structures that result from similar development from the same or similar embryonic tissues. Homologous organs include the human arm, the foreleg of a horse, the wing of a bat, and the flipper of a porpoise. All are modified limbs that develop from similar embryonic tissues, even though each has a different function. Structures or organs must have both structural and developmental similarities in order to be homologous. Structures with only the same function may not be homologous, but instead indicate evolutionary parallelism. Wings of birds and insects are functionally similar, but are not homologous, as they develop from different embryonic structures.

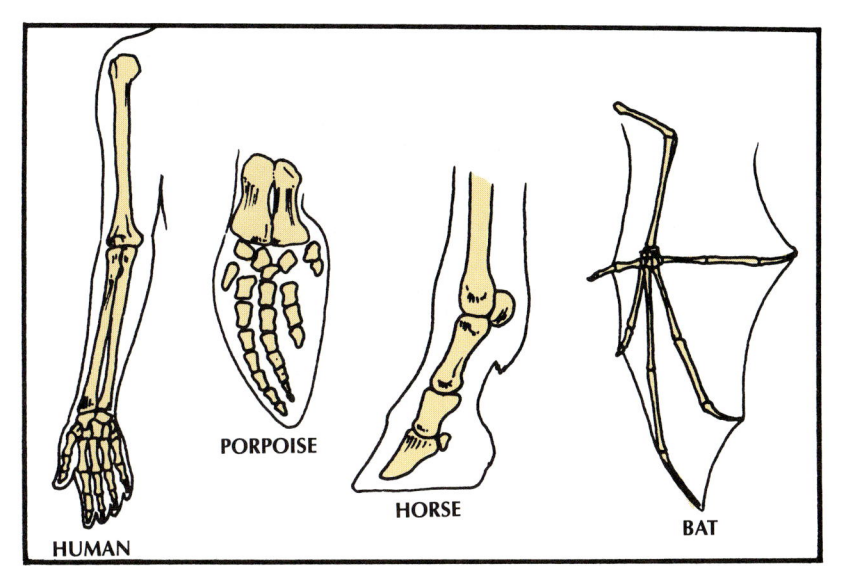

Comparative embryology is the study of the development of the embryos of different kinds of organisms. The similarities found among embryos of different species indicate that evolution has occurred. Scientists have discovered that the early embryos of certain animals are so similar that it is difficult to tell them apart. For example, the embryos of a shark and a chicken look almost the same in their early stages of development. It is only in the later stages of development that differences begin to appear. These similarities indicate that sharks and chickens may have had a common ancestor. Of course, sharks and chickens eventually develop into very different organisms. This fact suggests that these animals have changed, or evolved over a very long period of time.

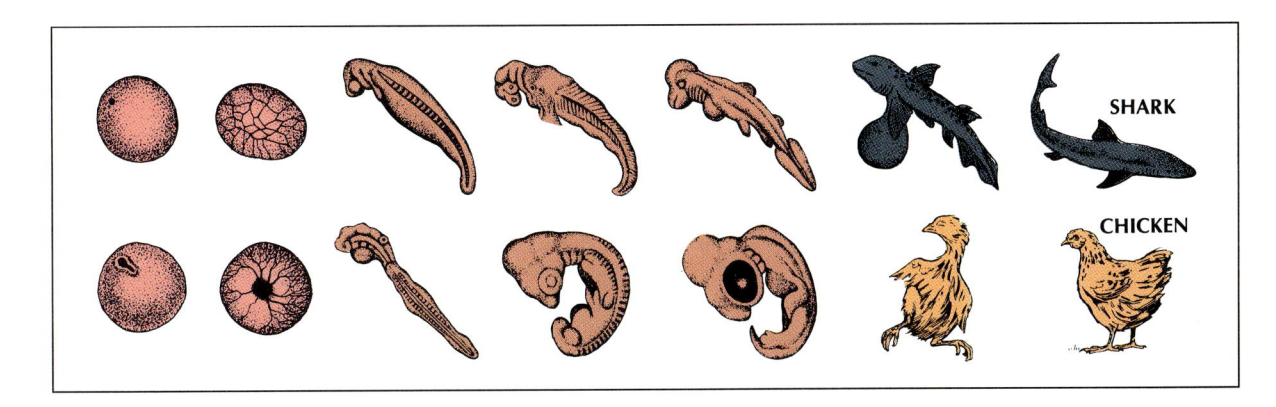

PERSPECTIVES
frontiers

Improving a Theory

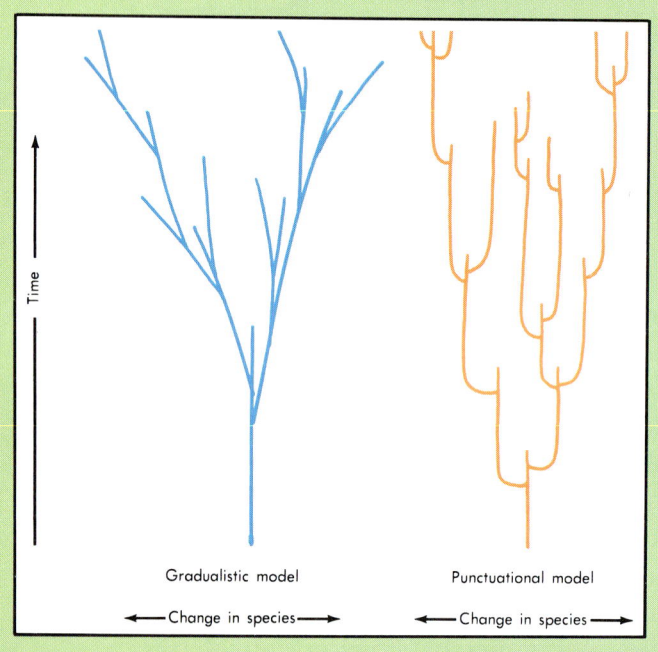

Gradualistic model — Change in species
Punctuational model — Change in species
Time

Charles Darwin originally proposed a theory for evolution based on the hypothesis that species gradually changed over long periods of time and developed into new species. His theory, called gradualism, was named for the speed at which species might develop. One problem with gradualism, which Darwin could not explain, is that fossil remains of many organisms do not show gradual changes. Instead the fossils provide evidence for very sudden changes in species. This evidence clearly contradicts gradualism.

Recently, numerous paleontologists and biologists hypothesized that some types of animals did not change for long periods of time and then suddenly evolved into different species. Scientists have called this type of evolution the punctuational model, as opposed to Darwin's gradualistic model. The speed at which a new species develops may range from only one generation, or in less than a year, to 50 000 years. Even the latter figure is considered fast when compared to the millions of years that Darwin believed to be necessary for evolution. The figures on this page illustrate species change in both the gradualistic and punctuational models of evolution. For a change to occur in as fast as one generation, a mutation would have to occur. However, scientists feel that it is unlikely that such a single mutation could occur and not be harmful to the organism. (If you recall, mutuations may be harmful or helpful.)

The punctuational model of evolution is one possible way to explain the fossil evidence. However, the discoveries of science continue. As more information is gathered, theories of evolution will be modified. Further studies are needed to establish one evolutionary model that will include all the theories of the past.

main ideas   Chapter Review

1. The atmosphere of early earth may have been much different from what it is today. 21:1
2. Changes in the environment may result in the appearance of new species and disappearance of old species. 21:2
3. Darwin's theory of evolution by natural selection states that a slow, gradual change in living things occurred over a long period of time. 21:2
4. According to Darwin's theory, those species suited to the environment survive and produce offspring. 21:2, 21:4, 21:5
5. Fossils provide evidence of changes in living things. 21:3
6. Many species that lived in the past do not exist today. Many species living today were not always present. 21:3
7. Evidence indicates that living things are still changing. 21:5
8. Mutations may have caused the changes in species. 21:6

vocabulary

Define each of the following words or terms.

amino acids extinct protein
evolution natural selection

study questions

DO NOT WRITE IN THIS BOOK.

A. True or False

Determine whether each of the following sentences is true or false. If the sentence is false, rewrite it to make it true.

F 1. Charles Darwin developed his theory of evolution ~~while sailing on a ship~~. 20 years after his explorations
F 2. Darwin's theory of natural selection was ~~never~~ published.
T 3. New varieties of plants and animals are developed through the process of selection.
T 4. Darwin and Wallace developed the same theory.
T 5. Some scientists believe the first life was formed in the oceans.
T 6. Some scientists believe that life came from nonliving substances.

F 7. Changes in living things over the years have generally been ~~fast and dramatic~~. slow and gradual

F 8. The early atmosphere of earth is believed to have been ~~the same as~~ it is today. different from what

F 9. Scientists have ~~yet to find any~~ evidence of natural selection. found

T 10. Through the years, certain chemical insecticides have become less effective in controlling flies.

B. Multiple Choice

Choose the word or phrase that completes correctly each of the following sentences.

1. An animal that is suited to its surroundings is likely to (*survive, move away, not survive*).
2. Natural selection results in a species (*mutating, changing, becoming less suited to the environment*).
3. (*Amino acids, sugars, mutations*) are the building blocks of proteins.
4. Dinosaurs and (*lobsters, birds, trilobites*) are most closely related.
5. A mutation (*always, sometimes, never*) increases an animal's chance for survival.
6. The early atmosphere of the earth may not have contained (*hydrogen, methane, oxygen*).
7. The number of species that have lived in the past is (*greater than, less than, the same as*) the number of species alive today.
8. Fossils show that today's horse is (*larger, smaller, the same*) size as its ancestors.
9. Evolution by natural selection is a (*theory, myth, scientific law*).
10. Mutations are caused by (*changes in genes, evolution, natural selection*).

C. Completion

Complete each of the following sentences with a word or phrase that will make the sentence correct.

1. The first life on Earth may have begun in the oceans .
2. Evolution means that over a period of many years living things changed .
3. One evidence of natural selection can be seen in the remains of living things called fossils .

4. __Fossil__ records show that species have changed during millions of years.
5. Many changes in living things may have come about through changes in the earth's __climate__.
6. The first living material most likely came from __nonliving__ substances.
7. Flies that are resistant to certain insecticides are more plentiful today than in 1945. The survival of these flies is an example of _____.   natural selection, survival of the fittest
8. The short grass in a hay field has less chance to survive than the __tall__ grass.
9. In all species of living things there is a(n) _____ of offspring.   oversupply
10. A(n) __extinct__ species is one that is no longer present.

D. How and Why   See Teacher's Guide at the front of this book.
1. How did Darwin's voyage lead to the development of his theory of evolution?
2. What are the main parts of Darwin's theory of evolution by natural selection?
3. Why are more bacteria resistant to antibiotics today than in 1960?
4. What natural protection do a polar bear and a rabbit have?
5. The dodo is an extinct bird. It had a large, heavy body and small wings. Using these two facts, why do you think the dodo became extinct?

challenges

1. Charles Darwin made contributions to biology other than the theory of evolution by natural selection. Use library sources to learn about his other biological work. Report to the class on his accomplishments.

interesting reading

Freedman, Russell, *They Lived With the Dinosaurs*. New York, NY: Holiday House, 1980.

Ricciuti, Edward R., *Older Than the Dinosaurs: The Origin and Rise of the Mammals*. New York, NY: Crowell, 1980.

Taylor, Ron, *The Story of Evolution*. New York, NY: Warwick Press, 1981.

Descent and Change

SIDE ROADS

Biomes

A biome, or life zone, is an area containing certain plants and animals. Climate determines the kinds of organisms found in each biome. There are seven major biomes on Earth.

In the tundra biome, winters are long and cold with temperatures as low as −40°C. Summers are also cool, with temperatures between 0°C and 15°C. The tundra receives little precipitation—between 10 and 15 cm per year. The growing season may only last two months. Although the surface thaws in the summer, a layer of frozen soil below, called permafrost, remains year round.

In the tundra biome, permafrost and cold winds prevent plants from growing large. Trees are stunted and many plants form cushionlike mats close to the ground. A few migratory birds and rodents, along with animals such as caribou and the arctic fox, live in the tundra.

The taiga is the largest land biome. The growing season is a little longer than that of the tundra. The ground thaws in the warm seasons. Temperatures range between −24°C in the winter and 22°C in the summer. The average precipitation for the year is 35 to 40 cm. Evergreens thrive in this cool, moist climate. Conifer and aspen forests provide food and shelter for animals such as elk, moose, lynx, beaver, ducks, and geese. The trees of the taiga also provide much of the lumber for human use.

A temperate deciduous forest biome has four distinct seasons. Rainfall averages 65 to 150 cm per year. The temperature range is between −24°C and 38°C. Deciduous forests of oak, maple, or hickory offer food and shelter to many woodland animals such as foxes, rabbits, and squirrels. Many animals hibernate or migrate during the winter when food is not available. Much of the world's deciduous forestland has been cut down for its valuable timber and to expose its rich soil for farming.

In the grassland biome, rainfall varies between 25 and 75 cm per year. Temperatures range from 0°C to 25°C. Grasses are the major plants that survive the low temperatures and dry winds. In this region, large crops of grain are grown. Grazing animals such as bison and pronghorn, and burrowing animals such as prairie dogs, are found in grasslands. In North America, coyotes are the most common grassland predators. Cattle and sheep are the primary domesticated animals raised in the grassland biome.

In the tropical savannahs, there is a yearly rainfall of 100 to 150 cm, but most of it occurs during a short, rainy season. The rest of the year is dry and plagued with frequent fires. Temperatures are high, averaging between 20°C and 30°C. Grasses and thin trees are found in this biome. Many of the animals, such as gazelle and zebra, graze in the grasslands. Kangaroos are also found in some savannahs. Predators such as lions and leopards can be found stalking their prey.

Rain falls nearly every day in the tropical rain forest biome, averaging 200 to 225 cm per year. Temperatures remain about 25°C most of the time. Plants form dense jungles in this warm, moist climate. There are no seasonal changes. Plants such as vines and mahogany trees, and animals such as parrots and monkeys are common. Fruits and termites are important members of the food webs. Other life forms are too abundant to count.

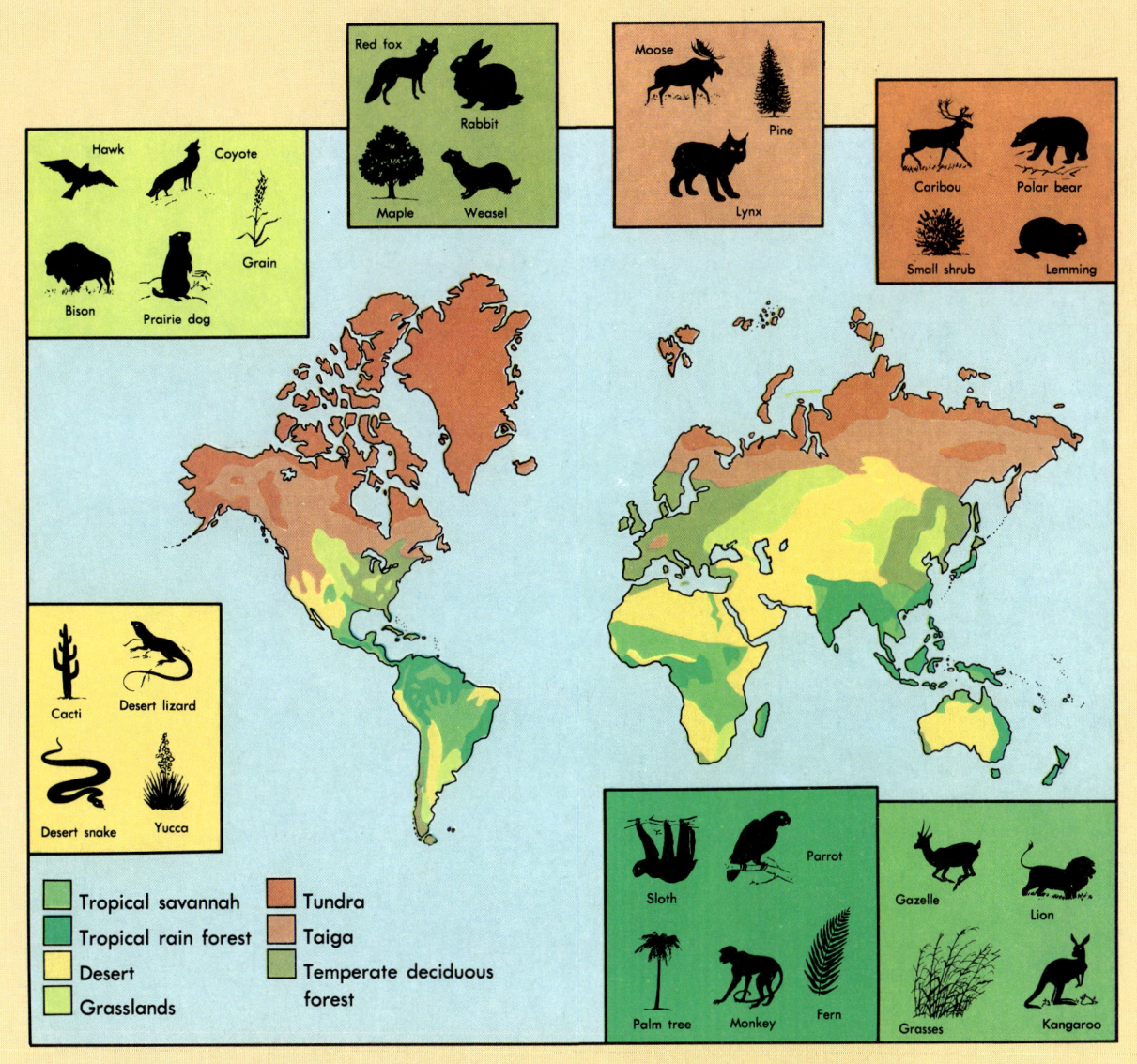

The desert biome has high temperatures during the day followed by cool nights. The temperatures range from 10°C to 38°C. The rainfall is less than 25 cm per year. Plants such as cacti and yuccas have adapted to this hot, dry region. Most desert animals, such as snakes and kangaroo rats, sleep during the heat of the day and are active at night. Lizards can be observed in the shade of thin trees and shrubs, scampering after an occasional insect.

Biome classifications are based on their climates, soils, plants, and animals. Some biomes are very important to humans, but have been severely damaged. Other biomes offer very little for human use. By continuing to study the biomes, we can learn more about how they can be used to our benefit. However, we must also be careful that we do not destroy them.

A resource is something that can be used for the support of human life. A resource also improves the quality of life. Grinnell Glacier in Glacier Park, Montana may be considered as a resource. How might Glacier Park improve your life? What are some other resources that support life on Earth? Why must these resources be preserved?

Larry Burton

Conservation
unit 6

Soil, air, and water are the most important resources on the earth. All three are necessary for maintaining life. However, the quality of our lives is determined by how well the resources are managed. How can conservation of soil, water, and air be best achieved? What steps can be taken to better manage these resources in the future?

Robert McKenzie/
Tom Stack & Associates

Soil, Water, and Air Conservation chapter 22

Introducing the chapter: Exhibit a variety of materials such as soil, wood, coal, iron, water, oil, aluminum foil, copper, and rock. Ask students to name practical uses for each. Have students explain which materials are natural resources and which materials are derived from natural resources. Present the content in Section 22:1. Afterwards, have students classify each material as renewable or nonrenewable.

22:1 Conservation of Natural Resources

Natural resources are those things in the environment that are useful to people. Soil, water, and air are three valuable natural resources. Natural resources such as lakes, forests, and seashores are enjoyed by many. We use these resources for recreation to promote good physical and mental health. Natural resources such as forests, coal, iron, natural gas, and minerals have economic value. They are needed for construction, heating, and transportation. Solar energy is one of the most important of all natural resources.

Natural resources can be classified as either renewable or nonrenewable. Renewable resources are those which can be used and replaced. Forests and other living things are examples. A nonrenewable resource is one which cannot be replaced. Examples are coal, air, and natural gas.

More and more people are taking an interest in conservation (kahn sur VAY shun). Conservation is the wise and careful use of natural resources. A person who practices conservation uses natural resources with as little waste as possible. The aim is to leave an adequate supply of resources for the future.

Have students read Section 22:1 and answer the reading guide questions. Then proceed to the discussion suggestion that follows.

GOAL: You will learn the methods used in the conservation of natural resources such as soil, water, and air.

Define natural resources.

Enrichment: Discuss the practical importance of conservation in people's lives. Ask students to list ways in which people can save money through conservation.

Examples: using less gasoline, paper drives to raise money for school projects, using less electricity.

What is conservation of natural resources?

Table 22–1.
Some Careers in Conservation

Forest Ranger	Protection and care of state and federal forests
Air Quality Engineer	Design and installation of air pollution control equipment
Ecologist	Scientific study of the relationship between living things and the environment
Soil Conservationist	Planning and use of soil conservation methods
Water Quality Technician	Tests water for pollution

FIGURE 22–1. Trees are a renewable resource (a). Fossil fuels cannot be replaced in a person's lifetime (b).

How is soil made?

Students in urban areas may feel they are far removed from problems related to soil and water. Use of films and filmstrips on these topics help clarify and relate the important ideas.

Explain the difference between topsoil and subsoil.

22:2 Soil

Soil did not exist when the earth was first formed. Instead, soil was made from solid rock that was broken into particles by weathering. Formation of soil is a slow, continuous process. It takes between 200 and 400 years to form 1 cm of soil.

Soil is made of two layers. The upper layer is **topsoil,** and the lower layer is **subsoil.** Topsoil is the part in which most plants grow. Topsoil contains rock particles, humus, and many kinds of living organisms. **Humus** is the material formed from the decay of dead plants and animals. Humus makes soil more crumbly and increases its water-holding

476 *Soil, Water, and Air Conservation*

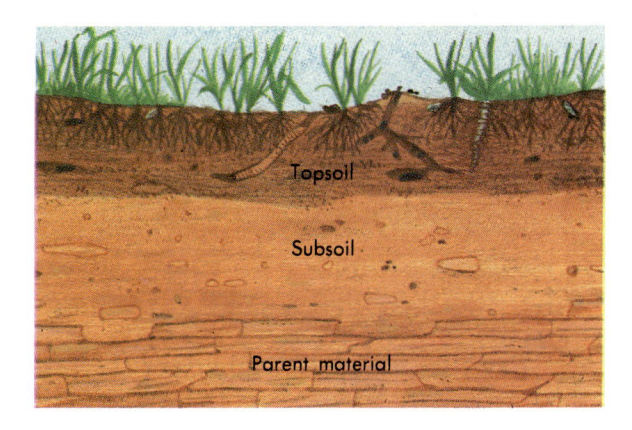

capacity. The fertility of topsoil depends on the amount of minerals, humus, and living things present. Bacteria, molds, yeasts, earthworms, and insects are some of the living things found in topsoil. Subsoil consists only of rock particles of various sizes. Subsoil is lighter in color than topsoil because it does not contain humus.

Soils in different areas vary in color, composition, and depth. The properties of soil affect the kinds of plants that will grow. Soils rich in humus are dark in color. They are usually the most fertile for raising crops. Some farm soils are naturally acidic. Lime is spread on the soil to decrease the acidity. Adding fertilizer to soil increases its fertility. **Fertilizer** is used to increase the organic matter and mineral nutrients in the soil.

FIGURE 22-2. Topsoil is fertile because of the organic material it contains. Layers of subsoil are lighter in color and contain mostly materials derived from the weathered parent rock.

Natural mineral composition of soil is determined by the kinds of rocks from which the soil is formed. Some soils are naturally deficient in minerals such as iron that certain plants need.

See Teacher's Guide.   Activity 22-1

activity

GROWING PLANTS IN DIFFERENT SOILS

(1) Soak 12 bean seeds in water overnight. (2) Label three flowerpots A, B, and C. (3) Fill pot A two-thirds full of topsoil, pot B two-thirds full of subsoil, and pot C two-thirds full of sand. (4) Plant 4 bean seeds in each pot. (5) Place the pots in a warm, sunny location and water them regularly. Be sure to water the three pots equally. (6) When the beans sprout, measure and record the height of each bean plant once a day for two weeks. (7) Record the appearance of each plant. (8) Determine an average growth per day for each pot. In which pots did the plants grow the most? The least? Explain your answers.

Place a saucer or pie pan under each pot to catch excess water. The pot with the sand may have to be watered more often because water soaks through the sand more rapidly and moist sand dries out more rapidly.

topsoil—most, sand—least

Topsoil contains mineral elements available to plants. Also, it retains moisture more than subsoil and sand.

22:2 Soil 477

22:3 Soil Erosion and Mineral Loss

What causes soil erosion?

Nearly 3 000 000 metric tons of topsoil are lost by erosion each year in the United States. **Rainwater loosens and washes away topsoil. Winds dry out topsoil and blow it away.** In many places, soil erosion exceeds the rate of soil formation.

Rainwater may either run off the surface or soak into the ground. Runoff water has enough force to carry soil particles. Even a slow flow of water will easily wash away loose soil. The steeper the slope, the faster the runoff water travels and the greater the erosion.

Erosion is a natural process that has always occurred. The rate of erosion is increased when soil cover such as trees and grass is stripped from the land.

Plants are the best protection against soil erosion. Grass stems and leaves cover soil surface. Runoff moves over the protective carpet of grass and causes little soil erosion. Also, plant roots hold the soil particles together. In a forest, the trunks of trees and exposed roots slow the speed of runoff. When water travels slowly, it has less carrying force.

FIGURE 22-3. Ground covers such as grasses are planted on the slopes near highways to prevent erosion (a). Erosion can be a problem between rows of crops (b).

Erosion is heaviest where row crops, such as corn, tomatoes, and beans, are planted on steep slopes. Row crops are grown in rows with exposed soil between the rows. The exposed soil makes pathways for runoff. If the field is steep, the runoff travels swiftly and the rate of soil erosion is high.

Crops remove minerals from the soil faster than they can be restored by natural processes.

Topsoil is ruined in ways other than erosion. Growing the same crops year after year removes soil minerals. Minerals in the soil are mainly those minerals present in the rock from which the soil came. The removal of minerals decreases soil fertility. To keep the soil fertile, these minerals must be returned to the soil.

Rainfall also reduces the mineral content of topsoil. Water soaking into the soil dissolves minerals and carries them deep into the subsoil. Here they are of little benefit to growing plants. This removal of minerals from topsoil by water is called **leaching**. Organic matter in the soil helps to prevent leaching by absorbing water and holding the topsoil.

Some minerals such as nitrates are very soluble in water.

What is leaching and why is it harmful?

activity
See Teacher's Guide.

SOIL DRAINAGE Activity 22-2

(1) Obtain three funnels of equal size and label them A, B, and C. Support each one over a beaker. (2) Place a small wad of cotton in the bottom of each funnel. (3) Fill funnel A half-full with dry sand. Fill funnel B half-full with dry clay. Fill funnel C half-full with dry loam. See Figure 22–4. Loam is a kind of topsoil that is rich in humus. (4) Pour 250 mL of water into each funnel. (5) Record the amount of water that runs into each of the beakers in 10 minutes. Which soil allows the most water to pass through? sand Which soil would allow the most leaching? sand

FIGURE 22–4.

22:4 Soil Conservation

Soil conservation is important to both city and rural dwellers. We all depend on soil for food. By conserving topsoil, a farmer may obtain high crop yields. At the same time, the condition of the soil is maintained for future use. Soil conservation has helped many farmers improve soil fertility. As a result, even better crop yields are obtained.

What are good soil conservation methods? Recall that soil erosion is greatest on steep, uncovered land, and humus is important to the fertility and water-holding capacity of topsoil. Therefore, anything done to keep soil covered, reduce rapid runoff, and increase the amount of humus aids soil conservation.

Plowing back and forth across a sloped field is called **contour plowing**. Contour plowing produces horizontal ridges in the soil which reduces runoff.

Point out that the price of food is related to the size of crop yields. If agricultural production goes down due to loss of topsoil, food prices may increase. Emphasize this basic principle of soil conservation.

A compost pile stores all sorts of vegetation over time to allow the material to decay. Adding the decayed material to topsoil increases the fertility. Leaves, grass, garbage, and manure are materials that can be composted.

Describe four methods used to conserve topsoil.

22:4 Soil Conservation 479

FIGURE 22–5. Contour plowing is a good conservation practice (a). Terraces keep the soil from eroding away and allow farmers to grow crops on very steep slopes (b).

FIGURE 22–6. Root nodes on this clover plant remove nitrogen from the air and return it to the soil.

Contour plowing can reduce soil erosion by nearly 50 percent. Very steep slopes are sometimes **terraced.** A terraced slope looks like a flight of very wide steps. Carving the land in this way is expensive. It is profitable only if the soil is very fertile and a valuable crop is grown. Citrus orchards and grapes are often grown on terraced hillsides.

Crop rotation is another soil conservation method. In crop rotation, one crop is grown one year and a different crop is grown the next year. Different crops vary in the kinds and amounts of minerals taken from the soil. By changing the crops that are grown, the nutrient loss can be reduced. Minerals are replaced by the decay of organic matter and by adding mineral fertilizer.

Grasses and legumes (LEG yewmz), such as alfalfa and clover, help guard against erosion. They have many uses in soil conservation programs. For example, some farmers plant strips of these plants between strips of row crops. This practice is called **strip-cropping.** Runoff is decreased by the covered strips of soil. This method allows a sloping field to be planted in row crops, yet prevents erosion.

Many farmers plant oats or other grains after the main crop is harvested in the fall. The grain grows a few centimeters high before the first frost. It serves as a ground-cover crop during rainy winter months. In the spring, the grain plants are plowed under. They increase the humus content and supply the nitrogen to the soil. A crop that is grown as a cover and plowed under before maturity is called a green manure crop.

Soil, Water, and Air Conservation

Wind erosion is also a problem in soil conservation. **Shelter belts** are very effective against wind erosion. Shelter belts are thick rows of trees that break the wind. They reduce wind erosion by breaking the wind's speed.

Winds blow away loose topsoil and can create dust storms.

See Teacher's Guide.   Activity 22-3

MODEL OF A CONTOUR PLOWED FIELD

(1) Put a 2 cm thick layer of plaster of paris in a small baking dish or pie pan. Spread the plaster until it evenly covers the pan. Smooth the surface to make it flat. (2) With the wide end of a pencil, make a series of parallel grooves across the plaster. The grooves should be about 0.5 cm deep and 1 cm apart. Allow the plaster to harden for one day. (3) Hold the pan over an empty pan or sink and tilt it slightly so the grooves point down. Use a sprinkling can to sprinkle water on the surface of the plaster. Does it run off or stay on the plaster? *runs off* (4) Repeat this procedure with the grooves horizontal. What happens to the water? Compare your observation with rainfall on a contour plowed field. *water stays in*

Water trapped in furrows soaks into the ground.

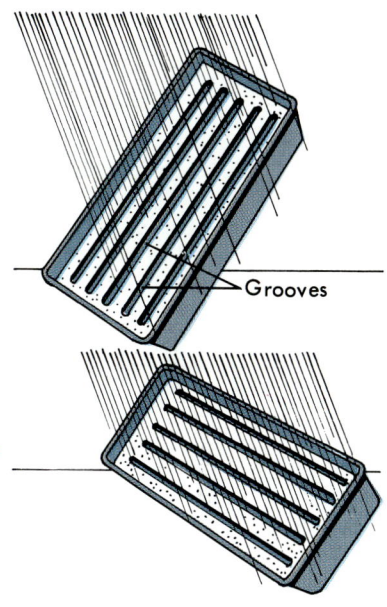

FIGURE 22-7.

22:5 Water Resources

Water is needed by all plant and animal life. People require vast amounts of water. The average American uses about 250 L of water every day. Also, many industries need large quantities of water to make their products. For example, it takes about 250 000 L of water to produce one metric ton of steel.

People obtain water in many ways. The water supplies of many cities come from rivers or lakes. Chicago gets its water from Lake Michigan, one of the Great Lakes. Los Angeles obtains water from the Colorado River. Other cities get water from underground springs or from wells.

Water is also obtained from reservoirs (REZ urv worz). Reservoirs serve as basins to collect rainwater. Much of New York City's water supply travels through pipes from reservoirs in rural areas more than 160 km away.

Students may wish to do research to find out how their local water supplies are obtained.

How is water obtained for use by people?

22:6 Water Conservation

List the goals of water conservation.

Water conservation has three basic goals. The first is to increase the amount of **groundwater,** or water that soaks into the ground, and improve the quality of water that is collected and stored. The second is to clean polluted water and to keep unpolluted water clean and usable. The third is to stop the wasteful use of water.

Water conservation can decrease the consumer's water bill. You may wish to exhibit a water bill in class and explain how the charges are calculated. Also, point out that consumers pay for the water used in agriculture and industry in the prices of the products they buy.

Groundwater supplies can be increased by conserving soil and reducing runoff. Planting grass and trees, contour plowing, terracing, and strip-cropping are effective in increasing the water-holding capacity of the ground. Dams and storage reservoirs prevent flooding and trap water that does not soak into the ground.

You may want to explain how CO_2 and H_2O combine in the air to form a weak carbonic acid which is "natural" acid rain.

Water pollution is a problem in many streams, lakes, and rivers. In some places, industrial wastes are dumped into nearby water. Large cities sometimes release untreated sewage into lakes and oceans. This sewage often contains nitrogen and phosphorus compounds that act as fertilizers for algae. Decay of algae at the end of the summer growing season uses oxygen. If there is a large increase in the amount of algae in a lake or pond, fish may die from a lack of oxygen. Dumping large amounts of wastes into bodies of water kills both animal and plant life. Disease-causing organisms thrive on such polluted water.

FIGURE 22-8. Pollution can result when topsoil, sewage, and other debris are washed into the water supply during a flood (a). Pine Flat Dam on King's River, CA, prevents flooding and provides a reservoir for recreation (b).

a

b

Bureau of Reclamation

U.S. Army

Soil, Water, and Air Conservation

Sewage treatment plants help to keep water clean. In a sewage treatment plant, the sewage is stored in large tanks. Here it is stirred constantly as air passes through it. Harmful compounds are changed by the air into harmless substances. When the process is completed, the sewage may be put into river and ocean waters without harmful effects. Also, purified water separated from the sewage can be used for irrigation.

Much of our water supply is wasted every day. In flushing a toilet or taking a shower, more water is used than is really needed. Decreasing the waste of water aids water conservation.

How can pure water supplies be increased?

Untreated sewage dumped into streams reduces the oxygen content because the oxygen in the water oxidizes the sewage. Thus, there is not sufficient oxygen for fish and plant life to survive.

You may wish to take the class to a water treatment plant in your area.

22:7 Water Sources for the Future

Water conservation practices will not solve all water problems. Many parts of the world simply do not have enough water for drinking and for growing crops. In some areas, thousands of people die from lack of water and food. New sources of fresh water are desperately needed now and in the future.

Changing ocean water to fresh water is a possible solution to the water problem of some arid regions. Ocean water cannot be used directly for drinking or for irrigating land. The ocean's salts kill plants and ruin the soil for crops. However, with the salts removed, this water could supply all of our needs for irrigating land and drinking. **One way that seawater can be changed to fresh water is by heating the water so that it evaporates. The salt is left behind, and the evaporated water is collected.**

One imaginative idea is to tow icebergs to arid regions and use the melting ice as a source of fresh water.

The water cycle is a natural distillation process powered by solar energy.

Explain how evaporation produces pure water.

FIGURE 22–9. Seawater is changed to fresh water at this desalination plant.

FIGURE 22–10. Polluted air contains acids and other substances which can damage property as well as affect a person's health.

Name four substances that pollute air.

What is photochemical smog?

FIGURE 22–11. Sometimes polluted air is trapped near the surface by a temperature inversion. A layer of warm air prevents the air near the surface from rising.

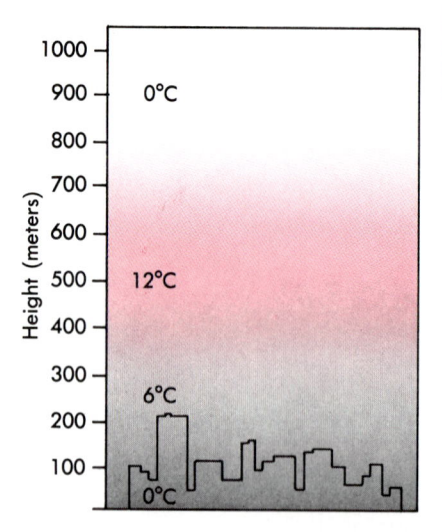

22:8 Air Pollution

If you live in or near a large city, you may have noticed that the air is sometimes hazy and unpleasant to breathe. Air that contains material from smokestacks, industrial waste, and motor vehicles is polluted air. Polluted air contains chemicals that can harm your health. These chemicals can also kill plants, reduce crop yields, and harm animals.

The two major air pollutants in the United States are carbon monoxide and sulfur dioxide. **Carbon monoxide** is a colorless, odorless gas. Ninety-five percent of all carbon monoxide in cities comes from the engines of motor vehicles. An excess of carbon monoxide in the air can cause suffocation and death.

Sulfur dioxide is a colorless gas that has an unpleasant odor. Most sulfur dioxide pollution results from the burning of oil and coal that contain sulfur. Sulfur burns to form the compound sulfur dioxide. Sulfur dioxide in air dissolves in rainwater forming **acid rain.** Acid rain can harm and kill plants. Fish living in lakes and streams may be killed by certain minerals dissolved by acid rainwater and washed into the lake or river. Acid rain destroys metal, concrete, marble, and paint on buildings and cars.

Nitrogen oxides, hydrocarbons (hi druh KAR bunz), and small particles are other substances that pollute the air. They are products of burning fuels. A mixture of smoke and waste gases is called **smog.** "Photochemical" refers to a chemical change that occurs in the presence of light. Photochemical smog often irritates the linings of the nose and throat. It may also cause the eyes to water and sting.

How can air pollution be reduced? One way is to remove solid, unburned particles formed when fuels are burned. Filters placed in smokestacks can remove these particles. Another way is to burn fuels more completely. The production of waste gases is reduced in complete combustion. Many industries burn natural gas instead of coal or oil. Natural gas burns more completely. It releases fewer wastes into the air, thus producing less pollution.

Soil, Water, and Air Conservation

Electric generating plants use large amounts of coal and oil. They release much sulfur dioxide into our air. Air pollution could be reduced if a different fuel were used to produce electricity.

Hydrocarbons, carbon monoxide, and nitrogen oxides are the main pollutants in automobile exhausts. Some devices have been developed which reduce auto exhaust pollutants. One device increases the amount of air in the gasoline-air mixture used by the engine. Gasoline burning is more complete and fewer pollutants are produced. Another device, called a catalytic converter is attached to an exhaust pipe. This converter changes carbon monoxide to nonpoisonous carbon dioxide.

Many scientists are studying alternatives to the gasoline engine. Engines that burn hydrogen gas do not produce pollutants. Turbine engines burn fuel more completely than a gasoline or diesel engine. Electric automobiles operating on rechargeable batteries do not produce waste gases and other pollutants.

U.S. Department of Energy

FIGURE 22-12. The use of electric cars in cities helps reduce pollution.

How can air pollution be reduced?

Most air pollution is produced by natural sources such as volcanoes and plant life. However, the pollution most harmful to people is that produced by cars and industries in urban areas.

Place the jar where it is secure and will not be disturbed. If a microscope is available you may wish to remove the tape and have students observe it under the microscope to see the pollution particles.

activity
See Teacher's Guide.   Activity 22-4

DETECTING AIR POLLUTION

Pollution particles may be detected with a glass jar (screw cap), block of wood, and cellophane or masking tape. (1) Unscrew the cap and nail it, inside up, to the wood. (2) Wrap some tape around the glass jar with the sticky side exposed to the air. (3) Screw the jar back into the cap. (4) Select a site where the jar will be exposed to breezes from all directions. Try to place it a few meters above the ground or on a rooftop. (5) Label the four sides of the tape **N, S, E,** and **W** to correspond to the compass points. (6) At the end of one week, remove the jar and observe the tape. Darkening of the tape shows many particles have been trapped. From which direction did most of the pollution particles arrive?

Darkest side of the tape points to the direction.

FIGURE 22-13.

making sure
See Teacher's Guide.

1. What are the two major air pollutants in the United States? How are these pollutants produced?

22:8 **Air Pollution**   485

See Teacher's Guide.

PERSPECTIVES

skills

Using Graphs

Graphs are used often to show and explain scientific information. The study of science involves your understanding many facts and seeing how they are related. The ability to interpret graphs makes it possible for you to acquire much information quickly. The *bar graph* below shows how land is used in a typical community.

1. What part of the community uses more land than any other. How many hectares are used?
2. What part of the community uses the least amount of land?
3. Determine the total number of hectares used for all activities.

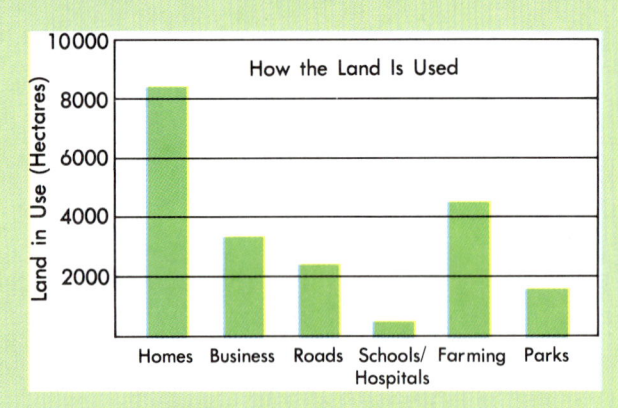

Line graphs show change such as growth, development over a time period, or a process. The following graph shows changes in air pollution levels for a time period of a typical day in a city.

1. How would you account for the highest amount of pollution to occur around 5:00 P.M.?
2. What do you think causes the first peak of pollution around 9:00 A.M.?
3. When does the lowest level of pollution occur? Why?

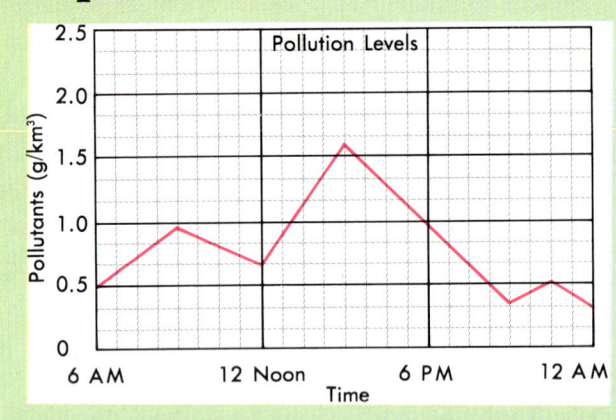

Circle graphs show the parts of a whole in the form of percentages. Even without percentages given, you can see the relative areas given to each part of the circle graph. The circle graph below shows water usage for a typical household.

1. In a household, what two uses of water amount to about two thirds of the total?
2. If a typical household uses 1020 L of water per day, how many liters would be used for bathing?
3. Suppose you had to conserve water and you were requested to cut the amount you used by 40 L. Using the graph, list all the ways by which you might be able to decrease the amount of water you use.

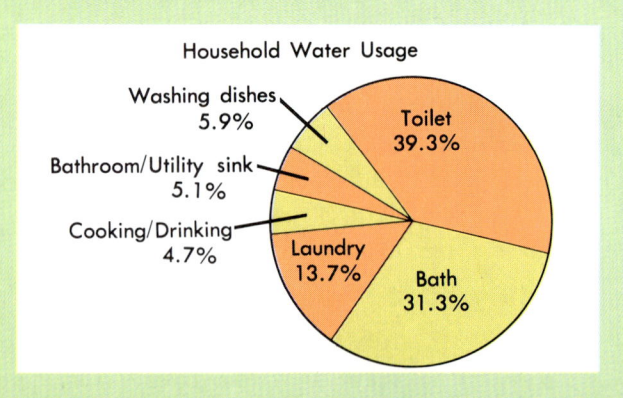

main ideas

Chapter Review

1. Soil, water, and air are three important natural resources. 22:1
2. Topsoil is made by the weathering of rock and the decay of organic materials. 22:2
3. Erosion is a natural process. The rate of erosion is high where soil is exposed to rain and wind. 22:3
4. Soil conservation practices reduce the erosion and leaching of topsoil. 22:4
5. The water supplies for many cities come from rivers, lakes, underground springs, wells, or reservoirs. 22:5
6. Water conservation practices can help provide an adequate supply of pure, fresh water. 22:6
7. Heat energy is needed to convert seawater to fresh water. 22:7
8. The air of some large cities is polluted through the addition of dust, vapor, and fumes by industries and motor vehicles. 22:8
9. Air pollution can be reduced through the use of devices that remove pollutants from smokestacks and engine exhausts. 22:8

vocabulary

Define each of the following words or terms.

acid rain	fertilizer	smog
carbon monoxide	groundwater	strip-cropping
conservation	humus	subsoil
contour plowing	leaching	terracing
crop rotation	shelter belt	topsoil

study questions

DO NOT WRITE IN THIS BOOK.

A. True or False

Determine whether each of the following sentences is true or false. If the sentence is false, rewrite it to make it true.

1. Natural resources such as minerals and forests have no economic value.
2. Weathering changes rock to soil.

F 3. It takes about ~~ten~~ years to form 1 cm of topsoil.   200–400
F 4. ~~Subsoil and topsoil are both~~ rich in humus.   Topsoil is
T 5. Humus increases the ability of soil to hold water.
T 6. Topsoil is removed by erosion.
T 7. Water conservation is closely related to soil conservation.
T 8. Water erosion is greater on slopes than on flat land.
F 9. Most carbon monoxide is produced ~~during the burning of oil and coal.~~   by automobile engines
F 10. ~~Increasing~~ the runoff of rainwater on farm fields is good for water conservation.   Decreasing

B. Multiple Choice

Choose the word or phrase that completes correctly each of the following sentences.

1. (*Strip-cropping, Fertilizing, Contour plowing*) increases the mineral content of soil.
2. Soil erosion is likely to be greatest in a field planted in (*corn, clover, grass*).
3. (*A tree, Coal, Air*) is a renewable resource.
4. (*Subsoil, Humus, Topsoil*), consists only of rock particles.
5. The best protection against soil erosion is (*grass, corn, trees*).
6. Groundwater is increased through (*contour plowing, leaching, fertilizing*).
7. Soil nutrients are returned to the soil by (*strip-cropping, leaching, fertilizing*).
8. Erosion is heaviest in (*cover crops, row crops, strip-cropping*).
9. A mixture of smoke and waste gas is called (*sulfur dioxide, smog, carbon monoxide gas*).
10. Industrial air pollution can be reduced by burning (*natural gas, coal, oil*).

C. Completion

Complete each of the following sentences with a word or phrase that will make the sentence correct.

1. Dumping untreated sewage into a river causes the water to become _____.   polluted
2. Water may be obtained from _____ which serve as basins to collect rainwater.   reservoirs, lakes, rivers

3. The wise and careful use of natural resources is _____. conservation
4. __Lime__ is spread on soil to decrease its acidity.
5. Alfalfa and _____ are examples of legumes. Answers may include clover, peas, etc.
6. Plowing that follows the natural slope of the land is __contour__ plowing.
7. _____ dissolves in rain to produce acid rain. sulfur dioxide
8. _____, _____, and _____ are three ways of preventing floods. Dams, reservoirs, planting grass and trees, contour plowing, terracing
9. _____ is one method of purifying ocean water. Desalination
10. Most of the air pollution in our cities comes from _____ and _____ in the air. carbon dioxide, sulfur dioxide, monoxide

D. How and Why   See Teacher's Guide at the front of this book.
1. Why is soil conservation important to people who live in cities?
2. What causes soil erosion?
3. How do contour plowing and strip-cropping prevent erosion?
4. How does crop rotation increase fertility?
5. What can be done in homes to decrease the waste of water?

challenges

1. Obtain a soil testing kit and test the acidity of a variety of soil samples.
2. Visit your local water treatment plant to learn how your community obtains its water.
3. Obtain information about organic gardening. Learn how this method increases soil fertility.
4. Obtain a career pamphlet that describes jobs in conservation.

interesting reading

Hahn, James and Lynn. *Environmental Careers.* New York: Franklin Watts, 1976.

Berger, Melvin, *The New Earth Book: Our Changing Planet.* New York: Crowell, 1980.

National Geographic Society, "An Atlas of Energy Resources," *National Geographic.* February 1981, pp. 58-69.

SIDE ROADS
Using Fires to grow Forests

Bob McKeever/Tom Stack & Associates

In the past, Smokey the Bear's rule on the prevention of forest fires was strictly followed. But recently, forest rangers have begun to use small controlled fires to stimulate tree growth. Scientists noticed that the giant redwood trees of California were not regenerating, and smaller trees were growing poorly. Studies showed that redwood seeds grow only in bare soil, formed by falling trees, landslides, erosion, and fires.

Without fires, tree branches and pine needles accumulate on the forest floor and cover the bare soil. By using controlled fires, all ground cover could be burned off, allowing tree seedlings to grow.

Further observations revealed that the insect and fungi populations that harm trees were increasing. Forests were becoming more susceptible to disease. Controlled fires have been used to eradicate harmful insects and fungi, resulting in a healthier forest. Such fires also eliminate dense layers of ground cover that provide fuel for large uncontrollable forest fires.

A fire can burn through a forest in three ways. A ground fire burns beneath the forest floor and consumes buried plant remains. A surface fire burns across the top of the ground and feeds on dead material and small plants. A crown fire starts on the forest floor, and climbs to the tops of trees. A crown fire can totally destroy a forest.

As scientists studied the effects of fires on the redwood forests, they found that small surface fires were needed for the overall success and continuation of the forests. Redwood was not the only tree species that required fires for regeneration. The jack pines and lodge pole pines both require heat from fires for their cones to release seeds.

Foresters have calculated that controlled fires reduce the damage caused by large uncontrolled fires by as much as 90 percent. The renewed growth of plant life provides a better source of food for wild animals. Such fires do not necessarily destroy nutrients in the soil or harm streams and rivers in the burned regions.

It has been estimated that major fires sweep North America about every 100 years. History also shows that some groups of American Indians continuously set prairies and forests on fire. Such fires were used to flush out wild animals for hunting and to minimize undergrowth in which enemies could hide.

Forest rangers will have to decide how much burning should be done, and in what areas. Using controlled fires to regenerate forest growth will become more and more popular in the future. The key term, however, is controlled. Campers and hikers should continue to adhere to the strict rule of Smokey the Bear and put out all forest fires.

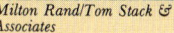

Deer and forests are natural resources. Human changes in the natural resources can be either helpful or harmful. Because of some changes in the past, some resources may cease to exist. What can be done to ensure adequate supplies of natural resources in the future? What are some good forest and wildlife conservation practices?

Forest and Wildlife Conservation chapter 23

Introducing the chapter: Show a film on forest or wildlife conservation such as a film in the life science list, page 35T. Review and discuss the main points made by the film. Have students choose a topic and prepare a report on some aspect of wildlife conservation.

23:1 Forest Resources

Forests cover about 2.5 million km² of the United States—about one third of the total land area of the country. Federal, state, and local governments own about 0.8 million km² of forest. Private industry owns about 0.9 million km². Farm woods make the rest.

Forests fill many needs. Besides having a natural beauty, they are the habitat for a wide variety of wildlife. Forests offer recreation and a supply of raw materials for many products. Lumber, plywood, timber, and paper are the major wood products.

Forest fires destroy the trees, topsoil, and animals that live in the forest. Losing a forest destroys the habitat in which the wild animals live. Fires also cause air and water pollution. After a fire in Paradise Canyon, California, one class took a field trip to see its effects. They talked with the forest rangers who managed the forest where the blaze occurred.

GOAL: You will learn the methods used in the conservation of forest and wildlife resources.

List three reasons why forests are an important resource.

Point out the aesthetic value of forest and wildlife. The study of forest and wildlife conservation will be greatly enriched through the showing of appropriate films and filmstrips.

493

FIGURE 23–1. After the trees are gone, the soil can be quickly eroded by rain.

How do forest fires harm a watershed?

Airplanes and helicopters are often used to reseed land after a forest fire.

The students learned that many trees had been destroyed. Wild animals had been killed or driven from the burned area. Also, much of the humus in the topsoil had been destroyed. Brush and trees on a watershed had been destroyed. A **watershed** is a region from which all runoff water drains into the same main body of water. For example, all the land that sheds its water into a certain pond is the watershed for the pond. The large drainage basin of the Mississippi River is another example of a watershed.

Loss of grasses, brush, and trees exposed the forest soil to erosion. Without the soil's protective cover, topsoil was being washed into the water supply. When the autumn rains came, the soil was lost as muddy runoff water. Bodies of water within the watershed became polluted with mud. Many plants and animals living in the polluted water died. Some flooding occurred because of the increased runoff water. Floodwaters carried soil into small reservoirs and even into streets in a nearby town.

Steps were taken by foresters to solve the problems created by the forest fire. A forester is a person whose job is to care for a forest. Foresters planted mustard and rye grass seed. These grasses grow very quickly and prevent further wind and water erosion. Tree seedlings were also planted. However, at least 20 years is needed to fully restore a forest. Meanwhile, the wildlife and the beauty of a forest are lost.

FIGURE 23–2. Lumber mills cut timber into boards for houses and other building uses.

activity

A WATERSHED

Activity 23-1

Objectives: To model and observe a watershed

See Teacher's Guide.

Materials

aquarium
metric ruler
pieces of flat rock about 5 cm in length
plants, small
putty
sand
sod or moss
sprinkling can

Grasslike carpeting such as Astroturf and plastic plants may be used for a permanent model. Puncture small holes in the carpet so the water will soak through it.

Procedure

1. Cover the bottom of the aquarium with pieces of flat rock.
2. Place a second layer of rocks over two thirds of the first layer as shown in Figure 23–3. Place a third layer of rocks over two thirds of the second layer.
3. Seal the spaces between the rocks on the top and sides next to the glass with the putty.
4. Add sand to the end of the aquarium with the three complete rock layers. Fill the end three-fourths full of sand.
5. Shape a hill from the sand so that the foot of the hill is 2.5 cm tall. Form a small lake on top of the hill.
6. Uncover the bedrock at the foot of the hill to expose part of the ledge.
7. Cover the sand with the sod or moss and plant the small plants in it.
8. Sprinkle water into the lake on the hill until it is full.
9. Observe and record what happens to the water.
10. Fill the lake again and sprinkle water on the hillside. Observe and record what happens to the water and the landscape.

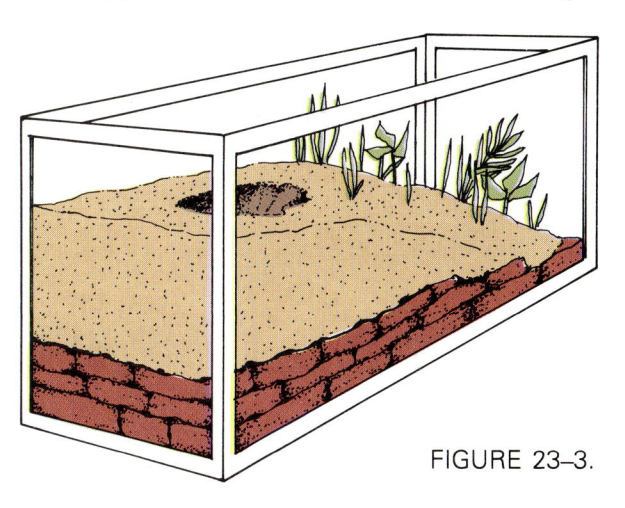

FIGURE 23–3.

Observations and Data

Record observations from Steps 9 and 10.

See Teacher's Guide.

Questions and Conclusions

1. What happened to the water that you added to the lake in Step 8?
2. What happened to the water that you added in Step 10?
3. From your observations how do you think streams are formed?
4. From your observations how do you think lakes are formed?
5. What will happen to the lake on top of the hill during very dry weather?
6. What area of your model is the watershed?
7. Into what area of your model does the watershed drain?

23:1 Forest Resources

FIGURE 23-4. Clear cutting leaves an area open to wind and rain erosion. Many years are necessary for a clear-cut area to reseed itself.

How has the amount of forest land been reduced?

New varieties of trees are being developed that are more disease resistant and grow more rapidly.

23:2 Forest Conservation

At one time, nearly half of our country was covered with forests. Most of the forests were cleared for farmland during the days of early settlement. The original timber stands of the Northeast and Midwest are now gone. They were cut, burned, grazed, and plowed until few trees remained. If these practices had continued, forests would now be scarce. Today, most of our lumber comes from the Far West. Forest resources must be conserved to fill the demand for wood. Forest conservation includes the careful regulation of tree growth in a forest and the wise use of wood as a natural resource.

How were forestry practices of the past unwise? Lumbering methods were much different years ago than they are today. What is known as clear cutting was practiced. **Clear cutting** is a process in which all trees are removed from a very large area. No young trees are left and no older trees remain to reseed the area. Only bare stumps remain when the loggers finish.

Forest conservation practices in the United States began in 1905. President Theodore Roosevelt signed a bill that created the United States Forest Service. The Forest Service is part of the United States Department of Agriculture. It operates national forests and offers guidance to those who manage public and private forests. The agency also employs scientists and technicians to search for and test new and improved conservation practices.

FIGURE 23-5. Much wood is used for the making of paper products.

496 Forest and Wildlife Conservation

activity

WOOD PRODUCTS

See Teacher's Guide.

Activity 23-2

Your class should be divided into small groups (4 to 5 students per group). Each group should then make a list of as many wood products as they can. After 15 minutes, find out which group(s) listed the most products. Combine the lists from the groups and tally the total number of wood products listed by the class. How long was your group's list? What were some of the most unusual products named by the class?

Examples of wood products include lumber, shingles, plywood, turpentine, charcoal, paper, cellulose, sponges, cardboard, wood alcohol.

23:3 Forest Conservation Practices

Forest conservation is the use of different methods for preserving and increasing the number and size of trees in forests. Many forests are grown as crops which are harvested from time to time. Periodically the mature trees are harvested, and young trees are planted on the cut over land. The young trees grow to replace those that were removed. Successful forest conservation involves several practices. These practices are described below.

Preventing forest fires is one practice. Forest fires can be caused by lightning bolts or volcanic eruptions. However, about 90 percent of all forest fires are caused in some way by people. Most fires are the result of carelessness.

What is forest conservation?

Call attention to the fact that forest conservation practices protect and develop watersheds, thereby increasing water resources.

What causes forest fires?

FIGURE 23–6. A forest fire warden watches for any signs of smoke or fire (a). Each year, insect pests destroy valuable forests (b).

a

b

Grant Heilman Photography

Rich Brommer

23:3 Forest Conservation Practices

National and state forests are supervised by forest rangers. From their high towers and airplanes, the rangers watch for smoke. As soon as smoke or fire is spotted, they determine its exact location. Then, fire fighters and their equipment are brought to the scene. If the fire is very large or far from main roads, airborne equipment may be used.

Unwanted trees in a forest are removed through **improvement cutting.** Crooked, aged, and diseased trees, as well as trees of less desirable species are cut. This practice makes room for the growth of healthy, more valuable trees. Improvement cutting increases the lumber yield and improve its quality.

In **selective cutting,** only mature trees are cut. They are carefully selected and marked before the lumbering begins. Younger trees are left for future cutting. When selective cutting is practiced, a good crop of trees is obtained every few years.

In **block cutting,** a section of trees is cut from the forest. Trees that remain around the block will reseed the cut area. In time, the cut area will be restored. Sometimes the cut block is planted with seedlings to speed the regrowth.

Reforestation is the renewing of the forest by seeding or by planting young trees. Some lumbering companies grow their own young trees. This is done for the reforestation of lands from which they have cut trees. Many state governments grow young trees in nurseries for reforestation.

Fighting harmful insects and disease is another practice. Harmful insects and diseases cause widespread destruction of forests. The American chestnut tree was once a valuable timber tree from the Northeast. Most of these trees have been destroyed by a disease accidentally introduced from Asia in 1892.

Dutch elm disease is another problem. It has been spreading through the Midwest for years. A bark beetle helps to spread the disease. One known treatment is to cut and destroy an infected elm tree as soon as the disease is found. This method can slow the spread of the disease.

Define: improvement cutting, selective cutting, block cutting, reforestation.

How may insects and disease harm forests?

23:4 Vanishing Wildlife

Forests are homes for some wildlife. Wildlife is also found in deserts, grasslands, rivers, lakes, and farm fields. Wildlife includes all untamed animals and plants. Destruction of habitats and overhunting can reduce wildlife numbers. In some cases, a species may be completely killed off. A species that no longer exists is said to be extinct. We may be able to find substitutes for wood, coal, and oil, but we can never replace an extinct species.

One example of an extinct species is the passenger pigeon. The last one died in the Cincinnati Zoo in 1914. At one time, there were more than 2 billion passenger pigeons in the United States. John Audubon, the famous naturalist, reported seeing flying flocks of these birds so thick that they blocked the sun. Thousands of these birds roosted together in trees at night. Could their extinction have been prevented?

Passenger pigeons were the victims of hunters. The birds were easy prey because of their roosting habits. Hunters came at night and killed hundreds of passenger pigeons at a time. Sacks full of them were taken away to be sold the next day. Farmers also killed hundreds of the birds and left them on the ground as animal food.

Finally, there were only a few passenger pigeons left. Perhaps a disease caused the death of those that managed to survive hunters. Whatever the cause, the passenger pigeon is now extinct.

Some other species, such as the bald eagle and the whooping crane, are near extinction. These are endangered species. An **endangered species** is one whose numbers are so small it could easily become extinct. If its members are not protected, the endangered species is likely to disappear. It is estimated that there are only a few dozen whooping cranes in the United States. How long will they remain? These birds could become extinct although they are protected by law. They could be wiped out by disease, harmful effects of pesticides, or illegal hunting.

What is an endangered species?

FIGURE 23-7. Three endangered animal species are the Florida manatee (a), the Santa Cruz salamander (b), and the black-tailed prairie dog (c).

Table 23-1.
Some Endangered Species in North America

California condor	Indiana bat
Eastern couger	Shortnose sturgeon
Eastern timber wolf	Arctic peregrine falcon
Mississippi sandhill crane	Utah prairie dog
Gray whale	Southern bald eagle
Florida manatee	Whooping crane

Wildlife are suited to their habitat. If the environment is changed and a habitat is destroyed, the numbers of wildlife species will decrease or be changed.

Every plant and animal is affected by its surroundings. If the environment changes, a species may decrease in number or completely disappear from a region. Yet, some other species may increase in number due to the change. For instance, years ago the snowshoe rabbit thrived in the Northeastern forests. When the forests were cleared for farming, the snowshoe rabbit disappeared from the region. Surprisingly, the number of cottontail rabbits increased. They were better able to survive in farm fields than in the forest. Clearing the land for farming caused a decrease in snowshoe rabbits and an increase in cottontail rabbits.

23:5 Wildlife Resources

Explain why wildlife is a valuable resource.

Various types of wildlife resources benefit people. For example, fish provide us with food. Game fish in lakes, rivers, and streams also provide the enjoyment of fishing. Waterfowl, big game, and upland game (rabbits, quail, grouse, and pheasants) provide sport for hunters. Money spent for hunting and fishing helps support many small communities. Wildlife adds to the beauty of the field and forest. Many people think wildlife is worth preserving just for its natural beauty. What do you think?

How can wildlife help control pests?

Wildlife can help control pests. Hawks, owl, crows, and snakes eat mice and rats. They also reduce the number of insects. One scientist found more than 200 caterpillars in a crow's stomach. This gives you some idea of the usefulness of wildlife insect-eaters.

500 *Forest and Wildlife Conservation*

Grant Heilman Photography

Ohio Department of Natural Resources

a

b

FIGURE 23–8. Clearing land for development destroys wildlife habitats (a). Keeping our waters clear and clean aids wildlife (b).

23:6 Wildlife Conservation

Destruction of wildlife has come about largely because of changes in natural habitats and unwise hunting practices. The changes in natural habitats have mostly been caused by humans. People have cleared forests, plowed grasslands, burned underbrush, and drained swamps. Each of these practices has reduced the number of wild plants and animals. Water pollution has made many of our lakes, rivers, and streams unfit for life. The hunting of wolves, eagles, foxes, and other species has also reduced wildlife resources.

Wildlife conservation helps ensure that all wildlife species will continue to exist in future years. Wildlife resources may be conserved in many ways. Both city and country dwellers can help. Bird feeders are one kind of conservation practice. Seeds are put in the feeders for birds to eat. This is very helpful to them in winter when food is scarce.

One of the best things that can be done for wildlife is the planting of shrubs and trees. These provide natural cover for animals. Many trees and bushes produce nuts and fruits that provide food for wildlife. Unfortunately, most American farmers like to keep their fences free of bushes and young trees. Some farmers poison or clip them each year to prevent further growth. Cleared fence rows are free from wildlife as well as brush.

How are wildlife habitats harmed or destroyed?

Describe three wildlife conservation practices.

FIGURE 23–9. Woodlots and brushy fence rows between fields of crops provide good habitats for wildlife.

Kansas Department of Economic Development

23:6 Wildlife Conservation

Wildlife conservation is based on principles of ecology.

Wildlife conservation requires the maintenance of an environment suited for breeding, feeding, and protection against natural enemies. Also, the environment must be kept free of harmful pollutants such as chemical pesticides that kill wildlife or reduce their ability to reproduce.

Each state has game laws to protect wild animals. These laws restrict the months in which certain animals may be hunted or trapped. They also limit the number of animals that may be killed by each hunter. In some areas, fishing is allowed only during certain seasons. Also, the size, weight, and number of fish that may be kept are regulated. To be effective, game laws must be enforced.

State conservation departments raise game birds and fish for stocking fields and streams. Trout, bass, and other fish species are raised in fish hatcheries. They are then shipped in special trucks to lakes and streams where they are released. In some states, young pheasants are raised and sold to sports clubs for stocking the land on which members hunt.

activity   See Teacher's Guide.   Activity 23-3

CONSERVATION WITH A BIRD FEEDER

Construct a wooden bird feeder. Set it outside your classroom window or near your school building. Place one type of seed in the feeder. Keep a record of the number and kinds of birds that visit it.

If possible, put another bird feeder near the first one. Place a different type of food in it. Record the number and kinds of birds that visit this feeder. How do the two feeders compare in the kinds and numbers of birds?

Remind students that if they set up winter bird feeders it is important to continue to supply the feeder throughout the winter since the birds will come to depend on it as a source of food.

making sure   *reforestation, allow brush to grow where possible, supply feed when necessary,*

1. What are the methods used by people to help conserve wildlife? *reduce use of harmful pesticides, game laws, stocking fields, lakes, and streams*

23:7 Wildlife Refuges

What is a wildlife refuge?

Wildlife refuges are areas in which wild animals and their habitats are protected. Special efforts are made to provide water, protective cover, and winter feeding within a refuge. In some refuges, no hunting or trapping is allowed. In others, hunting may be limited. Wildlife refuges are often owned and managed by national or state governments. However, some refuges are owned and controlled by private groups.

502   **Forest and Wildlife Conservation**

Yellowstone National Park in Wyoming became the first federal wildlife refuge in 1894. The killing of wildlife there is forbidden. The park, along with its protected resources, is a credit to the efforts of the early conservationists. The federal government now operates 356 wildlife refuges under the National Wildlife Refuge System. These refuges are maintained by the United States Fish and Wildlife Service. They cover about 30 million acres of land.

The Migratory Bird Treaty between the United States and Canada aids conservation of migratory birds. Migratory birds are those that travel north in summer and south in winter. **Migratory bird refuges** have been created to maintain and increase these bird populations. Under a law called the Duck Stamp Act, every duck and goose hunter over 16 years of age must buy a special "duck-stamp" each year. Money collected from the purchase of the stamps are used to maintain waterfowl refuge areas.

Permanent wilderness areas are regions untouched by civilization. Many of these areas have been set up in the West. The aim is to preserve them in the wild state so that people can enjoy them now and in the future. They are "forever wild." Although people can camp and hike in these areas, no roads, motor vehicles, or power lines are allowed. People enter only on foot or horseback.

Enrichment: Have students discuss the pros and cons of maintaining wild life refuges.

Enrichment: Ask any students who have visited a state or national park to describe their experiences. Discuss the care needed to preserve these places for future generations.

Explain the purpose of permanent wilderness areas.

FIGURE 23–10. The Everglades in Florida is protected by federal law. No hunting or trapping is allowed (a). Bird sanctuaries provide food and cover for migratory flocks (b).

23:7 Wildlife Refuges

PERSPECTIVES
people

Grizzly Bears and Humans

The grizzly bear is considered by many scientists to be an endangered species. As humans colonized North America, the grizzly bear was gradually eradicated from most of its natural habitats. Now the grizzly population is slowly rising, and some experts have upgraded the status of the species from endangered to threatened. A threatened species is one that could become extinct in the foreseeable future if steps are not taken to preserve it.

Dr. Charles Jonkel, a professor at the University of Montana, studies the grizzly bear. He is learning about many different aspects of the bear, in hopes of solving some of the problems that occur between bears and people.

Dr. Jonkel pays special attention to the habitats of the grizzly bear. He wants to know what areas the bears inhabit and how humans use the same habitats. As the number of human fatalities increases from encounters with the grizzly bear, Dr. Jonkel is trying to develop bear repellants.

Johnny Johnson

A repellant is a chemical that could be sprayed around a tent or cabin that would keep a bear away. Dr. Jonkel and his students have tested loud noises, strobe lights, electric fences, moth balls, skunk spray, and human urine as bear repellants. Evidence suggests that urine from male humans is an effective deterrent to grizzly encounters. Another effective repellant is a mixture of synthetic skunk musk and Mace, a spray sold in stores to ward off attackers.

Dr. Jonkel cautions that the slight rise in grizzly bear numbers is probably due to humans being educated about the bear. He warns, however, that the grizzlys' habitats are rapidly being destroyed, and the bear's population will probably begin to decline again. People are also learning that feeding bears can result in an attack and possibly fatality.

From his studies, Dr. Jonkel has been able to make many recommendations. He suggests that when timber in a bear's territory is cut, small areas with some cover, such as shrubs and small trees, should be left. Road construction should also be kept to a minimum. Dr. Jonkel recommends that intersections, when necessary, should be built with sharp curves at the crossroads. The curves would lessen the chance of human-grizzly encounter. He believes that if people see fewer bears, there would be fewer disturbances.

Many times an area is logged for timber at the time of the year when bears are feeding. In this case, lumbermen should wait and cut the trees after the bears have moved elsewhere.

It will be important to see how foresters and scientists manage the grizzly bear population. Work by Dr. Jonkel and his students indicates that it may be possible for humans and bears to co-exist.

main ideas — Chapter Review

1. Forests and wildlife are two important natural resources.	23:1, 23:4
2. Forests may be destroyed by fires, insects, disease, and unwise lumbering practices.	23:1, 23:2
3. Forests are valuable because of their beauty, their role in preventing erosion, and as a source of raw materials.	23:1, 23:2
4. Good forest conservation ensures an adequate future supply of forest products.	23:3
5. Some wildlife species are extinct. Others are endangered species and are near extinction.	23:4
6. Food, pest control, and natural beauty are some ways wildlife resources benefit people.	23:5
7. Civilization tends to drive away many wildlife species.	23:6
8. Wildlife conservation protects game animals, game fish, birds, and other wildlife species.	23:6
9. Wildlife requires protective cover, food, and clean water for survival.	23:7

vocabulary

Define each of the following words or terms.

block cutting improvement reforestation
clear cutting cutting selective cutting
endangered migratory bird refuge watershed
 species permanent wildlife refuge
game laws wilderness area

study questions

DO NOT WRITE IN THIS BOOK.

A. True or False

Determine whether each of the following sentences is true or false.
If the sentence is false, rewrite it to make it true.

T 1. Forest fires are harmful to soil, water, and wildlife.
F 2. Loss of ground cover during a forest fire ~~prevents~~ erosion. increases
T 3. Forests aid soil and water conservation.
F 4. Reforestation ~~destroys~~ forests. renews
T 5. Lightning can cause a forest fire.
T 6. Forest land may be easily eroded after a forest fire.

F 7. Clear cutting is ~~a good~~ forest conservation practice.   an unwise
T 8. Shrubs and trees are important for wildlife conservation.
T 9. Improvement cuttings are made to remove dead and diseased trees.
F 10. Block cutting is ~~harmful~~ to forest conservation.   helpful

B. Multiple Choice

Choose the word or phrase that completes correctly each of the following sentences.

1. About (*90, 20, 10*) percent of forest fires in the United States are caused in some way by people.
2. The (*elm, American chestnut, white pine*) is a tree species that has been almost totally destroyed by disease.
3. The (*bald eagle, bison, passenger pigeon*) is a species that has become extinct in the last 100 years.
4. The (*snowshoe rabbit, cottontail rabbit, whooping crane*) is near extinction.
5. It takes at least (*20, 100, 200*) years before a forest is fully restored after a forest fire.
6. The United States Forest Service was created in (*1892, 1894, 1905*).
7. A species that no longer exists is said to be (*protected, endangered, extinct*).
8. A region from which all runoff water drains into a certain body of water is a(n) (*refuge, reservoir, watershed*).
9. (*Block cutting, Improvement cutting, Selective cutting*) involves the careful choosing and marking of mature trees before cutting.
10. Clearing the land in the Northeastern forests causes an increase in (*snowshoe rabbits, bears, cottontail rabbits*).

C. Completion

Complete each of the following sentences with a word or phrase that will make the sentence correct.

1. Today, most of our lumber comes from the __western__ part of the United States.
2. State __game__ laws protect wildlife from too much hunting.
3. The large drainage basin of the Mississippi River is an example of a __watershed__.

4. _____ and _____ are two conservation practices that help wildlife. *game laws, stocking fields and streams*
5. The _____ operates national forests and employs forest conservationists. *U.S. Forest Service*
6. The process of removing unwanted trees from a forest is called _____. *improvement cutting*
7. Areas in which wild animals and their habitats are protected are wildlife _____. *refuges*
8. Two endangered species are the _____ and _____. *Answers will vary.*
9. _____ birds are those that travel north in summer and south in winter. *migratory*
10. The passenger pigeon was an animal that decreased in numbers until it became _____. *extinct*

D. How and Why *See Teacher's Guide at the front of this book.*
1. What important products are obtained from forests?
2. How may forests be preserved for future use?
3. How may wildlife be preserved for future generations?
4. How do wildlife refuges help to preserve wild animals?
5. Some states pay hunters to kill crows and hawks. Is this wise? Explain your answer.

challenges

1. Obtain information on the food preferences of wild birds. List the food habits of birds that make them valuable to agriculture. Explain why birds need greater energy than most other animals.
2. Obtain a book on trees from a library. Learn to identify 10 trees in your area. You can locate trees planted on street and in parks. List the uses of these trees.

interesting reading

Burt, Olive W., *Rescued: America's Endangered Wildlife on the Comeback Trail.* New York: Messner, 1980.

Sage, Bryan, *Antarctic Wildlife.* New York: Facts on File, 1982.

Poynter, Margaret, *Wildland Fire Fighting.* New York: Atheneum, 1982.

Appendix A

Scientific Notation

Scientific notation greatly simplifies the handling of large and small numbers. They are shortened by expressing decimal places as powers of ten. A power of ten is the number of times a number is multiplied by ten. The number 6 000 is written as 6×10^3 or $6 \times 10 \times 10 \times 10$.

A number is written in scientific notation by moving the decimal point until a single digit is to the left of the decimal point. The number of places the decimal point moved is the exponent of the power of ten. For a number larger than one, the decimal point is moved left, and the exponent is positive. For a number smaller than one, the decimal point is moved right, and the exponent is negative. The number 0.002 is written 2×10^{-3} or 2 multiplied by 1/10 three times—$2 \times 1/10 \times 1/10 \times 1/10$.

Example:
The estimated volume of water in the Pacific Ocean is 700 000 000 000 m³. What is the volume written in scientific notation?
Solution
Step 1: Write the number and the unit.

$$700\ 000\ 000\ 000\ m^3$$

Step 2: Move the decimal point until a single digit is to the left of it.

$$7.00\ 000\ 000\ 000\ m^3$$

Step 3: Count the number of places you moved the decimal point. Use that number as the exponent.

$$700\ 000\ 000\ 000\ m^3 = 7.0 \times 10^{11}\ m^3$$

$$\text{The volume of water} = 7 \times 10^{11}\ m^3$$

Example:
One second is 0.000 011 5 day. How is it written in scientific notation?
Solution
Step 1: Write the number and unit.

$$0.000\ 011\ 5\ \text{day}$$

Step 2: Move the decimal point until a single digit is to the left of it.

$$0\ 000\ 01.1\ 5\ \text{day}$$

Step 3: Count the number of places you moved the decimal point and use that number as the exponent.

$$0.000\ 011\ 5\ \text{day} = 1.15 \times 10^{-5}\ \text{day}$$

$$1\ s = 1.15 \times 10^{-5}\ \text{day}$$

Appendix B

Science Classroom Safety

The science classroom is a safe place in which to perform activities if you are careful. You must assume responsibility for the safety of yourself and your classmates. Here are some safety rules to help guide you in protecting yourself and others from injury.

1. Do not perform activities that are unauthorized. Always obtain your teacher's permission.
2. Study your assignment. If you are in doubt about a procedure, ask your teacher for help.
3. Use the safety equipment provided for you. Know the location of the fire extinguisher, safety shower, fire blanket, and first aid kit.
4. Safety glasses and safety apron should be worn when any activity calls for heating, pouring, or mixing of chemicals.
5. Report any accident, injury, or incorrect procedure to your teacher at once.
6. Smother fires with a towel. If clothing should catch fire, smother it with a blanket or coat or quench it under a safety shower. **NEVER RUN.**
7. Handle chemicals and bend glassware only under the direction of your teacher. If you spill acid or another corrosive chemical, wash it off immediately with water. Never taste any chemical substance or draw poisonous materials into a glass tube with your mouth. Never inhale chemicals. Keep combustible materials away from open flames.
8. Place broken glass and solid substances in designated containers. Keep insoluble waste material out of the sink.
9. When your activity is completed, be sure to turn off the water and gas and disconnect electrical connections. Clean your work area. Return all materials and apparatus to their proper places.

First Aid

1. Report all accidents or injuries to your teacher at once.
2. Know where and how to report an accident or injury. Know the location of the phone and fire alarm, and where to locate the nurse.
3. All cuts and bruises should be treated as directed by the instructions included in your first aid kit and should then be reported to a nurse or physician.
4. In case of severe bleeding, apply pressure or a compress directly to the wound. **GET MEDICAL ATTENTION IMMEDIATELY.**
5. If any substance is spilled or gets into your eyes, wash them with plenty of water and notify your teacher for additional aid.
6. Minor burns should be immersed in cold water immediately. In cases of severe burns, **NOTIFY YOUR TEACHER AT ONCE.**
7. In case of fainting or collapse, give the person fresh air and recline him/her so that the head is lower than the body. **NOTIFY YOUR TEACHER AT ONCE.** Mouth-to-mouth resuscitation may be necessary. Call a nurse or physician.
8. In case of poisoning, **NOTIFY YOUR TEACHER WHO WILL CALL A PHYSICIAN AT ONCE.** Note the suspected poisoning agent.
9. If any solution, acid, or base is spilled on you or your desk, wash the area with plenty of water at once. Baking soda (sodium bicarbonate) may be used on acid burns and boric acid on base burns. **NOTIFY YOUR TEACHER AT ONCE.**

Appendix C Weather Tables

Table C-1. Relative Humidity (%).

Dry Bulb °C	\multicolumn{10}{c}{Difference between wet and dry bulb readings in Celsius degrees}									
	1	2	3	4	5	6	7	8	9	10
10	88	77	66	55	44	34	24	15	6	
11	89	78	67	56	46	36	27	18	9	
12	89	78	68	58	48	39	29	21	12	
13	89	79	69	59	50	41	32	22	15	7
14	90	79	70	60	51	42	34	26	18	10
15	90	80	71	61	53	44	36	27	20	13
16	90	81	71	63	54	46	38	30	23	15
17	90	81	72	64	55	47	40	32	25	18
18	91	82	73	65	57	49	41	34	27	20
19	91	82	74	65	58	50	43	36	29	22
20	91	83	74	67	59	53	46	39	32	26
21	91	83	75	67	60	53	46	39	32	26
22	92	83	76	68	61	54	47	40	34	28
23	92	84	76	69	62	55	48	42	36	30
24	92	84	77	69	62	56	49	43	37	31
25	92	84	77	70	63	57	50	44	39	33

Table C-2. Wind Force

Terms used by U. S. National Weather Service	km per hour	Specifications for use on land	Beaufort number
Calm	<1	Calm, smoke rises vertically	0
Light air	2–5	Direction of wind shown by smoke draft	1
Light breeze	6–12	Wind felt on face; leaves rustle	2
Gentle breeze	13–20	Leaves and small twigs in constant motion	3
Moderate breeze	21–29	Raises dust and loose paper	4
Fresh breeze	30–39	Small trees in leaf begin to sway	5
Strong breeze	40–50	Large branches in motion	6
Moderate gale	51–61	Whole trees in motion	7
Fresh gale	62–74	Breaks twigs off trees	8
Strong gale	75–87	Slight structural damage occurs	9
Whole gale	88–101	Trees uprooted; much structural damage	10
Violent storm	102–120	Rarely experienced; widespread damage	11
Hurricane	>120	Devastation occurs	12

Appendix D

The Microscope and Its Use

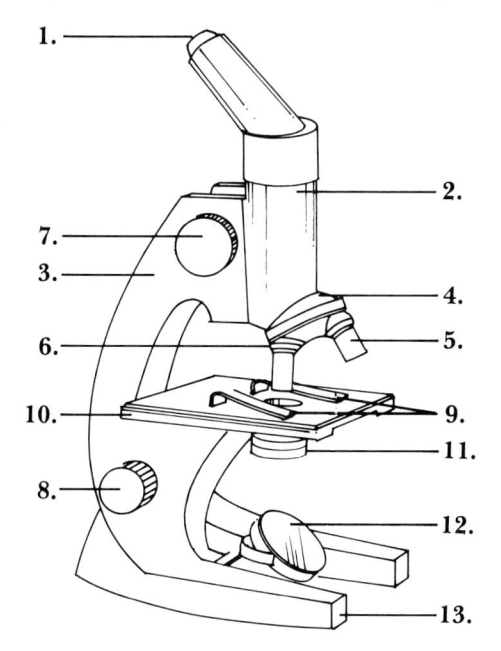

1. Eyepiece
2. Body tube
3. Arm
4. Revolving nosepiece
5. Low power objective lens
6. High power objective lens
7. Coarse adjustment lens
8. Fine adjustment knob
9. Stage clips
10. Stage
11. Diaphragm
12. Mirror
13. Base

These procedures should always be followed when using the microscope.
1. Always carry the microscope with both hands. Hold the arm with one hand. Place the other hand beneath the base.
2. Place the microscope on the table gently with the arm toward you and the stage facing a light source. The top of the table should be cleared of other objects.
3. Look through the eyepiece and adjust the diaphragm so that the greatest amount of light comes through the opening in the stage. The circle of light is called the field of view.
4. Turn the nosepiece so that the low power objective lens (10×) clicks into place.
5. Always focus first with the coarse adjustment and the low power objective lens. Raise the body tube by turning the coarse adjustment knob.
6. Turn the nosepiece until the high power objective lens clicks into place. Use only the fine adjustment with this lens. There will be less light coming through the opening in the stage. Do not raise the body tube.
7. Be sure to keep your fingers from touching the lenses.
8. Use only special lens paper to clean the lenses.
9. Before putting the microscope away, always turn the low power objective into place over the stage.
10. Raise the body tube until the low power objective is about two or three centimeters from the stage.

Appendix E
Classification of Living Organisms

	Phylum	Examples
Monera Kingdom	Schizomycophyta	Bacteria
	Cyanophyta	Blue–green algae
Protista Kingdom	Euglenophyta	Euglenas
	Chrysophyta	Golden algae, diatoms
	Pyrrophyta	Dinoflagellates
	Sarcodina	Amoebas, foraminifera
	Ciliophora	Ciliates, paramecia
	Mastigophora	Flagellates
	Sporozoa	Sporozoans
	Myxomycota	Slime molds
Fungi Kingdom	Zygomycota	Sporangium fungi—bread mold
	Basidiomycota	Club fungi—mushrooms, shelf fungi, rusts, smuts
	Ascomycota	Sac fungi—yeasts, pencillium mold
Plant Kingdom	Phaeophyta	Brown algae
	Chlorophyta	Green algae
	Rhodophyta	Red algae
	Bryophyta	Bryophytes—liverworts, mosses
	Tracheophyta	Vascular plants
	Lycopsida	Club mosses
	Psilopsida	Psilopods
	Sphenopsida	Horsetails
	Pteropsida	Ferns, conifers, flowering plants
Animal Kingdom	Porifera	Sponges
	Coelenterata	Coelenterates—jellyfish, coral
	Platyhelminthes	Flatworms—tapeworms, planaria
	Nematoda	Roundworms—hook worms
	Annelida	Segmented worms—earthworms, leeches
	Mollusca	Mollusks—snails, clams
	Arthropoda	Arthropods—insects, spiders, crustaceans
	Echinodermata	Echinoderms—starfish, sea urchins
	Chordata	Chordates—fish, amphibians, reptiles, birds, mammals

Appendix F

Measuring with the International System (SI)

The International System (SI) of measurement is accepted as the standard for measurement throughout most of the world. Four base units of the International System are listed in Table F–1.

Table F–1. SI Base Units

Measurement	Unit	Symbol
Length	Meter	m
Mass	Kilogram	kg
Time	Second	s
Temperature	Kelvin	K

Other measurement units are combinations of the base units or are considered supplementary units. Celsius temperature is a supplementary unit. The Celsius scale (°C) has 100 equal graduations between the freezing temperature (0°C) and the boiling temperature of water (100°C). The following relationship exists between the Celsius and kelvin temperature scales:

$$K = °C + 273$$

Larger and smaller units of measurement in SI are obtained by multiplying or dividing the base unit by some multiple of ten. The new unit is named by adding a prefix to the name of the base unit. Examples are given in Table F–2.

Table F–2. Common SI Prefixes

Prefix	Symbol	Multiplier	Prefix	Symbol	Multiplier
Greater than 1			Less than 1		
Mega-	M	1 000 000	Deci-	d	.1
Kilo-	k	1 000	Centi-	c	.01
Hecto-	h	100	Milli-	m	.001
Deka-	da	10	Micro-	μ	.000 001

Several units derived from the base units of SI are listed below.

Table F–3. Units Derived from SI Units

Measurement	Unit	Symbol	Expressed in base units
Energy	Joule	J	$kg \cdot m^2/s^2$
Force	Newton	N	$kg \cdot m/s^2$
Power	Watt	W	$kg \cdot m^2/s^3$ (J/s)
Pressure	Pascal	Pa	$kg/m \cdot s^2$ (N/m²)

Glossary

The glossary contains all of the major science terms of the text and their definitions. Below is a pronunciation key to help you use these terms. The word or term will be given in boldface type. If necessary, the pronunciation will follow the term in parenthesis.

PRONUNCIATION GUIDE

a . . . back (BAK)
er . . . care, fair (KER, FER)
ay . . . day (DAY)
ah . . . father (FAHTH ur)
ar . . . car (KAR)
ow . . . flower, loud (FLOW ur, LOWD)
e . . . less (LES)
ee . . . leaf (LEEF)
ih . . . trip (TRIHP)
i(i + con + e) . . . idea, life (i DEE uh, LIFE)
oh . . . go (GOH)
aw . . . soft (SAWFT)
or . . . orbit (OR but)
oy . . . coin (KOYN)

oo . . . foot (FOOT)
yoo . . . pure (PYOOR)
ew . . . food (FEWD)
yew . . . few (FYEW)
uh(u + con) . . . comma, mother (KAHM uh, MUHTH ur)
sh . . . shelf (SHELF)
ch . . . nature (NAY chur)
g . . . gift (GIHFT)
j . . . gem, edge (JEM, EJ)
ing . . . sing (SING)
zh . . . vision (VIHZH un)
k . . . cake (KAYK)
s . . . seed, cent (SEED, SENT)
z . . . zone, raise (ZOHN, RAYZ)

A

acceleration (ak sel uh RAY shun): the rate at which an object changes speed

acceleration of gravity: the rate of acceleration of a falling body; 9.8 m/s² near Earth

acid rain: rain that contains weak acids produced when air pollutants such as sulfur dioxide react with water in the air

action force: the force that produces movement in one direction

actual mechanical advantage (A.M.A.): the number of times that a machine multiplies the effort force; the resistance force divided by the effort force

air mass: a huge body of air with definite moisture and temperature properties covering a large portion of the earth's surface

alloy: a mixture of two or more metals

amino (uh MEE noh) **acids:** the building blocks of protein

amoeba (uh MEE buh): a one-celled organism that moves by changing its shape; member of the protist kingdom

amphibians: vertebrates that have moist skin and no scales; live part of their lives in water and part on land

amplitude (AM pluh tewd): distance a wave rises or falls from its rest position; height of a wave

angiosperm (AN jee uh spurm): a type of seed plant that has flowers and produces seeds inside a fruit

anticline (ANT ih kline): an upward fold of layered rocks

anticyclone: an area of air in which the pressure is higher than the air around it

Archimedes' principle: the buoyant force is equal to the weight of the fluid that is displaced by an object; when the weight of the displaced fluid is equal to the weight of the object, the object floats

area: the number of square units required to cover an area

arthropod: a phylum of animals that have segmented bodies, jointed legs, and exoskeletons; insects, spiders, crabs, shrimp, and lobsters are examples of arthropods

atmosphere: the air that surrounds Earth; extends 900 km about Earth's surface

atrium: part of the heart; receives blood returning from the lungs; pushes blood into the ventricle

B

bacteria: one-celled organisms with cell walls

barometer (buh RAHM ut ur): an instrument that measures air pressure

Bernoulli's principle: the pressure in a moving stream of fluid is less than the pressure in the fluid around it

bile: a digestive juice produced by the liver; aids in the breakdown of fat

bird: a warm-blooded vertebrate that is covered with feathers and lays eggs

blending: the inheritance of a trait that is a combination of the traits present in the parents

block cutting: a forest conservation method in which a section of tress is cut from a forest; a method of improvement cutting

Boyle's law: the pressure of a gas increases as its volume decreases, provided temperature remains constant

breed: a group of plants or animals that have similar traits and a common ancestor

buoyancy (BOY uhn see): the upward force exerted by a fluid on objects on or within the fluid; equal to the mass of the fluid displaced by the object

C

cambium: the growth tissue of a plant stem; produces xylem and phloem cells

carbon monoxide: a colorless, odorless gas that is toxic to humans; an air pollutant

cartilage (KART uh lihj): dense, strong, rubbery tissue present in most joints; lacks the minerals that make bones hard

cell: the basic unit of structure and function in all living organisms

centripetal (sen TRIHP ut al) **force:** the force that keeps an object moving in a circular path; acts towards the center of the circle

Charles' law: the volume of a gas increases when temperature is increased; if the volume is constant, the pressure of a gas increases when temperature is increased

chemical change: change in the chemical properties of a substance such that new substances with different properties are produced

chlorophyll: a green substance in plant cells that is used to make food

chromosome: thin, threadlike bodies that are formed by the chromatin of a nucleus; contain genes that control the inheritance of different traits

clear cutting: the removal of all trees from a very large area; a method of lumbering

climate: the average weather for a region over a period of years

climax community: the end result of succession in a community

cloud ceiling: the altitude at which the cloud cover becomes broken or overcast

coefficient: the number in front of a chemical symbol or formula; used to balance an equation

cold front: a boundary that forms when a cold air mass moves against a warm air mass

community: all living things in a certain area

compound: a substance formed when two or more elements are chemically joined

condensation: the change from a gas to a liquid; opposite of evaporation

conservation (kahn sur VAY shun): wise and careful use of natural resources

conservation of momentum: momentum cannot be created or destroyed

consumers: animals in a food web that eat plants or other animals; cannot make food

continental margin: the portion of the seafloor that lies next to the continents

continental shelf: a gently sloping portion of the ocean floor that borders the coastlines of a continent

contour line: a line drawn on a map joining all points of Earth's surface having the same elevation

contour plowing: plowing back and forth across a sloped field; produces horizontal ridges in the soil that reduce runoff; a soil conservation method

control: part of an experiment that is held constant

convection current: upward and downward movements of air in the atmosphere that are caused by density differences

core: the center section of Earth; consists of a solid inner region and liquid outer region

covalent (koh VAY lunt) **bond:** the chemical bond between atoms in which electrons are shared

crop rotation: changing the crops that are grown in given fields from year to year; a soil conservation method

crossbreed: a mixed breed; cross between two different breeds

crust: the outer layer of Earth; a thin, rocky skin around the mantle; consists of all the continents and ocean floors of Earth's surface

cubic meter (m^3): the basic unit of volume in SI

cyclone: a region of low air pressure caused by rising air at the center of the region

cytoplasm (SITE uh plaz um): the living, jellylike material found between the cell membrane and nucleus

D

deceleration (dee sel uh RAY shun): negative acceleration

declination: the angle formed with the geographic north pole when a compass points to the magnetic north pole

decomposer: an organism that causes the decay or breakdown of dead plants and animals

density (DEN sut ee): mass per unit of volume; expressed in g/mL or g/cm³

descent and change: living things may change as they pass from one generation to the next

dew point: the temperature of air at which water vapor first begins to condense

diffusion (dihf YEW zhun): movement of particles from regions of higher concentration to regions of lower concentration of that material

dominant (DAHM uh nunt) **species:** groups of living things that are present in greatest numbers

dominant trait: the trait that determines how a plant or animal will appear

drag: the force of air against any object that moves through it

E

ecology (ih KAHL uh jee): the study of relationships between organisms and their environments

efficiency (ih FISH uhn see): comparison of a machine's work output to the work input

effort force: the force applied to a lever

electron (ih LEK trahn): a negatively charged particle that makes up part of an atom; has a very small mass and is located outside the nucleus

element: a substance that cannot be broken down into simpler substances by a chemical change

embryo (EM bree oh): an organism in its early stages of life

emigration (ehm ih GRAY shun): movement out of a community

endangered species: a species of plants or animals whose numbers are so small that it could easily become extinct

energy: the capacity to do work

energy pyramid: a structure that shows how energy is used in a food chain; each successive level of the pyramid has less energy

engine: a machine that produces mechanical power

environment (in VI ruhn ment): sum total of the surroundings of an organism or population

epicenter: the point on Earth's surface directly above the focus, or exact location, of the earthquake center

epidermis: the outer, protective layer of a leaf

era: a major division in the history of Earth

erosion (ih ROH zhuhn): the process by which rocks and soil are worn away by wind, moving water, and moving ice

evaporation: the change from a liquid into a gas; opposite of condensation

evolution: a change in the hereditary features shown by a population of organisms from one generation to the next

excretion: the process by which wastes are removed from an organism

exoskeleton: a skeleton on the outside of an animal's body; protects the soft, internal body parts

exosphere (EK so sfihr): the layer of the atmosphere farthest from Earth; begins at about 600 km and extends outward

experiment: a method used to seek an answer to a question or to test a hypothesis

extinct: no longer existing

F

fault: a break in Earth's crust along which movement occurs

fertilization: the joining of egg and sperm cells

fertilizer: substance used to increase the organic matter and mineral nutrients in soil; makes soil more fertile and aids plant growth

fission: a process by which bateria and other living organisms reproduce; division into two parts

flatworm: a type of worm, flat in shape; planarian and tapeworm are examples

fluid: a substance that can move and change shape without separating

focus: true center of an earthquake

food chain: a series of animals feeding on plants and/or other animals

food web: a complex feeding system that contains more than one chain

force (F): any push or pull on an object

formula: an abbreviation for writing the name of a compound

fossil: preserved remains or traces of plants and and animals that lived long ago; any evidence of past life

fossil fuel: energy sources that come from the preserved remains of once living things; coal, oil, gasoline, and natural gas are fossil fuels

friction (FRIHCK shun): a force that slows down or prevents motion

fulcrum: the point at which a lever rotates

fungi: organisms such as yeasts, molds, and mushrooms that do not contain chlorophyll and cannot make their own food

G

game laws: laws that protect wild animals from exploitation and hunting

gas: that state of matter that has mass and occupies space, but does not have definite shape or volume

gene: a body in a chromosome that controls the inheritance of different traits; a unit of inheritance

genus: a group of closely related species

geology (jee AHL uh jee): the scientific study of Earth and its structure, formation, and history

geothermal (jee oh THUR mul) **energy:** energy that is generated from heat within Earth

germination: the early growth of an embryo plant

green manure crop: a crop that is grown as a cover and plowed under before maturity

groundwater: water in the ground

gymnosperm (JIHM nuh spurm): a type of seed plant where seeds are not enclosed in a fruit

H

habitat: the place where a species of plant or animal normally lives

heredity: the passing of traits or characteristics from parents to offspring

host: the organism that is lived on by a parasite

humus: material formed from the decay of dead plants and animals; found in topsoil

hurricane: a tropical cyclone; a low pressure area that forms in the trade winds, over regions of warm ocean water

hybrid: offspring from a cross between organisms that have different traits; contains the combined traits of two or more organisms

hybrid vigor: hardier and more disease resistant; a feature of crossbred plants and animals

hydroelectric (hy droh ih LEK trik) **energy:** energy that is generated by moving water

hypothesis (hi PAHTH uh sus): a possible solution for a problem; formed after studying facts and ideas relating to the problem

I

ideal mechanical advantage (I.M.A.): the number of times that a machine multiplies the effort force, not including the friction and weight of the lever; the effort arm length divided by the resistance arm length

immigration: movement into the community

improvement cutting: the removal of unwanted trees in a forest; a forest conservation practice

inclined plane: a slanted surface which may be used for raising objects to higher places; a simple machine

index fossil: a fossil used to identify specific rock layers; a guide fossil

inertia (ihn UR shuh): the tendency of matter to stay at rest or in motion, unless acted on by a force

International Date Line: the opposite side of Earth from the prime meridian; the 180° line of longitude

International System of Units (SI): the modern form of the metric system

invertebrate: animals that do not have a backbone or internal skeleton

ion: an atom with an electric charge; produced by the loss or gain of one or more electrons of an atom

ionic (i AHN ink) **bond:** a chemical bond formed by transferring electrons from one atom to another

isobar: a line on a weather map that connects points of equal air pressure

J

jet stream: swift, forceful winds that blow from west to east in the upper troposphere

joule (JEWL) **(J):** the metric unit used to measure work and energy; equivalent to newton-meter (N · m)

K

kilogram (kg): the basic unit of mass in SI; 1000 grams

kinetic energy: the energy of motion

kingdom: one of five main groups in the classification of all living organisms

L

large intestine: an organ through which wastes move to be excreted from the body

latitude (LAT uh tewd): imaginary lines on a map or globe of Earth that run east and west parallel to the equator

lava: magma that reaches Earth's surface; turns into rock upon cooling

law of conservation of energy: energy can change forms, but cannot be created or destroyed under ordinary conditions

law of dominance: theory that a trait from one parent may cover or mask a trait from another parent

law of uniformity: past changes in Earth were caused by processes that still exist today

leaching: the removal of mineral nutrients from topsoil by water; reduces fertility of soil

lever: a bar that is free to move about a fulcrum; a simple machine

lichen (LI kun): an alga and fungus living together; example of symbiosis

lift: the upward force on the wings of an airplane that allows it to fly; the unequal pressure pushing upward, due to the shape of the wing

liquid: the state of matter that has definite volume, but not definite shape

liter (L): the basic unit of volume in SI; used to measure liquids

liver: the digestive organ that produces bile

longitude (LAHN juh tewd): imaginary lines on a map or globe of Earth that run north and south through the poles; also called meridians

longshore current: a wave that hits the shore at an angle and produces a flow of water along the shore

M

machine: a device that can make work easier by changing the direction or amount of force

magma: molten rock material within Earth

magnetic north pole: a location about 1670 km south of the geographic north pole approximately along the 90° line of longitude; near Bathhurst Island of northern Canada

magnetic south pole: a location about 2670 km north of the geographic south pole approximately along the 140° line of longitude; near the coast of Antarctica

mammals: warm-blooded vertebrates that have hair and feed their young with milk

mantle: the region that surrounds Earth's outer core; consists mainly of solid rock that is somewhat fluid in areas

mass: the amount of matter in an object

meridians (muh RIHD ee unz): imaginary lines on a map or globe of Earth that run north and south through the poles; also called lines of longitude

mesosphere (MEZ uh sfihr): the middle layer of the atmosphere; lies above the stratosphere and extends about 50 to 80 km above Earth

metabolism: the sum of all chemical processes in living things that keep them alive

metamorphosis (met uh MOR fuh sus): series of changes in the lives of animals such as insects and frogs

meteorology (meet ee uh RAHL uh jee): the scientific study of weather

meter (m): the basic unit of length in SI

microclimate: climate in a small area

microorganism: a living organism so small that it can be seen only with a microscope

migratory bird refuge: an area created to maintain the population of birds that travel seasonally

mineral: a naturally occuring, crystalline substance that has a definite chemical composition

mixture: a substance that contains two or more elements or compounds that are mixed together but not chemically joined

molecule: particles that form when atoms are joined by covalent bonds

mollusk: soft-bodied animal that may or may not have shell or shell-like covering; snails, oysters and clams are examples of mollusks

momentum (moh MENT uhm): quantity of motion; the product of the mass of an object and its speed

monera: one of the five kingdoms of living organisms

mutation (myew TAY shun): a change in a gene or chromosome that causes a change in an inherited trait

N

natural resources: things in the environment that are useful to people

natural selection: the process by which organisms best adapted to their environments are able to survive and produce fertile offspring

neutron: a subatomic particle located in the nucleus that has no electric charge

newton (N): a unit of force; amount of force needed to accelerate one kilogram at a rate of one meter per second squared

nuclear fission: the process of splitting an atomic nucleus into two or more parts; energy is released

nuclear fusion: the process of combining two or more atomic nuclei to form one larger nucleus; opposite of fission, but energy is also released

nucleus: (1) the central part of an atom; (2) the control center of a cell

O

occluded front: a region where a cold front approaches and overtakes a warm front; characteristic of both warm and cold fronts

ocean-basin floor: mostly flat regions of the ocean floor along with mid-ocean ridges and ocean trenches

oceanography (oh shun AHG ruh fee): the scientific study of the oceans.

organ: groups of tissues working together to perform a function

organism: a whole and complete living thing

osmosis (ahs MOH sis): diffusion of water through a membrane

P

paleontology (pay lee ahn TAHL uh jee): the study of prehistoric living things through fossils

pancreas: a digestive organ that produces enzymes that pass into the small intestine

paramecium: a one-celled organism, shaped like a shoe, that moves by the beating action of short, hairlike structures called cilia; a member of the protist kingdom

parasite: an organism that lives on or in another organism; a type of consumer

period: (1) a time division of an era: (2) time between the passage of two successive crests of a wave

permanent wilderness area: regions untouched by civilization; an area that preserves the wild state of nature

phloem (FLO em): plant tissue that transports food substances downward

photosynthesis: the process by which green plants absorb solar energy to produce food for the growth and repair of tissues

physical change: the change in a substance from one form to another without a change in composition

pistils: reproductive parts of the flower that contain the ovary

plankton: a type of tiny animal and plant life in the sea; the first link in the oceanic food chain

plasma: the state of matter that is extremely hot and composed of electrical properties

plate: a large section of Earth's crust that is slowly moving in a specific direction

plate tectonics: the theory explaining formation of Earth's surface structures by movement of large crustal plates

pollen: tiny grains produced by the stamen of a flower that contain nuclei for reproduction

pollination: the transfer of pollen from stamen to pistil; occurs within the same flower or between flowers

pollution (puh LEW shun): waste products in the air, water, and on land

population (pahp yah LAY shun): the total number of any one species living in a community

potential energy: the energy of position, or stored energy

power: the rate at which work is done; the amount of work per unit of time

precipitation: moisture that falls from the atmosphere

predator: mammals that eat flesh of other animals; usually kill their prey

pressure: the amount of force per unit area; the unit of pressure is the pascal (Pa)

prey: an animal that a predator uses for food

prime meridian: the line of longitude that passes through Greenwich, England; 0° longitude

producer: an organism in a food web that makes its own food

protein: the basic material of living organisms; made up of amino acids

protist: one of the five kingdoms of living organisms; includes bacteria and other microscopic organisms

proton: a positively charged particle that makes up part of the atom; located in the nucleus of the atom

protoplasm: the "living material" of a cell

pulley: a wheel that turns on an axle; a simple machine

purebred: a member of a breed

R

radial symmetry: the form and shape of a body whose similar parts extend out from the center

radioactive dating: a method for finding the age of a rock, based on the rate of decay of certain radioactive elements

reaction force: the force equal in size and opposite in direction to an action force

recessive trait: a hidden or masked trait that is present

reduction division: a type of cell division in which the chromosome number of the dividing cells is reduced by half

reforestation: the renewing of a forest by seeding or planting young trees

relative humidity: the percent of water vapor in air, based on the amount the air can hold at a given temperature

reptiles: vertebrate animals that are cold-blooded, breathe air, and live mainly on land

resistance (rih ZIHS tunts) **force:** the force that is overcome when work is done

respiration: the release of energy from the breakdown of food within the cell; process through which cells receive oxygen and release carbon dioxide

revolution: the movement of Earth in a curved path or orbit around the sun; the movement of any body in a circular path around a central point

rhizomes: an underground stem of a plant

rip current: a flow of water that occurs when longshore currents go back to sea

rotation: the turning of Earth about its axis; turning motion of any body on its axis

roundworm: a type of worm, rounded and tubelike in shape; hookworms are roundworms

S

salinity (say LIHN ut ee): the amount of dissolved salts in a given mass of water; a measure of water's saltiness

scavenger (SKAV un jur): an animal that feeds on dead animals; a type of consumer

screw: a circular inclined plane; a simple machine

seismic sea wave: a wave produced by an earthquake

seismograph (SIZE muh graf): a device used to record shock waves sent out by earthquakes

selection: a process used to improve plant and animal breeds; choosing only certain individuals for reproduction

selective cutting: a forest conservation method where only mature trees are cut; a type of improvement cutting

sepals: green, leaflike structures on the underside of a flower

shelter belt: thick rows of trees that break the wind and prevent the erosion of soil

small intestine: a tubelike organ where most of food digestion and absorption takes place

smog: a mixture of smoke and waste gases; product of air pollution

society: a group of individuals of the same species that live together in an organized way

solar energy: energy that comes from the sun

solid: the state of matter that has definite volume and definite shape

specialized cell: a cell that performs a particular job

species: a single, distinct group of living things

speed: the distance an object travels per unit of time; refers to how fast an object travels per unit of time

spinal cord: a long bundle of nerves that connects the brain to other parts of the body

spiracle: tiny opening in exoskeleton of an insect; allows air to enter the body

sporozoan (spor uh ZOH un): a one-celled organism that obtains food from the organism in which it lives; member of the protist kingdom

stamens: reproductive parts of a flower that produce pollen

stimulus: anything in the environment that causes a change in the behavior of a living thing

stoma: small, porelike structure of a plant's epidermis

stratosphere (STRAT uh sfihr): the second layer of the atmosphere; lies above the troposphere and extends 11 to 50 km above Earth

strip-cropping: planting strips of grasses and legumes between strips of row crops to prevent erosion; a soil conservation method

sublimation (sub luh MAY shun): the physical change in a substance from a solid to a gas or a gas to a solid

subscript: the small number in a formula that shows how many atoms are in a molecule of a substance

subsoil: the lower layer of soil; underneath topsoil

succession: the process of slow, gradual change in a community

surface tension: forces that hold particles in a liquid's surface together

surge: a mass of water associated with a storm that produces an abnormal rise in sea level along the seacoast

symbiosis (sihm bee OH sus): two organisms living together in common effort so that both will benefit

symbol: a shorthand method or abbreviation for writing the name of an element

syncline (SIHN kline): a downward fold of layered rocks

system: organs that work together to perform one main function

T

temperature: the average kinetic energy of the molecules of a substance

terminal speed: the speed reached when air resistance against an object is equal to the pull of gravity; the point at which a falling body stops accelerating

terracing: a method of contour plowing on very steep slopes; results in very wide steps; a soil conservation method

territory: a certain area defended by animals

thermosphere (THUR muh sfihr): the fourth layer of the atmosphere; lies above the mesosphere and extends from 80 km to about 600 km above Earth

tide: a periodic rise and fall of the ocean along a seacoast

time zone: a geographical region within which the same standard time is used

tissue: a group of cells performing the same function

topographic (tahp uh GRAF ihk) **map:** a map of Earth's surface that shows the landscape, including the heights of mountains and depths of valleys

topsoil: the uppermost layer of soil

tornado: a violent storm characterized by a funnel-shaped cloud and associated with very strong thunderstorms; a region of whirling wind that circles around a low pressure center about 60 km in diameter

transpiration: the loss of water through the leaves of a plant

troposphere (TROHP uh sfihr): the first layer of the atmosphere; extends from the ground to an average height of 11 km above Earth

tuber: a rhizome that is enlarged because of the storage of food in its cells

U

unconformity: a buried erosion surface separating two rock masses of which the older was eroded before the younger was deposited

undertow: the flow of water back to the ocean, underneath the incoming waves

V

ventricle: part of the heart; receives blood from the atrium; pushes blood through vessels into the lungs and other body parts

vertebrate: an animal that has an internal skeleton

virus: the smallest living thing known; made of a DNA or RNA core; can cause disease

volume: (1) the space occupied or filled by an object; (2) loudness of a sound

W

warm front: a boundary that forms when a warm air mass moves against a cold air mass; warm air slides forward and above the cold air

watershed: a region from which all runoff water drains into the same main body of water

watt (W): the unit of power; equivalent to 1 joule per second (J/s)

wavelength: distance between two corresponding points on consecutive waves; horizontal distance from one crest to another

wave period: the time between the passage of two successive wave crests

weather: condition of the atmosphere at a certain time and space

weathering: the breakdown of rock by the action of water, ice, plants, animals, and chemical changes

wedge: a type of inclined plane

weight: the measure of the force of gravity that Earth exerts on every object on or near its surface

wheel and axle: a wheel attached to an axle; a simple machine

wildlife refuge: an area in which wild animals and their habitats are protected

wind: the movement of air in the atmosphere

work (W): a force acting through a distance

X

xylem: plant tissue that transports water and dissolved minerals upward

Z

zygote (ZI goht): a fertilized egg cell

Index

Index

A

Acceleration, 111-113; rate of, 112
Acid rain, 484
Action force, 114; *act.*, 114; 115
Adult insect, 355
Africa, 173
African sleeping sickness, 308
Agar, 280
Aileron, 140; 141
Air, 25, 35, 36, 221-222; composition of, 221, 222; density of, 136; moisture in, 231-233; saturated, 232; *act.*, 10; 227; 228; 233; 235; *illus.*, 222; 223; 224; *table*, composition, 222
Air mass, 243-246; *illus.*, 244; 245; 246
Air pollution, 484, 485; *act.*, 485; *illus.*, 484
Air pressure, 224-225, 230, 247; affecting weather, 247-249; average, 251; *act.*, 225; 235; *illus.*, 224; 225; 248
Air resistance, 116
Air sac, *illus.*, 357
Airplane, 139-141; *illus.*, 140; 141
Albinism, 444; *act.*, 445
Alcohol, 302
Alfalfa, 405
Algae, 280, 317-318, 422; *illus.*, 318; 452
Algin, 280
Alligator, 375; *illus.*, 373
Alloy, 36
Alps Mountains, 173
American plate, 172
Amino acid, 451
Ammonite, 210; *illus.*, 210
Amoeba, 305-306; *act.*, 307; *illus.*, 305; 306; *table*, classification, 298

Amphibian, 368-373; *illus.*, 368; 369; 370; 371; 388
Amplitude, 274; *illus.*, 274
Anal spot, 306
Aneroid barometer, 224; 255; *illus.*, 225
Angiosperm, 321; *illus.*, 321
Animal, 411, 412, 417; amphibian, 368-373; arthropod, 353-358; bird, 375-377; classification of, 297-311, 341-379; cold blooded, 365, 368; domestic, 378; earthworm, 346-349; evolution of, 459-465; fish, 364-368; flatworm, 344-345; inherited traits of, 429-446; jellyfish, 343; mammal, 378-380; mollusk, 351-353; reptiles, 373-375; roundworm, 345; spiny-skinned, 350-351; sponge, 342-343; *act.*, birds, 377; blood circulation, 367; cell division, 440; earthworm, 349; fish, 368; frog, 372, 373; hydra, 344; insect, 358; *illus.*, bat, 379; bird, 376, 377; blending, 435; camouflage, 384, 385; cat, 364; crab, 353; earthworm, 346, 347, 348; fish, 365, 366, 367; flatworm, 345; frog, 369, 370, 371; insect, 354, 355, 356, 357; jellyfish, 343, 344; killer whale, 378; octopus, 352; parasite, 345; platypus, 379; reptiles, 373, 374; roundworms, 345; salamander, 369; scorpion, 354; sea urchin, 350; shark, 365; shrew, 378; snail, 352; snake, 342; sponge, 343; starfish, 350, 351; tadpole, 369; toad, 368; *table*, chromosomes, 437; classification, 298; fruit flies, 432

Animal breeding, 445-446; *illus.*, 445
Animal cell, 294; *illus.*, 294
Animal fossil, 200-213; *illus.*, 200; 201; 203
Annual, 324; *illus.*, 324
Annual growth ring, 325; *act.*, 327; *illus.*, 325
Antenna, *illus.*, 356; 357
Anticline, 184; *act.*, 185
Anticyclone, 248; *illus.*, 248
Anus, *illus.*, 346; 370
Apatite, *table*, 175
Appalachian Mountains, 184
Archaeopteryx, 211; *illus.*, 211
Archimedes' principle, 135
Area, 14; *act.*, 15; 16
Arthropod, 353-358; *illus.*, 353; 354; 355; 356; 357
Asexual reproduction, 301, 303, 320
Asia, 173
Atlantic Ocean, 173, 271
Atmosphere, 215, 222-223; heating of, 226, 229
Atom, 30-31
Atomic clock, 60; *illus.*, 60
Atrium, 371
Audubon, John, 499
Axis, 151
Axle, 91

B

Backbone, 363, 371
Bacteria, 300-302; *act.*, 302; *illus.*, 300; 301; 452; *table*, classification, 298
Bald eagle, *illus.*, 376
Barometer, 224-225; *act.*, 225
Bat, *illus.*, 379
Bay of Fundy, 277; *illus.*, 278
Beef tapeworm, 420; *illus.*, 421
Beetle, *illus.*, 355

Bernoulli's principle, 138-139; *act.,* 139
Bile, 366
Bird, 212, 375-377; *act.,* 377
Birthrate, 395, 396
Blending, 435; *act.,* 436; *illus.,* 435; 436
Block cutting, 498
Blood, 356, 366, 371
Blood vessel, *illus.,* 356
Blue-green algae, *illus.,* 452
Boiling, 51
Bone, 366, 374
Bony fish, 365, 366
Boundary community, 402
Boyle's Law, 134
Brachiopod, *illus.,* 207
Brain, 363, 371, 378; *illus.,* 347; 357
Bread mold, *illus.,* 303
Breaker, 275; *illus.,* 275
Breed, 446
Breeding, 445-446; *illus.,* 445; 446
Bristle, *illus.,* 346
Bud, 326; *illus.,* 326
Budding, 303
Bulb, 326; *illus.,* 326
Buoyancy, 134-137; *act.,* 137; *illus.,* 134; 135; 136; 137
Burning, 53; *act.,* 53
Butterfly, 355; *illus.,* 355

C

Calcite, *table,* 175; 176
Cambium, 325; *illus.,* 322; 325
Cambrian Period, 207; *table,* 204
Camouflage, 384-385
Carbon dioxide, 50, 215, 296, 328; *act.,* 331
Carbon monoxide, 484
Carboniferous Period, 208; *illus.,* 204; 209
Cardinal, *illus.,* 376
Career, in science, 6-8; *act.,* 8; *table,* 7; **conservation,** 476
Carrageenan, 280
Cartilage, 364
Cat, *illus.,* 364
Catalytic converter, 485
Caterpillar, 355

Cell, 293-295; animal, 294; guard, 328; plant, 294; structure of, 293-295; *illus.,* 294; *table,* protoplasm, 294
Cell activities, 295-297; *act.,* osmosis, 297; *illus.,* respiration, 296
Cell membrane, 293, 295, 305; *illus.,* 294
Cell wall, 294; *illus.,* 294
Cellulose, 294
Celsius temperature scale, 48; *table,* 48
Cenozoic Era, 204, 212-214; *table,* 204
Centimeter, 12
Centipede, 353
Central nervous system, 371
Centripetal force, 117-118
Chain reaction, 56
Charles' Law, 133
Chemical change, 52-53
Chemical compound, 32-33; formula of, 34; *act.,* 34; *table,* 34
Chemical equation, 54-55
Chemical formula, 34; *table,* 34
Chemical symbol, 29, 34; *table,* 29
Chlorine, 262
Chlorophyll, 317, 327
Chloroplast, 318; *illus.,* 294; 329
Chordate, 363
Chromatin, 436-437
Chromosome, 437, 438, 439, 440, 441, 444; *illus.,* 437; *table,* 437
Cilia, 306; *illus.,* 306
Circular motion, 117-119; *act.,* 119
Circulatory system, 341, 347, 350, 366, 371
Cirrus cloud, 234; *illus.,* 234
Class, 298
Classification, 297-311; *table,* kingdoms, 298
Clear cutting, 496
Climate, 254-255; *illus.,* 254; 255
Climax community, 400-401
Cloud, 234-235; precipitation from, 236; reflecting solar energy, 226; and thunderstorms, 246-247; types of, 234-235; *act.,* 235; *illus.,* 231; 234; 235
Cloud ceiling, 241
Coal, 208, 209
Cocoon, 348, 355
Coefficient, 54
Cold-blooded animal, 365, 368
Cold front, 245-246, 249; *illus.,* 246
Colorado River, 196
Community, 389-390, 419; boundary, 402; **climax,** 400-401; competition in, 397-399; and ecology, 404-405; food chain in, 411-424; population growth in, 390-397; *act.,* 393; 399; 401; *illus.,* boundary, 402; climax, 400; pond, 390; succession, 401; tidepool, 390
Compass, 162
Competition, 397-399
Compound, 32, 33; *table,* 33, 34
Compound eye, *illus.,* 356
Compound machine, 98-99
Condensation, 51; *act.,* 52; *illus.,* 231
Conglomerate, 182
Conifer, 321
Conjugation, 307; *illus.,* 307
Conservation, of the forest, 493-498; of natural resources, 475-485; *act.,* 502; pollution, 485; soil, 479, 481; *table,* careers, 476
Conservation of energy, law of, 75
Conservation of momentum, 121
Consumer, 414, 417, 420; *act.,* 415
Continent, 173; *act.,* 173
Continental margin, 268
Continental shelf, 268
Contour line, 157
Contour plowing, 479-480; *act.,* 481; *illus.,* 480
Control, 11
Convection current, 226; *act.,* 272; *illus.,* 226

Cooperation, 403
Core, 169
Corundum, *table,* 175; 176
Cotyledon, 333; *illus.,* 333
Cougar, *illus.,* 299
Covalent bond, 33; *illus.,* in water, 33
Crab, *illus.,* 353
Crayfish, *illus.,* 353
Crest, 275
Cretaceous Period, 211; *table,* 204
Crop, *illus.,* 346
Crop rotation, 480
Crossbred, 446
Crust, 170, 174, 180, 184-185
Crustacean, 353; *illus.,* 353
Crystal, 180
Cubic meter, 15
Cumulonimbus cloud, 235, 246-247; *illus.,* 235; 246; 247
Cumulus cloud, 234; *illus.,* 234
Cuticle, 327; *illus.,* 328
Cyclone, 247-248, 249; *illus.,* 248
Cyst, 306
Cytoplasm, 293, 294, 305; *illus.,* 294

D

Darwin, Charles, 453-454; *illus.,* 454
Darwin's theory, 453-455; main points of, 461
Dead Sea, 263
Deceleration, 112-113
Declination, 162
Decomposer, 414; *illus.,* 414
Density, 26-28, 224, 226; of seawater, 264-266, 272; *act.,* 28; 137; 266; *illus.,* 135; *table,* 27
Desalination plant, *illus.,* 483
Desert cactus, *illus.,* 455
Devonian Period, 208; *table,* 204
Dew point, 231; *act.,* 233
Diamond, *table,* 175
Diatom, *illus.,* 318
Diesel engine, 78

Diffusion, 295-296, 305; *illus.,* 296
Digestion, 292, 305
Digestive system, 342, 346, 347, 350, 356, 366, 370
Dinosaur, 210 211; *illus.,* 210; 211
Disease, 309, 345, 393, 396, 397, 498; African sleeping sickness, 308; Dutch elm, 498; malaria, 309; viral, 310
Division of labor, 403
DNA, 295, 311, 438, 444; *illus.,* 295
Dolomite, 182, 183
Dominant species, 391, 400
Dominant trait, 430-438
Drag, 141, 275
Drift bottle, 272
Drone, 403
Dry ice, 50
Duck Stamp Act, 503
Duck-billed platypus, 379; *illus.,* 379
Dutch elm disease, 498

E

Ear, 371
Eardrum, 371
Earth, age of, 216; as a magnet, 161-163; mapping of, 156-158; mass of, 150; motion of, 151-152; shape of, 149-150; size of, 149-150; structure of, 169-171; *act.,* mapping, 156, 157, 158; motion, 154; shadows, 152; *illus.,* 148; 150; 151; 160; 161; 163; 170; *table,* 152
Earthquake, 172, 186-188, 279; *illus.,* 186; 187; *table,* 186
Earth's future, 214-215
Earthworm, 346-349; *act.,* 349; *illus.,* 346; 347; 348; 349
Easterlies, 227; *illus.,* 227
Ecologist, 404
Ecology, 404-405
Efficiency, of engine, 78; of simple machines, 94-97
Effort force, 84, 85, 86, 91, 93

Egg, 320, 331, 348, 355, 358, 368, 374-375, 439, 440, 442, 443; *illus.,* 439; 440
Egg nucleus, 332
Ekman current meter, 272
Electric generator, 57
Electron, 30; and bonding, 32, 33; *table,* 31
Element, 29; *table,* 29; 33
Elevator, 140
Embryo, 332, 333, 442, 443; *illus.,* 442
Emigration, 395
Endangered species, 499; *table,* 500
Endodermis, *illus.,* 322
Endosperm, 333, 334
Endospore, 301
Energy, 45-47; for cell activities, 292, 296; conservation of, 75; from digestion, 292, 305; efficiency of engine, 78; in food web, 412; heat, 47-49, 53, 128; kinetic, 45-46, 75, 133, 278; and photosynthesis, 329; potential, 45, 75; solar, 58-59, 226; sources of, 56-59, 278; and work, 75; *act.,* 47; 49; 59; *table,* 45
Energy level, 30, 31; and bonding, 32, 33; *table,* 31
Energy pyramid, 416-418; *act.,* 418; *illus.,* 417
Engine, 77-78; Hero's, 114; *illus.,* 78
Entomologist, 6-7
Environment, 392, 394, 404-405, 454, 455, 458, 462; affecting roots, 323; *act.,* 394
Enzyme, 346, 370
Eohippus, 212; *illus.,* 212
Epicenter, 188; *illus.,* 188
Epidermis, 327; *illus.,* 328
Equator, 153, 155
Era, 204-214; *illus.,* 205; *table,* 204
Eratosthenes, 149-150
Erosion, 178, 181, 197-198, 478-479; *act.,* 479
Esophagus, *illus.,* 346
Euglena, 308; *illus.,* 309
Eurasian plate, 172, 184

Europe, 173
Evaporation, 51, 231; *act.,* 52; *illus.,* 231
Everglades, *illus.,* 503
Evolution, main points, 461; theory of, 451-465; *illus.,* 460; 461
Excretion, 347
Exoskeleton, 353, 364
Exosphere, 223; *illus.,* 223
Experiment, 9-11; controlled, 11
Extinct, 456
Eye, 371

F

Fall, 153
Family, 298
Fault, 184-185, 186
Faulting, *illus.,* 184
Feldspar, 175
Felis concolor, 299
Felis paradalis, 299
Felis sylvestris, 299
Fern, 320; *illus.,* 320
Fertilization, 332, 348, 358, 368, 377, 440; *illus.,* 440
Fertilized egg cell, 442, 443
Fertilizer, 477
Fish, 364-368; *act.,* 367; 368; *illus.,* 365; 366; 367
Fission, 306, 307, 317; *illus.,* 306; 307
Flagella, 300, 308; *illus.,* 300
Flagellate, 308; *illus.,* 309
Flatworm, 344-345; *illus.,* 345
Flight, 140-141
Flounder, *illus.,* 365
Flower, 331-333; 431; complete, 331; incomplete, 331; *illus.,* 331; 332
Fluid, 127-138; Bernoulli's principle of, 138-139; buoyancy of, 136; used for cooling, 128; *act.,* 129; 137; 139
Fluorite, *table,* 175; 176
Focus, 188
Fog, *illus.,* 234
Folding, *illus.,* 184
Food, 296, 305, 306, 318, 322, 327, 328, 333, 395, 396

Food chain, 411-424; *act.,* 415; *illus.,* 412; 413; 414; 419; 420; 421
Food vacuole, 305, 306; *illus.,* 306
Food waste, 347
Food web, 413; *act.,* 415; *illus.,* 413
Force, 67-73, 82, 112, 113, 131; action, 114; buoyancy, 134-135; centripetal, 117-118; effort, 84, 85, 86, 91, 93; of friction, 72; geologic, 197, 198; lift, 139-140; reaction, 114; resistance, 84, 85, 86, 93; and work, 73-75; *act.,* 68; 70; 114; 115; buoyancy, 137; friction, 72; *table,* 69
Forest conservation, 496-498
Forest resource, 493-498; *act.,* 495; 497; *illus.,* 494; 496; 497
Formula, 34; *table,* 34
Fossil, 57, 197, 200-203, 205, 206, 207, 208, 209, 210, 211, 212; *act.,* fossil formation, 202; pertrified wood, 201; *illus.,* 200; 201; 203; 207; 208; 209; 210; 211; 212; *table,* geologic history, 204
Fossil evidence, 456-457
Fossil fuel, 56, 57-58, 209, 215
Fossil record, 202-203
Foucault pendulum, 151; *illus.,* 151
Freezing, 50
Friction, 70-73; *act.,* 72
Frog, 368-373; reduction division in, 439-440; *act.,* 372; 373; *illus.,* 369; 370; 371; 388; eggs, 369; heart, 371
Front, 245-246; cold, 245-246; occluded, 246; warm, 245; *act.,* 252; *illus.,* 245; 246
Fruit, 321, 333
Fruit fly, *act.,* 432; *illus.,* 430
Fuel, 208, 209, 215; efficiency, 78; fossil, 56, 57-58, 209, 215
Fulcrum, 84
Fungi, 302-304, 414, 422; *act.,* yeast, 304; *illus.,* 303; 414; *table,* classification, 298
Fusion, 56, 58; *illus.,* 56

G

Galapagos Islands, *illus.,* 453
Galena, *table,* 176
Gas, 25, 36; *act.,* 10; *table,* 26
Gas laws, 133-134
Gasoline engine, 77-78
Geese, *illus.,* 376
Gene, 311, 437, 438, 440, 444
Genus, 298, 299
Geologic clock, 199
Geologic time, 204-215; Cenozoic Era, 212-213; earth future, 214-215; Mesozoic Era, 210-211; Paleozoic Era, 206-209; Precambrian Era, 205-206; *illus.,* 205; 207; 208; 209; 210; 211; 212; 213; 214
Geologist, 6
Geology, 169
Geothermal energy, 57
Germination, 334; *act.,* 334; 335
Gill, 366, 367
Gill cover, 366
Gizzard, *illus.,* 346; 356
Glaciation, *illus.,* 213; 214
Glacier, 198, 213, 214; *illus.,* 254; 456
Glucose, 329
Gold rush, 211
Graduated cylinder, 15; *act.,* 28
Gram, 17
Grand Canyon, 195-198, 203; *illus.,* 196; 198
Grand Teton Mountains, 184
Granite, 175
Graphite, 205-206
Grasshopper, 35-358; *illus.,* 356; 357; 410
Gravity, 68-69, 115-116, 150; affecting roots, 323; *illus.,* 69; *table,* 115
Grease, 71
Great Lakes, *illus.,* 214
Green manure crop, 480
Greenhouse effect, *illus.,* 226
Ground water, *illus.,* 328
Guard cell, 328; *illus.,* 328
Guide fossil, 203

Gulf stream, 271
Gullet, *illus.*, 306
Gymnosperm, 321
Gypsum, *table*, 175; 176

H

Habitat, 391-392
Hail, 236
Heart, 366, 375; *illus.*, 356
Heat, *act.*, 10
Heat energy, 47-49, 51, 53, 128; *act.*, 10; 49; 59
Helium, 136
Hemoglobin, 371
Herbaceous stem, 324
Heredity, 429-446; and blending, 435-436; and evolution, 451-463; and genes, 436-438; hybrid, 433-434; mutations, 444, 462-463; and reduction division, 439-441; *act.*, cell division, 441; fruit flies, 432; hybrid, 436; mutation, 445; probability, 435; *illus.*, 430; 431; 438; 439; 440; 441; 443; 444; 445; blending, 435, 436; chromosomes, 437; hybrid, 433, 434; *table*, chromosomes, 437
Hero's engine, 114; *illus.*, 114
Hibernation, 369
Hickory, 391
Himalaya Mountains, 173
Honeybee, 403, 422
Hookworm, *illus.*, 345
Hormone, 371
Horned lizard, *illus.*, 455
Horse, *illus.*, evolution of, 460
Host, 344, 420
Humidity, 231-233, 243; *act.*, 233; 235; *illus.*, 231; *table*, 232
Humus, 476, 477, 480
Hurricane, 248-249
Hybrid, 431
Hybrid cross, 431; 433-434; *act.*, 436
Hybrid plant, *illus.*, 431
Hybrid vigor, 446

Hydra, *act.*, 344
Hydraulic lift, 131
Hydroelectric power, 57
Hydrogen, 34
Hygrometer, 232
Hypothesis, 8-9

I

Ice, 50; *act.*, 178
Ice age, 213, 214-215
Ideal mechanical advantage (I.M.A.), 87, 88, 89, 91, 93
Igneous rock, 174, 179-180; *act.*, 180; *illus.*, 179
Iguana, *illus.*, 373
Immigration, 395
Improvement cutting, 498
Inclined plane, 83, 91-94; *act.*, 92; *illus.*, 91; 93
Index fossil, 203
India, 173
Indian Ocean, 173
Inertia, 107-108, 118, 119; *act.*, 108; 119; *illus.*, 118
Inherited trait, 429-446; blended, 435-436; dominant, 430-438; and evolution, 451-463; and genes, 436-438; hybrid, 433-434; and mutations, 444, 462-463; and reduction division, 439-441; *act.*, cell division, 441; fruit flies, 432; hybrid, 436; mutation, 445; probability, 435; *illus.*, 430; 431; 438; 439; 440; 441; 443; 444; 445; blending, 435, 436; chromosomes, 437; hybrid, 433, 434; *table*, chromosomes, 437
Insect, 354-358; *act.*, 358; *illus.*, 355; 356; 357
International Date Line, 161
International System of Units, 12
International Time Zones, 159
Intestines, 370-371; *illus.*, 346; 370
Invertebrate, 363, 364
Ion, 33, 223

Ionic bond, 33
Ionosphere, 223; *illus.*, 223
Iron, 281; *illus.*, 206
Isobar, 251
Isotope, 216
Italy, 173

J

Jellyfish, 343; *illus.*, 207; 343; 344
Jet stream, 227, 228; *illus.*, 228
Joule, 74
Jurassic Period, 211; *table*, 204

K

Kelp, *illus.*, 318
Killer whale, *illus.*, 378
Kilogram, 17
Kilometer, 12
Kinetic energy, 45-46, 75, 133, 278; *act.*, 47
Kingdom, 298; *table*, classification, 298

L

Land breeze, 229; *illus.*, 229
Large intestine, 371
Larva, 355; *illus.*, 355
Latitude, 155
Lava, 189, 190
Law of conservation of energy, 75
Law of dominance, 430-438
Law of uniform change, 197
Leaf, 327-329; *act.*, 329; 330; 331; *illus.*, 326; 327; 328; 329
Left atrium, 371
Legume, 405
Length, 12-14; *act.*, 16
Lever, 84-89; *act.*, 85; *illus.*, 84; 86; 87; 88
Lichen, 422-423; *act.*, 423
"Life code", 295
Life span, 291
Lift, 139, 140
Lightning, 247
Limestone, 175, 182, 183

Lines of force, 163
Liquid, 25; pressure in, 129-131; *act.*, 129; 132; 137; *table*, 26
Liter, 15
Liver, 370
Liverwort, 318-319; *act.*, 319; *illus.*, 319
Living organism, 297; basic unit of, 293-295; characteristics, 297-311; in communities, 389-406; evolution of, 451-463; *act.*, 293; *table*, kingdoms, 298
Local winds, 229-230; *illus.*, 229
Longitude, 155
Longshore current, 276
Lubricant, 71
Lung, 368

M

Machine, 83-99; compound, 98-99; efficiency of, 94-97; ideal, 96; simple, 83-97; *act.*, 85; 90; 92; *illus.*, 84; 86; 87; 89; 93. *See also* Engine
Magma, 189, 190
Magnet, 161-163; *illus.*, 163
Magnetic field, 162, 163; *illus.*, 163
Magnetic north pole, 161, *illus.*, 163
Magnetic south pole, 161; *illus.*, 161
Magnetometer, 162
Malaria, 309
Mammal, 212, 378-379; *illus.*, 364; 378; 379
Mammary gland, 379
Mammoth, 212
Manganese, 281
Mantle, 170, 172
Map, 156-158; *act.*, 156; 157; 158; *illus.*, 156; 157; 158; 159; 160; 161
Marble, 183, 205
Marchantia, 319; *act.*, 319; *illus.*, 319

Mass, 17, 26, 108, 112, 113, 120, 135; of earth, 150; measurement of, 69
Mass spectrometer, 216
Mastodon, 212
Matter, 23-26; chemical change in, 52-53; inertia of, 107-108; nuclear change in, 56; physical change in, 50-52; properties of, 23; states of, 25-26; *act.*, 24; *table*, 26
Measurement, 12-17; of area, 14, 15; of density, 26, 27; of force, 67-71; of length, 12-14; of liquid pressure, 129-132; of mass, 17, 69; of speed, 109-113, 116, 117; of temperature, 48; of volume, 15; of weight, 68, 69; *act.*, 15; 16; 68; 70; 116; 117; 132; *table*, 12
Mechanical advantage, 86-89; actual, 86; ideal, 87, 88, 89, 91, 93; of levers, 86-89; of pulley, 89; of wedge, 93; of wheel and axle, 91
Melting, 50
Mendel, Gregor, 430-436
Mercury barometer, 224; *illus.*, 224; 225
Meridian, 155
Mesosphere, 223; *illus.*, 223
Mesozoic Era, 204, 210-211; *table*, 204
Metabolism, 292, 369
Metamorphic rock, 174, 183; *illus.*, 183
Metamorphosis, 355; *act.*, 373
Meteorite, 216
Meteorology, 251
Meter, 12
Metric system, 12-17; *act.*, 15; 16; *table*, 12
Microclimate, 255
Microorganism, 300
Mid-Atlantic Ridge, 173, 266
Migration, 376-377
Migratory Bird Treaty, 501
Milk, 379
Miller, Stanley, 451
Millimeter, 12
Millipede, 353

Mineral, 175, 206, 281, 478-479; *act.*, 176; *illus.*, 175; *table*, 175; 176
Mixture, 35-37; *act.*, 36; 37; 52
Mold, 303; *illus.*, 303; *table*, classification, 298
Molecule, 33; *act.*, 34
Mollusk, 351-353; *illus.*, 352
Molt, 358
Momentum, 120-121
Monera, 298, 300-302; *act.*, 302; *illus.*, 300; 301; *table*, classification, 298
Monument Valley, *illus.*, 168
Moon, 215, 216
Mosquito, *illus.*, 419
Moss, 318-319; *illus.*, 319
Motion, 107-122; circular, 117-119; linear, 107-117; *act.*, 108; 114; 115; 116; 117; 119
Mount St. Helens, 190; 490; 491
Mountain, 173, 229
Mountain breeze, 229
Mountain goat, *illus.*, 386; 387
Mouth, *illus.*, 346
Mushroom, 303-304; *table*, classification, 298
Mutation, 444; and evolution, 462-464; *act.*, 445; *illus.*, 444; 462; 463

N

National Weather Service, 251
National Wildlife Refuge System, 503
Natural resource, conservation of, 475-485
Natural selection, 454, 457-459
Nerve, 363, 371; *illus.*, 357
Nerve center, *illus.*, 357
Nerve cord, *illus.*, 357
Nervous system, 347, 350, 357
Neutron, 30
Newton, 67
Nitrogen, 221
Nitrogen cycle, 405
Nitrogen-fixing bacteria, 405

Nodule, 281, 282, 405
Nonrenewable resource, 475
North America, 172, 173
North American plate, 184, 185
North Pole, 69, 155, 161, 162
Nuclear change, 56
Nuclear fission, 56, 58
Nuclear fusion, 56, 58
Nuclear membrane, 293; *illus.,* 294
Nuclear power plant, 58
Nuclear reactor, 58, 128
Nucleus, 30, 293, 295, 307; *illus.,* 294
Nymph, 358

O

Oak, 391
Obsidian, 190
Occluded front, 246; *illus.,* 246
Ocean, 261-282; density of, 264-266, 272; life in, 269-270; temperature of, 264-265; *act.,* 264; 266; 272; 276; 277; *illus.,* 260; 262; 263; 265; 267; 268; 270; 271; 273; 275; 276; 278; *table,* composition of, 262
Ocean current, 271-272; *act.,* 272; *illus.,* 271; 272; 276
Ocean resources, 280-282
Ocean trench, 266
Ocean wave, 273-277; seismic, 279-280; *act.,* 276; 277; *illus.,* 273; 274; 275; 276
Ocean-basin floor, 268
Oceanography, 261
Ocelot, *illus.,* 299
Octopus, 352-353; *illus.,* 352
Oil, 71
Oral groove, 306; *illus.,* 306
Order, 298
Ordovician Period, 207; *table,* 204
Ore deposit, 206
Organ, 321, 341
Organism, 297
Orthoclase, *table,* 175; 176
Osmosis, 296, 322; *act.,* 297

Ovary, 331, 332, 333, 348; *illus.,* 331
Ovipositor, 358
Ovule, 332; *illus.,* 331
Oxygen, 221, 296, 305, 371, 482
Oxygen-carbon dioxide cycle, 404-406; *illus.,* 405
Oyster, 263, 352

P

Pacific Ocean, 282
Pacific plate, 172, 185
Painted Desert, 210
Paleontology, 200
Paleozoic Era, 204, 206-209; *illus.,* 207; 208; 209; *table,* 204
Pancreas, 370
Paramecia, 306-307, 398; *act.,* 307; *illus.,* 306; 307; *table,* classification, 298
Parasite, 344-345, 420; *illus.,* 345; 420; 421
Passenger pigeon, 499
Pea, *illus.,* 431
Pendulum, 151
Peppered moth, 460; *illus.,* 461
Perch, 366; *illus.,* 366; 367
Perennial, 325; *illus.,* 324
Period, 204; *table,* 204
Permanent wilderness area, 501
Permian Period, 209; *table,* 204
Petal, 331
Petrified wood, 200; *act.,* 201
Petroleum, 209
Pharynx, 347
Phloem, 325; *illus.,* 322; 325
Photochemical smog, 484
Photosynthesis, 329-331; *act.,* 330; 331
Phylum, 298
Physical change, 50-52; *act.,* 52
Pine cone, *illus.,* 321
Pistil, 331, 332, 333, 431; *illus.,* 331

Pith, *illus.,* 325
Planarian, 345
Plankton, 269, 270; *illus.,* 269
Plant, 411, 412, 416; classification of, 297-311; evolution of, 457-458; fern, 320; flower, 331-333; inherited traits of, 429-446; leaf of, 327-329; liverwort, 318-319; moss, 318-319; root of, 322-323; seed, 321, 333-335; stems of, 324-326; *act.,* blending, 436; cell division, 441; hybrid, 436; leaf, 328, 329; liverwort, 319; mutation, 445; photosynthesis, 330, 331; roots, 323; seed, 334, 335, 455; stems, 326; *illus.,* algae, 318; angiosperms, 321; annuals, 324; blending, 435, 436; cactus cop, 336; fern, 320; flower, 331, 332; hybrid, 431, 433, 434; leaf, 327, 328; liverwort, 319; moss, 319; perennials, 324; roots, 322, 323; seed, 333, 334, 335; stem, 325, 326; *table,* chromosomes, 437; classification, 298
Plant breeding, 445-446; *illus.,* 446
Plant cell, 294; *illus.,* 294; 329
Plant fossil, 200-203, 208, 209, 211, 212; *act.,* 201; 202; *illus.,* 201
Plasma, 26; *table,* 26
Plate, 171, 172
Plate tectonics, 171-173; *act.,* 173; *illus.,* 171; 172; 173
Platypus, 379
Polar climate, 254
Polar easterlies, 227
Pollen, 331, 332
Pollination, 322, 446; *illus.,* 446
Pollution, air, 484,-485; water, 482-483; *act.,* 485; *illus.,* 58; 484
Pond community, 389, 401; *act.,* 415; *illus.,* 390; 401
Population, 390-394, 404, 419; *act.,* 393; *illus.,* 391

Population change, 394-399; *act.*, 394; 399; *illus.*, 394; 395
Portuguese man of war, *illus.*, 269
Potential energy, 46, 75
Power, 75-76; *act.*, 77
Precambrian Era, 204, 205-206; *table*, 204
Precambrian rock, 206
Precipitation, 231-233, 236; *illus.*, 231; 236
Predator, 379, 396, 418-419
Pressure, air, 224-225, 230, 247, 248, 249; in gases, 133-134; in liquids, 129-132, 138; *act.*, 132; 139; 225; 235; *illus.*, 224; 225
Prey, 418
Primary wave, 187
Prime meridian, 155, 161
Principle of uniformity, 197
Problem, 8
Producer, 414, 417; *act.*, 415
Product, 55
Protein, 451
Protist, 298, 305-309; *act.*, amoeba, 307; hay infusion, 307; *illus.*, 305; 306; 307; 309; *table*, classification, 298
Protococcus, 317-318; *illus.*, 318
Proton, 30
Protoplasm, 294; *table*, 294
Psychrometer, *act.*, 233; *illus.*, 232
Pulley, 89; *act.*, 90; *illus.*, 89; 90
Pumice, 190
Pupa, 355; *illus.*, 355
Purebred, 446

Q

Quartz, *table*, 175; 176
Quartzite, 183
Quaternary Period, 212-213; *table*, 204
Queen bee, 403

R

Radial symmetry, 350
Radiation, 223, 226
Radio wave, 223
Radioactive dating, 199, 216
Radioactive decay, 199
Radioactive element, 199
Rain, 236; acid, 484
Rance River, 278
Reactant, 54
Reaction force, 114; *act.*, 114; 115
Recessive trait, 431-438
Red blood cell, 371
Reduction division, 439-440; *act.*, 441; *illus.*, 439
Reforestation, 498
Reindeer moss, 423
Relative humidity, 232-233
Remoras, *illus.*, 423
Renewable resource, 475
Reproduction, 292, 301, 374, 377, 442-443
Reptile, 373-375; *illus.*, 373; 374
Resistance force, 84, 85, 86, 93
Respiration, 296, 302, 305; *illus.*, 296
Respiratory system, 357
Response, 292
Revolution, 152; *illus.*, 152
Rhizoid, 319
Rhizome, 325; *illus.*, 326
Right atrium, 371
Rip current, 276; *illus.*, 276
RNA, 295, 311
Rock, 174-185; age of, 198-199, 203, 216; erosion of, 178, 181, 197-198; igneous, 174, 179-180; metamorphic, 174, 183; sedimentary, 174, 180-183, 196, 197, 198, 203, 206; weathering of, 177-178, 196-197; *act.*, 176; 178; 180; 182; 183; 185; *illus.*, 175; 178; 179; 196; 197; 198; 206; 213; cycle, 174
Rock formation, 184-185
Root, 322-323; *act.*, 323; *illus.*, 322; 323
Root hair, 322; *illus.*, 322
Rotation, 151, 227; *illus.*, 151
Roundworm, 345; *illus.*, 345
Rudder, 140
Rust, 52, 55

S

Salamander, *illus.*, 369; 388
Salinity, 263, 265
Salt, 32, 261-263
San Andreas Fault, 185
Sandstone, 182, 183
Scavenger, 414; *illus.*, 414
Science, 5-6; career in, 6-8; method of, 8-11; using skills, 8-9; *act.*, 6; *table*, 7
Scoria, 190
Scorpion, 354; *illus.*, 354
Screw, 93-94
Sea anemone, 422; *illus.*, 422
Sea breeze, 229
Sea floor, 266-268; *illus.*, 267; 268
Sea life, 269-270; *illus.*, 269; 270
Sea urchin, *illus.*, 350
Seasons, 152-154; *act.*, 154; *illus.*, 153
Seawater, 261-266, 481; density of, 264-266, 272; temperature of, 264-265; *act.*, 264; 266; *table*, composition, 262
Seaweed, 280; *illus.*, 281
Secondary wave, 187
Sediment, 181; *act.*, 183
Sedimentary rock, 174, 180-183, 197, 198, 203, 206; *act.*, 182; 183; *illus.*, 180; 181; 196; 197
Seed, 321, 333-335; *act.*, 334; 335; 455; *illus.*, 333; 334; 335
Seed coat, 333
Seed dispersal, 335; *illus.*, 335
Seed leaf, 333
Seed plant, 321, 333-335
Seismic sea wave, 279-280
Seismograph, 186-187; *illus.*, 187
Selection, 445-446
Selective cutting, 498
Sepal, 331; *illus.*, 331
Sewage treatment plant, 483
Sex cell, 320, 439
Sex chromosome, 441
Sex determination, 441; *illus.*, 441

Sex organ, 348
Sexual reproduction, 320, 348
Shale, 182, 183
Shark, *illus.,* 365; 423
Shell, 352
Shelter belt, 481
Shoot, *illus.,* 326
Shrew, 378
SI unit, 12–17; *table,* 12
Sierra Nevada Mountains, 211
Silurian Period, 208; *table,* 204
Simple eye, 357; *illus.,* 356
Skeletal system, *illus.,* 364
Skeleton, 363, 364, 366, 374, 375
Slate, 183
Sleet, 236
Slime band, 348
Sling psychrometer, *act.,* 233; *illus.,* 232
Small intestine, 371
Smog, 484
Snail, 351, 352; *illus.,* 207; 352
Snake, 374; *illus.,* 342
Snow, 236
Society, 403
Sodium, 262
Sodium chloride, 32, 34
Soil conservation, 476–481; *act.,* 477; 479; 481
Soil erosion, 478–479; *act.,* 479
Solar energy, 58–59; *act.,* 59
Solar radiation, 226
Solid, 25; *table,* 26
Solution, 36; *act.,* 37
Sound, speed of, 122
Sounding, 266–267
South Pole, 69, 155, 161
Space Shuttle, *illus.,* 106
Species, 298, 299, 391, 400, 404, 453, 454, 456, 457; changes in, 459–462; endangered, 499; *table* 500
Speed, 109–113; average, 110; and Bernoulli's principle, 138–139; change of, 111–113; and drag force, 141; and momentum, 120–121; of sound, 122; terminal, 116; *act.,* 116; 117; *table,* 115
Sperm, 320, 348, 368, 439, 440; *illus.,* 439; 440
Spider, 354; *illus.,* 354

Spinal cord, 363, 371; *illus.,* 364
Spiny-skinned animal, 350–351; *illus.,* 350; 351
Spiracle, 357; *illus.,* 356; 357
Spirilla, *illus.,* 300
Spirogyra, *illus.,* 318
Sponge, 342–343; *illus.,* 207; 343
Spore, 303, 320
Spore case, 320; *illus.,* 309
Sporozoa, 309; *illus.,* 309
Spring, 154
Square centimeter, 14
Squid, 352–353
Stamen, 331, 332, 431; *illus.,* 331
Starch, 322, 329
Starfish, 350–351; *illus.,* 350; 351
Steel, 36
Stem, 324–326; modified, 331–333; *act.,* 326; 327; *illus.,* 325; 326
Stimulus, 292
Stoma, 327–328; *illus.,* 327; 328
Stomach, 370
Strata, 181; *illus.,* 181
Stratosphere, 223; *illus.,* 223
Stratus cloud, 235; *illus.,* 234
Strip-cropping, 480
Subatomic particle, 30
Sublimation, 50; *illus.,* 50
Subscript, 34
Subsoil, 476, 477
Succession, 400–401; *act.,* 401; *illus.,* 400; 401
Sulfur dioxide, 484
Summer, 153
Sun, 215, 226
Surface tension, 128
Surge, 280
Symbiosis, 421–423; *act.,* 423; *illus.,* 422; 423
Symbol, 29, 34
Syncline, 184; *act.,* 185
System, 341

T

Tadpole, 368–369; *illus.,* 369
Talc, *table,* 175; 176

Tapeworm, 344–345, 420; *illus.,* 345; 421
Temperate climate, 254
Temperate deciduous forest, *illus.,* 255
Temperature, 47–49, 170, 183, 190, 231, 232, 233, 243; of seawater, 264–265; *act.,* 49; *table,* 48
Tentacle, 343, 352, 422
Terminal speed, 116
Termite, 422
Terrace, 480
Territory, 398
Tertiary Period, 212; *table,* 204
Testes, 348
Theory, Darwin's, 453–455, 461
Thermosphere, 223; *illus.,* 223
Thorax, *illus.,* 356
Thunder, 247
"Thunderhead", 235
Thunderstorm, 235, 246–247; *illus.,* 247
Tidal wave, 279
Tide, 215, 277–278; *act.,* 278; *illus.,* 278
Time zone, 159–161; *illus.,* 159; 160
Tissue, 321, 325, 341; *illus.,* 325
Toad, 268, 369; *illus.,* 368
Topaz, *table,* 175; 176
Topographic map, 157; *act.,* 158; *illus.,* 158
Topsoil, 476, 477, 478, 479
Tornado, 249–250; *illus.,* 250
Trade winds, 227
Trait, blended, 435–436; dominant, 430–438; and evolution, 451–463; and genes, 436–438; hybrid, 433–434; and mutations, 444, 462, 463; and reduction division, 439–441; *act.,* cell division, 441; fruit flies, 432; hybrid, 436; mutation, 445; probability, 435; *illus.,* 430; 431; 438; 439; 440; 441; 443; blending, 435, 436; chromosomes, 437; hybrid, 433, 434; *table,* chromosomes, 437

Transpiration, 328
Trench, 172
Triassic Period, 210; *table,* 204
Trichina, 345
Trilobite, 203, 207, 456; fossil, 456
Tropic of Cancer, 153
Tropic of Capricorn, 154
Tropical climate, 254
Tropical rain forest, *illus.,* 255
Troposphere, 222-223; *illus.,* 223
Tsetse fly, 308
Tsunami, 279
Tuber, 326
Turtle, *illus.,* 373; 374
Twins, 443; *illus.,* 443
Typewriter, 98-99; *illus.,* 99
Typhoon, 249
Tyrannosaurus, 211; *illus.,* 211

U

Unconformity, 197; *illus.,* 197
Undertow, 275
Uniformity, principle of, 197
United States Forest Service, 496
Uranium, 56, 199

V

Vacuole, 305; *illus.,* 294; 306
Venom, 374
Vertebrate, 363, 368
Virus, 310-311; *illus.,* 310; 311
Volcano, 189-190; *illus.,* 146-147; 189; 190
Volume, 15, 26; *act.,* 16
Volvox, 308; *illus.,* 309

W

Wallace, Alfred, 454; *illus.,* 454
Warm front, 245; *illus.,* 245; 246
Waste material, 296, 306, 347
Water, 34, 36, 50, 55, 128, 231, 234, 236, 296, 305; affecting roots, 323; conservation of, 481-483; loss by plants, 328; *act.,* 129; 178; 328; 415. See *also* Seawater
Water resource, 481
Water vapor, 231-232; *table,* 232
Watershed, 494; *act.,* 495
Waterspout, *illus.,* 250
Watt, 75
Wave, ocean, 273-277; seismic, 279-280; *act.,* 276; 277; *illus.,* 273; 274; 275; 276
Wave period, 274
Wavelength, 274, 275; *illus.,* 274
Weather, 241-255; *act.,* 242; 252; 253; *illus.,* 240; 243; 244; 245; 246; 247; 248; 249; 250; 251; 253; 254; 255
Weather forecasting, 252-253, 256; *act.,* 253; *illus.,* 253
Weather map, 251; *act.,* 252; *illus.,* 251
Weathering, 177-178; of rocks, 196-197; *act.,* 178; *illus.,* 197
Wedge, 93; *illus.,* 93
Weight, 68-70; *act.,* 70; *table,* 69
Westerlies, 227; *illus.,* 227
Wheel and axle, 91; *illus.,* 91
White blood cell, 371

White Cliffs of Dover, 211
Wildlife, 499-503
Wildlife conservation, 501 502
Wildlife refuge, 502-503
Wildlife resources, 500
Wind, 227-230, 248, 250; *act.,* 230
Winter, 153, 154
Wolf, *illus.,* 419
Woody stem, 325
Work, 73-75; and energy, 75
Worker bee, 403
Wrench, 83

X

Xylem, 322, 325; *illus.,* 322; 325

Y

Yeast, 302-304; *act.,* 304; *table,* classification, 298
Yellowstone National Park, 503

Z

Zygote, 442

2 3 4 5 6 7 8 9 10 11 12 13 14 15—95 94 93 92 91 90 89 88 87 86